HEAT TRANSFER
Principles and Applications

logy

Prentice-Hall of India Private Limited
New Delhi-110001
2006

Rs. 275.00

HEAT TRANSFER: Principles and Applications
by Binay K. Dutta

ISBN-81-203-1625-8

The export rights of this book are vested solely with the publisher.

Sixth Printing ... ... **November, 2006**

Published by Asoke K. Ghosh, Prentice-Hall of India Private Limited, M-97, Connaught Circus, New Delhi-110001 and Printed by Meenakshi Art Printers, Delhi-110006.

Contents

CONTENTS vii

Preface

The venture of adding a new book on an old subject that has been amply treated and covered in a large number of texts at both introductory and advanced levels deserves an explanation. *Heat Transfer* is a subject common to a number of engineering disciplines such as chemical, nuclear, metallurgy, biochemical as well as to several branches of chemical technology dealing with polymers, ceramics, vegetable oils and fats, pharmaceuticals and fine chemicals, food processing, etc. However, because of the differences in the nature and areas of applications it is difficult to incorporate enough material in a single volume to specifically meet the requirements and demands of the students of different disciplines. As a matter of fact, most of the texts on the subject at the undergraduate level, some of which are of excellent quality within their scope, primarily deal with the fundamentals along with problems that aim at strengthening the theoretical understanding. Little emphasis is placed on introducing the students to industrial applications and practice. No doubt there are some very good books on applications and practice, generally meant for the professionals, that make good references but not texts for the undergraduates.

In the above context, the present text consisting of eleven chapters is an attempt to provide a balanced mix of the fundamental physical principles underlying the phenomenon of transport of heat, applications to the solution of practically relevant problems, some with a design orientation, basic principles of design, construction, operation and selection of heat exchangers and an overview of some of the recent developments in heat transfer equipment particularly for the chemical process industry. Chapter 1 is introductory and explains the importance of the subject through the example of a chemical process and briefly discusses the systems of units and their conversions. Chapters 2–7, 10 and 11 mainly deal with the fundamental principles of the three basic modes of heat transmission, namely conduction, convection and radiation, while Chapters 8 and 9 are devoted to applications and some aspects of process design principles and heat transfer practice.

One of the major difficulties that a junior undergraduate frequently encounters is how to deal with a physical problem in order to translate it into tractable mathematical equations that can be solved using simple analytical tools. Keeping this in mind, formulation and solution of simple real-life problems and their physical understanding have been given adequate emphasis particularly in Chapters 2 and 3. Chapter 2 explains the application of Fourier's law to steady state heat conduction problems in three common geometries in conjunction with the energy conservation principle. The steps of the mathematical formulation and solution have been illustrated with quite a few examples. The industrial practice of thermal insulation has been described very briefly.

The concept of heat transfer coefficient is probably the single most important concept so far as practical applications are concerned. Chapter 3 elaborates this concept and highlights its importance through a large number of solved problems ranging from the conceptual types to those involving direct application such as the determination of the optimum insulation thickness. The physical basis of the mathematical equations, and particularly that of the boundary conditions, has been explained at each step. Forced convection heat transfer has been presented in Chapter 4 and natural or free convection in Chapter 5. Besides qualitative treatment of the phenomena and description of the dimensionless groups in heat transfer and the common strategies to correlate them, the objective has been to familiarize the students with the more important and tested correlations available in the literature for the estimation of heat transfer coefficient in diverse physical situations. Selection and use of the correlations have been illustrated with carefully designed examples.

Boiling and condensation have been discussed in Chapter 6. While the thermophysical principles underlying the phenomena and the regimes of boiling and condensation have been explained, a pretty large number of useful correlations along with the ranges of their applicability for the estimation of boiling or condensation heat transfer flux or coefficient have been cited. Basic principles of radiation and the method of calculation of the rate of radiation heat exchange are dealt with in Chapter 7. The mathematical tools for the analysis of radiation and radiation heat exchange are often quite involved and need a grasp of 3-D geometry. Some glimpses of the theoretical approaches, for example those for the determination of view factors, have been explained with a few examples. The principles and methodology of calculation of rates of radiation exchanges in practically relevant situations have been discussed in great detail.

Chapters 8 and 9 are devoted to heat exchange equipment and evaporators. The constructional features, material selection, scientific principles and methodology of process design calculations, and some aspects of operational features and maintenance of the shell-and-tube heat exchanger have been presented. Introduction of compact heat exchangers has revolutionized the heat exchange equipment during the last two decades. The more important equipment and devices of this category, for example the plate heat exchanger and the spiral-plate and the spiral-tube heat exchangers, and their applications and selection have been described. Illustrative pictures of some of the latest equipment of these types from global manufacturers have been included. Evaporators constitute a class of heat exchange equipment used in many industries including chemical, biochemical, food and metallurgical industries. Construction and operation of evaporators of conventional and novel designs (for example, falling film, rising film, plate type, etc.) have been reviewed and the selection of evaporators has been discussed. Evaporator auxiliaries (vacuum equipment, steam traps, mist eliminators, etc.) have also been described. Short questions formulated on these chapters, particularly those on operation, selection and troubleshooting, aim at providing some exposure to the common problems in the industrial practice of heat exchange. Excellent articles on heat exchange equipment, auxiliaries and practice are frequently published in technical journals such as Chemical Engineering, Chemical Engineering Progress, Hydrocarbon Processing, etc. Many of these articles, including the very recent ones, have been cited in the references.

Chapters 10 and 11 deal with some theoretical aspects of unsteady state and multidimensional heat conduction and convective heat transfer in boundary layer flow. These topics prepare the students for advanced courses in heat transfer and help them develop the skill of theoretical analysis of complex heat transfer problems. Some of the mathematical tools and concepts that the students frequently apply mechanically (for example those behind the techniques of separation of variables and similarity solution) have been elaborated and used for the solution of typical problems. The material contained in these chapters, in addition to that in Chapters 2 and 3, will help the students develop the skill of model-based solution of practical problems. However, a discussion of numerical techniques and their applications, though extremely useful, has been kept out of the scope of the book.

Sufficient care has been taken to formulate the problems for both examples and exercises. Most of the problems characterize real-life situations that are likely to provoke the imagination of the students. In many cases, a preamble in the problem statement gives its industrial relevance and connection. A collection of short questions has been appended to each chapter to help the students assess their physical and technical understanding. Though the book is primarily designed to meet the requirements of an undergraduate course on heat transfer, it will be of help to the practising engineers in the need of a brush-up of the fundamentals and bits of technical information.

For the solution of many of the problems, thermophysical properties of fluids and their mixtures are necessary. These properties are not always readily available in the literature and may have to be estimated using suitable equations and correlations. Sources of such equations and correlations have been mentioned and their uses have been illustrated in the course of solution of some of the problems (e.g. Example 6.3).

SI units are generally used throughout the book. However, there are some intentional departures because process design engineers often prefer 'kcal' to 'watt' and 'kg/cm^2' to 'bar'. In addition, the English Engineering System of units is still used by many. This is why purely SI units have not been strictly adhered to in order to familiarize the students with other units used in the engineering practice. A unit conversion table has been provided for ready reference.

I take this opportunity to express my indebtedness to many of my students, friends and former students in the industries and other organizations, who all helped me in connection with the preparation of the text. Admittedly, I got educated through interactions with many of my former students by sharing their professional experiences whenever an opportunity arose. In particular, I must mention the names of Dr. Narayan Bandyopadhyay, Satyen Chakraborty, Dulal Hui, Swapan Saha, Tarun Ghosh, Abhijitbikas Pal, Sahajamal Biswas, Soumendranath Datta and Prabir Das. Dr. Bandyopadhyay and Mr. Chakraborty also reviewed a part of the manuscript and gave useful suggestions for improvement. Mr. Kalyan Basu, who is one of the finest chemical engineers I have ever met, helped me a lot on many occasions and devoted his time to review the chapter on evaporation and evaporators. I am also indebted to Dr. Chandana Saha of Boston for her help on many occasions. I also acknowledge the help that I got from the staff members of the departmental and of central libraries of the University College of Science and Technology, Calcutta.

A number of organizations in India and abroad helped me by supplying and allowing to incorporate and reproduce in the book information, printed materials, pictures and photographs. The persons of these organizations whom I came in touch with and got help from are Samuel R. Wharry, Jr. (AMETEK, Wilmington, Delaware), H. Kumar (APV, Goldsboro, North Carolina), Louis F. LaPosa (Swenson Process Equipment, Harvey, Illinois), Samuel W. Croll III (Croll Reynolds, Westfield, New Jersey), Roger F. Krueger (KEMCO, Houston, Texas), Pete Mickus (Tranter, Inc., Wichita Falls, Texas), Eva Samuelsson (Alfa Laval, Lund, Sweden) and Mr. S.K. Chatterjee (Lloyd Insulations, Calcutta). I express my sincere thanks and gratitude to all of them.

It has been a great pleasure working with the editorial and production staff of Prentice Hall of India. I am thankful to my publisher for bringing out the book in its present shape. Finally, the encouragement and inspiration that I got from my wife, Ratna, and my daughter, Joyita, have been invaluable to me in the course of preparation of the manuscript. Words of thanks or gratitude are not sufficient to express my feelings towards them.

Binay K. Dutta

Notations

There are not enough symbols in the English and Greek alphabets to allow the use of each letter for representing just one quantity. Consequently, some symbols have been used to denote more than one entity, but their use should be clear from the context. In the following list of symbols used in the text, all the subscripts made use of for a given symbol have not been shown as their meaning should again be clear from the context.

Description	Symbol	Unit
a	heat transfer area per unit length of each pass in a heat exchanger	m^2
a_s	shell-side flow area, heat exchanger	m^2
A	area of heat transfer or exchange	m^2
b	breadth, of a fin	m
B	baffle spacing, heat exchanger	m
Bi	Biot number, hl/k, ha/k (a = radius)	—
c	velocity of light, 2.998×10^8	m/s
c'	clearance between two adjacent tubes (Fig. 8.12)	m
c_p	specific heat	kJ/kg °C
C_1, C_2, C_3, C_4	integration constants	—
C_D	drag coefficient for skin friction	—
C_{sf}	constant, Eq. (6.6)	—
d	diameter, characteristic length	m
D_H	hydranlic or equivalent shell diameter	m
D_S	inside diameter of a shell	m
E	emissive power of a body	W/m^2
E_b	hemispherical total emissive power of a blackbody	W/m^2
$E_{b\lambda}$	monochromatic emissive power or spectral emissive power of a blackbody	$W/m^2\ \mu m$
$E'_{b\lambda}$	directional spectral emissive power	W/m^2 sr
f	Fanning's friction factor	—
F	factor that modifies 'liquid only Reynolds number' Re_l to a two-phase Reynolds number	—

$F_{b,\,(\lambda_1 - \lambda_2)}$	fraction of the radiant emission from a blackbody lying in the wavelength range λ_1 to λ_2	—
F_{ij}	view factor or shape factor for the surface pair i–j	—
$F_{i\epsilon}$	emissivity factor of surface i	—
F_D	drag force per unit breadth of the plate	N/m
F_T	LMTD correction factor	—
Fo	Fourier number, $\dfrac{\alpha t}{l^2}$, $\dfrac{\alpha t}{a^2}$ (a = radius)	—
g_x	component of the acceleration due to gravity in the x-direction	m/s^2
G	mass velocity or flow rate	kg/m^2 s
	irradiation	W/m^2
Gr$_l$	local Grashof number	—
Gr$_L$	Grashof number, based on characteristic length L	—
Gz	Graetz number	—
h	heat transfer coefficient	W/m^2 °C, kcal/h m^2 °C
h	Planck's constant, 6.6256×10^{-34}	J s
H_s	enthalpy of solution	kJ/kg
i	enthalpy of a stream	kJ/kg
I_b	total intensity of radiation	W/m^2 sr
$I_{b\lambda}$	spectral radiation intensity	W/m^2 μm sr
I_λ	intensity of radiation of wavelength λ	W/m^2
j_H	Colburn factor	—
J	radiosity	W/m^2
Ja	Jacob number, $c_{pl}\,(T_v - T_w)/L_v$	—
k	thermal conductivity	W/m °C, kcal/h m °C
k_B	Boltzmann constant, 1.3805×10^{-23}	J/K
l	thickness, length	m
L	pipe/tube length, characteristic length, plate height/length	m
L_e	main beam length	m
L_v	molar heat of vaporization	kJ/kg-mol
m	mass	kg
m'	flow rate of a stream	kg/s
	rate of condensation per unit breadth	kg/m s
n	insulation life	yr
	number of tube passes	—

N	number of shell passes	—
	number of tubes	—
	number of walls in an enclosure	—
N_b	number of baffles	—
N_t	number of tubes in a shell	—
Nu	Nusselt number, hd/k	—
$\overline{\text{Nu}}$	average Nusselt number	—
Nu′	modified Nusselt number or condensation number	—
Nu_x	local Nusselt number	—
NTU	number of transfer units	—
p	perimeter, of a fin	m
	pressure	kPa, bar, atm
	tube pitch	m
Pe	Pectlet number, $\rho V l c_p / k$	—
Pr	Prandtl number, $c_p \mu / k$	—
Pr_l	liquid Prandtl number	—
ΔP_r	return loss	kg/m²
ΔP_s	shell-side pressure drop	kg/m²
ΔP_t	tube-side pressure drop	kg/m²
q	heat flux	W/m²
	radiant heat flux	W/m²
Q	rate of heat transfer	W, kcal/h
Q'	rate of electrical heat generation	W
Q_a	rate of radiation absorption by a gas	W
Q_g	rate of radiant emission by a gas	W
Q_i	total amount of radiation emitted by the surface i	W
$Q_{i,\,\text{net}}$	net rate of loss of radiant energy from the ith surface	W
Q_{ij}	rate of radiant energy transfer from surface i to surface j	W
$(Q_{ij})_{\text{net}}$	net rate of radiant energy loss from the ith surface because of radiation exchange with the jth surface	W
$Q_{ij,\,\text{net}}$	net rate of exchange of radiation between the surfaces i and j	W
r	radius, radial position	m
r'	dimensionless radius	—
$\bar{r}$	dimensionless radial position	—
r_H	hydraulic radius	m
R	capacity ratio, Eq. (8.27)	—
	heat transfer resistance	°C/W
	radius of a cylinder or sphere	m
	thermal resistance	°C/W
	universal gas constant	kJ/K kmol
R'	electrical resistance	ohm

R_d	heat transfer resistance due to dirt factor	m^2 °C/W
Ra_l	Rayleigh number	—
Re	Reynolds number, $\rho V d/\mu$	—
Re_f	film Reynolds number	—
Re_l	plate Reynolds number, Eq. (11.45), liquid only Reynolds number	—
Re_x	local Reynolds number, $\rho V x/\mu$	—
Re_L	Reynolds number based on characteristic length L	—
s	specific gravity	—
S	boiling suppression factor	
	solar constant	W/m^2
S_D	diagonal pitch, of a tube bank	m
S_L	longitudinal pitch	m
S_T	transverse pitch	m
St	Stanton number, $hV/\rho c_p$	—
t	time	s
T	temperature	°C, K
$\overline{T}$	dimensionless temperature	—
ΔT	temperature difference or driving force	°C, K
ΔT_{cm}	corrected mean driving force	°C
ΔT_m	log mean temperature difference (LMTD)	°C
$T_{\delta l}$	temperature at the edge of the laminar sub-layer	°C
u	x-component of velocity	m/s
u'	fluctuating velocity component in the x-direction	m/s
$\overline{u'^2}$	mean square velocity of the fluctuating velocity, x-component	$(m/s)^2$
u^+	dimensionless velocity, Eq. (11.153)	—
u_m	mean velocity of a fluid in a pipe	m/s
U, U_0	overall clean heat transfer coefficient	W/m^2 °C, kcal/h m^2 °C
U_d	overall heat transfer coefficient including the dirt factor	W/m^2 °C, kcal/h m^2 °C
v	transverse velocity component	m/s
	y-component of velocity	m/s
v'	fluctuating velocity component in the y-direction	m/s
V	average velocity	m/s
	linear fluid velocity	m/s
$\mathbf{V}$	velocity vector	m/s
V_o	superficial fluid velocity	m/s
V_∞	free stream velocity	m/s

w	flow rate of a stream	kg/s
	mass fraction vapour (or quality) in a liquid	—
	weight fraction of solute in a solution	—
	width, of a fin	m
	z-component of velocity	m/s
w'	fluctuating velocity component in the z-direction	m/s
W_c	cold fluid flow rate	kg/s
W_h	hot fluid flow rate	kg/s
x	distance (from the leading edge of a plate)	m
	longitudinal coordinate	—
$\bar{x}$	dimensionless axial position	—
y	transverse coordinate	—
$y+$	dimensionless distance from the immersed plate, $\left(y\sqrt{v}\right)\sqrt{\tau_w/\rho}$	—
z	axial position in a cylinder	m
$\alpha \ (= k/\rho c_p)$	thermal diffusivity	m²/s
α	absorptivity	—
β	coefficient of volumetric expansion	K⁻¹, °C⁻¹
δ	fluid film thickness	m
δ_l	thickness of the laminar sub-layer	m
	boundary layer thickness	m
	condensate film thickness	m
δ_T	thermal boundary layer thickness	m
$\delta(x)$	local momentum boundary layer thickness	m
ε	bed porosity	—
	emissivity	—
ε_H	eddy diffusivity of heat	m²/s
ε_M	turbulent or eddy diffusivity of momentum	m²/s
ε_g	emissivity of gas	—
ε_λ	spectral emissivity (a function of λ)	—
η	fin efficiency	—
	heat exchanger effectiveness	—
	similarity variable, $x/2\sqrt{\tau}$	—
	similarity variable, Eq. (11.29)	—
θ	cone angle (in the spherical coordinate system)	radians
	polar angle, dimensionless variable	—
Θ	eigenfunctions	—
λ	wavelength	Å (= 10⁻¹⁰ m)
		nm (= 10⁻⁹ m)
		μm (= 10⁻⁶ m)
	latent heat	kJ/kg
λ_n	eigenvalues	—
μ, μ_w	viscosity, wall viscosity	kg/m s
μ_M	eddy viscosity	kg/m s

ν	momentum diffusivity, μ/ρ (or kinematic viscosity)	m²/s
	frequency	s⁻¹
ξ	$(ph/kbw)^{1/2}$	m⁻¹
	similarity variable, $C\dfrac{y}{x^{1/4}}$	—
ρ	density	kg/m³
	reflectivity	—
σ	Stefan-Boltzmann constant	5.669×10^{-8} W/m² K⁴
	surface tension	N/m, dyne/m
τ	momentum flux (or shear stress)	N/m²
	transmissivity	—
τ_c	temperature ratio, Eq. (8.26)	—
ϕ	polar angle (in the spherical coordinate system)	radians
ϕ_t	dimensionless viscosity ratio	—
ϕ_s	viscosity correction factor for the shell-side fluid	—
φ	viscosity correction factor	—
Ψ	stream function	—
Ψ_v	volumetric rate of heat generation	W/m³
ω	solid angle	sr

1

Introduction

Heating or cooling of materials is an indispensable part of processing, production, fabrication or shopfloor jobs in engineering practice. Hardly any job or operation can be identified that does not involve heating or cooling of a body or a material at some stage or other. Whenever a material or a fluid stream has to be heated or cooled, or whenever a material has to be stored in a hot or cold condition, we have to see that it is done in the most economic way possible so that the loss or gain of heat during the process is minimium.

In this connection it will be worthwhile to cite a practical example of a process that involves heating or cooling of materials, or fluid streams, in a number of steps. Consider the process of manufacturing nitric acid by catalytic air-oxidation of ammonia (Fig. 1.1). Liquid ammonia is the raw material. It is common practice to store liquid ammonia at atmospheric pressure and at a temperature of about –33°C. A refrigeration unit is used to keep the liquid cool (or, in other

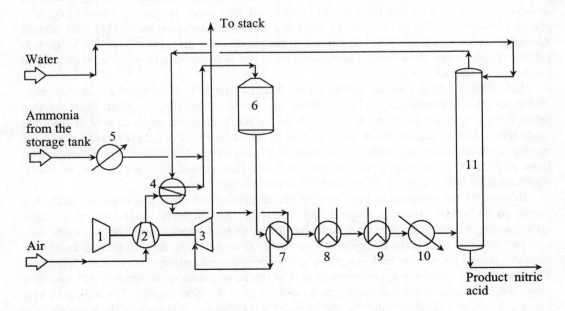

Fig. 1.1 A simplified flow diagram of the nitric acid plant.
(1) Steam turbine (2) multistage air compressor with interstage cooler (3) tail-gas turbine (4) tail-gas preheater (5) ammonia vaporizer (6) reactor (this also includes a waste-heat boiler) (7) tail-gas preheater (8) economizer (9) boiler feed water preheater (10) gas cooler and (11) absorption tower.

words, to continuously remove the heat that enters the storage tank from the ambient air). The storage tank is properly insulated to substantially reduce the transfer of heat from the outside air to the liquid ammonia. Liquid ammonia from the tank is pumped to a *vaporizer* (5) where heat supplied by steam (or any other hot fluid as available) converts the liquid to a vapour.

Compressed air is used for the oxidation of ammonia. Air is usually compressed in a two-stage centrifugal compressor (2) provided with an *interstage cooler*. When compressed, the temperature of a gas increases. Hot air leaving the first stage of the compressor is cooled in a *heat exchanger* by cold water. The cooled gas moves to the second stage of the compressor and exchanges heat with the 'tail gas' (4) from the absorption tower (11). Ammonia vapour is mixed with this compressed air in an appropriate proportion and fed to the reactor (6). The reaction products attain a temperature of around 800°C. This hot product stream, therefore, contains a huge amount of energy. This energy is recovered in a *waste heat boiler* which is a part of the reactor unit, in which transfer of heat from the hot gas to the boiling water occurs, and thus a large quantity of steam is generated. Even after the gas leaves the waste heat boiler, its temperature remains at about 275°C and has to be further reduced before the gas reaches the absorption tower (the gas containing the oxides of nitrogen is absorbed by water in a number of absorption towers in succession to produce nitric acid). Cooling of this gas from 275°C is performed in a series of *heat exchangers* (7–10). In the first heat exchanger (7), heat transfer from the hot gas occurs to the 'tail gas' leaving the final absorption tower. The heated tail gas is then fed to a turbine (3) that drives the air compressor. The second heat exchanger (8) acts as an *economizer*, and the third one (9) is used to heat boiler feed water. In this way, most of the heat energy is recovered. The gas then flows through a series of absorption towers (11) after passing through a *cooler condenser* (10). Absorption of the gas by water gives nitric acid. The absorption process is exothermic. As a result, each absorption tower is required to be provided with a liquid *cooler* (i.e. a heat exchanger).

This simple example shows how important heat transfer is in a chemical process industry. Numerous other examples can be cited. Consider the case of a thermal power plant. There are high rating boilers in which *heat transfer* occurs from the hot combustion gases as well as from the white-hot refractory furnace walls to water boiling inside an array of tubes. Steam leaving the turbine is condensed in huge surface condensers. In the hot rolling mill of an integrated steel plant, large metal blocks are shaped into sheets which are then cooled by copious flow of water. Application of heat transfer is more varied in a nuclear power plant.

Because of the ubiquity of heat transfer phenomena in industrial processes, its study has found an important place in engineering curricula. An engineer should have a sound understanding of the basic principles and applications of heat transfer. How do we *define* heat transfer? Heat transfer is the transport of heat energy from one point in a medium to another or from one medium to another in the presence of a temperature gradient or a temperature difference. The temperature difference between two points in the same medium, or between two mediums which are in thermal contact, is known as the *driving force* for heat transfer. The subject of heat transfer involves a detailed study of the physical mechanism of transport of heat energy, the methods of calculation or prediction of rates of transport of heat energy in various practical situations, and the applications of the theoretical principles to sizing, design and improvement of related equipment used in practice.

It is pertinent here to point out the difference between heat transfer and thermodynamics. Heat transfer is a rate process—the principles of heat transfer enable us to calculate the rate of transport of thermal energy. Classical thermodynamics, on the other hand, deals with processes and systems at equilibrium. The theoretical amount of energy to be supplied (or removed) in order to change a system from one equilibrium state to another is dictated by thermodynamics. But thermodynamics is silent about how quickly this change of equilibrium state can be effected.

1.1 MODES OF HEAT TRANSFER

Three modes of heat transfer are recognized—conduction, convection and radiation. The first two modes appear to be dominant in many practical fields. However, radiation is the most important mode of heat transfer at high temperatures.

1.1.1 Conduction

Conduction heat transfer is an atomic or molecular process. It occurs in the presence of a temperature difference and is not accompanied by any macroscopic or bulk motion in the medium. Conduction is the only mode of heat transfer in a solid medium. It may also occur in a stagnant liquid, or a gaseous medium. The basic law of conduction, the Fourier's law, is explained in Chapter 2, and a few situations of steady state conduction are analyzed. Conduction heat transfer plays a major role in heat loss from furnaces, hot pipe lines (e.g. steam pipes), and process vessels and equipment. It is also important in the process of heat gain by a fluid stored in an insulated tank at a cryogenic temperature. In fact, conduction has a role to play in virtually all of the heat transfer equipment, but in many cases it may not have a governing role.

1.1.2 Convection

Existence of motion or a velocity field in a liquid or a gaseous medium greatly enhances the rate of heat transfer. Convection means the transport of heat energy by way of displacement of fluid elements from one point to another point which is at a different temperature. Convection may be of two types—*forced convection* and *free* or *natural convection*. Forced convection occurs when motion in the medium is caused by an external mechanical agency such as a pump, a blower, an agitator, etc. or by an externally imposed pressure gradient. Free or natural convection occurs when motion in the medium is created by an adverse density gradient, as a result of temperature difference. This happens if the temperature of a fluid at a lower level becomes higher than that at an upper level.

While calculation of the rate of heat transfer by conduction is based on the Fourier's law, the effect of fluid motion on convective heat transfer rate cannot be taken into account by using any similar law. An empirical *heat transfer coefficient*, which is based on the *phenomenological observation* that *heat flux is proportional to the temperature driving force*, is defined instead. The major component of the study of convective heat transfer involves the development and application of correlations for the heat transfer coefficient in diverse physical situations.

1.1.3 Radiation

A body at a temperature above absolute zero always emits energy in the form of electromagnetic waves. The rate of release of such energy is proportional to the fourth power of the absolute temperature of the body. This phenomenon is called radiation and the basic governing law is known as the *Stefan-Boltzmann law*. Ordinarily, the contribution of the radiative component to the total rate of heat transfer from a body becomes significant, if the temperature of the body is sufficiently high. All bodies, however, are not equally good emitters of radiation. A standard or perfect emitter is called a *blackbody*, and acts as a reference with which any other body may be compared in respect of its effectiveness as an emitter. This property of a body is called its *emissivity*. Another quantity which comes into play while calculating the rate of radiative heat exchange between two bodies is called the *view factor*. It takes into account the fraction of the area of a body which can be 'seen' by another body. The study of radiative heat transfer involves the basic principles of radiation and their applications to computations of heat transfer rates in different geometrical configurations of physical systems.

In systems of practical interest, heat transfer mostly occurs by a combination of two or all of the above three modes. Consider, for example, the case of heat loss from a furnace wall. The inner surface of the wall gains heat from the flame and the combustion gases by both radiation and convection. Heat transfer through the refractory and the insulating brick layers of the furnace wall occurs by conduction. Finally, the transfer of heat from the outer surface of the furnace to the ambient occurs by convection and radiation. Boiling of liquids and condensation of vapours, which involve a change of phase, are other ramifications of heat transfer that occur mainly by convection.

1.2   HEAT TRANSFER EQUIPMENT

Heat transfer equipment is designed and built by utilizing the basic principles and correlations of heat transfer along with other relevant physical and engineering principles. The objective is to achieve the desired rate of heat transfer from one medium to another in the best possible way. Most of the heat transfer equipment comes under the general name of *heat exchangers,* and includes *reboilers, condensers, vaporizers* as well. An engineer, particularly a chemical or a mechanical engineer, must have a basic understanding of the design methodology, construction and operation of heat exchangers.

1.3   SYSTEMS OF MEASUREMENT, UNITS AND DIMENSIONS

In this section we give a brief overview of the *units* and *dimensions* of physical quantities and also of the interrelations among the different units used for measurement of such quantities. However, we will generally use the SI units throughout this book.

A *unit* is a 'standard of measurement' of a physical quantity. *Dimension* of a quantity indicates how a physical quantity is related to the basic or fundamental quantities. A system of measurement is based on a few fundamental or basic quantities and their units, which are called the 'basic units'. All other quantities are obtainable by some kind of combination of the basic quantities, and are called 'derived quantities'. The unit of a derived quantity is obtainable in terms of the basic units in any system of measurement.

Until recently, the CGS (Centimetre-Gram-Second) and the FPS (Foot-Pound-Second) systems were most widely used, particularly in scientific literature. The fundamental quantities and their dimensions in these systems are shown in Table 1.1. A few derived quantities of both these systems are also shown in the table. In the study of heat transfer, we will consider these four fundamental quantities—mass, length, time and temperature as the basic units.

Table 1.1 Relevant fundamental and derived quantities of the CGS and the FPS systems

Quantity	Dimension		Unit	
	CGS	FPS	CGS	FPS
Fundamental				
Mass	M	M	g	lb
Length	L	L	cm	ft
Time	t	t	s	s
Temperature	T	T	°C	°F
Derived				
Area	L^2	L^2	cm^2	ft^2
Density	$M L^{-3}$	$M L^{-3}$	g/cm^3	lbm/ft^3
Velocity	$L t^{-1}$	$L t^{-1}$	cm/s	ft/s
Force	$M L t^{-2}$	$M L t^{-2}$	$g\ cm/s^2$ (dyne)	$lb\ ft/s^2$ (poundal)

1.3.1 English Engineering System

This system of measurement and units has been largely used in engineering calculations all over the world (Table 1.2). It differs from the FPS system in that force is considered as a fundamental quantity here whereas in the FPS system, force is a derived quantity (see Table 1.1). So this system uses five fundamental quantities (i.e. mass, length, time, temperature and force, leaving aside the quantities used for electrical measurements).

Table 1.2 Units and dimensions in the English Engineering System

Quantity	Dimension	Unit
Fundamental		
Mass	M	lbm
Length	L	ft
Time	t	s
Temperature	T	°F
Force	F	lbf
Derived		
Energy	F L	ft-lbf
Viscosity	$M L^{-1} t^{-1}$	lbm/ft s

Let us consider Newton's second law of motion that relates the force acting on a body to its mass and acceleration.

$$F = k'ma \tag{1.1}$$

where

F is the force

m is the mass

a is the acceleration

k' is a constant.

In both the FPS and the CGS systems, we define unit force as the amount of force which acting on a body of unit mass, produces unit acceleration. A unit force is then 1 lb ft/s^2 (FPS) or 1 g cm/s^2 (CGS). This definition reduces k' in Eq. (1.1) to unity.

In the English Engineering system, force being a fundamental quantity, unit force or lbf is defined as the amount of force which, when acting on a body of mass m of 1 lbm produces an acceleration of 32.17 ft/s^2. In this system, k' in Eq. (1.1) is replaced by $1/g_c$. That is,

$$F = \frac{1}{g_c} ma \tag{1.2}$$

Putting the values and units of the different quantities in Eq. (1.2), we have

$$1 \text{ lbf} = \frac{1}{g_c} (1 \text{ lbm})(32.17 \text{ ft/s}^2)$$

or

$$g_c = 32.17 \frac{\text{lbm ft}}{\text{lbf s}^2} \tag{1.3}$$

It may be recalled that 32.17 ft/s^2 is the average value of the acceleration due to gravity g over the earth's surface. Though the value of g_c and the average value of g are *numerically equal*, *they have different units*. While g is the acceleration due to gravity, g_c is known as the 'Newton's law conversion factor' or 'the gravitational conversion factor'. The following simple example explains this point of the English system.

The weight of a body of mass m is $W = mg/g_c$ (lbf). The potential energy of a body of mass m at an elevation of h ft is mgh/g_c (ft-lbf), and the kinetic energy of the same body when moving at a velocity v is $mv^2/2g_c$ (ft-lbf).

The unit of heat energy in the English system is Btu as it is in the FPS system.

1.3.2 The International System (SI) of Units (The Système Internationale (SI) de Unitès)

The International System of units was recommended in 1960 at the Eleventh General Conference on Weights and Measures. Since then, it has gradually replaced the English system beginning from the late sixties. Nevertheless, the English system is still preferred by some professionals, because many design data, charts and correlations are readily available in this system.

Some important basic units of SI and also a few derived units are listed in Table 1.3. Conversion of the unit of a quantity from the English system to SI and vice versa, is often required in engineering calculations. Conversion of some of the quantities relevant to our purpose is given in Table 1.4. In engineering practice relating to heat transfer, the quantities 'kcal', 'm', 'h' and '°C' are commonly used.

1.4 EXAMPLES OF UNIT CONVERSION

(i) The coefficient of viscosity of water is 1 cP at 20°C. Calculate its value both in English and SI units.

$$1 \text{ cP} = 0.01 \text{ poise} = 0.01 \frac{g}{cm \text{ s}} = \frac{(0.01)(10^{-3} \text{ kg})}{(10^{-2} \text{m})(\text{s})} = 0.001 \frac{kg}{m \text{ s}} \text{ in SI unit.}$$

Table 1.3 Relevant units and dimensions in the SI

Quantity	Dimension	Unit
Fundamental or basic units*		
Mass	M	kg
Length	L	m
Time	t	s
Temperature	T	K (kelvin)
Derived units		
Force	$M\ L\ t^{-2}$	$\dfrac{kg\ m}{s^2}$ = newton (N)
Work	$M\ L^2\ t^{-2}$	$\dfrac{kg\ m^2}{s^2}$ = N m = joule (J)
Specific heat	$L^2\ t^{-2}\ T^{-1}$	$\dfrac{kg\ m^2}{s^2\ kg\ K} = \dfrac{J}{kg\ K}$
Power	$M\ L^2\ t^{-3}$	$\dfrac{kg\ m^2}{s^3}$ = watt, W (J/s)

* The other fundamental or basic units of SI are:
  electric current: ampere (A); amount of substance: mole (mol); and luminous intensity: candela (cd).

$$1\ cP = 0.01\ poise = 0.01\ \frac{g}{cm\ s} = \frac{(0.01)\left(\dfrac{1}{453.6}lbm\right)}{\left(\dfrac{1}{30.48}ft\right)\left(\dfrac{1}{3600}h\right)} = 2.42\ \frac{lbm}{ft\ h}$$

(ii) **The thermal conductivity of magnesite at 600°C is 6.7×10^{-3} cal/cm s °C. Find its value both in the English and the SI units.**

$$6.7 \times 10^{-3}\ \frac{cal}{cm\ s\ °C} = \frac{(6.7 \times 10^{-3})\left(\dfrac{1}{252}Btu\right)}{\left(\dfrac{1}{3600}h\right)\left(\dfrac{1}{30.48}ft\right)\left(\dfrac{9}{5}°F\right)} = 1.62\ \frac{Btu}{h\ ft\ °F}$$

$$6.7 \times 10^{-3}\ \frac{cal}{cm\ s\ °C} = \frac{(6.7 \times 10^{-3})(4.187\ W\ s)}{(s)\left(\dfrac{1}{100}m\right)(K)} = 2.805\ \frac{W}{m\ K}$$

Table 1.4   Units and conversions of some important physical quantities

Quantity	SI	English	CGS	SI → English	English → SI	Other units/conversion
Mass	kg	lbm	g	1 kg = 2.2046 lbm	1 lbm = 0.4536 kg	1 lbm = 453.6 g
Length	m	ft	cm	1 m = 3.2808 ft	1 ft = 0.3048 m	1 ft = 30.48 cm
Area	m²	ft²	cm²	1 m² = 10.7639 ft²	1 ft² = 0.0929 m²	1 ft² = 929 cm²
Volume	m³	ft³	cm³	1 m³ = 35.3134 ft³	1 ft³ = 0.028317 m³	1 ft³ = 28317 cm³
Density	kg/m³	lbm/ft³	g/cm³	1 kg/m³ = 0.06243 lbm/ft³	1 lbm/ft³ = 16.018 kg/m³	1 lbm/ft³ = 16018 g/cm³
Force	N (kg m/s²)	lbf	dyne (= 10⁻⁵ N)	1 N = 0.2248 lbf	1 lbf = 4.4482 N	1 lbf = 4.4482 × 10⁵ dyne
Pressure	N/m² (= pascal, Pa) (= 1.4504 × 10⁻⁴ lbf/inch²)	lbf/ft² lbf/inch²	dyne/cm²	1 N/m² = 1 Pa (= 2.0885 × 10⁻² lbf/ft² = 6.8946 × 10³ N/m²)	1 lbf/ft² = 47.881 N/m² 1 lbf/inch² = 1psi	1 atm = 14.7 psi = 1.013 × 10⁵ N/m² = 1.013 bar = 1.033 kg cm² 1 kg/cm² = 0.97 atm
Power	W	ft-lbf/s	dyne-cm/s	1 W = 0.737 ft-lbf/s = 1.34 × 10⁻³ hp	1 ft-lbf/s	1 hp = 746 W = 641.6 kcal/h
Heat	J (= N m)	Btu	cal	1 J = 9.4783 × 10⁻⁴ Btu	1 Btu = 1055.04 J	1 cal = 4.187 J (or W s) 1 Btu = 252 cal
Heat flux	W/m²	Btu/(h ft²)	cal/(s cm²)	1 W/m² = 0.317 Btu/(h ft²)	1 Btu/(h ft²) = 3.154 W (or J/s)	1 Btu/(h ft²) = 2.712 kcal/(h m²)
Heat flow rate	W	Btu/h	cal/s	1 W = 3.4121 Btu/h	1 Btu/h = 0.2931 W	1 Btu/h = 0.252 kcal/h
Heat transfer coefficient	W/(m² K)	Btu/(h ft² °F)	cal/(s cm² °C)	1 W/(m² K) = 0.1761 Btu/(h ft² °F)	1 Btu/(h ft² °F) = 5.678 W/(m² K)	1 Btu/(h ft² °F) = 4.88 kcal/(h m² °C)
Specific heat	J/(kg K)	Btu/(lbm °F)	cal/(g °C)	1 J/(kg K) = 2.3884 × 10⁻⁴ Btu/(lbm °F)	1 Btu/(lbm °F) = 4187 J/(kg K)	
Thermal conductivity	W/(m K)	Btu/(h ft °F)	cal/(s cm °C)	1 W/(m K) = 0.5778 Btu/(h ft °F)	1 Btu/(h ft °F) = 1.7307 W/(m K)	1 Btu/(h ft °F) = 1.488 kcal/(h m °C)
Viscosity	kg/(m s) (= Pa s)	lbm/(ft h)	g/(cm s) (= poise)	1 kg/(m s) = 2419.2 lbm/(ft h)	1 lbm/(ft h) = 4.133 × 10⁻⁴ kg/(m s)	1 poise = 2.42 lbm/(ft h) = 0.1 kg/(m s)

Note: The brackets in the denominators of the above units have been dropped in subsequent chapters for ease of convenience.

2

Steady State Conduction in One Dimension

Conduction of heat means transport of heat energy in a medium from a region at a higher temperature to a region at a lower temperature without any macroscopic motion in the medium. The difference in temperature between the regions causes the flow of heat and is called the *temperature driving force*. Heat conduction is also called *diffusion of heat*.

The mechanism of heat conduction in a medium depends upon the state of the medium, i.e. whether it is a solid, a liquid or a gas. In a solid, the molecular motion is restricted to vibrations about an equilibrium position. In the presence of a temperature gradient heat energy is transferred from one molecule to a neighbouring molecule through molecular vibrations. In metals, however, conduction of heat occurs more through the drift of free electrons than by molecular vibrations. The motion of free electrons in metals is similar to that of molecules in a gas (free electrons are often referred to as *electron gas*), and this is why a material having good electrical conductivity also possesses good thermal conductivity.

In a gaseous medium, conduction of heat occurs through collisions of molecules having more thermal energy (i.e. faster moving) with molecules having less thermal or kinetic energy (slower moving). While similar phenomenon is partly responsible for heat conduction in a liquid, there are other factors too which involve intermolecular forces in the liquid.

Transport of heat in a solid occurs only by conduction. In this chapter we will describe the basic law of heat conduction and discuss its applications to heat transfer calculations in single- or multi-layer solids of three common geometries—plane wall, cylindrical and spherical. We will also consider problems which involve generation of heat (e.g. a nuclear fuel) in solids having one of the above geometries.

2.1 THE BASIC LAW OF HEAT CONDUCTION—FOURIER'S LAW

The basic law of heat conduction in a medium was established by J.B.J. Fourier, a physicist, in 1822 from his experimental data on the rate of heat flow. The law states that *if two plane parallel surfaces each having an area A are separated by a distance l and are maintained at temperatures T_1 and T_2, respectively ($T_1 > T_2$) (Fig. 2.1), the rate of heat conduction Q at steady state through the wall is given by*

$$Q = kA \frac{T_1 - T_2}{l} \tag{2.1}$$

where k is called the thermal conductivity of the solid and is assumed to be constant throughout the wall.

9

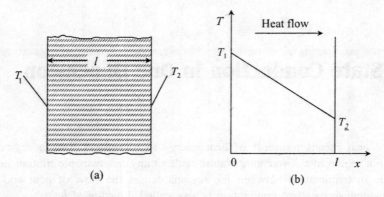

Fig. 2.1 Steady state conduction of heat in a plane wall: (a) the wall and (b) the temperature profile.

In the differential form the Fourier's law is expressed as

$$q_x = -k \frac{dT}{dx} \qquad (2.2)$$

where q_x is the heat flux (i.e. the rate of heat conduction in the x-direction per unit area normal to the x-direction), and dT/dx is the temperature gradient in the x-direction. Heat flow in the direction of decreasing temperature is a positive quantity, therefore dT/dx is negative. The negative sign in Eq. (2.2) makes both sides of it consistent with respect to sign. The units of the various quantities in Eqs. (2.1) and (2.2) are: Q, W (watt) [or, kcal/s, Btu/h, etc.]; k, W/m °C [or, cal/s cm °C, Btu/h ft °F, etc.]; q_x, W/m^2 [or, kcal/s cm^2, Btu/h ft^2, etc.]; dT/dx, °C/m [or, K/m, °F/ft, etc.]. The quantities q_x and dT/dx are vector quantities.

Equation (2.1) is the integrated form of Eq. (2.2) over the thickness l of the wall for an area A.

2.2 THERMAL CONDUCTIVITY

Thermal conductivity is a fundamental property of a material that gives a measure of the effectivity of the material in transmitting heat through it. Besides its chemical constitution, characteristics of a material (solid, liquid or gas), nature of the solid state (crystalline or amorphous), and physical conditions (temperature, pressure) have significant effects on thermal conductivity.

A material in the crystalline state has a higher thermal conductivity than that of the same material in the amorphous state. The orderly arrangement of the atoms in a crystalline solid allows faster transmission of thermal energy through vibrations of the crystal lattices. The thermal conductivity of a substance depends to some extent on temperature. For metals, thermal conductivity generally decreases with an increase in temperature (aluminium is an exception). A metal in the pure state has the maximum thermal conductivity. The thermal conductivity decreases with increasing amounts of impurities in a metal. Most nonmetallic substances are poor conductors of heat (Tye, 1969).

Solid substances having low thermal conductivities are called *thermal insulators*. Such a substance can be applied to the surface of a metal to reduce its rate of exchange of heat energy with the surroundings. Besides low thermal conductivity, an insulator must be chemically inert, available in a form suitable for application on a surface, and should be light and cheap. Commercial insulators are ceramic (e.g. insulating bricks), diatomaceous (e.g. rock wool) or polymeric (e.g. expanded polyurethane, expanded polystyrene, etc.) materials. These materials are either highly porous, or have a high void volume filled with an inert gas. A gas being a very poor conductor of heat, a high void volume fraction makes an insulator more effective in reducing the flow of heat. In fact, the presence of entrapped gas is more responsible for the insulating property than for the 'true' thermal conductivity of the material. The thermal conductivity of an insulating material increases with temperature.

It is relevant here to point out the difference between an insulating material and a refractory. The latter is a substance capable of withstanding high temperatures without physical deterioration, but need not necessarily have a low thermal conductivity like an insulator. However in most cases, low thermal conductivity is a desirable characteristic of refractory materials. A refractory material must have chemical resistance to the environment to which it is exposed. A furnace, for example, is provided with an innermost layer of refractory bricks, one or two layers of insulating bricks, and often with an outer layer of ordinary bricks. Common examples of refractory materials are silica, magnesite, alumina, fire-clay, sillimanite (Al_2SiO_5), chromite ($FeO.Cr_2O_3$), zirconia (ZrO_2), etc. For most refractories, the thermal conductivity increases with increase in temperature (exception: magnesite bricks).

Thermal conductivities of liquids generally decrease with increase in temperature (exception: glycerine, water, etc. over certain ranges of temperature), but are nearly independent of pressure. Leaving aside liquid metals, water has the highest thermal conductivity of all liquids. Thermal conductivities of gases increase with temperature and decrease with molecular weight; pressure dependence becomes significant only at high pressures. Among the gases, hydrogen has the highest thermal conductivity.

Generally, gases have lower thermal conductivities than those of liquids, and liquids have conductivties lower than those of solids. For example, at the ambient temperature air has a thermal conductivity of 0.0262 W/m K (0.0151 Btu/h ft °F), water has a thermal conductivity of 0.63 W/m K (0.364 Btu/h ft °F), whereas pure silver has a conductivity of 410 W/m K (237 Btu/h ft °F).

In heat transfer calculations, it is often satisfactory and sufficient to take the thermal conductivity value of a substance at the average temperature of the material or medium. However, a linear or a quadratic equation as given below can be used to describe thermal conductivity as a function of temperature.

$$k = k_0(1 + aT + bT^2) \tag{2.3}$$

where a and b are constant coefficients, and k_0 is the thermal conductivity at $T = 0$ K.

If the thermal conductivity of a solid is the same in all directions, the material is called *isotropic*. But there are some materials in which the conductivity depends upon the direction as well. Such materials are called *anisotropic*. For example, the thermal conductivity of wood in the direction of the grains is different from that in the transverse direction. So wood is an anisotropic material.

Thermal conductivities of some substances are given below in Table 2.1.

Table 2.1 Thermophysical properties of some common materials

Material	Temperature (°C)	Density (kg/m^3)	Specific heat (kJ/kg °C)	Thermal conductivity (W/m °C)
Non-metals				
Building bricks	20	1600	0.84	0.69
Fireclay brick (fired at 1450°C)	500	2300	0.96	1.28
Magnesite	1200	—	—	1.90
Window glass	20	2700	0.84	0.78
Asbestos sheet	59	—		0.166
Corkboard (density = 160 kg/m^3)	30	—		0.043
Wool	30	130–200		0.036
Metals				
Carbon steel (0.5% C)	100	7800	0.465	52
Stainless steel (316)	27	8240	0.468	13.4
Stainless steel (304)	27	7900	0.477	14.9
Copper (pure)	27	8933	0.385	401
Aluminium (pure)	27	2702	0.903	237
Silver (99.9%)	20	10,525	0.234	407
Zinc (pure)	20	7144	0.384	112.2

2.3 STEADY STATE CONDUCTION OF HEAT THROUGH A COMPOSITE SOLID

For a 'single-layered' plane wall, the rate of heat conduction Q can be calculated directly from Eq. (2.1), if the surface area A, the wall thickness l, the thermal conductivity k, and the temperature difference are known. We can extend the same equation to the case of a composite or 'multi-layered' wall.

Let us consider a composite wall consisting of three layers (Fig. 2.2) of materials 1, 2 and 3, having thicknesses l_1, l_2 and l_3, and thermal conductivities k_1, k_2 and k_3, respectively. The boundary temperatures of the different layers are shown in Fig. 2.2.

The area of heat conduction A is constant. Therefore, the rates of heat flow at steady state[1] through the individual layers are equal ($Q_1 = Q_2 = Q_3 = Q$, say). If the thermal conductivities of the layers are independent of temperature, the temperature distribution in each layer must be

---

[1] The term *steady state* or *steady state heat transfer* will occur frequently in our discussion. It will be useful to explain steady state and to differentiate it from equilibrium state. 'Steady state' means the state of the system that does not change with time. In the case of heat transfer, in particular, it means that both the temperature and the heat flux at any location remain constant or invariant with respect to time. For example, if we consider the plane wall in Fig. 2.1, the flow of heat occurs at steady state at the rate given by Eq. (2.1) if the temperatures T_1 and T_2 at the surfaces have been maintained for a sufficiently long time. But the system is surely not at thermal equilibrium because, in that case, the temperature of the wall must have been the same throughout and the rate of heat flow through the wall zero.

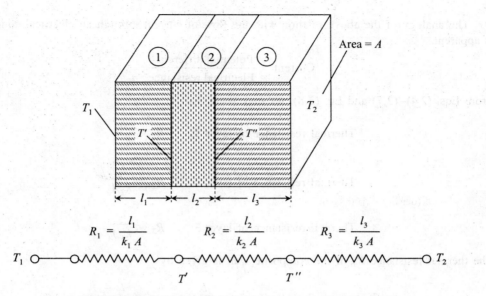

Fig. 2.2 Heat conduction in a composite wall and its electrical analogue.

linear (but of different slopes). The rates of heat flow through the walls as given by Fourier's law are as follows:

Layer 1:
$$Q = \frac{k_1 A(T_1 - T')}{l_1}; \qquad T_1 - T' = Q\frac{l_1}{k_1 A} \qquad (2.4)$$

Layer 2:
$$Q = \frac{k_2 A(T' - T'')}{l_2}; \qquad T' - T'' = Q\frac{l_2}{k_2 A} \qquad (2.5)$$

Layer 3:
$$Q = \frac{k_3 A(T'' - T_2)}{l_3}; \qquad T'' - T_2 = Q\frac{l_3}{k_3 A} \qquad (2.6)$$

Adding Eqs. (2.4), (2.5) and (2.6), we have

$$T_1 - T_2 = Q\left(\frac{l_1}{k_1 A} + \frac{l_2}{k_2 A} + \frac{l_3}{k_3 A}\right)$$

or

$$Q = \frac{T_1 - T_2}{\dfrac{l_1}{k_1 A} + \dfrac{l_2}{k_2 A} + \dfrac{l_3}{k_3 A}} \qquad (2.7)$$

In Eq. (2.7), $(T_1 - T_2)$ is the overall temperature driving force that causes a rate of heat transfer Q. This equation, in general, means

$$\text{Rate of heat transfer} = \frac{\text{Temperature driving force}}{\text{Thermal resistance}} \qquad (2.8)$$

The analogy of the above relation with the flow of current through an electrical conductor is apparent.

$$\text{Current} = \frac{\text{Potential difference}}{\text{Electrical resistance}}$$

From Eqs. (2.4)–(2.7) and Eq. (2.8), we may write

Thermal resistance of layer 1: $R_1 = \dfrac{l_1}{k_1 A}$ (2.9a)

Thermal resistance of layer 2: $R_2 = \dfrac{l_2}{k_2 A}$ (2.9b)

Thermal resistance of layer 3: $R_3 = \dfrac{l_3}{k_3 A}$ (2.9c)

The thermal resistance R_T of the composite wall is given by

$$R_T = R_1 + R_2 + R_3 = \frac{l_1}{k_1 A} + \frac{l_2}{k_2 A} + \frac{l_3}{k_3 A} = \frac{T_1 - T_2}{Q}$$

It means that thermal resistances in series are additive as in the case of electrical resistances in series. The electrical analogue of heat conduction through the wall is also shown in Fig. 2.2.

Example 2.1   The wall of a cold storage consists of three layers—an outer layer of ordinary bricks, 25 cm thick, a middle layer of cork, 10 cm thick, and an inner layer of cement, 6 cm thick. The thermal conductivities of the materials are—brick: 0.7, cork: 0.043, and cement: 0.72 W/m °C. The temperature of the outer surface of the wall is 30°C, and that of the inner is –15°C. Calculate (a) the steady state rate of heat gain per unit area of the wall, (b) the temperatures at the interfaces of the composite wall, and (c) the percentages of the total heat transfer resistance offered by the individual layers. What additional thickness of cork should be provided to make the rate of heat transfer 30% less than the present value?

SOLUTION   Let the three layers be layer 1: brick; layer 2: cork; and layer 3: cement. If we consider unit area of the wall (1 m^2), then from Eq. (2.9), we get

Thermal resistance of the brick layer, $R_1 = \dfrac{l_1}{k_1 A} = \dfrac{0.25 \text{ m}}{(0.7 \text{ W/m °C})(1 \text{ m}^2)} = 0.357 \text{ °C/W}$

Thermal resistance of the cork layer, $R_2 = \dfrac{l_2}{k_2 A} = \dfrac{0.10 \text{ m}}{(0.043 \text{ W/m °C})(1 \text{ m}^2)} = 2.326 \text{ °C/W}$

Thermal resistance of the cement layer, $R_3 = \dfrac{l_3}{k_3 A} = \dfrac{0.06 \text{ m}}{(0.72 \text{ W/m °C})(1 \text{ m}^2)} = 0.083 \text{ °C/W}$

Total thermal resistance = $R_1 + R_2 + R_3$ = 0.357 + 2.326 + 0.083 = 2.766 °C/W

Temperature driving force = 30 – (–15) = 45°C

(a) Rate of heat gain by the wall, $Q = \dfrac{45°C}{2.766 \ °C/W} = \boxed{16.27 \ W}$

(b) From Eq. (2.4) and Fig. 2.2, the temperature drop across the brick layer

$$T_1 - T' = Q \frac{l_1}{k_1 A} = (16.27)(0.357) = 5.8°C$$

Now $T_1 = 30°C$; therefore, $T' = 30°C - 5.8°C = \boxed{24.2°C}$ (interface temperature between brick and cork).

Similarly, $T' - T'' = Q \dfrac{l_2}{k_2 A} = (16.27) (2.326) = 37.8°C$

Therefore,

$T'' = 24.2 - 37.8 = \boxed{-13.6°C}$ (interface temperature between cork and cement).

(c) Per cent thermal resistance offered by the brick layer $= \dfrac{0.357}{2.766} \times 100 = \boxed{12.9\%}$

$$\text{cork layer} = \frac{2.326}{2.766} \times 100 = \boxed{84.1\%}$$

$$\text{cement layer} = \frac{0.083}{2.766} \times 100 = \boxed{3.0\%}$$

Second part

Desired rate of heat flow (30% less than the present value) $= (16.27)(0.7) = 11.39 \ W$

Required thermal resistance $= \dfrac{\text{Driving force}}{\text{Rate of heat flow}} = \dfrac{45}{11.39} = 3.95 \ °C/W$

Additional thermal resistance to be provided $= 3.95 - 2.766 = 1.184 \ °C/W$

Additional thickness of cork to be provided $= (1.184)(0.043)(1) = \boxed{5.1 \ cm}$

Example 2.2 The walls of a house in a cold region consist of three layers—an outer brickwork of 15 cm thickness and an inner wooden panel of 1.2 cm thickness. The intermediate layer is made of an insulating material 7 cm thick. The thermal conductivities of the brick and the wood used are 0.70 W/m °C and 0.18 W/m °C, respectively. The inside and outside temperatures of the composite wall are 21°C and –15°C, respectively. If the layer of insulation offers twice the thermal resistance of the brick wall, calculate (a) the rate of heat loss per unit area of the wall and (b) the thermal conductivity of the insulating material.

SOLUTION Consider 1 m² of the wall area.

(a) If l = thickness of a layer, A = area, and k = thermal conductivity, then, thermal resistance $= \dfrac{l}{kA}$

In the given case, $A = 1.0 \ m^2$

Thermal resistance of the layer of brick (l = 0.15 m; k = 0.70 W/m °C)

$$= \frac{0.15}{(0.70)(1.0)} = 0.2143 \text{ °C/W}$$

Thermal resistance of the layer of wood (l = 0.012 m; k = 0.18 W/m °C)

$$= \frac{0.012}{(0.18)(1.0)} = 0.0667 \text{ °C/W}$$

Thermal resistance of the insulating layer = (2)(0.2143) = 0.4286 °C/W

Total thermal resistance = 0.2143 + 0.4286 + 0.0667 = 0.7096 °C/W

Temperature driving force = 21 − (−15) = 36°C

$$\text{Rate of heat loss} = \frac{36°C}{0.7096 \text{ °C/W}} = \boxed{50.7 \text{ W}}$$

(b) For the insulating layer, l = 0.07 m; A = 1.0 m²; thermal resistance = $\dfrac{l}{kA}$ = 0.4286 °C/W. Therefore,

$$\text{Thermal conductivity, } k = \frac{0.07}{(1.0)(0.4286)} = \boxed{0.1633 \text{ W/m °C}}$$

2.4 STEADY STATE HEAT CONDUCTION THROUGH A VARIABLE AREA

In the case of a plane wall the area for heat flow is constant. However, there are solids of other geometries in which the area for heat flow is variable. Two common geometries of practical importance are cylindrical and spherical, in both of which the area depends upon the radius or the radial position. Here, we develop equations for heat flow in cylindrical and spherical bodies.[1]

2.4.1 The Cylinder

Let us consider a hollow cylinder of inside radius r_i, outside radius r_o and length L (Fig. 2.3). The inner and outer curved surfaces are maintained at temperatures T_i and T_o, respectively. Heat flow occurs in the radial direction. The area of heat flow varies from $2\pi r_i L$ (inside) to $2\pi r_o L$ (outside).

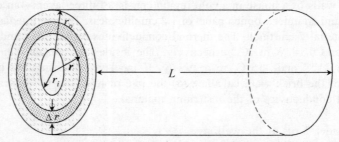

Fig. 2.3 Radial heat conduction through a thin cylindrical shell.

---

[1]Detailed analysis of conduction heat transfer in solids of various shapes is available in the literature (see, Schneider, 1955; Myers, 1971).

We can make a 'heat balance' over a thin cylindrical shell of inside radius r and thickness Δr as shown in Fig. 2.3.

Rate of heat input to the thin shell at the radial position (r) = (Area)(Flux) = $(2\pi rL) q_r|_r$

Here q_r denotes the radial heat flux*, and the notation ' $|_r$' means that the quantity is evaluated at the radial position r.

Similarly,

Rate of heat output from the shell at the radial position $r + \Delta r = (2\pi rL) q_r|_{r+\Delta r}$**

At steady state, there cannot be any accumulation of heat in the element, and so the rate of heat input must be equal to the rate of heat output. Therefore,

$$(2\pi rL) q_r|_{r+\Delta r} = (2\pi rL) q_r|_r$$

or

$$\frac{(2\pi rL)q_r|_{r+\Delta r} - (2\pi rL)q_r|_r}{\Delta r} = 0$$

Taking the limit $\Delta r \to 0$ and invoking the definition of the derivative of a function, we may write,

$$\frac{d}{dr}(2\pi rLq_r) = 0$$

or

$$\frac{d}{dr}(rq_r) = 0 \tag{2.10}$$

Integrating, we get

$$r\,q_r = \text{constant} = C_1 \text{ (say)} \tag{2.11}$$

The radial heat flux*** is expressed as

$$q_r = -k\frac{dT}{dr}$$

Then,

$$r\left(-k\frac{dT}{dr}\right) = C_1$$

or

$$\frac{dT}{dr} = -\frac{C_1}{k}\frac{1}{r}$$

Integrating again, we have

$$T = -\frac{C_1}{k}\ln r + C_2 \tag{2.12}$$

---

* The suffix r in q_r refers to the radial direction, and not to the radial position.

** This means that the entire quantity $(2\pi rL) q_r$ is evaluated at the radial position $r + \Delta r$.

*** We have assumed that heat enters the thin shell at r and leaves it at $r + \Delta r$, i.e. T decreases with r. In conformity with this, dT/dr has to be negative. But heat flows in the direction of decreasing temperature, and the heat flux is a positive quantity here. The negative sign is necessary in order to maintain consistency of sign [see Eq. (2.2)].

here C_2 is an integration constant. The values of C_1 and C_2 can be determined by using the 'boundary conditions', that is, the known temperatures at the boundaries r_i and r_o of the solid.

$$T = T_i \quad \text{at} \quad r = r_i, \qquad T = T_o \quad \text{at} \quad r = r_o$$

Substituting in Eq. (2.12), we get

$$T_i = -\frac{C_1}{k} \ln r_i + C_2 \qquad \text{and} \qquad T_o = -\frac{C_1}{k} \ln r_o + C_2$$

The above two equations can be solved for C_1 and C_2.

$$C_1 = \frac{(T_i - T_o)k}{\ln (r_o/r_i)} \qquad \text{and} \qquad C_2 = T_i + \frac{(T_i - T_o) \ln r_i}{\ln (r_o/r_i)}$$

Substituting for C_1 and C_2 in Eq. (2.12), we get

$$T = T_i - \frac{(T_i - T_o)}{\ln (r_o/r_i)} \ln (r/r_i) \tag{2.13}$$

Equation (2.13) gives the temperature distribution in the cylinder. The rate of heat flow can be easily obtained from Eq. (2.11) which essentially means that the product of area and radial heat flux is a constant.

Rate of heat flow is given by

$$Q = 2\pi r L q_r = 2\pi L C_1 = \frac{2\pi L k(T_i - T_o)}{\ln (r_o/r_i)} \tag{2.14}$$

Comparing Eq. (2.14) with Eq. (2.8), we get

$$\text{Driving force} = (T_i - T_o); \qquad \text{Thermal resistance} = \frac{\ln (r_o/r_i)}{2\pi L k}$$

We may thus rewrite Eq. (2.14) in the form

$$Q = k \, 2\pi \frac{(r_o - r_i)}{\ln (r_o/r_i)} L \frac{(T_i - T_o)}{(r_o - r_i)} \tag{2.15}$$

The distance through which conduction occurs is the thickness of the wall of the cylinder, i.e. $r_o - r_i$. If we compare Eq. (2.15) with Eq. (2.1), the rate of heat conduction is found to be the same as that through a plane wall of thickness $(r_o - r_i)$ and area $2\pi \, [(r_o - r_i)/\ln (r_o/r_i)] \, L$. This area is called the 'log mean area of the cylinder' because it is calculated on the basis of the log mean cylinder radius, r_M. Therefore,

$$r_M = \frac{r_o - r_i}{\ln (r_o/r_i)}$$

Now we consider heat conduction through a composite cylindrical wall consisting of three layers denoted by 1, 2 and 3 having thermal conductivities k_1, k_2 and k_3 and having inner radii r_i, r' and r'', respectively. The outer radius of the composite cylinder is r_o. The temperatures at these radial positions are T_i, T', T'' and T_o, respectively. A cross-section of the assembly is shown in Fig. 2.4.

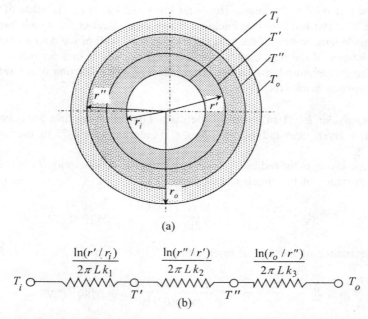

(a)

$$\frac{\ln(r'/r_i)}{2\pi L k_1} \qquad \frac{\ln(r''/r')}{2\pi L k_2} \qquad \frac{\ln(r_o/r'')}{2\pi L k_3}$$

$T_i \circ\!\!-\!\!\bigwedge\!\!\bigwedge\!\!\bigwedge\!\!-\!\!\circ\!\!-\!\!\bigwedge\!\!\bigwedge\!\!\bigwedge\!\!-\!\!\circ\!\!-\!\!\bigwedge\!\!\bigwedge\!\!\bigwedge\!\!-\!\!\circ\, T_o$

$\qquad\qquad\quad T' \qquad\qquad\qquad T''$

(b)

Fig. 2.4 (a) Cross-section of the composite cylinder and (b) the electrical analogue of the heat transfer resistances.

The rates of heat flow through the individual layers (which will be equal at steady state) are given by

$$Q = \frac{2\pi L k_1 (T_i - T')}{\ln(r'/r_i)} = \frac{2\pi L k_2 (T' - T'')}{\ln(r''/r')} = \frac{2\pi L k_3 (T'' - T_o)}{\ln(r_o/r'')}$$

or

$$T_i - T' = \frac{Q\ln(r'/r_i)}{2\pi L k_1}; \qquad T' - T'' = \frac{Q\ln(r''/r')}{2\pi L k_2}; \qquad T'' - T_o = \frac{Q\ln(r_o/r'')}{2\pi L k_3} \qquad (2.16)$$

Adding the above equations and rearranging, we get

$$Q = \frac{T_i - T_o}{\dfrac{\ln(r'/r_i)}{2\pi L k_1} + \dfrac{\ln(r''/r')}{2\pi L k_2} + \dfrac{\ln(r_o/r'')}{2\pi L k_3}} \qquad (2.17)$$

Equation (2.17) is of the form of Eq. (2.8). The total thermal resistance of the composite cylinder, given by the denominator of Eq. (2.17), is the sum of the thermal resistances of the individual layers. The overall temperature driving force is $T_i - T_o$. The electrical analogue of the problem is also shown in Fig. 2.4. The rate of heat conduction can be calculated if the cylinder dimensions, thermal conductivities of the layers and the driving force are known. Conversely, the temperature T' and T'' can be obtained from the given or the calculated value of Q.

Example 2.3 A cylindrical hot gas duct, 0.5 m inside radius, has an inner layer of fireclay bricks

(k = 1.3 W/m °C) of 0.27 m thickness. The outer layer, 0.14 m thick, is made of a special brick (k = 0.92 W/m °C). The brickwork is enclosed by an outer steel cover which has a temperature of 65°C. The inside temperature of the composite cylindrical wall of the duct is 400°C. Neglecting the thermal resistance of the steel cover, calculate the rate of heat loss per metre of the duct and also the interface temperature between the ceramic layers. What fraction of the total resistance is offered by the special brick layer?

SOLUTION   Consider L = 1 m length of the duct. The following data are given:

Fireclay brick layer: inner radius, r_i = 0.5 m, r' = outer radius = 0.77 m, thermal conductivity, k_1 = 1.3 W/m °C

Special brick layer: outer radius, r_o = 0.91 m, thermal conductivity, k_2 = 0.92 W/m °C.

Heat transfer resistance of the fireclay brick layer

$$R_1 = \frac{\ln (r'/r_i)}{2\pi L k_1} = \frac{\ln (0.77/0.5)}{2\pi(1)(1.3)} = 0.0529 \text{ °C/W}$$

Heat transfer resistance of the special brick layer

$$R_2 = \frac{\ln (r_o/r')}{2\pi L k_2} = \frac{\ln (0.91/0.77)}{2\pi(1)(0.92)} = 0.0289 \text{ °C/W}$$

Total resistance, $R_1 + R_2$ = 0.0529 + 0.0289 = 0.0818 °C/W

Driving force = $T_i - T_o$ = 400 – 65 = 335°C

Rate of heat loss, $Q = \dfrac{T_i - T_o}{R_1 + R_2} = \dfrac{335}{0.0818} = \boxed{4095 \text{ W}}$

Interface temperature T' is given by, $Q = \dfrac{T_i - T'}{R_1}$

or

$$4095 = \frac{400 - T'}{0.0529}$$

or

$$T' = \boxed{183°C}$$

Fractional resistance offered by the special brick layer = $\dfrac{R_2}{R_1 + R_2} = \dfrac{0.0289}{0.0818} = \boxed{0.353}$

Example 2.4   A tapered stainless steel rod, perfectly insulated on the curved surface, has end diameters of 0.06 m and 0.12 m respectively, and is 0.2 m long. The thicker end is fixed to a hot wall and the thinner end is maintained at 30°C. The steady state rate of heat loss through the rod is found to be 50 W. The thermal conductivity of stainless steel is 15 W/m °C. Calculate (a) the hot end temperature, (b) the temperature gradients at both the ends, and (c) the temperature at a section of the rod 0.15 m away from the cold end.

SOLUTION   This is a case of steady state heat conduction through a variable area. The system is shown in Fig. 2.5. We first determine the axial temperature distribution in the tapered solid.

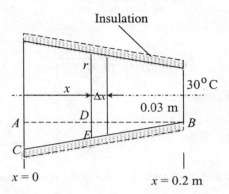

Fig. 2.5   A tapered rod (Example 2.4).

Because the curved surface of the rod is perfectly insulated, heat conduction occurs in the axial direction (x-direction) only (i.e. the temperature at any cross-section is uniform). Let us choose the origin at the centre of the thicker end of the rod. If r is the radius of the rod at any axial position x, and q_x is the axial heat flux, we have

$$\text{Rate of heat input at } x = \pi r^2 \; q_x|_x$$

$$\text{Rate of heat output at } x + \Delta x = \pi r^2 \; q_x|_{x + \Delta x}$$

$$\text{Rate of heat accumulation} = 0$$

Therefore, by a steady state heat balance, we get

$$\pi r^2 \; q_x|_x - \pi r^2 \; q_x|_{x + \Delta x} = 0$$

Dividing by Δx throughout and taking the limit $\Delta x \to 0$, we get

$$\frac{d}{dx}(\pi r^2 q_x) = 0$$

Integrating and putting the expression for the axial heat flux, we have

$$-\pi r^2 k \frac{dT}{dx} = C_1 \text{ (constant)} \tag{i}$$

The local radius r depends on the axial position x and should be expressed in terms of x before Eq. (i) is further integrated. The relation between r and x may be obtained from the similar triangles ABC and DBE.

$$AC = 0.03 \text{ m}; \qquad AB = 0.2 \text{ m}; \qquad DB = (0.2 - x) \text{ m}; \qquad \frac{DE}{AC} = \frac{DB}{AB}$$

$$DE = \frac{(0.03)(0.2 - x)}{0.2} = 0.03 - 0.15x; \text{ and } r = 0.03 + DE = 0.06 - 0.15x$$

Substituting for r in Eq. (i) and integrating, we get

$$T = -\frac{C_1}{\pi k} \int \frac{dx}{(0.06 - 0.15x)^2} + C_2$$

or

$$T = -\frac{C_1}{(\pi k)(0.15)} \frac{1}{(0.06 - 0.15x)} + C_2 \tag{ii}$$

where C_2 is the integration constant.

Rate of heat input at $x = 0$ is $\quad -\left[\pi r^2 k \frac{dT}{dx}\right]_{x=0} = C_1$

But the rate of heat input at $x = 0$ (which is the same as the rate of heat loss through the rod) $= 50$ W. Therefore,

$$C_1 = 50$$

Also, at $x = 0.2$ m, $T = 30°C$; thermal conductivity, $k = 15$ W/m °C
Substituting all these values in Eq. (ii), we get

$$C_2 = 265.8$$

Putting the values of C_1 and C_2 in Eq. (ii), the axial temperature profile in the tapered cylinder is obtained as

$$T = 265.8 - \frac{7.07}{0.06 - 0.15x} \tag{iii}$$

(a) At the hot end, $x = 0$, and the corresponding temperature is

$$T = 265.8 - \frac{7.07}{0.06} = \boxed{148°C}$$

(b) From Eq. (i), we have

$$\frac{dT}{dx} = -\frac{C_1}{\pi r^2 k}$$

At $x = 0$, $\quad r = 0.06$. Therefore,

$$\frac{dT}{dx} = \frac{-50}{(\pi)(0.06)^2(15)} = \boxed{-294.7 \text{ °C/m}}$$

This is the temperature gradient at the hot end.

Similarly, the temperature gradient at the cold end ($x = 0.2$ m) can be found to be $\boxed{-1179 \text{ °C/m}}$.

(c) At a section 0.15 m away from the cold end, $x = 0.2 - 0.15 = 0.05$. The temperature at this point can be calculated by putting this value of x in Eq. (iii) which gives $T = 131°C$.

2.4.2 The Sphere

The temperature distribution in a hollow sphere or the rate of heat transfer through it can be determined by following the same procedure as used in the case of a cylinder.

Let us consider a hollow sphere of thermal conductivity k and inner and outer radii r_i and r_o respectively, with the corresponding surface temperatures T_i and T_o. Considering a thin spherical shell of inner radius r and thickness Δr (Fig. 2.6), we may write

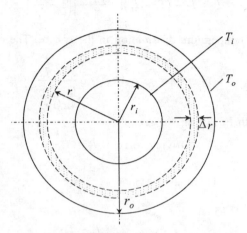

Fig. 2.6 Cross-section of the spherical shell.

$$\text{Rate of heat input at } r = 4\pi r^2 \ q_r|_r$$

$$\text{Rate of heat output at } r + \Delta r = 4\pi r^2 \ q_r|_{r+\Delta r}$$

Because the rate of heat accumulation is zero at steady state, the heat balance equation is

$$4\pi r^2 \ q_r|_r - 4\pi r^2 \ q_r|_{r+\Delta r} = 0$$

or

$$\frac{4\pi r^2 q_r|_{r+\Delta r} - 4\pi r^2 q_r|_r}{\Delta r} = 0$$

Taking limit $\Delta r \to 0$, we get

$$-\frac{d}{dr}(r^2 q_r) = 0$$

or

$$r^2 \ q_r = C_1 \qquad (2.18)$$

Putting $q_r = -k \dfrac{dT}{dr}$, and integrating, we get

$$T = \frac{C_1}{kr} + C_2 \qquad (2.19)$$

The boundary conditions are: $T = T_i$ at $r = r_i$, and $T = T_o$ at $r = r_o$. $\qquad (2.20)$

Using these boundary conditions in Eq. (2.19), the integration constants can be found as

$$C_1 = k\,(T_i - T_o)\left(\frac{r_i r_o}{r_o - r_i}\right) \qquad \text{and} \qquad C_2 = T_i - (T_i - T_o)\left(\frac{r_o}{r_o - r_i}\right) \qquad (2.21)$$

Therefore,

$$T = T_i - (T_i - T_o)\left(\frac{r_o}{r_o - r_i}\right)\left(1 - \frac{r_i}{r}\right) \qquad (2.22)$$

Equation (2.22) gives the temperature distribution in the sphere. The corresponding rate of heat transfer is given by

$$Q = 4\pi r^2 q_r = 4\pi C_1 = \frac{T_i - T_o}{\dfrac{r_o - r_i}{4\pi k r_i r_o}} \qquad (2.23)$$

Comparing Eq. (2.23) with Eq. (2.8), we get

$$\text{Driving force} = T_i - T_o$$

$$\text{Thermal resistance} = \frac{r_o - r_i}{4\pi k r_i r_o}$$

If we rewrite Eq. (2.23) as

$$Q = k(4\pi\,r_i r_o)\,\frac{T_i - T_o}{r_o - r_i}$$

and compare it with Eq. (2.1), it appears that the rate of heat conduction through the spherical shell is the same as that through a plane wall of thickness $(r_o - r_i)$ and area $(4\pi r_i r_o)$. This is the area of the surface of a sphere of radius equal to the geometric mean radius of the given shell, i.e. $\sqrt{r_i r_o}$.

Similarly, if we consider a composite sphere having three layers of materials 1, 2 and 3, inner radii r_i, r' and r'', and thermal conductivities k_1, k_2 and k_3, respectively, and an outer radius r_o, (see Fig. 2.4), the rate of heat conduction through the composite sphere is given by

$$Q = \frac{T_i - T_o}{\dfrac{r' - r_i}{4\pi k_1 r' r_i} + \dfrac{r'' - r'}{4\pi k_2 r'' r'} + \dfrac{r_o - r''}{4\pi k_3 r_o r''}} \qquad (2.24)$$

Equation (2.24) can be derived by following the same procedure as used in the case of a composite cylinder (it is left to the student as an exercise). If all other quantities are known, Q can be calculated, and if Q is known, the interface temperatures, T' and T'' can be calculated.

Example 2.5 Polyurethane foam and expanded polystyrene are two important insulations for low temperature applications. Polyurethane can be impregnated with a flame retardant[1] and is

---

[1] Common flame retardants are additives like polychlorinated or polybrominated compounds (e.g. decabromo-diphenyl ether) used together with antimony oxide.

probably safer. It is desired to calculate the heat gain by a Horton sphere (a spherical vessel used for cryogenic storage) of 16 m diameter that contains liquid ammonia at 4°C. The tank is insulated with a 10 cm thick layer of polyurethane foam having a thermal conductivity of 0.02 W/m °C (0.0115 Btu/h ft^2 °F). The outer surface temperature of the insulation is 27°C. Also, calculate the refrigeration requirement of the vessel.

SOLUTION   The rate of heat transfer can be calculated by using Eq. (2.23) directly.

Given: T_i = inside temperature of the tank = 4°C (this is the same as the temperature of the inner surface of the insulation if the thermal resistance of the tank wall is neglected); T_o = outside temperature of the insulation = 27°C; r_i = 16/2 = 8 m; r_o = 8 m + 0.1 m = 8.1 m; k = 0.02 W/m °C.

$$Q = \frac{T_i - T_o}{(r_o - r_i)/4\pi k r_i r_o} = \frac{(4 - 27)(4\pi)(0.02)(8)(8.1)}{8.1 - 8} = \boxed{-3746 \text{ W}}$$

The negative sign of Q indicates that the *sphere does not lose heat, but it gains heat instead.* It may be seen that the rate of heat transfer is almost the same as that through a plane wall of thickness 0.1 m and an area equal to the 'arithmetic mean area' of the insulation. This is because the insulation thickness is much smaller than the radius of the sphere.

Refrigeration requirement

Refrigeration capacity is commonly expressed in 'tons'. One ton of refrigeration is equivalent to the removal of 12,000 Btu of heat per hour (i.e. 1 ton = 12,000 Btu/h = 3.516 kW).

Therefore, the required refrigeration capacity to maintain the temperature of the sphere

$$= \frac{3.746}{3.516} = 1.07 \text{ tons}$$

2.5   STEADY STATE HEAT CONDUCTION IN BODIES WITH HEAT SOURCES

We have so far discussed steady state heat conduction in planar, cylindrical and spherical geometries without being accompanied by any generation of heat. However, there are situations in which heat generation occurs in a conducting medium. A few common examples are—a nuclear fuel element irradiated by high energy neutrons to trigger nuclear fission, an electrical conductor in which heat generation occurs because of the flowing current, and a catalyst pellet in which heat generation occurs because of a chemical reaction taking place inside the pellet. Also, here, the thermal conductivity of the medium may depend on temperature (unlike that assumed in the previous analyses). In this section, we shall develop the differential equations for heat conduction in the above three common geometries taking into consideration heat generation and also temperature-dependence of the thermal conductivity. The heat generation rate may be a function of its position in the solid.

2.5.1 The Plane Wall

Let us consider a plane wall of thickness l, area A, and temperature-dependent thermal conductivity $k(T)$. The two surfaces are maintained at temperatures T_1 and T_2 (Fig. 2.7). Heat conduction occurs in the x-direction which is normal to the wall. The point $x = 0$ is positioned on the left surface of the wall.

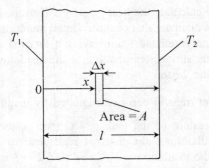

Fig. 2.7 An elementary volume in a plane wall.

We take an elementary volume (also called the *control volume*) in the solid of thickness Δx and area A (normal to the x-axis) placed at a distance x from the origin.

$$\text{Rate of heat input to the element at } x = Aq_x|_x$$

$$\text{Rate of heat output from the element at } x + \Delta x = Aq_x|_{x+\Delta x}$$

$$\text{Rate of heat generation in the element} = A\Delta x\psi_v(x)$$

Here $\psi_v(x)$ is the volumetric rate of heat generation (i.e. the amount of heat generated per unit time per unit volume of the solid) which is assumed to be a function of position, i.e. x. By the principle of conservation of energy, we have

$$\text{Rate of heat input} - \text{Rate of heat output} + \text{Rate of heat generation}$$
$$= \text{Rate of heat accumulation} \tag{2.25}$$

The accumulation term is zero at steady state. Therefore, we get

$$Aq_x|_x - Aq_x|_{x+\Delta x} + A\Delta x\psi_v(x) = 0$$

For a plane wall the area does not change with position. Dividing by $A\Delta x$ throughout, and taking limit $\Delta x \to 0$, we get

$$\frac{q_x|_x - q_x|_{x+\Delta x}}{\Delta x} + \psi_v(x) = 0$$

or

$$-\frac{dq_x}{dx} + \psi_v(x) = 0 \tag{2.26}$$

Substituting $q_x = -k(T)\dfrac{dT}{dx}$ in Eq. (2.26), we get

$$\frac{d}{dx}\left[k(T)\frac{dT}{dx}\right] + \psi_v(x) = 0 \tag{2.27}$$

The boundary conditions are

$$\text{At } x = 0, \qquad T = T_1 \tag{2.28a}$$

$$\text{At } x = l, \qquad T = T_2 \tag{2.28b}$$

Equation (2.27) is the governing differential equation for one-dimensional heat conduction in a plane wall with heat generation and temperature-dependent thermal conductivity. Solution of this equation subject to the boundary conditions, (2.28a) and (2.28b), gives the temperature distribution in the wall. The rate of heat loss from the wall can be determined from the temperature distribution. This approach of heat balance over a thin element (or a cylindrical or spherical shell) in order to develop the differential equation for temperature is called the *shell balance*.

In order to simplify the problem, we assume that the thermal conductivity $k(T)$ remains constant, i.e. independent of temperature. Also, the rate of heat generation $\psi_v(x)$ is assumed constant, i.e. $\psi_v(x)$ remains uniform throughout the wall. Equation (2.27) then becomes

$$\frac{d^2T}{dx^2} = -\frac{\psi_v}{k} \tag{2.29}$$

Integrating twice, we get

$$T = -\frac{\psi_v}{2k}x^2 + C_1 x + C_2 \tag{2.30}$$

Using the boundary conditions, Eqs. (2.28a) and (2.28b), we get

$$C_1 = \frac{T_2 - T_1}{l} + \frac{\psi_v}{2k}l \quad \text{and} \quad C_2 = T_1$$

Substitution of the expressions for the integration constants C_1 and C_2 in Eq. (2.30) yields the temperature distribution in the wall, that can be expressed as

$$T = T_1 + (T_2 - T_1)\frac{x}{l} + \frac{\psi_v x}{2k}(l - x) \tag{2.31}$$

It is clear from the above equation that the temperature distribution (i.e. the plot of temperature vs. position) in the wall is parabolic.

The rates of heat loss from the surfaces can be determined from Eq. (2.31).

The rate of heat loss from unit area of the surface at $x = 0$ is

$$-[q_x]_{x=0} = \left[k\frac{dT}{dx}\right]_{x=0} = \frac{k}{l}(T_2 - T_1) + \frac{\psi_v l}{2} \tag{2.32}$$

The rate of heat loss from unit area of the surface at $x = l$ is

$$[q_x]_{x=l} = -\left[k\frac{dT}{dx}\right]_{x=l} = \frac{\psi_v l}{2} - \frac{k}{l}(T_2 - T_1) \tag{2.33}$$

It is important to understand why the negative sign has been put on the LHS of the Eq. (2.32). Heat flux in the direction of positive x is denoted by q_x. But heat *loss* from the surface at $x = 0$ occurs (see Fig. 2.7) in the opposite direction (i.e. in the direction of decreasing x). Inclusion of the negative sign is necessary to make both sides of Eq. (2.32) compatible with respect to sign.

The combined heat flux from both the surfaces is

$$-[q_x]_{x=0} + [q_x]_{x=l} = \psi_v l \tag{2.34}$$

Now, the volume of solid per unit area of the wall $(1)(l) = l$. The rate of heat generation in this volume is $\psi_v l$, and this amount of heat must be conducted out from the two surfaces together in order to maintain the steady state condition. This physical reasoning is substantiated by Eq. (2.34).

Example 2.6 Two ends of a 5 cm diameter and 50 cm long aluminum rod, with the curved surface perfectly insulated, are maintained at 30°C and 300°C, respectively. The temperature-dependent thermal conductivity of the metal is given by: $k = 202 + 0.0545T$, (W/m °C), where T is in °C. Calculate the temperature gradient at each end of the rod, and the temperature midway in the rod at steady state.

SOLUTION This is a situation of steady state heat conduction in one dimension (because the curved surface is insulated, the temperature at any cross-section of the rod will be uniform) with (i) a constant heat transfer area, and (ii) temperature-dependent thermal conductivity. Heat flow occurs in the axial direction only.

The differential equation for heat conduction in this case can be directly written from Eq. (2.27), after putting $\psi_v = 0$. Thus, we get

$$\frac{d}{dx}\left[k(T)\frac{dT}{dx}\right] = 0 \quad (x \text{ is in metre, } T \text{ is in °C})$$

On integration, we get

$$k(T)\frac{dT}{dx} = C_1$$

Substituting for $k(T)$ and integrating again, we get

$$202T + \frac{0.0545T^2}{2} = C_1 x + C_2 \tag{i}$$

where C_1 and C_2 are integration constants, which can be determined by using the following boundary conditions:

$$\text{At } x = 0 \qquad T = 30$$
$$\text{At } x = 0.5 \qquad T = 300$$

Therefore,

$$C_1 = 113{,}930 \qquad \text{and} \qquad C_2 = 6084.5$$

The temperature distribution in the rod is given by

$$1.35 \times 10^{-4}\ T^2 + T = 564x + 30.1 \tag{ii}$$

Calculation of the heat fluxes

The steady state heat flux in the x-direction is

$$-k(T)\ dT/dx = -C_1 = -113{,}930 \text{ W/m}^2$$

which does not depend upon the axial position. Hence it is the same at both the ends.

Why is the heat flux negative? We have chosen the axial coordinate such that the temperature increases with the distance x. So, in reality, heat conduction occurs in the negative x-direction. Hence the negative sign of the flux.

In order to calculate the mid-plane temperature, we put

$$x = \frac{0.5}{2} = 0.25 \text{ m}$$

in the temperature distribution given by Eq. (ii), when we get

$$T = \boxed{167.3°C}$$

It may be noted that if the thermal conductivity was taken as $k = 202$ (constant), the mid-plane temperature would have been

$$\frac{30 + 300}{2} = 165°C$$

Temperature gradients at the ends of the rod

At $x = 0$, $T = 30°C$, $k(T) = 202 + (0.0545)(30) = 203.6$ W/m °C

$$\text{Temperature gradient} = \frac{dT}{dx} = \frac{C_1}{k(T)} = \frac{113{,}930 \text{ W}/\text{m}^2}{203.6 \text{ W}/\text{m °C}} = \boxed{560 \text{ °C}/\text{m}}$$

Similarly, at $x = 0.5$ m,

$$\frac{dT}{dx} = \boxed{521.8 \text{ °C}/\text{m}}$$

Example 2.7 The steady state temperature distribution in a 0.3 m thick plane wall is

$$T = 600 + 2500x - 12{,}000x^2$$

where T is in °C and x is in metre measured from one surface of the wall. One-dimensional steady state heat conduction occurs in the wall along the x-direction. The thermal conductivity of the material of the wall is 23.5 W/m °C.

(a) What are the surface temperatures and the average temperature of the wall?

(b) Calculate the maximum temperature in the wall and its location.

(c) Calculate the heat fluxes at the surfaces.

(d) Do you think that there is heat generation in the wall? If so, what is the average volumetric rate of heat generation? Is the rate of heat generation uniform?

SOLUTION Let us fix the origin ($x = 0$) on the left surface of the wall (see Fig. 2.7).

(a) At the surface $x = 0$, temperature $T_1 = 600 + (2500)(0) - 12{,}000(0)^2 = \boxed{600°C}$

At the surface $x = 0.3$ m, $T_2 = 600 + (2500)(0.3) - (12{,}000)(0.3)^2 = \boxed{270°C}$

Average temperature of the wall,

$$T_{av} = \frac{1}{l}\int_0^l T(x)\,dx = \frac{1}{0.3}\int_0^{0.3}(600 + 2500x - 12{,}000x^2)\,dx = \boxed{615°C}$$

(b) The maximum temperature occurs at a point where $dT/dx = 0$. That is

$$2500 - 24{,}000x = 0, \text{ or } x = \boxed{0.104 \text{ m}}$$

Therefore, the maximum temperature is

$$T_{max} = 600 + (2500)(0.104) - (12{,}000)(0.104)^2 = \boxed{730°C}$$

(c) Temperature gradient in the wall, $dT/dx = 2500 - 24{,}000x$

Heat flux at the left surface ($x = 0$) $= -k[dT/dx]_{x=0} = -(23.5)(2500) = \boxed{-58{,}750 \text{ W}/\text{m}^2}$

Heat flux at the right surface ($x = 0.3$ m) $= -k[dT/dx]_{x=0.3}$

$$= -(23.5)[2500 - (24{,}000)(0.3)]$$

$$= \boxed{110{,}450 \text{ W}/\text{m}^2}$$

(d) We see in part (c) above that the heat flux at the left surface is negative. This means that heat flow occurs opposite to the x-direction (i.e. away from the wall). On the right surface, the heat flux is positive, i.e. heat flow occurs from the wall. Because heat is conducted out of both the surfaces, there must be a source of heat in the wall. So there is heat generation in the wall.

Total rate of heat loss from both the surfaces per unit wall area $= 110{,}450 - (-58{,}750) = 169{,}200$ W/m².

Volume of the wall per unit surface area $= 0.3$ m³/m².

So, the average volumetric rate of heat generation $= \dfrac{169{,}200}{0.3}$ W/m³ $= \boxed{564{,}000 \text{ W/m}^3}$

It is rather easy to check whether the rate of heat generation is uniform (or whether it is a function of position). Comparison of the temperature distribution given in the problem with Eq. (2.31) shows that the rate of heat generation in the wall is uniform.

Example 2.8 A rectangular block of material A ($k_A = 24$ W/m °C), 0.1 m thick, is sandwiched between two walls of metals B ($k_B = 230$ W/m °C) and C ($k_C = 200$ W/m °C) of thicknesses 0.12 m and 0.15 m, respectively. The outer surface temperature of wall B is 100°C and that of wall C is 150°C. Heat generation occurs in A at a uniform volumetric rate of 2.5×10^5 W/m³. Develop equations for steady state temperature distributions in the three layers and the maximum temperature in the assembly. Also, calculate the percentage of total heat conducted out through the wall B. Assume one-dimensional heat flow.

SOLUTION The composite wall is shown in Fig. 2.8. The origin, $x = 0$, is fixed on the outer surface of the wall B. If ψ_v is the volumetric rate of heat generation, the differential equations for

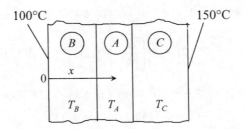

Fig. 2.8　Composite wall (Example 2.8)

temperatures in the walls may be written with the help of Eq. (2.29) and noting that there is no heat generation in the walls B and C [i.e. $\psi_v = 0$].

For the wall B,

$$\frac{d^2 T_B}{dx^2} = 0$$

which on integrating twice, we get

$$T_B = C_1 x + C_2 \tag{i}$$

Similarly for the wall C,

$$T_C = C_3 x + C_4 \tag{ii}$$

For the wall A,

$$\frac{d^2 T_A}{dx^2} = -\frac{\psi_v}{k}$$

or

$$T_A = -\frac{\psi_v}{2 k_A} x^2 + C_5 x + C_6 \tag{iii}$$

There are six integration constants (C_1 to C_6) in total. Their values are determined by using the appropriate boundary conditions. The temperatures at the outer surfaces of B ($x = 0$) and of C ($x = 0.12 + 0.1 + 0.15 = 0.37$ m) are given by

$$x = 0, \qquad T_B = 100 \tag{iv}$$

$$x = 0.37, \qquad T_C = 150 \tag{v}$$

Putting Eq. (iv) in Eq. (i), and Eq. (v) in Eq. (ii), we get

$$T_B = 100 + C_1 x \tag{vi}$$

$$T_C = 150 + C_3 (x - 0.37) \tag{vii}$$

Other conditions are obtained from the *continuity of temperature* and *continuity of heat flux* at the interfaces between B and A and between A and C.

$$x = 0.12 \ (B\text{-}A \text{ interface}), \qquad T_B = T_A \text{ (continuity of temperature)} \tag{viii}$$

and

$$-k_B \frac{dT_B}{dx} = -k_A \frac{dT_A}{dx} \text{ (continuity of heat flux)} \tag{ix}$$

From Eqs. (iii), (vi) and (viii), at $x = 0.12$, we get

$$100 + C_1(0.12) = -\left[\left(\frac{\psi_v}{2k_A}\right)(0.12)^2\right] + C_5(0.12) + C_6$$

Putting the values of ψ_v (= 2.5×10^5 W/m^3) and k_A (= 24 W/m °C), we get

$$C_1 = -1458.3 + C_5 + 8.333C_6 \tag{x}$$

From Eqs. (iii), (vi) and (ix), at $x = 0.12$, we get

$$-k_B C_1 = \psi_v(0.12) - C_5 k_A$$

Putting the values of k_A, k_B and ψ_v, we have

$$C_1 = -130.4 + 0.104C_5 \tag{xi}$$

Similarly, at $x = 0.22$ (A-C interface), $T_A = T_C$ \hfill (xii)

and

$$-k_A \frac{dT_A}{dx} = -k_C \frac{dT_C}{dx} \tag{xiii}$$

From (iii), (vii) and (xii), at $x = 0.22$, we have

$$-\left[\left(\frac{\psi_v}{2k_A}\right)(0.22)^2\right] + C_5(0.22) + C_6 = 150 + C_3(0.22 - 0.37)$$

or

$$C_6 = 402 - 0.22C_5 - 0.15C_3 \tag{xiv}$$

From (iii), (vii) and (xiii), we get

$$C_3 = 0.12C_5 - 275 \tag{xv}$$

Solving the simultaneous Eqs. (x), (xi), (xiv) and (xv), we get

$$C_1 = 96.6, \qquad C_3 = -14, \qquad C_5 = 2175, \qquad C_6 = -74.5$$

Putting the values of the constants and the values of the other quantities in Eqs. (iii), (vi) and (vii), we get

Temperature distribution in B: $\boxed{T_B = 100 + 96.6x}$ \hfill (xvi)

Temperature distribution in A: $\boxed{T_A = -5208x^2 + 2175x - 74.5}$ \hfill (xvii)

The temperature distribution in C: $\boxed{T_C = 155.2 - 14x}$ \hfill (xviii)

The maximum temperature will occur in A at a position determined from $dT_A/dx = 0$

or

$$-10,416x + 2175 = 0$$

or

$$\boxed{x = 0.209 \text{ m}} \text{ from the open surface of } B.$$

The maximum temperature,

$$T_{A, \, max} = -5208(0.209)^2 + 2175(0.209) - 74.5 = \boxed{152.6°C}$$

The last part is left to the student as an exercise.

2.5.2 The Cylinder

To analyse heat conduction in a cylinder with a heat source, we choose the case of an electrical wire carrying current. The wire of radius r_i has a layer of electrical insulation over it to a radius of r_o. The temperature of the outer surface of the insulation is T_o. The thermal conductivity of the material of the wire is k_m, and that of the insulation is k_c. The volumetric rate of heat generation in the wire ψ_v is assumed to remain uniform throughout the medium.

A longitudinal section of the insulated wire is shown in Fig. 2.9. In order to develop the governing differential equation for the system we make a heat balance over a control volume which is a thin cylindrical shell of inside diameter r and thickness Δr (see Section 2.4).

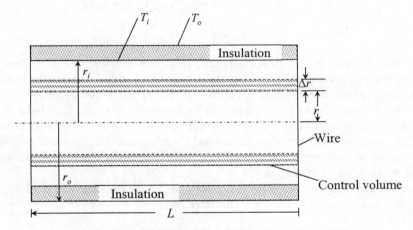

Fig. 2.9 The longitudinal section of a wire with insulation.

If the length of the wire is L and heat conduction occurs in the radial direction only,*

Rate of heat input at $r = (2\pi rL) \, q_r|_r$

Rate of heat output at $r + \Delta r = (2\pi rL) \, q_r|_{r + \Delta r}$

Rate of heat generation = (volume of the element) × (volumetric rate of heat generation)

$$= 2\pi r \Delta r L \psi_v$$

There is no accumulation of heat in the element at steady state. Therefore, by a heat balance [see Eq. (2.25)], we have

$$(2\pi rL) \, q_r|_r - (2\pi rL) \, q_r|_{r + \Delta r} + 2\pi r \Delta r L \psi_v = 0$$

---

*This assumption is correct if the length of the wire is much larger than its diameter. Similar assumption is commonly used in the analysis of heat conduction in a 'long cylinder'.

Dividing by $2\pi\Delta rL$ throughout and taking limit $\Delta r \to 0$, we get

$$-\frac{d}{dr}(rq_r) + r\psi_v = 0 \qquad (2.35)$$

Substituting, $q_r = -k_m\dfrac{dT_m}{dr}$, we get

$$\frac{d}{dr}\left(r\frac{dT_m}{dr}\right) + \frac{r}{k_m}\psi_v = 0 \qquad (2.36)$$

where T_m denotes the local temperature in the wire.

Equation (2.36) is the governing differential equation for temperature distribution in a cylinder with heat generation. In order to obtain the complete solution of the equation we need to specify the boundary conditions at $r = 0$ and at $r = r_i$. Thus,

$$\text{At } r = 0, \qquad \frac{dT_m}{dr} = 0 \qquad (\text{or, } T_m \text{ is finite}) \qquad (2.37a)$$

$$\text{At } r = r_i, \qquad T_m = T_c \qquad (2.37b)$$

where T_c stands for the insulation temperature. Boundary condition (2.37a) arises from the cylindrical geometry (that has the axis of symmetry at $r = 0$) of the system which demands that the temperature gradient must vanish at the axis of the cylinder. As it will be seen later, the alternative boundary condition (2.37a) (shown in the bracket)—at $r = 0$, T_m = finite (it is obvious that it is true)—can be used as an alternative condition to arrive at the same final solution.

Boundary condition (2.37b), although straightforward, is not directly useful because T_c is not known. To make use of this condition we need to write down the differential equation for temperature in the layer of insulation.

The temperature equation for the insulation will, in fact, be the same as Eq. (2.36) except that there is no heat generation (i.e. $\psi_v = 0$). Therefore, we have

$$\frac{d}{dr}\left(r\frac{dT_c}{dr}\right) = 0 \qquad (2.38)$$

where T_c is the local temperature in the insulation. The boundary condition at the outer surface is

$$r = r_o, \qquad T_c = T_o \qquad (2.39)$$

The second boundary condition on T_c at $r = r_i$ (i.e. at the other surface of the insulation where the radial position is the same as the radius of the wire) is exactly the same as Eq. (2.37b). This condition, in other words, expresses the continuity of temperature at the boundary between the wire and the insulation.

Now let us attempt to solve Eq. (2.36) in order to determine the steady state temperature distribution in the wire. Integrating this equation, we get

$$r\frac{dT_m}{dr} = -\frac{r^2\psi_v}{2k_m} + C_1 \qquad (2.40)$$

where C_1 is an integration constant. Boundary condition (2.37a) yields $C_1 = 0$. Dividing both sides of Eq. (2.40) by r and integrating again, we get

$$T_m = -\frac{r^2 \psi_v}{4k_m} + C_2 \qquad (2.41)$$

where C_2 is another integration constant.

Now we revert to Eq. (2.38) and integrate it twice to get

$$T_c = C_3 \ln r + C_4 \qquad (2.42)$$

where C_3 and C_4 are two other integration constants. Using the boundary condition (2.39), we get

$$C_4 = T_o - C_3 \ln r_o$$

Putting the above result in Eq. (2.42), we have

$$T_c = C_3 \ln r + T_o - C_3 \ln r_o = T_o + C_3 \ln (r/r_o) \qquad (2.43)$$

We still have another boundary condition (2.37b) at our disposal. Using this condition together with Eqs. (2.41) and (2.43), we can write

$$-\frac{r_i^2 \psi_v}{4k_m} + C_2 = T_o + C_3 \ln (r_i/r_o) \qquad (2.44)$$

The above equation contains two unknowns (C_2 and C_3), and we need one more condition in order to determine these integration constants. For this purpose, we invoke the condition of continuity of heat flux at the interface between the wire and the insulation. Thus at $r = r_i$, we have

$$-k_m \frac{dT_m}{dr} = -k_c \frac{dT_c}{dr} \qquad (2.45)$$

The physical significance of the above condition is: the rate at which heat is conducted from inside the wire to the interface at $r = r_i$ must be equal to the rate at which heat enters the insulation at the same location at steady state.

We now determine the gradients of the temperatures of the wire and of the insulation (dT_m/dr and dT_c/dr) at $r = r_i$ from Eqs. (2.40) and (2.42) and substitute these in Eq. (2.45) to obtain

$$\frac{r_i \psi_v}{2k_c} = -\frac{C_3}{r_i}$$

or

$$C_3 = \frac{r_i^2 \psi_v}{2k_c} \qquad (2.46)$$

Substituting for C_3 from Eq. (2.46) in Eq. (2.44), we can obtain C_2 as

$$C_2 = T_o + \frac{r_i^2 \psi_v}{4} \left[\frac{1}{k_m} - \frac{2}{k_c} \ln (r_i/r_o) \right] \qquad (2.47)$$

Substituting for C_2 and C_3 from Eqs. (2.47) and (2.46) in Eqs. (2.41) and (2.43), respectively, we get

$$T_m = T_o + \frac{\psi_v}{4k_m}(r_i^2 - r^2) - \frac{r_i^2 \psi_v}{2k_c} \ln (r_i/r_o) \tag{2.48}$$

$$T_c = T_o - \frac{r_i^2 \psi_v}{2k_c} \ln (r/r_o) \tag{2.49}$$

Equation (2.48) is the temperature distribution in the wire, and Eq. (2.49) is that in the insulation.

Example 2.9 A two-layer composite cylinder of inner diameter 15 cm and outer diameter 30 cm has moderate volumetric rates of heat generation 100 kW/m³ and 40 kW/m³ in the inner and the outer layers, respectively. The thickness of the two layers is equal. The temperature at the inside surface ($r_i = 0.075$ m) of the assembly is 100°C, and that at the outside surface ($r_o = 0.15$ m) is 200°C. Thermal conductivities of the materials are: $k_1 = 30$ W/m °C for the inner layer, and $k_2 = 10$ W/m °C for the outer layer. Determine (a) the temperature distributions in the individual layers, and (b) the maximum temperature in the cylinder and the radial position at which it occurs.

SOLUTION We have to write two separate differential equations for the temperature distributions in the two cylindrical layers. These have to be solved using the given conditions and the conditions of the continuity of temperature and of the heat flux at the interface between the cylindrical shells. Heat flow occurs in the radial direction at steady state.

(a) The differential equation governing steady state radial temperature distribution with heat generation in the cylindrical geometry has been developed in Section 2.5.2. We will, therefore start directly with Eq. (2.35).

Temperature equation for the inner layer (layer 1) is

$$\frac{d}{dr}\left(r \frac{dT_1}{dr}\right) + \frac{r}{k_1}\psi_{v1} = 0$$

Integrating twice, we get

$$T_1 = -\frac{r^2}{4k_1}\psi_{v1} + A_1 \ln r + B_1 \tag{i}$$

where A_1 and B_1 are integration constants.

Similarly, the temperature equation for the outer layer (layer 2) is

$$T_2 = -\frac{r^2}{4k_2}\psi_{v2} + A_2 \ln r + B_2 \tag{ii}$$

where A_2 and B_2 are integration constants.

Given: $k_1 = 30$ W/m °C; $\psi_{v1} = 100$ kW/m³; $r_i =$ inside radius $= 0.075$ m; $k_2 = 10$ W/m °C; $\psi_{v2} = 40$ kW/m³; $r_o =$ outer radius $= 0.15$ m; r' (interface position) $= (0.075 + 0.15)/2 = 0.1125$ m

Boundary conditions: at the inner surface, $r = r_i = 0.075$ m, $T_1 = 100$°C

at the outer surface, $r = r_o = 0.15$ m, $T_2 = 200$°C

Using the boundary condition at $r = r_i$ in Eq. (i), we get

$$100 = \frac{-(0.075)^2}{(4)(30)}(10^5) + A_1 \ln(0.075) + B_1$$

or

$$B_1 - 2.59A_1 = 104.69 \qquad \text{(iii)}$$

Similarly, using the boundary condition at $r = r_o$ in Eq. (ii), we get

$$B_2 - 1.897A_2 = 222.5 \qquad \text{(iv)}$$

In order to determine the temperature profiles in the layers completely, all the constants A_1, B_1, A_2, and B_2 need to be evaluated. We have two more conditions at our disposal—the continuity of temperature and the continuity of heat flux at the interface, $r = r'$. Equating the temperatures of the layers at the interface ($r = r'$) and putting the values of different quantities, we get

$$2.185(A_1 - A_2) + (B_2 - B_1) = 2.11 \qquad \text{(v)}$$

Similarly, equating the heat fluxes in the layers at the interface and putting the values of various quantities, we get

$$A_1 - 0.333A_2 = 12.66 \qquad \text{(vi)}$$

Solving Eqs. (iii), (iv), (v) and (vi), we get

$$A_1 = 100, \ B_1 = 364, \ A_2 = 261, \text{ and } B_2 = 718.$$

Using these values of the constants in Eqs. (i) and (ii) and putting the values of all other quantities, we get

Temperature distribution in the inner layer,

$$\boxed{T_1 = 364 + 100 \ln r - 833.3 \, r^2}$$

Temperature distribution in the outer layer,

$$\boxed{T_2 = 718 + 261 \ln r - 1000 \, r^2}$$

(b) In order to calculate the maximum temperature in layer 1, we put $dT_1/dr = 0$. That is,

$$\frac{100}{r} - 2(833.3)r = 0$$

or

$$r = 0.245 \text{ m}$$

This radial position does not fall within layer 1. Therefore, no temperature maximum occurs within this layer. Similarly, it can be shown that no temperature maximum occurs within layer 2 either. So the maximum temperature occurs at the *outer boundary* ($r_o = 0.15$ m), and its value is $\boxed{200^\circ C}$.

2.5.3 The Sphere

Let us consider a solid sphere of radius r_o with 'skin temperature' T_o in which heat is being generated at a volumetric rate of $\psi_v(r)$, r being any radial position in the sphere. The thermal conductivity of the sphere, $k(T)$, is a function of temperature. Considering a thin spherical shell of thickness Δr (see Fig. 2.6), we have

$$\text{Rate of heat input to the shell at } r = 4\pi r^2 \ q_r|_r$$

$$\text{Rate of heat output from the shell at } r + \Delta r = 4\pi r^2 \ q_r|_{r+\Delta r}$$

$$\text{Rate of heat generation in the shell} = 4\pi r^2 \ \Delta r \ \psi_v(r)$$

$$\text{Rate of heat accumulation in the shell at steady state} = 0$$

By making a steady state heat balance [see Eq.(2.25)], we get

$$4\pi r^2 \ q_r|_r - 4\pi r^2 \ q_r|_{r+\Delta r} + 4\pi r^2 \ \Delta r \ \psi_v(r) = 0$$

Dividing both sides by $4\pi\Delta r$ throughout and taking limit $\Delta r \to 0$, we have

$$-\frac{d}{dr}\left(r^2 q_r\right) + r^2\psi_v(r) = 0 \tag{2.50}$$

Substituting for the local heat flux, $q_r = -k(T)\dfrac{dT}{dr}$, Eq. (2.50) reduces to

$$\frac{d}{dr}\left[r^2 k(T)\frac{dT}{dr}\right] + r^2\psi_v(r) = 0 \tag{2.51}$$

The boundary conditions are:

$$\text{At } r = 0, \qquad \frac{dT}{dr} = 0 \qquad (T = \text{finite}) \tag{2.52a}$$

$$\text{At } r = r_o, \qquad T = T_o \tag{2.52b}$$

The first boundary condition arises out of spherical symmetry. If, for the sake of simplification, we assume $k(T) = \text{constant}$, and $\psi_v(r) = \text{constant}$, Eq. (2.51) reduces to

$$\frac{d}{dr}\left(r^2\frac{dT}{dr}\right) = -\frac{r^2\psi_v}{k} \tag{2.53}$$

Integrating, we get

$$r^2\frac{dT}{dr} = -\frac{r^3\psi_v}{3k} + C_1$$

where C_1 is an integration constant. By using the boundary condition (2.52a), $C_1 = 0$.
Integrating once more, we get

$$T = -\frac{r^2\psi_v}{6k} + C_2 \tag{2.54}$$

where C_2 is another integration constant which can be determined by using the boundary condition (2.52b).

$$T_o = -\frac{r_o^2 \psi_v}{6k} + C_2$$

or

$$C_2 = T_o + \frac{r_o^2 \psi_v}{6k} \tag{2.55}$$

Substituting for C_2 in Eq. (2.54), we get

$$T = T_o + \frac{\psi_v}{6k}(r_o^2 - r^2) \tag{2.56}$$

Equation (2.56) is the temperature distribution in the sphere which is parabolic. The maximum temperature occurs at the centre. Thus, we have

$$T_{\text{max}} = T_o + \frac{\psi_v r_o^2}{6k} \tag{2.57}$$

2.6 AVERAGE TEMPERATURE OF A SOLID

Besides temperature distribution and the maximum value of temperature in a solid, another quantity of importance is the average temperature. The equations for the determination of average temperatures from temperature distributions in solid bodies of the above three geometries—plane wall, cylindrical and spherical—are developed below.

The plane wall

Consider again the plane wall (thickness = l; density = ρ; and specific heat = c_p) shown in Fig. 2.7. Assume a temperature distribution $T = T(x)$ in the wall. If A is the area of the wall, the amount of heat energy contained in an element of thickness dx is $(A\,dx)\rho c_p T$. The total amount of heat energy in the wall can be obtained by integrating this quantity. On the other hand, if T_{av} is the average temperature of the wall, the total amount of heat energy in it is $(A\,l)\rho c_p T_{\text{av}}$. Therefore, we have

$$Al\rho c_p T_{\text{av}} = \int_0^l (A dx)\rho c_p T(x) = A\rho c_p \int_0^l T(x)dx$$

or

$$T_{\text{av}} = \frac{1}{l}\int_0^l T(x)dx \tag{2.58}$$

The cylinder

In the case of a solid cylinder of length L and radius R (see Fig. 2.3), the volume of an elementary cylindrical shell is $2\pi r dr L$. If the radial temperature distribution in the cylinder is $T = T(r)$, the average temperature is given by

$$(\pi R^2 L)\rho c_p T_{\text{av}} = \int_0^R (2\pi r\, dr\, L)\rho c_p T(r) = 2\pi L\rho c_p \int_0^R rT(r)dr$$

Thus

$$T_{\text{av}} = \frac{2}{R^2}\int_0^R rT(r)dr \tag{2.59}$$

The sphere

Similarly, by integrating the amount of thermal energy contained in a thin spherical shell of thickness dr, it is easy to determine the average temperature of the sphere which has a temperature distribution of $T(r)$. Thus, here, we have

$$T_{av} = \frac{3}{R^3} \int_0^R r^2 T(r) dr \qquad (2.60)$$

2.7 APPLICATION OF CONDUCTION CALCULATION

Heat conduction calculation is very important in a variety of situations. In some systems the rate of heat conduction is required to be as low as possible, whereas in some others a high rate of heat flow is desirable. For example, all furnaces are heavily insulated in multiple layers to minimize heat loss and to maintain a high temperature inside. The calculation of the rate of heat loss from a rectangular furnace can be done by using Eq. (2.7). Various process equipment, reactors, steam pipes and cryogenic systems need to be well-insulated against loss or gain of heat (Mollory, 1969; Harrison and Pelanne, 1977). Besides its most conspicuous function of reducing heat loss, thermal insulation on a hot pipe reduces the pipe stress (Peng and Peng, 1998). Electrical wires are always kept insulated [polyvinyl chloride (PVC) is a common electrical insulator used], and this layer of insulation also acts as a thermal insulator. If the steady state wire temperature is high, degradation of the insulation may occur. Thus, conduction calculation is important in electrical insulation design. Conduction calculation in nuclear fuel elements is also important to determine the size of the fuel element to be used so that the maximum temperature in it does not exceed the allowable limit. Determination of insulation thickness on steam pipes and various process equipment is a routine part of process design calculations. Important insulating materials used over different temperature ranges are listed in Table 2.2. The Petroleum Conservation Research Association (PCRA) gives an excellent account of the properties and applications of industrial thermal insulations.

Table 2.2 Common insulations used for process equipment and pipelines

Insulation material	Temperature range °C	Thermal conductivity (10^{-3} W/m °C)	Thermal conductivity (10^{-3} kcal/h m² °C)	Density (kg/m³)	Applications
1. Perlite	−200 to 40	1–2	0.85–1.7	60–140	Cryogenic services
2. Polyurethane foam	−170 to 110	16–20	14–17	32	Tanks, vessels (cold/hot)
3. Polyurethane foam	−180 to −150	16–20	14–17	25–50	Pipes and fittings
4. Expanded polystyrene	−100 to 40	22–25	19–22	20–50	Chilled vessels
5. Fibre-glass blanket	−160 to 230	24–86	21–75	10–50	Chillers, tanks, vessels
6. Fibre-glass blanket for wrapping	−80 to 290	22–80	20–70	10–50	Pipes and fittings (hot/cold)
7. Fibre-glass board	20 to 450	32–52	27–50	25–100	Boilers, tanks, heat exchangers
8. Calcium silicate, board/block	230 to 1000	32–85	27–50	25–100	Boilers, breechings, chimney liners
9. Mineral fibre block	Up to 1100	50–130	43–112	210	Boilers and tanks
10. Fibre glass mats	60 to 360	30–55	26–47	10–50	Vessels, tanks
11. Mineral fibre blankets	Up to 750	37–80	32–69	125	Vessels, pipes
12. Mineral wool blankets	450 to 1000	50–130	43–112	175–290	Vessels, pipes

(a) A 'Rockloyd' pre-formed pipe section (made from bonded rockwool).

(b) A 'Rockloyd' mattress (made from light resin-bonded rock fibres, with wire mesh machine stitch on one side; used to insulate boilers, vessels, etc.).

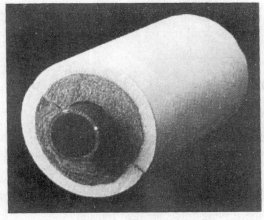

(c) A bilayer pre-formed insulation (inner layer—rockwool, outer layer—polyurethane foam) on a cold surface.

(d) Pre-formed pieces of polyurethane foam (different shapes).

(e) Measurement of insulation surface temperature by a digital contact thermometer.

Fig. 2.10   Thermal insulations [Courtesy: Llyod Insulations (India), Calcutta].

Common thermal insulators can be divided into four groups: (i) granular, (ii) fibrous, (iii) cellular, and (iv) reflective. The insulating property of materials of the first three groups is mainly due to the air or gas entrapped in the matrix (for granular and cellular materials) or between the fibres. Calcium silicate is a common granular insulating material. Mineral wools (e.g. rockwool, glass wool, etc.) are common fibrous insulations. A reflective insulation consists of a packed layer of low emissivity metal foils (e.g. aluminium) and is used for cryogenic applications.

Selection and application of insulation depend upon a number of factors such as temperature of a body, its surface characteristics, fire resistance, etc. 'Pre-formed' insulations (Fig. 2.10) of various shapes may be used depending upon the shape of the body to be insulated. A bonding material like phenol-formaldehyde resin may be used for fabrication of rockwool pre-formed insulation. For example, two half-cylindrical pre-formed pieces may be put on a pipe section and then held at place by bands or straps. A protective cover, commonly an aluminium foil (called 'cladding'), is used to prevent damage or weathering of the insulation. Mineral wool 'mats' made by holding a layer of the material between two wire nets by stitching may be applied to a flat surface or a vessel wall. For cryogenic service, a coating of a suitable polymer formulation is applied on the outer surface of the insulation (e.g. polyurethane foam) to form a vapour barrier that prevents penetration of water vapour in the insulation. Guidelines on the method of application of insulations are given in IS 7413 (1981). Several tips on application of insulations are suggested by Irwin (1991).

SHORT QUESTIONS

1. What is the effect of impurities on the thermal conductivity of a metal?

2. Give the approximate ranges of thermal conductivities of common solids, liquids and gases. Name a solid of low thermal conductivity and a liquid (not a metal) of high thermal conductivity.

3. How does thermal conductivity for solids, liquids and gases depend upon temperature?

4. How is thermal resistance defined? What is the driving force for heat transfer?

5. What types of quantities (vector or scalar) are the following: temperature, heat flux, thermal conductivity, temperature gradient?

6. A substance has a temperature-dependent thermal conductivity k(W/m °C) $= 0.45 + 8.1 \times 10^{-4}T + bT^2$, where $b = 6.2 \times 10^{-7}$, and T is in °C. What is the unit of the coefficient b? Find the thermal conductivity of the material at 1050°F. Also, calculate its average thermal conductivity over the temperature range 300°F to 1050°F.

7. The temperature distribution in a wall, 0.25 m thick, is given as T (°C)$= 250 + bx - cx^2$, where x (in metre) is the position of a point with respect to the surface of the wall which is at a higher temperature. The thermal conductivity of the wall is 5.95 W/m °C. If the rate of heat loss from the surface at $x = 0.25$ is 500 W/m^2, and the surface at $x = 0$ is insulated, calculate the average temperature of the wall.

8. Write down the expressions for thermal resistances of a wall, an annular cylinder and a spherical shell. How does thermal resistance depend upon the area of heat transfer?

9. The condition of continuity of temperature [e.g. Eq. (2.37b)] is frequently used at the surface of contact of two mediums. In what kind of situation is this condition not valid?

10. Which type of insulation do you recommend for the following applications? Insulation of: (i) an atmospheric ammonia storage tank at $-33°C$, (ii) a boiler, (iii) a steam pipe carrying saturated steam at $200°C$, (iv) a liquid air distillation column, (v) a SO_2 converter at $500°C$, (vi) a storage tank of molten tar at $65°C$, and (vii) a pipe carrying chilled methanol at $-50°C$.

11. Answer the following questions by referring to Fig. 2.11.

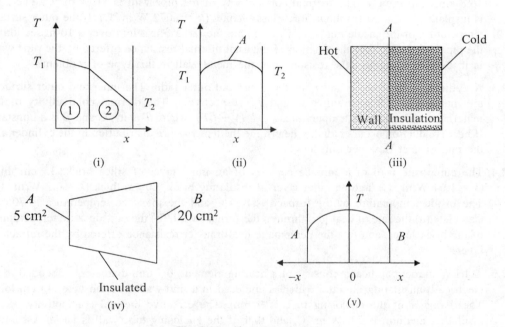

Fig. 2.11   Short Question 11.

(i) The steady state temperature distribution in a two-layered composite wall is shown in Fig. (i). Which layer does have higher thermal conductivity?

(ii) Steady state temperature distribution in a plane wall is shown in Fig. (ii). What is the heat flux at position A?

(iii) A 'perfectly insulated' wall is shown in Fig. (iii). What is the temperature gradient at the plane AA?

(iv) Figure (iv) shows a tapered solid block insulated on the curved surface. If heat flux at the face A is 100 W/m^2, what is the flux at the face B? Assume steady state condition and no heat generation.

(v) Steady state temperature distribution in a plane wall is shown in Fig. (v). What are the signs of the temperature gradients at surfaces A and B?

PROBLEMS

2.1 An annealing chamber has a composite wall made of a 17 cm thick firebrick layer (k = 1.1 W/m °C) and a 13 cm thick ordinary brick layer (k = 0.70 W/m °C). The inside and outside surface temperatures of the walls are 400°C and 45°C, respectively. Calculate the heat loss from 25 m^2 of furnace wall. Also, determine the temperature between the ordinary brick and the firebrick layers.

2.2 A 15 cm schedule 40 steam main carries saturated steam at 10.7 bar (gauge), and the temperature is 190°C. The inside and outside diameters of the pipe are 15.4 cm and 16.8 cm, respectively. The thermal conductivity of the pipe wall is 51 W/m °C. The pipe is insulated with a 10 cm thick fibre glass blanket (k = 0.072 W/m °C). If the outer surface temperature of the insulation is 41°C, calculate the rate of heat loss over a 10 m section of the pipe. Also, calculate the fraction of the total thermal resistance offered by the pipe wall. Is it justified to neglect the resistance of the metal wall in this type of problem?

2.3 A cylindrical shell has r_i and r_o as the inner and outer radii. The inner and outer surfaces are maintained at temperatures T_i and T_o, respectively. The thermal conductivity of the material is a function of temperature, $k = k_o(1 + \beta T)$, where T is in °C and β is a constant. If heat conduction occurs radially, determine the temperature distribution in the cylinder and the rate of heat flow per unit length.

2.4 The composite wall of a furnace consists of an inner layer of silica brick, 15 cm thick (k = 1.04 W/m °C), and an outer layer of insulating brick, 20 cm thick (k = 0.2 W/m °C). The inside temperature of the furnace is 800°C and the interface temperature is 705°C. Calculate (a) the rate of heat loss through the furnace wall, (b) the outside 'skin temperature' of the brick layer, and (c) the percentage heat transfer resistance offered by the refractory layer.

2.5 A 1 kW immersion heater consists of a heating element, 0.7 mm diameter, embedded in an electrically insulating medium which is enclosed in a thin-walled coiled tube, 40 cm long. The diameter of the enclosing tube is 9 mm. The effective thermal conductivity of the insulating medium is 2.0 W/m °C, and that of the enclosing tube wall is large. Assuming that the heating element is placed exactly centrally in the tube, calculate the maximum temperature of the element if the heater is being used to boil water. Also, assume that the surface temperature of the heater is 110°C (water boiling on a hot surface always has a degree of superheat).

[*Hint:* Consider the problem as one of steady state heat conduction through a cylindrical wall having r_i = 0.35 mm, r_o = 4.5 mm, L = 0.4 m, k = 2 W/m °C, Q = 1000 W, and T_o = 110°C.]

2.6 Maleic anhydride is a bulk organic chemical (for example, it is a raw material for the manufacture of unsaturated polyester resins) obtained by catalytic air-oxidation of benzene. In a 20 tpd (tons per day) maleic anhydride plant, molten crude maleic anhydride is stored at 60°C in a 5 m diameter and 3.5 m high cylindrical stainless steel tank, from where it is pumped to the refining column. The tank has a 15 cm thick insulation of glass fibre mat. However, some heat loss occurs from the insulated tank and steam heating (by steam pipes placed within the tank) of the liquid is done to compensate for the heat loss. If the outside wall temperature of the tank is 40°C, calculate the rate at which low pressure steam at 2.0

bar gauge pressure is required to be supplied to the heating pipes. Assume that the steam is saturated and only latent heat of condensation is available for heating. Consider heat loss from the cylindrical and top surfaces only. Neglect heat loss from the bottom of the tank to the tank foundation. Obtain the latent heat of condensation of steam from the steam table. The thermal conductivity of the insulation is 0.45 W/m °C.

[*Hint:* Calculate the total rate of heat loss from the top 'flat' surface and the curved surface. Divide it by the latent heat of steam. Neglect the resistance of the stainless steel wall.]

2.7 An ice-box of inner dimensions 1 m × 0.7 m × 0.6 m has a 6.5 cm thick layer of thermocol on it as insulation. It contains 12 kg of ice. If the outer surface temperature of the box is 20°C, calculate the time required for the ice to melt. The thermal conductivity of the insulation layer is 0.0355 W/m °C. Assume that this layer virtually offers all the heat transfer resistance. State any other assumption you make.

2.8 A 3 m diameter furnace operating at 1000°C has a 25 cm inner lining of refractory bricks followed by a 15 cm layer of bonded rockwool insulation. On the outside, the furnace is covered with a 6 mm carbon steel plate. If the outer wall temperature (skin temperature) is 50°C, calculate the rate of heat loss and the boundary temperature between the layers of the refractory and bonded wool. The thermal conductivity of the refractory lining is 1.5 W/m °C and that of the bonded rockwool is a linear function of temperature, $k = 0.648 + 1.55 \times 10^{-4}T$, W/m °C (where T is in °C).

2.9 A wall has two layers of insulation on it. The layers are of equal thickness but the thermal conductivity of layer 1 is twice that of the layer 2. How much more thicker should layer 2 be if the heat flow is to be reduced by 20%, the total wall thickness and the driving force remaining unchanged? What would be the fractional reduction in heat flow if the material of layer 1 is replaced by the material of layer 2?

2.10 A 4 inch schedule 40 pipe carrying a hot liquid at 250°C has two layers of insulation of equal thickness, 2 inch each. The outside temperature of the insulation remaining the same (40°C), the rate of heat loss per foot of the pipe reduces from 136 Btu/h to 130 Btu/h if the outer insulation thickness is increased by 25%. Calculate the thermal conductivities of the two layers of insulation.

2.11 A hot chamber has an 8 cm thick inner layer of firebrick ($k = 1.04$ W/m °C) and a 13 cm outer layer of ordinary brick ($k = 0.69$ W/m °C). The inside and outside surface temperatures of the composite wall are 400°C and 75°C, respectively. Considering that the outer surface temperature is too high, it is decided to apply a 5 cm thick layer of plaster on the outer surface. On doing this, the outer skin temperature is reduced to 60°C and the rate of heat loss decreases by 250 W/m² of the wall area. Calculate the thermal conductivity of the layer of plaster.

2.12 A cylindrical shell has two layers of insulation of equal thickness. The outer radius of the assembly is twice the inner radius. Also, the inner layer (layer 1) has a thermal conductivity thrice that of the outer layer (layer 2). What will be the per cent change in heat flow if the layers are interchanged (i.e. when the material of the outer layer forms the inner layer and vice versa)? The thicknesses of the layers remain the same as before and the temperature driving force remains unchanged.

2.13 Consider a hot gas duct similar to the one of Example 2.3. The length of the duct is 30 m. Hot gases enter the duct at a rate of 2500 kg/h. The temperature of the gas decreases along the duct because of heat loss to the ambient. If the gas enters at 400°C, at what temperature does it leave? Assume that the inside wall temperature is the same as that of the gas at any location. The outside wall temperature remains constant at 65°C. The thermal resistance of the steel shell is negligible. The average specific heat of the gas is 1040 J/kg °C.

[*Hint:* Consider a small element of length dx of the duct over which the gas temperature changes by dT because of heat loss. If m is the gas flow rate, R the thermal resistance per metre length of the duct, T_1 the inlet temperature (400°C), T_2 the outlet temperature (to be calculated), and l the duct length (30 m), then $- m c_p dT = \{(T - T_o)/R\}dx$. On integration, $\ln \{(T_1 - T_o)/(T_2 - T_o)\} = l/(mc_pR)$.]

2.14 Perlite is a typical cryogenic insulator. It is made of a volcanic rock containing a high percentage of glass. When heated to a high temperature, the material softens and the moisture present in it vaporizes to form a cellular microstructure. It is common practice to fill the jacket outside a cryogenic chamber (like the cold box in a liquid air distillation unit) with perlite powder and then evacuate it to achieve very good thermal insulation.

A spherical cryogenic vessel of 10 m diameter stores liquid nitrogen at a temperature of −196°C. It is provided with a perlite-filled evacuated jacket to guard against heat gain. If the outside temperature of the jacket is assumed to be 30°C, calculate the width of the jacket that would limit the heat flux (based on the outer surface) to within 0.5 W/m². The effective thermal conductivity of the insulation is 1.73×10^{-3} W/m °C.

2.15 Consider a composite solid block of the shape of the frustrum of a cone with the curved surface perfectly insulated (Fig. 2.12). The block has two layers of thicknesses l_1 (layer 1) and l_2 (layer 2). The exposed surface of layer 1 (thermal conductivity, $k_1 = 5$ W/m °C) has a radius r_1 (= 0.18 m) and a temperature of T_i (= 200°C), and that of layer 2 (thermal conductivity, $k_2 = 2$ W/m °C) has a radius r_2 (= 0.08 m) and a temperature T_o (= 100°C). If $l_1 = 0.2$ m, and $l_2 = 0.15$ m, calculate (a) the temperature distributions in the two layers, (b) the rate of heat flow through the assembly, (c) the heat flux at the open end of layer 2, and (d) the temperature and heat flux at the surface of contact of the layers.

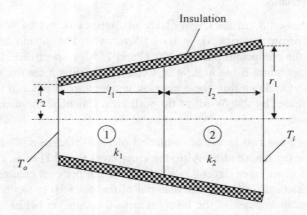

Fig. 2.12 Composite solid block (Problem 2.15).

2.16 A spherical vessel of 3 m inside diameter is made of AISI 316 stainless steel sheet of 9 mm thickness ($k = 14$ W/m °C). The inside temperature is $- 80$°C. The vessel is layered with a 10 cm thick polyurethane foam ($k = 0.02$ W/m °C) followed by a 15 cm outer layer of cork ($k = 0.045$ W/m °C). If the outside surface temperature is 30°C, calculate (a) the total thermal resistance of the insulated vessel wall, (b) the rate of heat flow to the vessel, (c) the temperature and heat flux at the interface between the polyurethane and the cork layers, and (d) the percentage error in calculation if the heat transfer resistance of the metal wall is neglected.

2.17 Consider a composite wall consisting of four different materials as shown in Fig. 2.13. The following data are given: $L_A = 0.1$ m, $L_C = 0.2$ m, $L_D = 0.12$ m, $H = 2$ m, $H_B = 1$ m, $k_A = 20$ W/m °C, $k_B = 10$ W/m °C, $k_C = 7$ W/m °C, $k_D = 25$ W/m °C, $T_1 = 120$°C, and $T_2 = 50$°C.

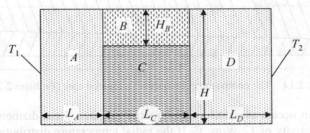

Fig. 2.13 Composite wall (Problem 2.17).

Calculate the rate of heat flow through the assembly per unit breadth. Assume one-dimensional heat flow only. Show the electric analogue of the problem.

[*Hint:* This heat transfer problem involves thermal resistances in a series-parallel configuration. If R_A, R_B, R_C and R_D are the thermal resistances of the individual slabs, the total resistance R_T, is given by

$$R_T = R_A + \frac{1}{1/R_B + 1/R_C} + R_D$$

The second term on the RHS is the sum of the thermal resistances of B and C in parallel. On the basis of unit breadth, $R_A = 0.0025$, $R_B = 0.02$, $R_C = 0.0286$, $R_D = 0.0024$ °C/W.]

2.18 In the case of calculation of the rate of heat transfer through a cylindrical wall of small thickness, the 'arithmetic mean area' of the wall can be used. Determine the ratio of the inner and the outer radii (r_o/r_i) of a cylindrical wall for which the use of the arithmetic mean area does not introduce more than 1% error in heat transfer calculation. Also, determine whether the use of the arithmetic mean area overestimates the heat transfer rate.

2.19 In order to measure the effective thermal conductivity of a powder, the material is packed in the annulus between two aluminium spheres. The outer diameter of the inner sphere is 60 mm, and the inner diameter of the outer sphere is 80 mm. The core of the assembly is electrically heated. In a steady state experiment, the power supply to the core is rated at 100 W when the temperatures of the aluminium surfaces are found to be 210°C and 60°C. What is the thermal conductivity of the sample?

2.20 Consider the composite cylinder made of a tapered core of material A and an annular tapered shell of material B as shown in Fig. 2.14. The outer diameter of the cylindrical composite body is 10 cm. The core diameters are 5 cm at end 1, and 8 cm at end 2. The length of the body is 1 m, and the end temperatures are $T_1 = 200°C$ and $T_2 = 60°C$. If the curved surface of the cylinder is perfectly insulated and the radial temperature distribution at any axial position is uniform, calculate the rate of heat flow through the cylinder. Given: $k_A = 10$ W/m K, $k_B = 20$ W/m K.

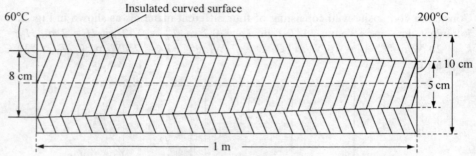

Fig. 2.14 The composite cylinder with a tapered core (Problem 2.20).

2.21 Heat generation occurs at a uniform volumetric rate in a 5 cm diameter cylinder having a thermal conductivity of 12 W/m °C. If the radial temperature distribution in the cylinder at steady state is given as: $T = 313.021 - 2.083 \times 10^4 r^2$ (T is in °C, and r in metres), determine (i) the surface and centreline temperatures of the cylinder, (ii) the volumetric rate of heat generation, and (iii) the average temperature of the cylinder. [*Hint:* Compare the given temperature distribution with Eq. (2.41) to calculate ψ_v.

2.22 A 30 cm thick wall of thermal conductivity 16 W/m °C has one surface (call it $x = 0$) maintained at a temperature 250°C and the opposite surface ($x = 0.3$ m) perfectly insulated. Heat generation occurs in the wall at a uniform volumetric rate of 150 kW/m³. Determine (a) the steady state temperature distribution in the wall, (b) the maximum wall temperature and its location, and (c) the average wall temperature.

   [*Hint:* The general form of the temperature distribution is given by Eq. (2.30). Use the boundary conditions $x = 0$, $T = 250$, $x = 0.3$, $dT/dx = 0$ (insulated surface), and obtain the values of C_1 and C_2.]

2.23 A solid composite sphere has an inner core (material 1) of radius r_1, and an outer shell (material 2) of outer radius r_2. The thermal conductivities of the materials are k_1 and k_2, respectively. If the corresponding rates of heat generations are ψ_{v1} and ψ_{v2}, and the surface temperature of the sphere is T_s, determine the steady state temperature profile in the composite sphere.

   [*Hint:* Two separate equations for the shell and the core have to be solved. Write down these equations based on Eq. (2.53). Satisfy the symmetry condition at $r = 0$, the surface condition at $r = r_2$, and the continuity of both temperature and heat flux at $r = r_1$.]

   [*Check:* if the materials are identical, i.e. $\psi_{v1} = \psi_{v2}$, and $k_1 = k_2$, both T_1 and T_2 reduce to the form of Eq. (2.56).]

REFERENCES AND FURTHER READING

Bureau of Indian Standards: IS 7413 (1981), Code of practice for the application and finishing of thermal insulation materials between 40°C and 700°C.

Glaser, P.E., I.A. Black, and P. Doherty, "Multilayer insulation", *Mech. Engg.,* August 1965, 23.

Harrison, M.R. and C.M. Pelanne, "Cost effective thermal insulation", *Chem. Engg.*, **84** (December, 1977), 62–76.

Hsu, S.T., *Engineering Heat Transfer*, D. Van Nostrand Co., New York, 1963.

Irwin, W., "Insulate intelligently", *Chem. Eng. Progr.*, **87** (May, 1991), 51–55.

Mollory, J.F., *Thermal Insulation*, Reinhold Book Corp., New York, 1969.

Myers, G.E., *Analytical Methods in Conduction Heat Transfer*, McGraw-Hill, New York, 1971.

Peng, L.C. and T.L. Peng, "Thermal insulation and pipe stress", *Hydrocarbon Processing,* May 1998, 111–113.

Petroleum Conservation Research Association, "Thermal insulation", *Publication No. 7*, PCRA, New Delhi.

Schneider, P.J., *Conduction Heat Transfer*, Addison-Wesley Pub. Co., Mass., 1955.

Tye, R.P. (Ed.), *Theory of the Thermal Conductivity of Solids*, Academic Press, London, 1969.

3

Heat Transfer Coefficient

The calculation of the rate of transport of heat in a solid body or medium is based on a well-defined simple law—the Fourier's law. While the law is equally valid for transport of heat in a fluid medium, there are very few physical situations in which heat transfer occurs in a fluid purely by conduction. For the Fourier's law to be applicable, the medium must be stagnant. But a fluid, in which flow of heat is occurring, rarely remains stagnant in practical situations. As a result, heat transfer in a fluid mostly occurs by convection. The use of the 'heat transfer coefficient' greatly helps us in calculating the rates of heat transfer by convection.

3.1 CONVECTIVE HEAT TRANSFER AND THE CONCEPT OF HEAT TRANSFER COEFFICIENT

Convection refers to transport of heat (or mass) in a fluid medium because of the motion of the fluid. Convective heat transfer is, therefore, associated with bulk motion in the fluid medium. *Bulk motion* or *macroscopic motion* or *bulk flow* can be easily distinguished from *molecular motion* or *microscopic motion*. Molecules in any medium, solid or fluid, are never static. In a solid, a molecule or an atom is in a vibrational state of motion, but has a mean position. Molecules or atoms of a fluid have translational motion in addition. As a result, the molecules are continually changing their positions. Any one position vacated by a molecule is instantaneously occupied by another molecule. This molecular or microscopic motion cannot be measured by a flow measuring device. On the other hand, bulk flow may be caused by a pressure gradient in a medium, and can be measured by a device such as venturi meter, orifice meter or pitot tube.

Let us consider a fluid having bulk (or macroscopic) motion. If the flow is laminar, the fluid elements or ensembles (a fluid element is conceived as a conglomerate of a large number of molecules, but of too small a size in the macroscopic scale) move in an orderly or streamline flow. In turbulent flow, the fluid elements move in a random manner while undergoing, on the average, a net displacement in the direction of flow. In either case, a moving fluid element carries with it heat energy as it moves from one point to another and thus contributes to the rate of heat transfer in the event of there being a temperature gradient.

When a fluid flows over a solid surface, its velocity is zero at the surface (no-slip condition) and increases gradually with distance from the surface, eventually attaining the *free-stream velocity*. Transport of heat to the fluid at or very near the surface occurs mainly by conduction because the fluid velocity is practically zero there. Since the velocity of fluid increases with distance from the surface, the role of convection becomes increasingly important. Irrespective of the nature of flow of the bulk fluid (laminar or turbulent), a fluid layer near the surface always remains laminar. The thickness of this layer decreases with increase in the velocity of the bulk fluid. This thin layer of fluid at the surface accounts for most of the resistance to heat transfer. At a higher velocity of the fluid, this resistance becomes smaller.

It is obvious that if this resistance to heat transfer is known, the rate of heat transfer to or from the surface can be calculated. Although this resistance can be calculated, at least in principle, if the velocity distribution of the fluid is clearly known, the exercise becomes very difficult in a great majority of practical flow situations. This is because either the velocity distribution is not well-understood, or the problem is formidable mathematically. A pragmatic approach to overcome the difficulty is to define a *heat transfer coefficient*, h, such that

$$Q = hA(T_s - T_o) \tag{3.1}$$

where Q is the rate of heat transfer (W), A is the area of heat transfer (m^2), T_s is the surface temperature, and T_o is the bulk fluid temperature. Written in words, Eq. (3.1) may be expressed as

Rate of heat transfer = (Heat transfer coefficient) × (Area) × (Temperature driving force)

$$\tag{3.2}$$

Heat transfer coefficient is a *phenomenological coefficient*. It is based on the observation that the rate of convective heat transfer is proportional to the area of heat transfer and the temperature driving force. The proportionality constant is called the heat transfer coefficient. That is,

$$Q \propto A \quad \text{and} \quad Q \propto T_s - T_o$$

or

$$Q = hA(T_s - T_o) = hA \, \Delta T$$

Therefore,

$$h = \frac{Q}{A(T_s - T_o)} = \frac{q}{T_s - T_o} = \frac{q}{\Delta T} \tag{3.3}$$

where q is the heat flux and ΔT the driving force. So, quantitatively, heat transfer coefficient is equal to the heat flux per unit driving force. Equation (3.1) is also known as the 'Newton's law of cooling'.

It should be remembered that the heat transfer coefficient combines in it the effects of velocity or motion in the medium and also the properties like thermal conductivity, viscosity, density of the medium, etc. Motion or velocity in a medium can be created by application of an external force. For example, an agitator creates motion in the liquid in a tank, or a pump creates the pressure necessary for flow of a fluid through a pipe. Heat transfer in such systems (for example, in an agitated vessel or in a pipe flow) is said to occur by *forced convection*. However, if the temperature of a particular layer of a fluid is higher than that of an upper layer, a bouyancy-induced motion is generated. Heat transfer augmented by this kind of motion is called *free convection heat transfer*. Forced convection heat transfer coefficients are generally much higher than their free convection counterparts.

Heat transfer coefficient is also called the *film coefficient*. This is because heat transfer at a phase boundary is sometimes visualized as occurring through a thin stagnant film adhering to the boundary or the interface between the phases. Beyond this film the fluid is assumed to be well-mixed. The change in the temperature of the fluid from the value at the surface to that at the bulk occurs across the film only, and there is no further temperature change beyond the film. The fluid in the film being 'stagnant' (i.e. it does not have any velocity), heat transfer through it occurs purely by conduction. The difference in the temperature between the interface and the bulk, i.e. the temperature drop across the film, is the driving force. The steady state temperature distribution in such a stagnant film is schematically shown in Fig. 3.1. The 'film thickness' is assumed to be

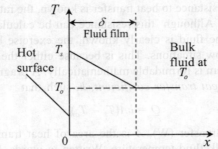

Fig. 3.1   Temperature profile in a stagnant film.
(Driving force = $T_s - T_o$).

δ. The temperature is plotted along the vertical axis and the distance from the hot surface along the horizontal axis. Temperature decreases linearly with distance because heat flow is assumed to occur at steady state by conduction (like steady state conduction through a solid wall). If k is the thermal conductivity of the fluid, the rate of heat transfer can be written directly by following the Fourier's law. Therefore, we have

$$Q = -kA\left(\frac{T_o - T_s}{\delta}\right) = \frac{k}{\delta}A(T_s - T_o) \tag{3.4}$$

where $(T_o - T_s)/\delta$ is the temperature gradient in the film. Comparing Eq. (3.1) with (3.4), we have

$$h = \frac{k}{\delta} \tag{3.5}$$

It is obvious that the heat transfer coefficient can be easily calculated using the above relation if k and δ are known. While the thermal conductivity (k) data are readily available, the film thickness δ is not known. Thus Eq. (3.5) does not help us to estimate the heat transfer coefficient. But heat transfer coefficient is important in the context of visualization of the convective heat transfer phenomenon as described in the *film theory of heat transfer*. Physical significance of δ is that it is the thickness of a stagnant layer of fluid that offers the same heat transfer resistance as that which actually exists in the fluid under the given hydrodynamic conditions. Because a stagnant film is a visualization rather than a reality, it is sometimes called a *fictitious film*. It is pertinent to mention here that the heat transfer coefficient at a surface is determined by the flow field and the thermophysical properties (density, viscosity, thermal conductivity, specific heat, etc.) of the fluid. The effects of all these factors and quantities are 'lumped' together to define the film thickness, δ.*

The inverse of the heat transfer coefficient, i.e. $1/h$, gives a measure of the heat transfer resistance.

The film theory of heat transfer does not help us in calculating the heat transfer coefficient h. But it has to be known before we can calculate the rate of heat transfer using Eq. (3.1). It may be noted that the other quantities (A, T_s and T_o) in Eq. (3.1) are either known or can be measured. Fortunately, a large number of correlations have been developed over the years by heat transfer researchers which are applicable to a variety of physical situations and geometries. The more

---

* The definition of mass transfer coefficient is also based on similar concepts.

important correlations will be discussed in the next chapter. Here we will consider a few physically realistic and practically important cases and problems involving convective heat transfer. Table 3.1 gives typical values of heat transfer coefficients (for both free and forced convection) for a number of systems.

Table 3.1 Values of heat transfer coefficients under typical conditions

Description	Heat transfer coefficients		
	$W/m^2 \; °C$	$kcal/h \; m^2 \; °C$	$Btu/h \; ft^2 \; °F$
1. Air flowing at a velocity of 12 m/s over a flat plate, 2 m in length, at 1 atm, 25°C (Laminar boundary layer flow of air)	23	19.5	4
2. Air flowing at a velocity of 10 m/s through a 1 inch (25 mm) pipe at 1 atm, 200°C (Turbulent pipe flow of air)	65	53.7	11
3. Water flowing at a velocity of 1.0 m/s through a 1-1/2 inch (38 mm) pipe at 25°C (Turbulent pipe flow of water)	3800	3200	650
4. Benzene flowing at a velocity of 1.0 m/s through a 1-1/2 inch (38 mm) pipe at 40°C (Turbulent pipe flow of an organic)	1200	980	200
5. A vertical plate, 1.5 m high, at a temperature of 65°C in contact with air at 25°C (Natural convection on a vertical plate in air)	4.5	4.0	0.8
6. A horizontal cylinder, 13 cm in diameter, at a temperature of 80°C in contact with air at 25°C (Natural convection on a horizontal cylinder in air)	6.0	5.0	1.0

Example 3.1 An ice-ball of initial diameter 0.06 m is suspended in a room at 30°C. The ice melts by absorbing heat from the ambient, the surface heat transfer coefficient being 11.4 W/m^2 °C. The air in the room is essentially dry. If the shape of the ball remains unchanged, calculate the time required for reduction in its volume by 40%. The density of ice is 929 kg/m^3 and its latent heat of fusion is 3.35×10^5 J/kg.

SOLUTION Let r be the radius and m the mass of the ice-ball at any time t. Also, let ρ be the density and λ be the latent heat of fusion of ice.

$$\text{Mass of the ice-ball, } m = \frac{4}{3}\pi r^3 \rho$$

$$\text{The rate of melting} = -\frac{dm}{dt} = -\frac{d}{dt}\left(\frac{4}{3}\pi r^3 \rho\right) = -4\pi r^2 \rho \frac{dr}{dt}$$

The mass m of the ice-ball decreases with time and dm/dt is negative. We put a negative sign before dm/dt, so that the rate of melting is expressed as a positive quantity.

$$\text{Rate of heat absorption by the ice-ball for melting} = -4\pi r^2 \rho \frac{dr}{dt}\lambda$$

Also, the rate of heat transfer from the ambient to the ice-ball

$$= \text{(area)(heat transfer coefficient)(driving force)} = 4\pi r^2 h(T_o - T_b)$$

where

h is the heat transfer coefficient

T_o is the ambient temperature

T_b is the temperature of the ice-ball.

By energy balance, we have

$$-4\pi r^2 \rho \frac{dr}{dt} \lambda = 4\pi r^2 h(T_o - T_b)$$

Integrating, we get

$$r = C_1 - \frac{h(T_o - T_b)}{\rho \lambda} t$$

where C_1 is an integration constant.

Putting the values of the quantities involved, we get

$$r = C_1 - \frac{(11.4)(30 - 0)}{(929)(3.35 \times 10^5)} t = C_1 - 1.1 \times 10^{-6} t$$

where the constant C_1 can be determined by using the condition

$$t = 0 \text{ (i.e. the initial time)}, r = 0.03 \text{ m}$$

Thus, we have

$$C_1 = 0.03 \quad \text{and} \quad r = 0.03 - 1.1 \times 10^{-6} t \tag{i}$$

If the volume of the ball is reduced by 40% of the original volume, we have

$$\frac{4}{3}\pi r^3 = 0.6 \times \frac{4}{3}\pi (0.03)^3$$

or

$$r = 0.0253 \text{ m}$$

The time required for melting can be calculated using Eq. (i)

$$0.0253 = 0.03 - 1.1 \times 10^{-6} t$$

Therefore,

$$t = 4267 \text{ s} = \boxed{1.184 \text{ h}}$$

Example 3.2 A 1 kW electric room heater has a coil of nichrom wire of diameter 0.574 mm and electrical resistance 4.167 ohm/m. If the temperature of the room remains constant at 21°C and the average heat transfer coefficient at the surface of the wire is 100 W/m^2 °C, calculate the time required for the heating coil, after it is switched on, to reach 63% of its steady state temperature rise. Assume that the wire itself offers negligible heat transfer resistance. The density of the material of the wire is 8920 kg/m^3, and its specific heat is 384 J/kg °C.

SOLUTION This is a problem of heating of a wire at unsteady state. The electrical heating capacity of the wire is 1000 W at an applied voltage of $V = 220$ V.

Total electrical resistance of the wire is given by, $1000 = V^2/R'$
or

$$R' = \frac{(220)^2}{1000} = 48.4 \text{ ohm}$$

Electrical resistance of the wire = 4.167 ohm/m

Therefore, the length of the wire, $l = \dfrac{48.4}{4.167} = 11.61$ m

Diameter of the wire, $d = 0.574$ mm

Area of the wire = Area of heat transfer

$$A = \pi d l = (\pi)(5.74 \times 10^{-4})(11.61) = 2.093 \times 10^{-2} \text{ m}^2$$

If T_f (°C) is the *steady state* or final temperature of the wire, we have

$$1000 = hA(T_f - T_o)$$

where T_o is the ambient temperature.

Therefore,

$$1000 = (100)(2.093 \times 10^{-2})(T_f - 21)$$

or

$$T_f = 499°C$$

The steady state temperature rise = 499 − 21 = 478°C

When 63% of this temperature rise is attained, the temperature of the wire at that moment is

$$21 + (0.63)(478) = 322°C$$

and the time required to reach this temperature of 322°C can be calculated by solving the following unsteady state heat balance equation.

If m is the mass of the wire, c_p its specific heat, and T the temperature of the wire at any time t, we have

$$\text{Rate of heat accumulation in the wire} = \frac{d}{dt}(mc_pT)$$

$$\text{Rate of heat loss} = hA(T - T_o)$$

where T_o is the ambient temperature.

Now, we write the unsteady state heat balance equation as

$$\frac{d}{dt}(mc_pT) = Q' - hA(T - T_o)$$

or

$$\frac{dT}{dt} = \frac{Q'}{mc_p} - \frac{hA}{mc_p}(T - T_o) \tag{i}$$

where Q' is the rate of electrical heat generation in the coil.

The initial condition is: $t = 0$, $T = T_o$. This means that the heater is at the ambient temperature when the switch is turned on.

The solution of Eq. (i) subject to the given initial condition is

$$T = T_o + \frac{Q'}{hA}\left[1 - \exp\left(\frac{hA}{mc_p}t\right)\right]$$ (ii)

Given: $T = 322°C$, $Q' = 1000$ W, $h = 100$ W/m^2 °C, $A = 2.093 \times 10^{-2}$ m^2, $T_o = 21°C$,

$$m = (\pi d^2/4)l\rho = 0.0268 \text{ kg, and } c_p = 384 \text{ J/kg °C}$$

Putting these values in Eq. (ii), we have

the required time, $t = \boxed{4.9 \text{ s}}$

Example 3.3   The steady state temperature distribution in a 0.2 m thick wall is known to be $T = 250 - 2750\, x^2$, where x is the position in the wall in metre, and T is in °C. The thermal conductivity of the material of the wall is 1.163 W/ m °C. The wall loses heat to an ambient at 30°C (Fig. 3.2).

(a) Calculate the heat transfer coefficient at the surface of the wall at $x = 0.2$ m.

(b) What can be said about the same at the other surface of the wall?

(c) What is the average volumetric rate of heat generation in the wall?

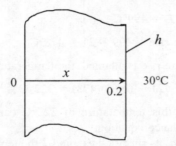

Fig. 3.2   Steady state temperature distribution in a thick wall (Example 3.3).

SOLUTION   (a) The rate of conduction of heat (at steady state) from inside the solid to the surface (at $x = 0.2$) must be equal to the rate of convective heat transfer from this surface to the ambient medium.
Therefore,

$$\text{at } x = 0.2, \quad -k\frac{dT}{dx} = h(T - T_o)$$

Now, at $x = 0.2$,

$$T = 250 - (2750)(0.2)^2 = 140°C$$

$$dT/dx = -2750(2x) = -2750(2)(0.2) = -1100 \text{ °C/m}$$

Given: $k = 1.163$ W/m °C

Therefore,

$$-(1.163)(-1100) = h(140 - 30)$$

or

$$h = \boxed{11.63 \text{ W}/\text{m}^2 \,^\circ\text{C}}$$

(b) At the other surface of the wall, at $x = 0$, $dT/dx = -2750\,(2)(x) = 0$. So, there is no heat flow at this surface of the wall, that is, the wall is insulated on one side.

(c) For unit surface area, volume of the wall = $(1 \text{ m}^2)(0.2 \text{ m}) = 0.2 \text{ m}^3$
Heat loss from unit area (from one side of the wall only; the other side does not lose heat)

$$= h(140 - 30)(1) = (11.63)(110) = 1279.3 \text{ W}$$

Average volumetric rate of heat generation in the wall $= (1279.3 \text{ W})/0.2 \text{ m}^3$

$$= \boxed{6396 \text{ W}/\text{m}^3}$$

3.2 OVERALL HEAT TRANSFER COEFFICIENT

Heat transfer from one fluid to another separated by an intervening solid wall is of great practical importance. Calculation of the rate of heat transfer in such a case is done by using an *overall heat transfer coefficient*. Two situations are considered below.

3.2.1 Heat Transfer between Fluids Separated by a Plane Wall

We will first consider heat transfer from one fluid to another separated by a plane solid wall of thickness l (Fig. 3.3). The temperatures of the bulk of the fluids on two sides of the wall are T_1

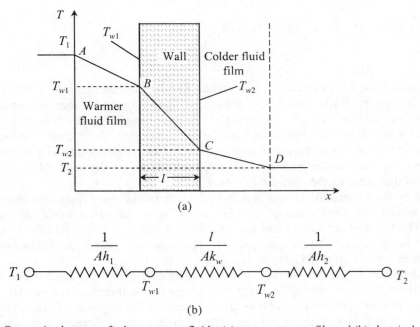

Fig. 3.3 Convective heat transfer between two fluids: (a) temperature profile and (b) electrical analogue.

and T_2 ($T_1 > T_2$), and the corresponding heat transfer coefficients between the wall and the fluids are h_1 and h_2, respectively. The temperatures at the fluid-wall interfaces are T_{w1} and T_{w2}. The thermal conductivity of the material of the wall is k_w. If the area of heat transfer is A, the rates of heat transfer on two sides of the wall are:

Rate of heat transfer from the hot fluid to the wall, $Q_1 = h_1 A(T_1 - T_{w1})$

Rate of heat transfer through the wall, $Q_w = k_w A(T_{w1} - T_{w2})/l$

Rate of heat transfer from the wall to the cold fluid, $Q_2 = h_2 A(T_{w2} - T_2)$

Because heat transfer is taking place at steady state through a constant area, we can write

$$Q_1 = Q_w = Q_2 = Q \text{ (say)}$$

Therefore,

$$T_1 - T_{w1} = \frac{Q}{Ah_1}; \quad T_{w1} - T_{w2} = \frac{Q}{A(k_w/l)}; \quad T_{w2} - T_2 = \frac{Q}{Ah_2} \tag{3.6}$$

Adding the above three relations, we get

$$T_1 - T_2 = \frac{Q}{A}\left(\frac{1}{h_1} + \frac{1}{k_w/l} + \frac{1}{h_2}\right) = \frac{Q}{UA} \tag{3.7}$$

where

$$\frac{1}{U} = \frac{1}{h_1} + \frac{1}{k_w/l} + \frac{1}{h_2}$$

Therefore,

$$Q = UA(T_1 - T_2) \tag{3.8}$$

Comparing Eq. (3.8) with Eq. (2.8), we have

$$\text{Total thermal resistance} = \frac{1}{UA} \tag{3.9}$$

It appears from Eq. (3.8) that the rate of heat transfer has been expressed in a form similar to Eq. (3.1). Here the driving force is the difference of temperature between the hot and the cold fluids, which is, in fact, the *overall driving force* for heat transfer. Here, the quantity U is called the *overall heat transfer coefficient*, and can be calculated if h_1, h_2, k_w, and l are known. The overall thermal resistance or heat transfer resistance, $1/UA$, is equal to the sum of the individual resistances—the heat transfer resistances offered by the 'film' of the hot fluid ($1/Ah_1$), the wall (l/Ak_w), and the 'film' of the cold fluid ($1/Ah_2$).

Based on the assumption of two hypothetical or 'fictitious' fluid films on either side of the wall, the temperature distributions in the films as well as in the wall are qualitatively shown in Fig. 3.3. The temperature distributions in all the three sections (*AB, BC* and *CD*) are predictably linear, but have different slopes in general. The electrical analogue of the problem showing the resistances in series is also depicted in Fig. 3.3.

The definition of the overall heat transfer coefficient given by Eq. (3.8) can be used to introduce the concept of *controlling resistance*. If one of the thermal resistance terms is much larger than the sum of the other two, it is called the controlling resistance. For example, if $1/Ah_1$ $\gg (l/Ak_w + 1/Ah_2)$, then $1/Ah_1$ is the controlling heat transfer resistance. Very often the heat transfer resistance offered by a metal wall (l/Ak_w) separating the two fluids is small (because k_w is large) compared to $1/Ah_1$ or $1/Ah_2$, and can be, therefore, neglected without causing an appreciable error in the calculation. Also, when heat transfer occurs between a gas phase and a liquid phase separated by a metal wall, the gas phase heat transfer resistance usually becomes significantly larger than the sum of the other two resistances (but may not necessarily be the controlling resistance). This is because the thermal conductivity of a gas is much smaller than that of a liquid.

3.2.2 Heat Transfer between Fluids Separated by a Cylindrical Wall

A well-known simple practical device for heat transfer from one fluid phase to another is the *double-pipe heat exchanger*. It consists of two concentric pipes properly fitted or welded with arrangements for pumping one of the fluids through the inner pipe and the other through the annular space. The fluids are thus brought in 'thermal contact' in order to achieve heat transfer. A counter-current double-pipe heat exchanger is schematically shown in Fig. 3.4.

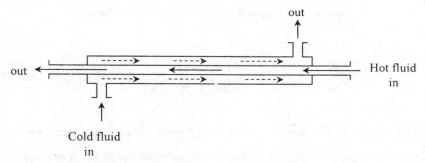

Fig. 3.4 Schematic of a counter-current double-pipe heat exchanger.

How to express the overall heat transfer coefficient for this or a similar device that involves heat transfer in a cylindrical geometry? It can be expected that the requisite equation will be a little different from Eqs. (3.7) and (3.8) because here the wall separating the fluids has a variable area along the radius. Let the inner and outer radii of the inner pipe be r_i and r_o respectively, and the corresponding heat transfer coefficients be h_i and h_o respectively. The inner surface temperature of the pipe wall is T_{wi} and that of the outer surface of the wall is T_{wo}. The bulk fluid temperatures are T_i and T_o ($T_i > T_o$). A cross-section of the device is shown in Fig. 3.5.

We consider a section of the assembly of length L over which the temperatures T_i and T_o are assumed to remain fairly constant. The following rates of heat transfer across the wall of the inner pipe can be written. The rates are equal at steady state.

The rate of heat transfer from the hot fluid to the inner surface (at $r = r_i$) is

$$Q = A_i h_i (T_i - T_{wi})$$

The rate of heat transfer through the pipe wall [see Eq. (2.14)] is

$$Q = \frac{2\pi L k_w (T_{wi} - T_{wo})}{\ln\ (r_o/r_i)}$$

The rate of heat transfer from the outer surface (at $r = r_o$) of the pipe to the cold fluid is

$$Q = A_o h_o (T_{wo} - T_o)$$

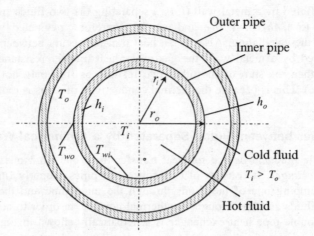

Fig. 3.5 Cross-section of a double-pipe heat exchanger.

Therefore, we have

$$T_i - T_{wi} = \frac{Q}{A_i h_i}, \quad T_{wi} - T_{wo} = \frac{Q}{2\pi L k_w / \ln(r_o/r_i)}, \quad T_{wo} - T_o = \frac{Q}{A_o h_o}$$

where

$A_i = 2\pi r_i L$ = area of heat transfer based on the inner radius of the inner pipe

$A_o = 2\pi r_o L$ = area of heat transfer based on the outer radius of the inner pipe

Adding the above three relations and rearranging, we get

$$Q = \frac{T_i - T_o}{\dfrac{1}{A_i h_i} + \dfrac{\ln(r_o/r_i)}{2\pi k_w L} + \dfrac{1}{A_o h_o}} \tag{3.10}$$

The inside and outside surface areas of the inner pipe are different. So, we should define *two* overall coefficients—an overall heat transfer coefficient U_i based on the inside surface area and another coefficient U_o based on the outside surface area. However, the rate of heat transfer and the driving force remain the same. Therefore, we have

$$Q = U_i A_i (T_i - T_o) = U_o A_o (T_i - T_o)$$

The above equation, when compared with Eq. (3.10), yields

$$U_i = \frac{1}{\dfrac{1}{h_i} + \dfrac{A_i \ln(r_o/r_i)}{2\pi k_w L} + \dfrac{A_i}{A_o h_o}} = \frac{1}{\dfrac{1}{h_i} + \dfrac{r_i \ln(r_o/r_i)}{k_w} + \dfrac{r_i}{r_o h_o}} \tag{3.11}$$

$$U_o = \frac{1}{\dfrac{A_o}{A_i h_i} + \dfrac{A_o \ln(r_o/r_i)}{2\pi k_w L} + \dfrac{1}{h_o}} = \frac{1}{\dfrac{r_o}{r_i h_i} + \dfrac{r_o \ln(r_o/r_i)}{k_w} + \dfrac{1}{h_o}} \tag{3.12}$$

The denominator of Eq. (3.10) and that of Eq. (3.11) or (3.12) is the sum of the three heat transfer resistances in series. The total thermal resistance R_T is related to U_i and U_o as follows:

$$Q = \frac{T_i - T_o}{R_T} = U_i A_i (T_i - T_o) = U_o A_o (T_i - T_o)$$

or

$$R_T = \frac{1}{U_i A_i} = \frac{1}{U_o A_o} \tag{3.13}$$

and

$$U_i A_i = U_o A_o \tag{3.14}$$

Example 3.4   A steam pipe of 97 mm inner diameter and 114 mm outer diameter (4 inch, schedule 80), is required to carry high pressure[⊕] saturated steam at 30 bar absolute pressure. It is covered by a layer of mineral wool* in order to reduce heat loss. Though an extra thick layer of insulation saves a lot of heat, it is expensive at the same time. As a standard practice, a design engineer usually allows the temperature at the outer surface of the insulation (also called the *skin temperature*) to remain at 15–25°C above the room temperature. In order to calculate the thickness of insulation of the above steam pipe, assume a skin temperature of 55°C. The ambient temperature is 30°C. The thermal conductivity of mineral wool may be taken as 0.1 W/m °C, and that of the pipe material (carbon steel) as 43 W/m °C. The external air-film coefficient for heat loss to the ambient is 8 W/m² °C. Calculate the thickness of insulation and the rate of heat loss per metre length of the pipe.

SOLUTION   We use the following notations: r_i = inner radius of the pipe, r' = outer radius of the pipe (this is the same as the inner radius of the layer of insulation), r_o = outer radius of the insulation, L = length of the pipe, k_c = thermal conductivity of the insulation, k_w = thermal conductivity of the pipe wall, T_i = steam temperature, T_s = skin temperature of the insulation, T_o = ambient temperature, and h = external heat transfer coefficient.

Thermal resistance of the insulation $= \dfrac{\ln (r_o/r')}{2\pi L k_c}$

Thermal resistance of the pipe wall $= \dfrac{\ln (r'/r_i)}{2\pi L k_w}$

Total thermal resistance of the pipe wall and the insulation $= \dfrac{1}{2\pi L} \left[\dfrac{\ln (r_o/r')}{k_c} + \dfrac{\ln (r'/r_i)}{k_w} \right]$

Driving force for heat conduction through the pipe wall and the insulation $= T_i - T_s$.

[⊕] Saturated steam at or below 3 kg/cm² gauge is generally called *low pressure steam*; that having a pressure between 3–12 kg/cm² is called *medium pressure steam*; above 12 kg/cm², it is called *high pressure steam*. But this classification should not be taken too rigidly.

* In practice, the insulation on a surface is covered or 'clad' with a thin aluminium sheet to protect the insulation against mechanical damage, weather and rain, spills and fire. The insulation not only helps save energy but also helps prevent possible burn injuries to the workmen from a hot surface (see also Section 2.6).

At steady state the rate of heat flow through the insulation and the outer air-film are equal under the driving force $(T_s - T_o)$ and the area $2\pi r_o L$. Therefore, we have

$$\frac{2\pi L(T_i - T_s)}{\ln (r_o/r')/k_c + \ln (r'/r_i)/k_w} = 2\pi r_o Lh(T_s - T_o)$$

Putting $r_i = 97/2 = 48.5$ mm, $r' = 114/2 = 57$ mm, $k_c = 0.1$ W/m °C, $k_w = 43$ W/m °C, $h = 8$ W/m² °C, $T_s = 55$°C, $T_i = 234$°C (which is the temperature of saturated steam at 30 bar absolute, as may be obtained from the steam table), $T_o = 30$°C, we get

$$\frac{234 - 55}{\ln(r_o/0.057)/0.1 + \ln(0.057/0.04850/43)} = r_o(8)(55 - 30)$$

This is a *transcendental equation* to be solved by trial or by any numerical technique. The solution is $r_o = 0.12$ m, and therefore the required thickness of insulation is

$$r_o - r' = (0.12 - 0.057) \text{ m} = \boxed{63 \text{ mm}}$$

The rate of heat loss from one metre length of the pipe

$$= 2\pi r_o Lh(T_s - T_o) = 2\pi(0.12)(1)8(55 - 30)$$

$$= 150.8 \text{ W}$$

Example 3.5 An 8% solution of alcohol from a fermenter is raised to 75°C in a heat exchanger before feeding it to the distillation column for concentration. A 50 mm (schedule 40 IPS) pipe is used to carry the hot liquid to the column. The pipe is insulated with a 4 cm thick layer of glass fibre blanket ($k = 0.068$ W/m °C). The liquid-side film coefficient is estimated to be 800 W/m² °C, and the outer air-film coefficient is 10 W/m² °C. The thermal conductivity of the liquid mixture (alcohol and water) can be calculated by using the Filippov equation $k_m = w_1 k_1 + w_2 k_2 - 0.72 w_1 w_2(|k_1 - k_2|)$, where w_1 and w_2 are the weight fractions of components 1 and 2 having thermal conductivities k_1 and k_2 respectively. The following data are also available: ambient air temperature = 28°C, actual pipe diameter: i.d. = 53 mm, o.d. = 60 mm. Thermal conductivity values are—pipe wall: 45 W/m °C, air: 0.0263 W/m °C, alcohol: 0.155 W/m °C, and water: 0.67 W/m °C. Calculate

    (a) the effective thicknesses of the air and the 'liquid films',

    (b) the overall heat transfer coefficient based on the i.d. of the pipe,

    (c) the overall heat transfer coefficient based on the o.d. of the insulation,

    (d) the percentage of the total resistance offered by the 'air-film',

    (e) the rate of heat loss per metre length of the pipe, and

    (f) the insulation 'skin temperature'.

SOLUTION Figure 3.6 shows a cross-section of the insulated pipe. Some of the notations used are also shown in the figure.

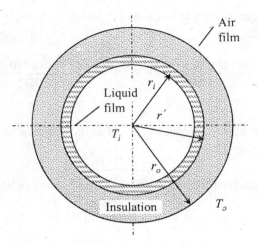

Fig. 3.6 Cross-section of the pipe (Example 3.5).

Given: inside radius of the pipe, $r_i = 0.053/2 = 0.0265$ m; outside radius of the pipe, $r' = 0.06/2 = 0.03$ m; outer radius of the insulation, $r_o = r' + 4$ cm $= 0.07$ m; inside temperature (bulk liquid), $T_i = 75°C$; ambient temperature, $T_o = 28°C$; $h_i =$ liquid-side heat transfer coefficient $= 800$ W/m^2 °C; and $h_o =$ outside heat transfer coefficient, 10 W/m^2 °C; liquid concentration: w_1 (weight fraction alcohol) $= 0.08$, w_2 (weight fraction of water) $= 0.92$.

(a) Thermal conductivity of the liquid mixture,

$$k_m = w_1 k_1 + w_2 k_2 - 0.72 w_1 w_2 (|k_1 - k_2|)$$

$$= (0.08)(0.155) + (0.92)(0.67) - (0.72)(0.08)(0.92)(|0.155 - 0.67|)$$

$$= 0.601 \text{ W/m °C}$$

Effective thickness of the 'liquid-film',

$$\delta_i = \frac{k_m}{h_i} = \frac{0.601 \text{ W/m °C}}{800 \text{ W/m}^2 \text{ °C}} = \boxed{0.75 \text{ mm}}$$

Effective thickness of the 'air-film',

$$\delta_o = \frac{k_a}{h_o} = \frac{0.0263 \text{ W/m °C}}{10 \text{ W/m}^2 \text{ °C}} = \boxed{2.6 \text{ mm}}$$

(b) Equation (3.11) may be used to calculate the overall heat transfer coefficient. There are four resistances in series. The calculations are based on 1 m length of the pipe ($L = 1$ m).

$$\frac{1}{U_i} = \frac{1}{h_i} + \frac{A_i \ln(r'/r_i)}{2\pi L k_w} + \frac{A_i \ln(r_o/r')}{2\pi L k_i} + \frac{A_i}{A_o h_o}$$

where

$k_w =$ thermal conductivity of the pipe wall $= 45$ W/m °C

$k_i =$ thermal conductivity of insulation $= 0.068$ W/m °C

Areas: $A_i = 2\pi r_i L = 2\pi(0.0265)(1) = 0.1665$ m^2, and $A_o = 2\pi r_o L = 2\pi(0.07)(1) = 0.4398$ m^2

Therefore, we obtain

$$\frac{1}{U_i} = \frac{1}{800} + \frac{(0.1665)\ln(0.03/0.0265)}{2\pi(1)(45)} + \frac{(0.1665)\ln(0.07/0.03)}{2\pi(1)(0.068)} + \frac{(0.1665)}{(0.4398)(10)}$$

$$= 0.3694$$

Thus the overall heat transfer coefficient based on the inside area,

$$U_i = \boxed{2.707 \text{ W/m}^2 \text{ °C}}$$

(c) If U_o is the overall heat transfer coefficient based on the outside area of the insulation, then

$$U_i A_i = U_o A_o \quad \text{[see Eq. (3.14)]}$$

Therefore, we get

$$U_o = \frac{U_i A_i}{A_o} = \frac{(2.707)(0.1665)}{0.4398}$$

$$= \boxed{1.025 \text{ W/m}^2 \text{ °C}}$$

(d) Total heat transfer resistance (based on inside area, 1 m pipe length),

$$= \frac{1}{U_i A_i} = \frac{0.3694}{0.1665} = \boxed{2.219 \text{ °C/W}}$$

Heat transfer resistance of the air-film $= \dfrac{1}{A_o h_o} = \dfrac{1}{(0.4398)(10)} = 0.2274$ °C/W

Per cent of the resistance offered by the air-film $= \dfrac{0.2274}{2.219} = \boxed{10.25\%}$

(e) Rate of heat loss per metre length of the pipe $= U_i A_i (T_i - T_o)$

$$= (2.707)(0.1665)(75 - 28) = \boxed{21.2 \text{ W}}$$

(f) If T_s is the outside surface temperature of the insulation (i.e. the insulation 'skin temperature'), we may write

$$U_o A_o (T_i - T_o) = h_o A_o (T_s - T_o)$$

or

$$(1.025)(75 - 28) = (10)(T_s - 28)$$

or

$$T_s = \boxed{32.8 °C}$$

Example 3.6 An insulated flat-headed cylindrical tank (i.d. = 1.5 m, height = 2.5 m, wall thickness = 0.006 m) is being used as a 'surge drum'* in a hot liquid line. The liquid being drawn

---

* A surge drum is sometimes provided in a fluid line in order to smoothen out any fluctuation in flow rate.

from the tank is at 80°C. The tank has a 4 cm thick insulation on it except on the bottom surface because it rests on a concrete foundation. The ambient temperature is 25°C. The heat transfer coefficients at the open surfaces are 5.5 W/m² °C for the flat top and 4 W/m² °C for the curved surface. Heat loss from the bottom of the tank can be neglected. The thermal conductivity of the insulation material is 0.05 W/m °C. If the average flow rate of liquid is 700 kg/h and its specific heat is 2000 J/kg °C, what should be the temperature of the liquid entering the tank? Assume that (i) steady state operating conditions exist, (ii) the liquid in the tank is well-mixed, and (iii) the heat transfer resistances of the 'liquid film' and of the tank wall are negligible. Also, calculate the insulation skin temperatures at the flat top surface and at the cylindrical surface.

SOLUTION   The total rate of heat loss from the tank is the sum of the heat losses from the cylindrical surface and the flat top. Referring to Fig. 3.7, we have the following values of the relevant quantities:

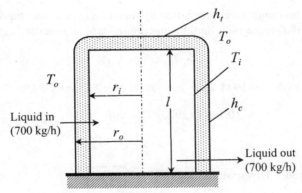

Fig. 3.7   Cross-section of the tank (Example 3.6).

Inner radius of the insulation, $r_i = 1.5/2 + 0.006 = 0.756$ m

Outer radius of the insulation, $r_o = 0.756 + 0.04 = 0.796$ m

Thermal conductivity of the insulation, $k_i = 0.05$ W/m °C

Heat transfer coefficient at the outer cylindrical surface, $h_c = 4$ W/m² °C

Heat transfer coefficient at the top flat surface, $h_t = 5.5$ W/m² °C

Height of the top of the insulation, $l = 2.5 + 0.006 + 0.04 = 2.546$ m

Temperature of the inner surface of the insulation (which is the same as the desired liquid temperature because the resistances of the liquid film and the metal wall are negligible), $T_i = 80$°C
Ambient temperature, $T_o = 25$°C
Heat transfer resistance of the cylindrical wall [see Eq. (3.10)]

$$= \frac{\ln (r_o/r_i)}{2\pi l k_i} + \frac{1}{2\pi r_o l h_c}$$

$$= \frac{\ln (0.796/0.756)}{2\pi(0.05)(2.546)} + \frac{1}{2\pi(0.796)(2.546)(4)}$$

$$= 0.0841 \text{ °C/W}$$

Heat transfer resistance of the flat insulated top [see Eq. (3.7)]

$$= \frac{1}{\pi r_o^2} \left[\frac{r_o - r_i}{k_i} + \frac{1}{h_t} \right]$$

$$= \frac{1}{\pi (0.796)^2} \left[\frac{0.04}{0.05} + \frac{1}{5.5} \right]$$

$$= 0.493 \, °C/W$$

Temperature driving force for heat transfer $= T_i - T_o = 80 - 25 = 55°C$

Total rate of heat loss, $Q = \dfrac{55}{0.0841} + \dfrac{55}{0.493} = 750 \, W$

This rate of heat loss must be compensated by maintaining the inlet liquid temperature higher than that in the tank. If T_l is the inlet liquid temperature, c_p is its specific heat, and m its flow rate, we have

$$mc_p (T_l - 80) = Q$$

Given: $m = 700 \, kg/h = 0.1944 \, kg/s$, $c_p = 2000 \, J/kg \, °C$ and $Q = 750 \, W$, we have

$$(0.1944)(2000)(T_l - 80) = 750$$

or

$$T_l = 82°C$$

Therefore, the inlet liquid temperature T_l should be $\boxed{82°C}$.

Insulation skin temperatures

Let T' be the insulation skin temperature at the flat top surface.

Rate of heat loss from the flat top surface $= 55/0.493 = 111.6 \, W = \pi r_o^2 \, h_t \, (T' - T_o)$ Therefore,

$$\pi (0.796)^2 (5.5)(T' - 25) = 111.6$$

or

$$T' = \boxed{35°C}$$

Similarly, the skin temperature of the cylindrical surface can be found to be $\boxed{38°C}$.

Example 3.7 A simple apparatus used for the experimental determination of vapour-liquid equilibrium data is the Othmer still. Heat is supplied to the still for vaporization of the liquid mixture; the vapour leaving the still is condensed and the condensate is recirculated to the still. After equilibrium is attained, samples of the liquid in the still and of the condensate are drawn and analyzed. The Othmer still is shown schematically in Fig. 3.8.

In the above apparatus, a 2.5 cm i.d. glass tube of 0.3 cm wall thickness is used as the still. A 2.5 m long nichrome wire (electrical resistance = 16.7 ohm/m) is wound on a 0.12 m section of the glass tube over which there is a 2.5 cm thick layer of insulation. The thermal conductivity of glass is 1.4 W/m °C, and that of the insulation is 0.041 W/m °C. The temperature of the

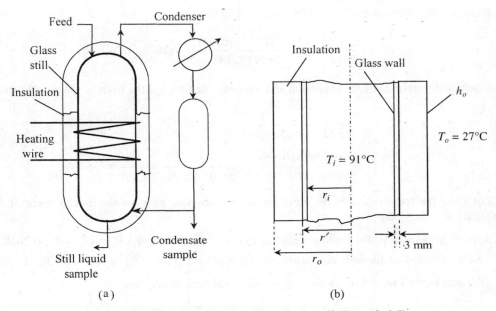

Fig. 3.8 Schematic diagram of the Othmer still (Example 3.7).

boiling liquid in the still is 91°C. The film coefficient of a boiling liquid is very large. The ambient temperature is 27°C, and the outside air-film coefficient is 5.8 W/m² °C. If power is supplied to the heating coil at 90 V, what is the rate of heat input to the boiling liquid at steady state?

SOLUTION   Resistance of the heating coil, $R' = (16.7 \text{ ohm/m})(2.5 \text{ m}) = 41.75$ ohm

$$\text{Rate of heat generation} = \frac{V^2}{R'} = \frac{(90)^2}{41.75} = 194 \text{ W}$$

A part of this heat flows to the boiling liquid and the other part is lost to the ambient.

There are three heat transfer resistances in series—those of the glass tube, the insulation and the outer air-film. These are calculated below. The values of the various quantities involved are:

Inner radius of the glass tube, $r_i = 0.0125$ m

Outer radius of the glass tube, $r' = 0.0155$ m

Outer radius of the insulation, $r_o = 0.0405$ m

Length of the still covered with the heating coil, $L = 0.12$ m

Thermal conductivity of glass, $k_g = 1.4$ W/m °C

Thermal conductivity of the insulation, $k_i = 0.041$ W/m °C

Outside heat transfer coefficient, $h_o = 5.8$ W/m² °C

Temperature at the inner surface of the glass tube, $T_i = 91$°C

Ambient temperature, $T_o = 27$°C

Heat transfer resistance of the glass wall,

$$R_g = \frac{\ln (r'/r_i)}{2\pi L k_g} = \frac{\ln (0.0155/0.0125)}{2\pi (0.12)(1.4)} = 0.2038 \text{ °C/W}$$

Combined resistance of the insulation and the outer air-film (on the basis of the outside area),

$$R_i + R_f = \frac{\ln (r_o/r')}{2\pi L k_i} + \frac{1}{2\pi r_o L h_o}$$

$$= \frac{\ln (0.0405/0.0155)}{2\pi (0.12)(0.041)} + \frac{1}{2\pi (0.0405)(0.12)(5.8)} = 36.71$$

Let T_s be the temperature of the outer surface of the glass tube (or the inner surface of the insulation) at steady state.

Rate of heat input to the boiling liquid in the still, $Q_1 = (T_s - T_i)/R_g = (T_s - 91)/0.2038$

Rate of heat loss through the insulation, $Q_2 = (T_s - T_o)/(R_i + R_f) = (T_s - 27)/36.71$

The total rate of heat flow or the rate of electrical heat generation,

$$Q_1 + Q_2 = \frac{T_s - 91}{0.2038} + \frac{T_s - 27}{36.71} = 194$$

or

$$T_s = 130°C$$

Therefore, we have

$$Q_1 = \frac{130 - 91}{0.2038} = \boxed{191.4 \text{ W}}$$

$$Q_2 = Q - Q_1 = 194 - 191.4 = 2.6 \text{ W}$$

Example 3.8 A 10 gauge electrical copper wire has a 1.3 mm thick rubber insulation on it. The life of the insulation depends, among other factors, on the temperature of the wire. It is desired to calculate the maximum allowable current in the wire so that the temperature of the insulation does not exceed 90°C. The ambient temperature is 30°C, and the outer air-film coefficient is 15 W/m² °C. The following data are available. Thermal conductivity: rubber (0.14 W/m °C), copper (380 W/m °C); electrical resistance of the copper wire (0.422 ohm/100 m).
  Determine

(a) the maximum allowable current, and

(b) the corresponding temperatures at the centre of the wire and at the outer surface of the insulation.

SOLUTION When current flows through a wire, heat generation occurs. This heat energy is conducted through the wire and then through the rubber insulation (there is no heat generation in the insulation), and is finally convected out to the ambient air through the outer 'air-film'. Because the wire is very long, heat conduction in the radial direction only needs to be considered. We will assume that steady state conditions prevail. By writing the appropriate shell balance equations, we will have two ordinary differential equations—one for the temperature distribution in the wire, and another for that in the insulation. These equations have to be solved by using the prevailing boundary conditions. The problem is similar to that described in Section 2.5 except that here the outside film coefficient is given instead of the outside surface temperature of the insulation. We refer to the Fig. 3.9 below and use the following notations:

T_m is the local temperature in the wire (it is a function of radial position)

T_c is the insulation temperature (also a function of radial position)

k_m is the thermal conductivity of copper

k_c is the thermal conductivity of the insulation

T_o is the ambient temperature

ψ_v is the volumetric rate of heat generation

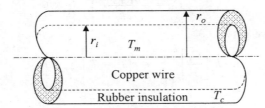

Fig. 3.9   Cross-section of the electrical wire (Example 3.8).

Applying Eq. (2.41) to the copper wire, we have

$$T_m = -\frac{r^2 \psi_v}{4k_m} + C_2 \qquad \text{(i)}$$

Applying Eq. (2.42) to the insulation, we have

$$T_c = C_3 \ln r + C_4 \qquad \text{(ii)}$$

In order to determine the integration constants C_2, C_3, and C_4, we need three boundary conditions. These are:

At the inner surface of the insulation, where $r = r_i$, we have

$$T_m = T_c \text{ (continuity of temperature)} \qquad \text{(iii)}$$

and

$$-k_m \frac{dT_m}{dr} = -k_c \frac{dT_c}{dr} \text{ (continuity of heat flux)} \qquad \text{(iv)}$$

At the outer surface of the insulation, where $r = r_o$, we have

$$-k_c \frac{dT_c}{dr} = h(T_c - T_o) \qquad \text{(v)}$$

What is the physical significance of the boundary condition (v)? It means that the rate at which heat is conducted from inside the insulation to the outer surface of it must be equal to the rate at which heat is convected away from this surface to the ambient. There cannot be any accumulation of heat at a surface even when the system is not at steady state.

If we put $r = r_i$ in Eqs. (i) and (ii) and use the boundary condition (iii), we get

$$-\frac{r_i^2 \psi_v}{4k_m} + C_2 = C_3 \ln r_i + C_4 \qquad \text{(vi)}$$

Differentiating (i) and (ii), putting $r = r_i$, and using (iv), we get

$$-k_m\left[-\frac{r_i \psi_v}{2k_m}\right] = -k_c\left[\frac{C_3}{r_i}\right]$$

or

$$C_3 = -\frac{r_i^2 \psi_v}{2k_c} \qquad \text{(vii)}$$

Similarly, using Eq. (v), we get

$$-k_c\left[\frac{C_3}{r}\right]_{r=r_o} = h[T_c - T_o]_{r=r_o}$$

or

$$-\frac{C_3 k_c}{r_o} = h(C_3 \ln r_o + C_4 - T_o)$$

Putting C_3 from (vii) we can solve the above equation for C_4 to obtain

$$C_4 = T_o + \left(\frac{r_i^2 \psi_v}{2k_c}\right)\left(\ln r_o + \frac{k_c}{hr_o}\right) \qquad \text{(viii)}$$

Again substituting for C_3 and C_4 from Eqs. (vii) and (viii) in Eq. (vi), we get

$$C_2 = T_o + \frac{r_i^2 \psi_v}{2}\left(\frac{1}{2k_m} + \frac{\ln(r_o/r_i)}{k_c} + \frac{1}{hr_o}\right) \qquad \text{(ix)}$$

The integration constants having been determined, the temperature distributions in the wire and in the insulation [Eqs. (i) and (ii)] can now be written as

$$T_m = T_o + \frac{r_i^2 \psi_v}{2}\left[\frac{1}{2k_m} + \frac{\ln(r_o/r_i)}{k_c} + \frac{1}{hr_o}\right] - \frac{r^2 \psi_v}{4k_m} \qquad \text{(x)}$$

$$T_c = T_o + \frac{r_i^2 \psi_v}{2k_c}\left[\ln\frac{r_o}{r} + \frac{k_c}{hr_o}\right] \qquad \text{(xi)}$$

Numerical calculations

(a) We shall first develop an expression for ψ_v. If ρ' = electrical resistance per unit length of the wire, r_i = radius of the wire, and I = current strength, then on the basis of unit length of the wire, we have

$$\text{Rate of electrical heat generation} = I^2\rho' \text{ W}$$

$$\text{Volume of the wire} = (\pi r_i^2)(1) = \pi r_i^2 \text{ m}^3$$

Therefore,

$$\text{Volumetric rate of heat generation, } \psi_v = (I^2\rho')/(\pi r_i^2) \text{ W/m}^3 \tag{xii}$$

The maximum temperature in the insulation occurs at $r = r_i$, where the insulation is in contact with the wire.

The various quantities involved are:

$(T_c)_{\max} = 90°C; \quad T_o = 30°C;$

r_i = radius of the 10 gauge wire = 1.3 mm = 1.3×10^{-3} m; r_o = 1.3 mm + 1.3 mm = 2.6×10^{-3} m

$k_c = 0.14$ W/m °C; $k_m = 380$ W/m °C; $h = 15$ W/m^2 °C; $\rho' = 0.422$ ohm/100 m = 4.22×10^{-3} ohm/m

Putting the numerical values in Eq. (xi), we get

$$(T_c)_{\max} = 90 = 30 + \frac{(1.3 \times 10^{-3})^2 \psi_v}{2(0.14)}\left[\ln\frac{2.6}{1.3} + \frac{0.14}{15(2.6 \times 10^{-3})}\right]$$

or

$$\psi_v = 2.3211 \times 10^6 \text{ W/m}^3$$

From Eq. (xii), we have

$$I = \left(\frac{\pi r_i^2 \psi_v}{\rho'}\right)^{1/2}$$

$$= \left[\frac{(2.3211 \times 10^6)(\pi)(1.3 \times 10^{-3})^2}{4.22 \times 10^{-3}}\right]^{1/2}$$

$$= \boxed{54.04 \text{ A}}$$

(b) Temperature at the centre of the wire can be obtained directly from Eq. (x) on putting $r = 0$.

$$[T_m]_{r=0} = 30 + \frac{(1.3 \times 10^{-3})^2(2.3211 \times 10^6)}{2}\left[\frac{1}{(2)(380)} + \frac{\ln(2.6/1.3)}{0.14} + \frac{1}{(15)(2.6 \times 10^{-3})}\right]$$

$$= \boxed{90.004 \text{ °C}}$$

This is also the maximum temperature of the wire.

This shows that the wire temperature remains virtually uniform at 90°C because of the high thermal conductivity of copper. The temperature at the outer surface of the insulation can be directly obtained by putting $r = r_o$ in Eq. (xi). That is,

$$[T_c]_{r=r_o} = 30 + \frac{(1.3 \times 10^{-3})^2 (2.3211 \times 10^6)}{2(0.14)} \times \frac{(0.14)}{(15)(2.6 \times 10^{-3})}$$

$$= \boxed{80.3°C}$$

It is thus seen that the skin temperature of the insulation will be quite high if the maximum allowable current flows.

Example 3.9 A heat generating slab A (thickness = 0.25 m, k_A = 15 W/m °C) is sandwiched between two other slabs B (thickness = 0.1 m, k_B = 10 W/m °C) and C (thickness = 0.15 m, k_C = 30 W/m °C) as shown in Fig. 3.10. There is no heat generation in slab B or C. The temperature distribution in slab A is known to be $T_A = 90 + 4500x - 11,000x^2$, where T is in °C

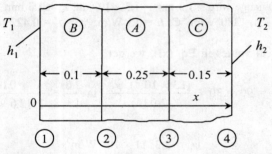

Fig. 3.10 Heat generating slab A sandwiched between slabs B and C (Example 3.9).

and x is the distance (in metres) from the left surface of B as shown in the figure. The wall B is in contact with a fluid at temperature T_1 = 40°C, the wall heat transfer coefficient being h_1. Similarly, the free surface of C loses heat to a medium at a temperature 35°C, and the surface heat transfer coefficient is h_2. Assume steady state condition.

(a) Calculate the temperatures at the surfaces of slab A. What is the maximum temperature in A and where does it occur?

(b) Determine the temperature gradients at both the surfaces of each of the slabs A, B and C. Comment on the signs of the gradients.

(c) Find the temperature profiles in slabs B and C. Also, calculate the values of the heat transfer coefficients h_1 and h_2.

SOLUTION (a) Temperature distribution in slab A,

$$T_A = 90 + 4500x - 11,000x^2$$

At the left surface of A, x = 0.1 m. Therefore,

$$[T_A]_{x=0.1} = 90 + 4500(0.1) - 11,000(0.1)^2 = \boxed{430°C}$$

Similarly, at the right surface, x = 0.1 + 0.25 = 0.35 m. Therefore,

$$[T_A]_{x=0.35} = \boxed{317.5°C}$$

At the point of maximum temperature in A, $dT_A/dx = 0$. Therefore, we have

$$4500 - (2)(11{,}000)x = 0$$

or

$$x = \boxed{0.2045\ \text{m}}$$

The maximum temperature in A, therefore, is

$$[T_A]_{\max} = 90 + (4500)(0.2045) - (11{,}000)(0.2045)^2 = \boxed{550.2^\circ\text{C}}$$

(b) *Temperature gradients in slab A*

At the interface (2), we have

$$x = 0.1 \quad \text{and} \quad [dT_A/dx]_{x=0.1} = 4500 - (2)(11{,}000)(0.1) = \boxed{2300\ ^\circ\text{C/m}}$$

At the interface (3), we have

$$x = 0.35 \quad \text{and} \quad [dT_A/dx]_{x=0.35} = 4500 - (2)(11{,}000)(0.35) = \boxed{-3200\ ^\circ\text{C/m}}$$

At $x = 0.1$, dT_A/dx is positive, i.e. temperature increases with x. So diffusion of heat occurs in the direction opposite to the x-direction. But at $x = 0.35$, dT_A/dx is negative and diffusion of heat occurs along the x-direction. So heat loss occurs from both the surfaces of A.

Temperature gradients in slabs B and C

By using the condition of continuity of heat flux, at the interface (2), we have

$$-k_B\left[\frac{dT_B}{dx}\right]_{x=0.1} = -k_A\left[\frac{dT_A}{dx}\right]_{x=0.1}$$

or

$$(10)\left[\frac{dT_B}{dx}\right]_{x=0.1} = (15)(2300)$$

Therefore,

$$\left[\frac{dT_B}{dx}\right]_{x=0.1} = \boxed{3450\ ^\circ\text{C/m}}$$

As there is no heat generation in slab B, the temperature profile in it must remain linear at steady state, and the temperature gradient shall remain uniform. So, at the interface (1), we have

$$\left[\frac{dT_B}{dx}\right]_{x=0} = \boxed{3450\ ^\circ\text{C/m}}$$

Similarly, the temperature gradients in C at the interfaces (3) and (4) can be found to be $\boxed{-1600\ ^\circ\text{C/m}}$.

(c) Steady state temperature distribution in a slab without heat generation is always linear.

For slab B, let $T_B = \beta_1^x + \beta_2$

At $x = 0.1$, $T_B = T_A = 430$ [from part (a)], and $dT_B/dx = 3450$

When $\beta_1 = 3450$, $\beta_2 = 85$, the temperature distribution in slab B is

$$T_B = 3450x + 85$$

Similarly, the temperature distribution in slab C may be found to be

$$T_C = 877.5 - 1600x$$

In order to calculate the surface heat transfer coefficients, we make use of the continuity of heat flux at the solid–fluid interface. That is,

$$h_1(T_1 - [T_B]_{x=0}) = -k_B \left[\frac{dT_B}{dx}\right]_{x=0}$$

or

$$h_1(40 - 85) = -(10)(3450)$$

Therefore,

$$h_1 = \boxed{766.6 \text{ W/m}^2 \, ^\circ\text{C}}$$

Similarly, the surface heat transfer coefficient at the other solid–fluid interface can be found to be

$$h_2 = \boxed{1129 \text{ W/m}^2 \, ^\circ\text{C}}$$

3.3 HEAT TRANSFER FROM EXTENDED SURFACES—THE FINS

Since thermal conductivities of gases are much smaller than those of liquids, the gas–phase heat transfer coefficient is also much smaller than the liquid–phase coefficient. Thus, if we make a heat transfer device for heat exchange between a gas and a liquid, the gas-side 'film' will offer most of the thermal resistance, and therefore a large heat transfer area will be required. A practical solution to the problem is to make an arrangement to increase the surface area of heat transfer on the gas side. This can be done by fixing attachments like rectangular metal strips or annular rings, called *fins*, to the surface of heat transfer (e.g. to the surface of a pipe). A cylindrical pipe with a ring-type fin is shown in Fig. 3.11. Because a fin increases the area available for heat transfer, a finned surface is also called an *extended surface*.

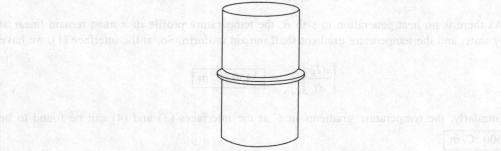

Fig. 3.11 A ring-type fin on a cylinder.

However, if a 'finned' pipe carries a hot fluid, the temperature at any point of the fin will be less than that of the pipe wall, thus providing a smaller driving force for heat transfer from the surface of the fin. Here we give a theoretical analysis to determine how effective a rectangular fin is in enhancing the rate of heat transfer.

We consider a thin rectangular plate that protrudes a distance l from a wall as shown in Fig. 3.12. This extended surface or fin has the length l, a thickness w and a breadth b. The temperature of the wall is T_w (which is also called the 'base temperature' of the fin), and that of

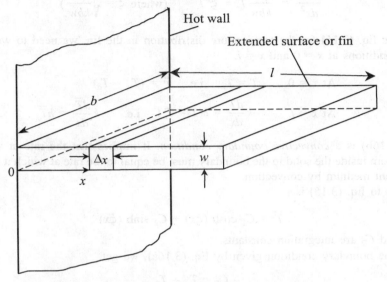

Fig. 3.12   Heat balance in a rectangular fin.

the ambient is T_o. Heat loss to the ambient occurs from both surfaces (top and bottom) of the fin. As the thickness of the fin is small, we assume that the temperature at any transverse section of it is uniform. Fin temperature varies only in the longitudinal direction and heat transfer occurs through the fin by conduction in the same direction.

Let us now consider a thin section of thickness Δx of the fin at a distance x from the base. For the breadth b of the fin, the available area of heat conduction is bw and the area of heat loss by convection from the exposed fin surface to the ambient is $p\Delta x$ [where p is the fin perimeter $= 2(b + w)$]. In order to make a heat balance over the small element, we identify the following heat input and output terms.

Rate of heat input at x	$= bw\ q_x\lvert_x$
Rate of heat output at $x + \Delta x$	$= bw\ q_x\lvert_{x+\Delta x}$
Rate of heat loss from the exposed surfaces	$= p\Delta x h(T - T_o)$
Rate of heat accumulation (at steady state)	$= 0$

where T is the local temperature of the fin and h is the surface heat transfer coefficient. Making a steady state heat balance, we have

$$bwq_x\lvert_x - bwq_x\lvert_{x+\Delta x} - p\Delta x h(T - T_o) = 0$$

Dividing by $bw\Delta x$ throughout and taking limit $\Delta x \to 0$, we get

$$-\frac{d}{dx}(q_x) = \frac{ph}{bw}(T - T_o)$$

Putting $q_x = -kdT/dx$, and defining $\overline{T} = T - T_o$, we have

$$\frac{d^2\overline{T}}{dx^2} = \frac{ph}{kbw}\overline{T} = \xi^2\overline{T} \qquad \text{(where } \xi = \sqrt{\frac{ph}{kbw}}) \qquad (3.15)$$

To solve Eq. (3.15) for the temperature distribution in the fin, we need to write down the boundary conditions at $x = 0$ and $x = l$.

$$\text{At } x = 0, \qquad T = T_w, \quad \text{i.e.} \quad \overline{T} = (T_w - T_o) \qquad (3.16a)$$

$$\text{At } x = l, \qquad -k\frac{dT}{dx} = h(T - T_o), \quad \text{i.e.} \quad -k\frac{d\overline{T}}{dx} = h\overline{T} \qquad (3.16b)$$

Equation (3.16b) is a *convective boundary condition*. It implies that the rate at which heat is conducted from inside the solid to the boundary must be equal to the rate at which it is transported to the ambient medium by convection.

Solution to Eq. (3.15) is

$$\overline{T} = C_1 \cosh(\xi x) + C_2 \sinh(\xi x) \qquad (3.17)$$

where C_1 and C_2 are integration constants.

Using the boundary condition given by Eq. (3.16a), we get

$$C_1 = T_w - T_o \qquad (3.18)$$

Differentiating $\overline{T}$ in Eq. (3.17) and using Eq. (3.16b), we get

$$-k[C_1\xi \sinh(\xi l) + C_2\xi \cosh(\xi l)] = h[C_1 \cosh(\xi l) + C_2 \sinh(\xi l)]$$

or

$$C_2 = -\frac{C_1\left[\dfrac{h}{k\xi}\cosh(\xi l) + \sinh(\xi l)\right]}{\cosh(\xi l) + \sinh(\xi l)} \qquad (3.19)$$

Substitution of the relations for C_1 and C_2 in Eq. (3.17) and simplification lead us to the final solution for the *dimensionless fin temperature* given below.

$$\frac{\overline{T}}{T_w - T_o} = \frac{T - T_o}{T_w - T_o} = \frac{\cosh\xi(l - x) + \dfrac{h}{k\xi}\sinh\xi(l - x)}{\cosh(\xi l) + \dfrac{h}{k\xi}\sinh(\xi l)} \qquad (3.20)$$

It may be seen from the solution [Eq. (3.20)] that the temperature decreases with distance x from the base. So the surface heat flux from the fin also decreases in the longitudinal direction because of a smaller temperature driving force. It is easy to visualize that if the thermal conductivity k of the material of the fin is extremely high, its thermal resistance will be negligibly small and the temperature will remain almost uniform throughout at T_w, which is the wall or base temperature. The maximum rate of heat transfer can be achieved under this condition, which is rather an ideal condition. It is, therefore, interesting and useful to calculate the actual rate of heat transfer and also its ratio to the maximum rate of heat transfer.

The actual rate of heat loss from the fin must be equal to the rate of heat input by conduction at the fin base from the wall at steady state. Therefore, we have

$$Q = (\text{cross-sectional area})(\text{heat flux at the fin base}) = (bw) \left[-k(dT/dx) \right]_{x=0}$$

The temperature gradient dT/dx can be determined from Eq. (3.20). Therefore, we get

$$Q = bw(T_w - T_o)(-k) \left[\frac{-\xi \sinh \xi (l - x) - \dfrac{h}{k\xi} \cdot \xi \cosh \xi (l-\xi)}{\cosh \xi l + \dfrac{h}{k\xi} \sinh \xi l} \right]_{x=0} \tag{3.21}$$

If the fin temperature remains uniform at T_w (i.e. the base temperature), the rate of heat loss is the maximum.

$$Q_{max} = (\text{total area})(\text{heat transfer coefficient})(\text{driving force})$$
$$= (pl + bw)(h)(T_w - T_o) \tag{3.22}$$

The fin efficiency η is defined as

$$\eta = \frac{\text{Actual rate of heat transfer}}{\substack{\text{Calculated rate of heat transfer if the temperature of the fin} \\ \text{remains uniform at the base value, } T_w}}$$

Therefore, from Eqs. (3.21) and (3.22) we get

$$\eta = \frac{Q}{Q_{max}} = \frac{bwk\xi}{h(pl + bw)} \frac{\sinh \xi l + \dfrac{h}{k\xi} \cosh \xi l}{\cosh \xi l + \dfrac{h}{k\xi} \sinh \xi l} \tag{3.23}$$

A simpler but nevertheless accurate expression for heat loss can be obtained by using an alternative form of the boundary condition Eq. (3.16b). If we assume that heat loss from the edge of the fin (at $x = l$) is negligibly small, we can write

$$x = l, \qquad -k\frac{dT}{dx} = 0 \tag{3.24}$$

Solution of Eq. (3.15) subject to boundary conditions (3.16a) and (3.24) can be obtained as

$$\frac{T - T_o}{T_w - T_o} = \frac{\cosh [\xi(l - x)]}{\cosh \xi l} \tag{3.25}$$

which is the longitudinal temperature distribution in the fin. The rate of heat loss from both the surfaces (neglecting that from the edge) is

$$Q = (bw)\left[-k\frac{dT}{dx}\right]_{x=0} = bwk\xi\frac{(T_w - T_o)\sinh \xi l}{\cosh \xi l} \qquad (3.26)$$

The fin efficiency is given by

$$\eta = \frac{Q}{Q_{max}} = \frac{bwk\xi}{plh(T_w - T_o)}\frac{(T_w - T_o)\sinh \xi l}{\cosh \xi l} = \frac{1}{\xi l}\tanh \xi l \qquad \left(\because \xi^2 = \frac{ph}{kbw}\right) \qquad (3.27)$$

If the fin thickness w is small, $p = 2(b + w) \approx 2b$, then $\xi^2 = 2h/kw$. It is easy to check the accuracy of Eq. (3.27) for a physically realistic situation. If we take $l = 5$ cm, $w = 4$ mm, $b = 1$ m, $k = 160$ W/m °C, $h = 50$ W/m °C, then $\xi = [ph/kbw]^{1/2} = 12.5$ m^{-1}, and $\xi l = 0.625$. From Eq. (3.23), $\eta = 0.88$; and from Eq. (3.27), $\eta = 0.887$. These values of η are very close to each other. Equation (3.27) is good enough for practical purposes.

Fins may be of different types: longitudinal fins with (a) rectangular, (b) triangular, (c) concave parabolic, (d) convex and parabolic, and (e) radial profile [Fig. 3.13(i)]. Other types include thin annular ring, cylindrical spine, conical spine, etc. External and internal longitudinal fins on pipes are shown in Fig. 3.13(ii). Fin efficiency may be calculated for each geometry of the fin. Choice of the shape and the material of a fin depend upon factors like the material cost, equipment volume, etc. A comprehensive treatment of heat transfer from fins and extended surfaces is available in Kern and Kraus (1972).

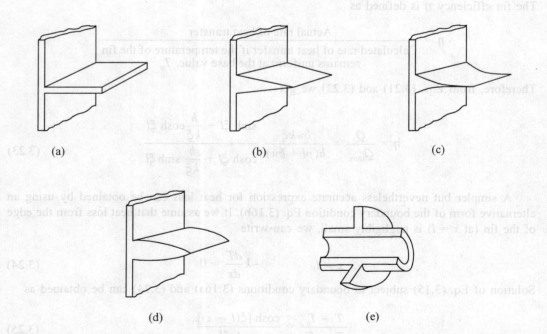

Fig. 3.13(i) A few common types of fins: (a) rectangular, (b) triangular, (c) concave parabolic, (d) convex parabolic, and (e) radial.

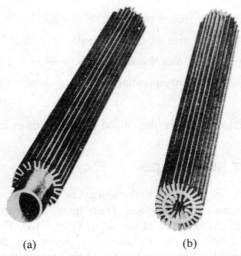

Fig. 3.13(ii) Longitudinal fins on pipes: (a) external and (b) both external and internal (Kakac, 1981).

Example 3.10 A 3 inch schedule 40 carbon steel pipe (actual i.d. = 78 mm, wall thickness = 5.5 mm) has eight longitudinal fins of 1.5 mm thickness. Each fin extends 30 mm from the pipe wall. The thermal conductivity of the fin material is 45 W/m °C. If the wall temperature, the ambient temperature, and the surface heat transfer coefficient are 150°C, 28°C, and 75 W/m^2 °C respectively, calculate the percentage increase in the rate of heat transfer for the finned tube over the plain tube.

SOLUTION The fins are of the shape of a rectangular plate. We use Eq. (3.27) to calculate the fin efficiency. It is assumed that $p = 2(b + w) \approx 2b$.

$$\xi = \sqrt{\frac{2h}{kw}} = \sqrt{\frac{(2)(75)}{(45)(0.0015)}} = 47.14, \qquad l = 30 \text{ mm} = 0.03 \text{ m}, \qquad \xi l = 1.414$$

$$\eta = (1/\xi l) \tanh(\xi l) = (1/1.414) \tanh (1.414) = 0.628$$

If we consider 1 m length of the pipe, then length of the fin = 1 m, breadth of the fin = 0.03 m, the area of a single fin = (2)(1)(0.03) = 0.06 m^2 [both sides included], and the total area of all the eight fins = (8)(0.06) = 0.48 m^2

The maximum rate of heat transfer from the fin surfaces (if the temperature all over the fin is the same as the base temperature, i.e. 150°C),

$$hA\Delta T = (75)(0.48)(150 - 28) = 4392 \text{ W}$$

Actual rate of heat transfer from the fins = $(\eta)(4392) = (0.628)(4392) = 2758$ W

Area of contact of a fin with the pipe wall = (1 m)(0.0015 m) = 0.0015 m^2

Area of contact of all the fins with the pipe wall = (0.0015)(8) = 0.012 m^2

Outer pipe radius = (78/2) mm + 5.5 mm = 0.0445 m

Area per metre = $2\pi(0.0445)(1) = 0.2796$ m^2

Free outside area of the finned pipe = 0.2796 − 0.012 = 0.2676 m²

Rate of heat transfer from this area of the pipe wall = (75)(0.2676)(150 − 28) = 2448 W

Total rate of heat transfer from the finned pipe = 2758 + 2448 = 5206 W

Rate of heat transfer from the corresponding 'unfinned' pipe = (75)(0.2796)(150 − 28)

$$= 2558 \text{ W}$$

Per cent increase in heat transfer for the finned pipe = (5206 − 2558)/2558 = $\boxed{103.5\%}$

3.4 THERMAL CONTACT RESISTANCE

When two flat solid surfaces are in contact, there may be scattered tiny gaps containing an entrapped gas (usually air) at the contact surface. These gaps, which prevent perfect contact of the surfaces, arise out of surface roughness of the solids (Fig. 3.14). The resulting heat transfer resistance at the interface is called the *thermal contact resistance*. If the surfaces are optically plane, a perfect contact is possible. There would be no gap, making the contact resistance zero. Thermal contact resistance at a solid–solid boundary is usually expressed in terms of a *contact heat transfer coefficient.*

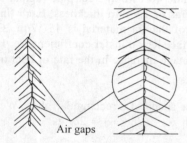

Air gaps

Fig. 3.14 Schematics of surfaces at contact.

Example 3.11 Pre-stressed multi-layered shells are often used for high pressure services (the urea reactor is a typical example) because of a smaller wall thickness required compared to a single-layer shell to withstand the same internal pressure. Such shells are often made by 'shrink fitting'. As such there is a possibility of existence of a thermal contact resistance at the interface between the two layers because of the entrapped gases in the crevices on the metal surfaces.

A two-layered cylindrical shell of high tensile steel (Cr-Mo-V) has an i.d. of 90 cm, and an o.d. of 110 cm. The layers are equal in thickness. Under steady operating conditions, the inside temperature of the shell is 180°C and that of the outer surface is 170°C, making a temperature drop of 10°C across the composite wall. The thermal conductivity of the alloy is 37 W/m °C, and the rate of heat loss through the wall is known to be 5.18 kW per metre length of the shell.

(a) Is there any thermal contact resistance at the interface between the layers?

(b) If so, calculate the contact resistance based on one metre length of the shell and also express it in terms of a contact heat transfer coefficient.

(c) Calculate the 'temperature jump' across the contact surface.

SOLUTION (a) If there is no contact resistance, the wall may be considered as a single-layer one for the purpose of heat transfer calculations because both the layers are made of the same material. The values of the relevant quantities are (see Fig. 3.15):

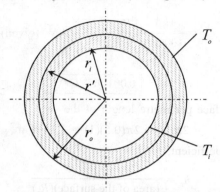

Fig. 3.15 Pre-stressed multi-layer shell (Example 3.11).

Inner radius of the shell, r_i = 0.45 m

Boundary between the layers, r' = 0.5 m

Outer radius of the shell, r_o = 0.55 m

Thermal conductivity, k = 37 W/m °C

Length of the shell, L = 1 m

Inside wall temperature, T_i = 180°C

Outside wall temperature, T_o = 170°C

The rate of heat transfer in the absence of any contact resistance,

$$Q = \frac{2\pi L k (T_i - T_o)}{\ln(r_o/r_i)}$$

$$= \frac{2\pi(1)(37)(180 - 170)}{\ln(0.55/0.45)}$$

$$= 11{,}585 \text{ W} = 11.585 \text{ kW}$$

The actual rate of heat loss (5.18 kW) is much less than this value. So, there is a thermal contact resistance at the interface between the layers.

(b) There are three heat transfer resistances in series. These are:

$$\text{Resistance of the inner layer} = R_1 = \frac{\ln(r'/r_i)}{2\pi L k} = \frac{\ln(0.50/0.45)}{2\pi(1)(37)} = 0.000453 \text{ °C/W}$$

$$\text{Resistance of the outer layer} = R_2 = \frac{\ln(r_o/r')}{2\pi L k} = \frac{\ln(0.55/0.50)}{2\pi(1)(37)} = 0.00041 \text{ °C/W}$$

If the contact resistance is R_c, we have

Total resistance = $0.000453 + 0.00041 + R_c = 0.000863 + R_c$

The rate of heat loss,

$$Q = \frac{180 - 170}{0.000863 + R_c} = 5180 \text{ W (given)}$$

or

$$R_c = \boxed{0.001067 \text{ °C/W}}$$

Area of the contact surface per metre length of the shell

$$= 2\pi r'L = 2\pi(0.5)(1) = 3.1416 \text{ m}^2$$

Contact heat transfer coefficient,

$$h_c = \frac{1}{(\text{area of the surface})(R_c)}$$

$$= \frac{1}{(3.1416)(0.001067)} = \boxed{298.3 \text{ W/m}^2 \text{ °C}}$$

(c) If ΔT is the 'temperature jump' at the contact surface at steady state, we have

$$Q = h_c(\text{area}) \, \Delta T = (298.3)(3.1416)(\Delta T) = 5180$$

or

$$\Delta T = \frac{5180}{(298.3)(3.1416)} = \boxed{5.5°C}$$

The physical meaning of the temperature jump is that, on an average, the outer surface of the inner layer will be 5.5°C above the inner surface of the outer layer.

3.5   CRITICAL INSULATION THICKNESS

Heat loss from an insulated cylindrical surface is pretty common in industrial operations, for example, heat loss from an insulated steam pipe. Here there are two major resistances in series— the resistance offered by the layer of insulation, and that offered by the 'gas-film' on the outside. While the use of a thicker layer of insulation increases the resistance to heat transfer, it also increases the outer radius and hence the area of heat transfer. Thus two factors, having opposite effects on the rate of heat transfer, come into play when the thickness of insulation on a pipe is increased. In some cases this results in a higher rate of heat transfer with increasing thickness of insulation. It is interesting to determine the insulation thickness for which the rate of heat transfer will be a maximum.

We consider a pipe (carrying steam, for example) with a layer of insulation on it as shown in Fig. 3.16. The inner radius of the layer of insulation and the corresponding temperature are r_i and T_i, respectively, and the outer radius of the insulation is r_o. The ambient temperature is T_o, the convective heat transfer coefficient outside the insulation is h_o, and the thermal conductivity of the insulating material is k_c. The heat transfer coefficient, on the inner side of the insulation, h_i, is assumed to be large. The rate of heat transfer from a section of length L of the insulated pipe can be written following Eq. (3.10).

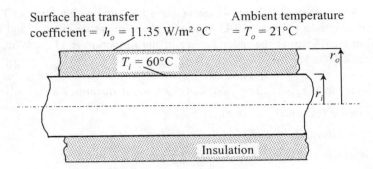

Fig. 3.16 Heat loss from an insulated pipe.

$$Q = \frac{T_i - T_o}{\dfrac{\ln(r_o/r_i)}{2\pi k_c L} + \dfrac{1}{A_o h_o}} = \frac{2\pi L(T_i - T_o)}{\dfrac{\ln(r_o/r_i)}{k_c} + \dfrac{1}{r_o h_o}} \qquad (3.28)$$

We want to determine the outer radius of insulation for which the heat loss will be a maximum. The condition for this is

$$\frac{dQ}{dr_o} = \frac{-2\pi L(T_i - T_o)\left(\dfrac{1}{k_c r_o} - \dfrac{1}{h_o r_o^2}\right)}{\dfrac{\ln(r_o/r_i)}{k_c} + \dfrac{1}{r_o h_o}} = 0$$

or

$$\frac{1}{k_c r_o} - \frac{1}{h_o r_o^2} = 0$$

or

$$r_{o,\,\text{crit}} = \frac{k_c}{h_o} \qquad (3.29)$$

It can be easily verified that the second derivative, d^2Q/dr_o^2, is negative and hence Eq. (3.29) gives the condition for the *maximum* rate of heat transfer. The quantity, $r_o - r_i$, is called the *critical insulation thickness*.

Once r_o is determined from Eq. (3.29), the maximum rate of heat loss can be calculated from Eq. (3.28). When an insulation is applied to a steam pipe, its thickness should be above the critical value. For an insulated electrical wire, however, the insulation thickness should be very near the critical value so that heat loss is maximum, and the insulation temperature remains low thus enhancing the life of the insulation.

It should, however, be noted that a critical insulation thickness may not always exist for an insulated pipe. For example, if the values of k_c and h_o are such that the ratio k_c/h_o $(= r_o)$ turns out to be less than r_i (which is the radius of the bare pipe), then the critical insulation thickness, $r_o - r_i$, becomes negative. This means that the critical insulation thickness *does not exist* in this case.

Example 3.12   A copper wire, 5.2 mm in diameter, is insulated with a layer of PVC (polyvinyl chloride) of thermal conductivity $k_c = 0.43$ W/m °C. The wire carries current, and its temperature is 60°C. The film coefficient at the outer surface of the insulation is 11.35 W/m² °C. Calculate the critical insulation thickness.

SOLUTION   Refer to Fig. 3.16. The values of the various quantities are:

Wire temperature (assumed constant), $T_i = 60$°C

Ambient temperature, $T_o = 21$°C

Inner radius of the insulation, $r_i = 5.2/2 = 2.6$ mm $= 2.6 \times 10^{-3}$ m

Surface heat transfer coefficient, $h_o = 11.35$ W/m² °C

Thermal conductivity of PVC, $k_c = 0.43$ W/m °C

The critical outer radius of insulation can be calculated directly from Eq. (3.29).

$$r_{o,\text{crit}} = \frac{k_c}{h_o} = \frac{0.43}{11.35} = 0.0378 \text{ m} = 37.8 \text{ mm}$$

The critical insulation thickness $= r_o - r_i = 37.8 - 2.6 = \boxed{35.2 \text{ mm}}$

The data given in the problem may be used to prepare a plot to show how the rate of heat loss through the layer of insulation changes with the thickness (or the o.d.) of the insulation. The rate of heat loss Q is given by Eq. (3.28). The values of Q per metre length of the insulated wire ($L = 1$ m) are calculated for different values of r_o. The plot of Q versus r_o (Fig. 3.17) shows a maximum that corresponds to the critical insulation thickness.

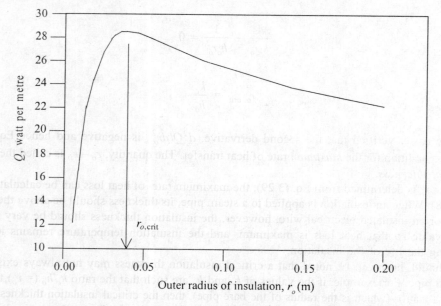

Fig. 3.17   Illustrative plot showing the critical insulation thickness (Example 3.12).

Example 3.13 A 15 cm (schedule 80, IPS) steam main has a 10 cm thick 'preformed' mineral fibre insulation on it. The thermal conductivity of the insulation is 0.035 W/m °C, and the heat transfer coefficient at the outer surface of the insulation is 10 W/m² °C. What is the critical insulation thickness?

SOLUTION The outer radius of the pipe corresponding to critical insulation thickness is [Eq. (3.29)]

$$r_o = \frac{k_c}{h} = \frac{0.035}{10} = 0.0035 \text{ m} = 3.5 \text{ cm}$$

However, the radius of the bare pipe (7.5 cm) is larger than the calculated critical value of the outer radius of insulation (3.5 cm). This is physically impossible. So, the critical insulation thickness does not exist in this case.

3.6 ECONOMIC (OR OPTIMUM) INSULATION THICKNESS

When the thickness of insulation on a hot surface is increased, the rate of heat loss is reduced (except below the critical insulation thickness, if it exists), i.e. the cost of energy lost is decreased. So there are two opposing factors that should be considered in determining the thickness of insulation. A thicker insulation saves more energy at the cost of higher investment on insulation. Similar is the case of applying insulation to a cold surface.

The economic or optimum insulation thickness is that for which the sum of the insulation cost and the cost of the heat loss is a minimum. A detailed calculation should take into consideration the fixed cost of insulation, variable costs like maintenance costs, service life of insulation, depreciation, taxes, etc.[1] Here we shall develop a simplified method of determination of the optimum insulation thickness. Factors like maintenance costs, depreciation, taxes, etc. will not be considered. The method is based on minimization of total *present cost* taking into consideration an interest rate. We shall take up the case of an insulated pipe carrying a hot fluid (e.g. steam). The various terms and quantities involved are as follows [see also Eq. (3.28)]:

Length of the pipe section = L, m

Outer radius of the insulation = r_o, m

Inner radius of the insulation (this is the same as the outer radius of the pipe) = r_i, m

Thermal conductivity of insulation = k_c, kcal/h m °C

Film coefficient at the outer surface of insulation = h_o, kcal/h m² °C

Pipe wall temperature (= temperature of the inner surface of the insulation) = T_i, °C

Ambient temperature = T_o, °C

Cost of insulation = C', Rs/m³

Insulation life = n, yr

Cost of heat energy = C_H, Rs/kcal

Interest rate = i, Re/(yr)(Re)

The inside heat transfer coefficient, h_i, is assumed to be large.

---

[1] The methodologies useful for cost estimation and economic analysis are discussed in Peters and Timmerhaus, 1991; Ulrich, 1984.

If $(r_o - r_i)$ is the insulation thickness and h_o the outer surface heat transfer coefficient, the overall heat transfer based on the inner radius of insulation is given by [see Eq. (3.11)]

$$U_i = \frac{1}{\dfrac{r_i}{k_c} \ln(r_o/r_i) + \dfrac{r_i}{r_o h_o}} \quad \text{kcal/h m}^2 \text{ °C} \tag{3.30}$$

Rate of heat loss (per year), $Q = (2\pi r_i L) \, U_i \, (T_i - T_o) \, (24)(300)$ kcal

$$= (2\pi r_i L) \, U_i \, (T_i - T_o) \, (7.2 \times 10^3) \text{ kcal}$$

(300 working days a year, on the average, is assumed).

Cost of heat loss $= QC_H$ Rs/yr

If i is the fractional annual compound interest rate (compounded annually), the total present value of heat loss P_1 over the service life of the insulation (n years) is given by

$$P_1 = \sum_{j=1}^{n} \frac{QC_H}{(1+i)^j} \quad \text{Rs} \tag{3.31}$$

Volume of insulation applied $= \pi(r_o^2 - r_i^2)L$ m^3

Present value of the insulation, $P_2 = \pi(r_o^2 - r_i^2) \, L \, C'$ Rs $\tag{3.32}$

The total present value or cost [from Eqs. (3.31) and (3.32)] is

$$P_T = \sum_{j=1}^{n} \frac{QC_H}{(1+i)^j} + \pi(r_o^2 - r_i^2)LC'$$

$$= C_H \left[\sum_{j=1}^{n} \frac{1}{(1+i)^j} \right] (2\pi r_i L) \left[\frac{1}{\dfrac{r_i}{k_c} \ln(r_o/r_i) + \dfrac{r_i}{r_o h_o}} \right] (T_i - T_o)(7.2 \times 10^3) + \pi(r_o^2 - r_i^2)LC' \tag{3.33}$$

The optimum insulation thickness is obtained by putting

$$\frac{dP_T}{dr_o} = 0 \tag{3.34}$$

The development of a general expression for r_o, and therefore for the optimum insulation thickness, $r_o - r_i$, based on the expression for P_T given by Eq. (3.33), is cumbersome. The application is illustrated in Example 3.14. The two components of the total cost, P_1 and P_2, and also the present value of total cost, P_T, are plotted against the insulation thickness, $r_o - r_i$, in Fig. 3.18. These are based on the data of Example 3.14. The present value of the total cost P_T passes through a minimum which corresponds to the optimum insulation thickness.

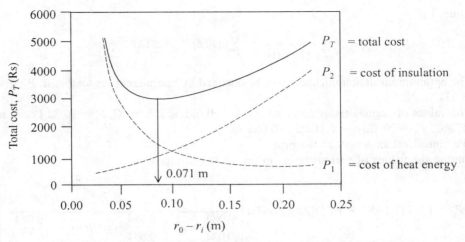

Fig. 3.18 Illustrative plot showing the optimum insulation thickness
(data taken from Example 3.14).

Example 3.14 A schedule 40 6-inch steam pipe line carries saturated steam at an absolute pressure of 8.2 bar. Calculate the optimum thickness of the bonded rockwool insulation for the steam pipe line. The following data are given:

Saturation temperature of steam at 8.2 bar (absolute) is 172°C

Ambient temperature = 20°C

Cost of steam = Rs 700 per ton

Latent heat of condensation of the steam at 172°C = 487 kcal/kg

Outer surface heat transfer coefficient = 10.32 kcal/h m^2 °C

Thermal conductivity of the insulation = 0.031 kcal/h m °C

Insulation cost (including the cost of installation, cladding, labour and supervision) is a function of the insulation diameter, $C' = (17,360 - 1.91 \times 10^4 \, d_i)$ Rs/m^3, over the range, $0.114 < d_i < 0.406$, where d_i (in metre) is the inner diameter of the insulation (or the o.d. of the pipe)

Service life of insulation is 5 years.

Interest rate = 0.18 Re/(yr)(Re)

The exact outer diameter of a schedule 40 6-inch n.b. (nominal bore) pipe is 0.168 m.

SOLUTION *Given:* $d_i = 0.168$ m; cost of insulation, $C' = 17,360 - 1.91 \times 10^4(0.168) = 15,750$ Rs/m^3.

Cost of energy: cost of steam = Rs 700 per ton; latent heat of steam = 487 kcal/kg
Therefore, cost of heat energy in the steam

$$C_H = \frac{\text{Rs } 700}{(1000 \text{ kg})(487 \text{ kcal/kg})} = 1.437 \times 10^{-3} \text{ Rs/kcal}$$

Interest rate, $i = 0.18$ Re/(yr)(Re);   insulation life, $n = 5$ yr

Therefore,

$$\sum_{j=1}^{5} \frac{1}{(1+i)^j} = \sum_{j=1}^{5} (1.18)^{-j} = 3.127$$

The optimum insulation thickness will be obtained by minimizing the total cost, P_T, given by Eq. (3.33).

The values of various terms involved are: $k_c = 0.031$ kcal/h m °C, $h_o = 10.32$ kcal/h m^2 °C, $T_i = 172$°C, $T_o = 20$°C, $r_i = 0.168/2 = 0.084$ m.

We consider 1 m length of the pipe.

Putting the values of various terms in Eq. (3.33), we have

$$P_T = (3.127)(1.437 \times 10^{-3})(2\pi)(0.084)L \left[\frac{1}{\frac{0.084}{0.031}\ln r_o - \frac{0.084}{0.031}\ln(0.084) + \frac{0.084}{10.32\,r_o}} \right] \times$$

$$(7.2 \times 10^3)(T_i - T_o) + \pi(r_o^2 - r_i^2)L\,(15{,}750)$$

$$= \frac{957.7}{\ln r_o + (0.0030/r_o) + 2.477} + 49{,}480\left(r_o^2 - r_i^2\right)$$

Now we put

$$\frac{dP_T}{dr_o} = -\frac{957.7[(1/r_o)-(0.0030/r_o^2)]}{[\ln r_o + (0.0030/r_o) + 2.477]^2} + 98{,}960\,r_o = 0$$

On simplification, we get

$$r_o\,[r_o \ln r_o + 2.477\,r_o + 0.0030]^2 - 0.00968\,(r_o - 0.0030) = 0$$

The solution of the above transcendental equation can be obtained graphically or numerically as

$$r_o = 0.155 \text{ m} = \text{outer radius of the insulation.}$$

Optimum insulation thickness $= r_o - r_i = 0.155 - 0.084 = 0.071$ m $= \boxed{71\,\text{mm}}$

Plots of the energy cost P_1, the insulation cost P_2, and the total cost P_T, based on 1 m length of the pipe are shown in Fig. 3.18.

$$P_1 = \frac{957.7}{\ln r_o + (0.0030/r_o) + 2.477}; \qquad P_2 = 49{,}480\left(r_o^2 - r_i^2\right); \qquad P_T = P_1 + P_2$$

The plot of P_T vs. $(r_o - r_i)$ shows a minimum corresponding to the optimum or economic insulation thickness.

Note: In actual practice about 75 mm thick rockwool insulation is applied to such a steam pipe. The value estimated here, theoretically, closely matches the thickness of insulation used in practice.

SHORT QUESTIONS

1. What do you mean by convection? How do you differentiate between free and forced convection?

2. Consider heat transfer from a solid surface to a flowing fluid. At what location does heat transfer to the fluid occur by conduction only? Draw the temperature profile in the liquid near the surface.

3. How is heat transfer coefficient defined? What is the physical significance of 'film thickness'? Mention the more important factors on which the heat transfer coefficient depends.

4. Explain the terms 'bulk fluid' and 'bulk temperature'.

5. How does the 'film thickness' change if the velocity of a fluid flowing over a surface increases?

6. What is 'Newton's law of cooling'?

7. Does a 'stagnant film' of a fluid really exist at a surface over which the fluid is flowing?

8. Why are the qualitative depictions of the temperature distributions in the films and in the wall in Fig. 3.3 linear but of different slopes?

9. What do you mean by 'controlling resistance' in heat transfer?

10. Consider heat exchange between a gas and a liquid phase separated by a wall. Which one of the side resistances is likely to control the rate of heat exchange?

11. Two overall coefficients, U_i and U_o, are defined in Eqs. (3.11) and (3.12). Which one of these coefficients is larger in magnitude and why?

12. How does a fin enhance heat transfer at a surface? Mention the common types of fins and discuss the comparative advantages of fins of different geometries.

13. Give a few practical and specific examples of use of fins in heat transfer.

14. Under what condition does the fin efficiency become nearly 100%?

15. What would be the nature of temperature distribution in a fin if the thermal conductivity of the fin material is very high?

16. What type of boundary condition is used at the fin edge?

17. What is 'critical insulation thickness'? Can you give a physical explanation of its existence? Does a critical thickness exist for every insulated cylindrical surface?

18. What is thermal contact resistance and what is its origin? What is its unit?

19. What do you mean by 'optimum insulation thickness'? What are the more important factors that should be taken into account while determining this thickness?

20. An aluminium ball, 7.5 cm in diameter, is cooled in an ambient at 25°C. When the temperature of the ball is 120°C, it is found to cool at the rate of 4°C per minute. Calculate the surface heat transfer coefficient. Given: $\rho = 2700$ kg/m^3, $c_p = 0.89$ kJ/kg °C.

21. One side of a 15 cm thick layer of insulation ($k = 0.75$ W/m °C) has a temperature of 200°C while the other loses heat to an ambient at 27°C at the rate of 600 W/m^2. Calculate the skin temperature of the insulation and the air-film coefficient.

PROBLEMS

3.1 A 2 kg copper ball at 200°C cools down in ambient air at 29°C. If it requires an hour for the ball to cool down to 35°C, calculate the average value of the surface heat transfer coefficient.

Because copper is an excellent conductor of heat, it may be assumed that the temperature in the ball remains uniform at any instant. For copper, density = 8950 kg/m³, specific heat = 0.383 kJ/kg °C.

[*Hint:* At any time t, $- mc_p \, (dT/dt) = Ah(T - T_o)$. Integrate this equation from $t = 0$, $T = 200°C$, to $t = 3600$ s, $T = 35°C$. Put the values of different quantities in the resulting expression and calculate h.]

[*Note:* The assumption that the temperature of the ball remains uniform everywhere in it at any instant leads to a *lumped parameter model* of the system. The value of the parameter (here temperature) is lumped (or put together and averaged) for the entire volume. If the variation of a 'parameter' with position in a body is significant, a *distributed parameter model* is used. The suggested method of solution of the problem is based on the lumped parameter model which is quite reasonable in view of the high thermal conductivity of copper.]

3.2 Annealing of a metal part is done by raising it to a high temperature followed by slow cooling to a lower temperature. A machine part weighing 5.7 kg and having the shape of a regular tetrahedron is heated to 1100 K and then cooled down in an annealing chamber under controlled conditions to 500 K. If the surface heat transfer coefficient is limited to 20 W/m² °C, and the ambient temperature in the chamber is 470 K, calculate the time of cooling of the metal part. Assume that cooling occurs from all the surfaces. The specific gravity of the material is 7.8, and its specific heat is 560 J/kg °C. Neglect any temperature variation within the solid at any instant.

3.3 A 0.35 m thick brick wall has one surface maintained at 500 K, and the other surface is in contact with air ($k = 0.027$ W/m °C) at a bulk temperature of 308 K. The area of the wall is 7.2 m². The rate of heat flow through the wall at steady state is known to be 2.75 kW. If the thermal conductivity of the brick in the wall is 0.81 W/m °C, calculate (a) the skin temperature of the outer surface, (b) the surface heat transfer coefficient, and (c) the thickness of the air-film.

3.4 A length of 12 gauge (2.05 mm dia) bare copper wire, connected to a power source, is kept in a room at 26°C. The circuit is activated and suddenly a current of 30 ampere starts flowing through it causing generation of heat. While the wire gets heated, it also dissipates heat into the room, the heat transfer coefficient being 18 W/m² °C. The wire is initially at the temperature of the room. What will be its steady state temperature? The resistance of the wire is 0.22 ohm/100 m, and 13.75 kJ heat is required to raise the temperature of 1 km length of wire by 1°C. The thermal conductivity of copper is known to be very high. Also, calculate the time required for attaining 63.2% of the steady state temperature rise.

3.5 A 10 cm (4-inch schedule 40 IPS) high pressure steam pipe carries saturated steam at 17 bar gauge pressure obtained from a waste heat boiler of an ethylene oxide plant. The actual i.d. of the pipe is 102 mm and its wall thickness is 6 mm ($k = 45$ W/m °C). The pipe is lagged with a 75 mm layer of glass wool ($k = 0.049$ W/m °C) which is held by a thin external cover of aluminium. The ambient air temperature is 32°C and the heat transfer coefficient at the outer surface is 14.2 W/m² °C. At this pressure, the temperature of the saturated steam is 206°C. Neglecting the steam-side heat transfer resistance, calculate (a) the thermal resistance offered by the insulation per metre pipe length, (b) the overall heat transfer coefficients based on the outside and the inside area, (c) the skin temperature of the aluminium cover, (d) the per cent of the total resistance offered by the air-film and the effective air-film thickness, and (e) the rate of heat loss from a 3 m section of the pipe.

3.6 A 6 m³ polymerization reactor has an outer jacket for heating or cooling as necessary. In a particular batch, 5 m³ of monomer at 24°C is charged into the reactor. The charge has to be heated to 80°C before the initiator is added to start the reaction. The liquid in the reactor is well-stirred and the liquid-side heat transfer coefficient is 400 W/m² °C. Hot oil flows through the jacket at a sufficiently high rate, and the average oil temperature is 125°C. The oil-side heat transfer coefficient is 300 W/m² °C. If the available heat transfer area is 10 m², calculate the overall heat transfer coefficient and the time required for heating the charge to the desired temperature. The relevant properties of the monomer are: density = 870 kg/ m³; specific heat = 1.81 kJ/kg °C, thermal conductivity = 0.25 W/m °C. Estimate the liquid 'film thickness' on the monomer side. What would be the heat transfer coefficient for water corresponding to this liquid 'film thickness'? Neglect the thermal resistance of the wall.

3.7 The steady state radial temperature profile in a 10 cm diameter solid sphere is known to be $T = 101.39 - 1.389 \times 10^3 r^2$ (T in °C, and r in metre). Its thermal conductivity is 12 W/m °C. The sphere is placed in an ambient at 30°C. (a) What is the maximum temperature in the sphere? (b) Do you think that there is heat generation in the sphere? If so, at what rate? (c) Calculate the film coefficient at the outer surface.

[*Hint:* Use the relation $-[kdT/dr]_{r = R} = h(T - T_o)_{r = R}$, where h is the film coefficient.]

3.8 The human body has literally hundreds of control systems in it to regulate various functions. So far as body temperature is concerned, it is maintained almost exactly at 37°C under normal conditions. Additional heat generation in the body compensates for extra heat loss when the body is exposed to a cold environment.

In order to calculate the heat loss from a human body, let us assume that the surface area of an average-size human body is equal to that of the curved surface of a cylinder, 1.70 m in height and 0.35 m in diameter. The heat transfer resistance of the skin is equal to that of 2 mm of skin tissues having a thermal conductivity of 0.37 W/m °C. The ambient temperature is 8°C on a cold wintry night. With wind blowing, the surface heat transfer coefficient is 18 W/m² °C. (a) Calculate the additional rate of heat generation in the body necessary to compensate for the heat loss under the given situation. (b) What will be the percentage decrease in heat loss if a person is clad in woollens 5 mm thick ($k = 0.04$ W/m °C)? (c) What would be the temperature of the skin surface of both the bare and the woollen-clad body?

3.9 A zero order exothermic chemical reaction occurs in a spherical catalyst. For this particular reactant-catalyst pair, the rate of reaction, and, therefore, the rate of heat liberation are virtually uniform in the catalyst pellet. Let ψ_v be the uniform volumetric heat generation rate. The pellet also loses heat to the surrounding fluid, the heat transfer coefficient being h. Other pertinent parameters are: radius of the pellet = r_o, and the thermal conductivity = k. Determine the steady state temperature distribution and also the average temperature of the catalyst pellet.

3.10 When the bulb of a mercury-in-glass thermometer is immersed in a hot liquid, heat transfer to mercury occurs through the glass wall. Very often, the heat transfer resistance is virtually offered by the glass wall only. If a thermometer bulb at temperature T_i is immersed in a medium at temperature T_o, the time taken by the thermometer to show a temperature change of 0.632 times the maximum change [i.e. to show a temperature $T_i + 0.632(T_o - T_i)$] is called its 'time constant'.

Consider a thermometer having a time constant of 20 s. The heat transfer resistance offered by the glass wall of the bulb is 50.5 °C/W. The thermometer is used to measure the temperature of a reactor wall by inserting the bulb in a close-fitting 'thermometer pocket' in the wall. An alert young engineer notices that now it takes 40 s to show 0.632 times the maximum temperature rise. He rightly concludes that this is due to the contact resistance between the thermometer bulb and the wall of the thermometer pocket. Calculate the value of this contact resistance.

3.11 A 12-gauge copper wire (diameter = 2.05 mm) has a current of 30 A flowing through it. It is covered with a 2.5 mm thick PVC ($k = 0.43$ W/m °C) insulation. Copper has a thermal conductivity of 380 W/m °C and the wire has an electrical resistance of 2.06 ohm/km. The outside air-film coefficient is 12 W/m^2 °C. It is desirable that the maximum temperature of the insulation be limited to 60°C. The ambient temperature is 29°C. (a) Is the current flowing above the allowable limit? (b) Find the maximum insulation temperature and its outer skin temperature? (c) What is the centre temperature of the wire? (d) What is the maximum allowable current if the critical insulation thickness is used?

3.12 Nitrogen gas at 1.3 bar absolute pressure is being heated by passing it through a 102 mm i.d. pipe at a rate of 0.7 kg/s. The temperature of the pipe wall is maintained constant at 140°C. Nitrogen ($c_p = 1.05$ kJ/kg °C) enters at 35°C. The heat transfer coefficient at the pipe wall is 130 W/m^2 °C. Determine the rise in the temperature of the gas over a 5 m section of the pipe.

[*Hint:* Make a differential heat balance over an elementary section of the pipe and integrate it.]

3.13 Demineralized water is heated from 110°F to 230°F by passing it through a 1-1/2 inch o.d. steam-jacketed tube (wall thickness = 0.083 inch) at a rate of 5000 lb/h. Steam at 320°F condenses on the outside of the tube. If the inside heat transfer coefficient is 250 Btu/h ft^2 °F and the heat transfer coefficient for steam condensing outside is 1800 Btu/h ft^2 °F, calculate the length of the tube required to achieve the temperature. The thermal conductivity of the tube wall is 25 Btu/h ft °F.

3.14 Many exothermic gas-phase catalytic reactions are carried out in reactors of 'shell-and-tube' type construction. Catalyst pellets are packed in a large number of tubes while a coolant is circulated through the shell in order to remove the heat of reaction. Proper control of the coolant flow rate and temperature is extremely important in order to prevent possible burn-out (for example, if the coolant flow rate is low) or quenching (if the coolant flow is more than necessary) of the reactor. In a particular reactor, 19 mm i.d. and 25 mm o.d. tubes are used to hold the catalyst. Heat generation due to chemical reaction occurs at a volumetric rate of 415 kW/m^3 of reactor volume. An eutectic mixture of sodium nitrate, sodium nitrite and potassium nitrate is used as the coolant and the outside heat transfer coefficient is 300 W/m^2 °C. (See Chapter 8 for a brief table of heat transfer media.) Assume that at any section of the reactor, the temperature in a tube is radially uniform except that there is a heat transfer resistance at the wall on the gas side (i.e. inside a tube) characterized by a film coefficient of 140 W/m^2 °C. The resistance offered by the tube wall may be safely neglected. If the bulk temperature in the tube is 400°C, what temperature of the coolant should be maintained in order to remove the heat of reaction under these set of conditions? What is the temperature of the tube wall?

[*Hint:* Take 1 m tube length as the basis. Calculate the rate of heat generation in this volume and then the heat flux based on the inside tube area. Determine the overall heat transfer coefficient on the same basis. The value of ΔT will follow.]

3.15 A thick copper cable, 9.5 mm in diameter, is insulated with a 6 mm layer of dielectric material having a thermal conductivity of 0.25 W/m °C. The outer surface temperature is maintained at 21°C. What current flow is allowed through the cable if the insulation temperature has to be within 60°C? The resistivity of copper is 1.72×10^{-6} ohm-cm and its thermal conductivity is 380 W/m °C. What is the centre temperature of the cable?

By what per cent will the allowable current be changed, if, instead of maintaining the surface temperature at 21°C, the cable is put in a medium at that temperature (i.e. 21°C) and the film coefficient at the insulation surface is 10 W/m^2 °C? What will be the outer surface temperature of the insulation in this case?

3.16 A 5 cm n.b. pipe (actual i.d. = 52 mm, o.d. = 60 mm) has eight rectangular longitudinal fins welded to the outer surface (Fig. 3.19). The fins are of 5 cm length and 1.6 mm thickness. The wall temperature of the pipe is 160°C. The finned tube loses heat to a flowing stream of air with a surface heat transfer coefficient of 115 W/m^2 °C. The thermal conductivity of the fin material is 46 W/m °C. Neglecting heat transfer from the edges of the fins, calculate (a) the fin efficiency, and (b) the rate of heat transfer to the air per metre length of the finned pipe.

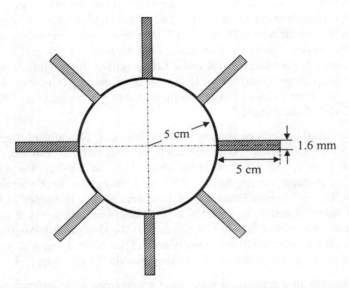

Fig. 3.19 A pipe with rectangular longitudinal fins (Problem 3.16).

3.17 A Horton sphere storing pressurized liquid ammonia at 4°C has a ladder on it for inspection and maintenance (Fig. 3.20). The ladder is fixed to the ends of a set of carbon steel rods, 38 mm diameter. The other ends of the rods are welded to the tank. A length of 0.35 m of each rod shows up outside the insulation of the tank. It can be assumed that heat loss from the surface of the rod outside the insulation is only important, and the heat transfer coefficient is 16 W/m² °C. The end of the rod to which the ladder is fixed may be taken to be at the ambient temperature (30°C). Radial temperature in the rod at any section is uniform and only the axial variation of temperature is to be considered. The thermal conductivity of the metal is 46 W/m °C. Calculate the rate of heat transfer through one such rod.

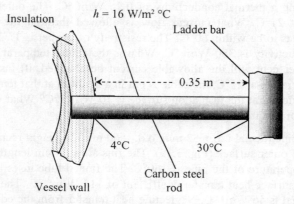

Fig. 3.20 Horton sphere (Problem 3.17).

3.18 A composite wall consists of material A (thickness = 15 cm, k_A = 10 W/m °C) and material B (thickness = 20 cm, k_B = 16 W/m °C). The exposed surface of A is in contact with a hot fluid at 150°C (heat transfer coefficient = 180 W/m² °C), and that of B is in contact with air at 38°C (film coefficient = 26 W/m² °C). The mid-plane temperature of A (i.e. 7.5 cm away from the exposed surface) at steady state is measured to be 130°C. (a) Is there any contact resistance at the junction of A and B? If so, what is its magnitude? (b) Calculate the temperature jump at the interface. (c) If there was no contact resistance, by what per cent the thickness of slab A should have been increased to get the same heat flux (keeping the thickness of slab B unchanged)?

3.19 The composite wall of an oven consists of three layers, A, B, and C of thicknesses 12 inch, 4 inch, and 8 inch and thermal conductivities 0.75, 0.08, and 0.50 Btu/h ft² °F, respectively. The inner layer is A, and it has an inside temperature of 2000°F. The outer layer (C) is exposed to an ambient temperature of 85°F and its surface heat transfer coefficient is 1.3 Btu/h ft² °F. It is estimated that there is a thermal contact resistance of 0.0021 h °F/Btu between the layers A and B, and 0.0025 h °F/Btu between the layers B and C (expressed on the basis of unit area of the wall). Calculate (i) the rate of heat loss from the oven per square foot of the wall area, (ii) the temperatures of the walls A and B at their junction, and the temperature jump. Draw the electrical analogue of the problem.

3.20 Long pipe sections in a plant need supporting arrangements. A common type of support, called the 'pipe clamp support' (Fig. 3.21), is used in the plant to hold a high pressure steam

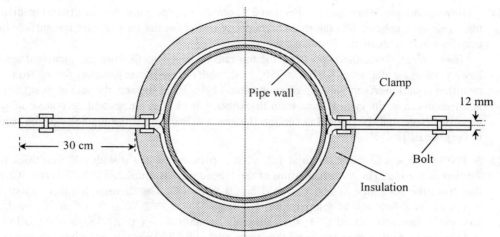

Fig. 3.21 Pipe clamp support (Problem 3.20).

pipe. The steam temperature is 240°C and the pipe is properly insulated. Two carbon steel slats (50 mm × 6 mm cross-section; thermal conductivity = 48 W/m °C) are used to make the support and are exposed to the ambient to a length of 30 cm on both sides as shown in the figure. Assume that: (i) heat loss from the clamp plates occurs only through the section exposed to the ambient air, (ii) although the clamp bars or slats are bolted, they are in contact with each other in the extended part, and for simplicity the bolt area can be neglected (this means that heat loss occurs through both sides of the combination of two steel plates of the given size, and the base temperature is 240°C), (iii) heat loss at the free ends of the clamp plates is negligible, and (iv) there is no transverse variation of temperature in a plate. Calculate the rate of heat loss through the clamp if the external air-film coefficient is 20 W/m² °C. The ambient temperature is 28°C.

3.21 Calculate the rate of heat loss from a circular fin, 4 inch o.d. and 0.05 inch thick, attached to a 1 inch diameter tube at 250°F. If the surface heat transfer coefficient is 1.9 Btu/h ft² °F, calculate the rate of heat loss through the fin. (The expression for the efficiency of a fin of circular geometry is available in Kern and Kraus, 1972.).

3.22 A wall at 200°F has an array of 'pin-fins' (i.e. thin cylindrical spikes protruding from the wall) of 0.25 inch diameter and 2.5 inch length arranged on 0.75 inch square pitch. If the ambient temperature is 90°F, the heat transfer coefficient is 1.8 Btu/h ft² °F, and the thermal conductivity of the material of the fin is 25 Btu/h ft °F, calculate the rate of cooling through the fins per square foot of the wall area.

3.23 A 0.25 inch o.d. tube is used to circulate Servotherm (a heat transfer oil from Indian Oil Corporation Ltd.) at 275°C through an experimental micro-reactor. The tube is lagged with asbestos (asbestos is a carcinogen!) of thermal conductivity 0.106 Btu/h ft °F. The heat transfer coefficient at the exposed surface of the insulation is 0.625 Btu/h ft² °F. If the ambient temperature is 30°C, calculate (on the basis of 10 ft length of the tube): (i) the rate of heat loss if the critical insulation thickness is used, (ii) the rate of heat loss without insulation (in this case take the bare surface heat transfer coefficient to be 2.5 Btu/h ft² °F), and (iii) the rate of heat loss if five times the critical insulation thickness is used. Also, prepare a plot to show the rate of heat loss as a function of the insulation thickness.

3.24 Following the procedure used in Section 3.5, obtain an expression for the critical insulation thickness on a sphere. The thermal conductivity of the insulation is k_c and the surface heat transfer coefficient of the insulated sphere is h.

[*Hint:* Write down the expression for the rate of heat loss Q from the insulated sphere having an outer diameter r_o. Put $dQ/dr_o = 0$, and obtain the expression for r_o from the resulting equation. As in the case of an insulated cylindrical surface, the rate of heat transfer is a maximum at the critical insulation thickness. Check that the second derivative of Q is negative. Note that the critical insulation thickness exists only if the diameter of the sphere is very small.]

3.25 A 100 mm pipe will carry steam at 160°C. It is planned to lag it with 100 mm thick pre-formed insulation. The outside air-film of the lagged pipe is estimated to be 5 W/m² °C, and that from the bare pipe at 160°C would be 20 W/m² °C. The thermal conductivity of the insulation is a function of temperature, $k = 0.9 + 3 \times 10^{-4}T$ W/m °C (T in °C), and the ambient temperature is 30°C. If the estimated insulation cost is Rs 15,000/m³ (including installation and other expenses) and the steam cost is Rs 750 per ton, calculate the pay-back period assuming a rate of interest of Re 0.18/(yr)(Re). Latent heat of steam is 1825 kJ/kg.

[*Hint:* Calculate the rate of heat loss, and hence the cost of steam condensed, from the bare pipe as well as the insulated pipe. Calculate the present value of the savings in energy achieved by putting the insulation over the service life of the insulation (say n years) and considering the given rate of interest compounded annually (see Section 3.6 and Example 3.14). Equate it to the cost of insulation and calculate the payback period, n.]

3.26 A thermos bottle, 75 mm i.d. and 200 mm in height, has a 1.6 mm double glass wall with an evacuated intervening space of 4 mm width. It has an outer plastic casing which also offers considerable resistance to heat transfer equivalent to that of a 5 mm thick glass plate. At a sufficiently high vacuum (so that the width of the intervening space is less than the mean free path of the molecules), the Knudsen equation may be used to calculate an 'equivalent film coefficient' for the evacuated space.

$$h_e = \frac{1}{2}\left[\frac{\alpha_i\alpha_o}{\alpha_o + (1 - \alpha_o)\alpha_i r_i/r_o}\right]\left(\frac{\gamma + 1}{\gamma - 1}\right)\sqrt{\frac{R}{2\pi TM}}$$

where α is a constant called the 'accommodation coefficient', r_i, r_o are the inner and outer radii of the evacuated space, respectively, γ is the ratio of specific heats of the gas, R is the gas constant, T is the temperature, K, and M is the molecular weight of the gas. If the residual gas in the evacuated space is air, α is approximately unity.

The bottle is filled with hot tea at 97°C. How long will it take for the tea to cool down to 60°C? Assume that tea has properties similar to those of water. The temperature of the liquid in the bottle remains uniform at any point of time (lumped parameter model). Four resistances in series have to be considered—those owing to two glass walls, that owing to the evacuated space, and the other owing to the plastic casing. Consider heat loss from the curved surface of the flask only. The termal conductivity of glass is 0.87 W/m °C. The outer skin temperature of the flask may be taken to be 30°C.

[*Note:* Double-walled tanks with the space between the walls filled with a thermal insulator but highly evacuated are used for cryogenic storage. For example, such a tank is used for ethylene storage at −100°C.]

REFERENCES AND FURTHER READING

Bird, R.B., W.E. Stewart, and E.N. Lightfoot, *Transport Phenomena*, John Wiley, New York, 1960.

Fletcher, L.S., "Recent advances in contact conductance heat transfer", *J. Heat Transfer*, **110** (Nov. 1988), 1059.

Incropera, F.P. and D.P. Dewitt, *Introduction to Heat Transfer*, 2nd ed., John Wiley, New York, 1996.

Jacob, M., *Heat Transfer*, Vol. 1, John Wiley, New York, 1957.

Kakac, S., A.E. Bergles, and F. Mayinger, *Heat Exchangers,* Hemisphere Publ. Corp., New York, 1981.

Kern, D.Q. and A.D. Kraus, *Extended Surface Heat Transfer*, McGraw-Hill Book Co., New York, 1972.

Ligi, J.J., "Processing with cryogenics", *Hydrocarbon Proc.*, April 1969, 93–96.

Menicatti, S., "Check tank insulation economics", *Hydrocarbon Proc.*, April 1969, 133–169.

Peters, M. and K.D. Timmerhaus, *Plant Design and Economics for Chemical Engineers*, 4th ed., McGraw-Hill Book Co., New York, 1991.

Sloane, B.A., "Maximise your energy savings with proper insulation", *Hydrocarbon Proc.*, October, 1992, 67–74.

Ulrich, G.D., *A Guide to Chemical Engineering Process Design and Economics*, John Wiley, New York, 1984.

4

Forced Convection

The concept and definitions of heat transfer coefficients, individual and overall, have been introduced in Chapter 3. We explained with examples how simple problems involving convective transport of heat can be solved if the *heat transfer coefficient* is known. But the question remains as to how we can obtain the heat transfer coefficient in a given situation. The heat transfer coefficient is not as simple a quantity as a fundamental property of a body or a system. It depends upon many factors—the physical properties of the fluid (like thermal conductivity, density, viscosity), the velocity field, the geometry of the system and its characteristic dimensions, etc. In convective heat transfer, transport of energy from one point to another occurs primarily through actual movement of the fluid elements or particles. This is why the flow field or the velocity field (the velocity components expressed as functions of three space coordinates and also time) tremendously influences the heat transfer coefficient. In a case of convective heat transfer it is not difficult to make a differential heat balance over an elementary volume considering all kinds of energy input, output, generation and accumulation terms in order to obtain a partial differential equation that governs the temperature distribution in the medium. But in most practical cases it becomes a formidable task to solve such equations for the calculation of heat transfer rates. Fortunately, a large number of empirical correlations have been developed over the years by heat transfer researchers. These correlations cover a great majority of situations encountered in industrial practice, and are used for the calculation of heat transfer coefficients. No doubt, the use of any correlation involves some error—that may be even 25–30% in some cases. However, in the absence of more reliable data or information in specific cases, these correlations are indispensable for the estimation of heat transfer coefficients.

In this chapter we will first qualitatively analyze convective heat transfer in a few simple systems in order to provide further understanding of this important mode of heat transfer. Later, we will discuss the more important heat transfer correlations and illustrate their uses. Convective heat transfer may be of two types—*forced convection* and *free* or *natural convection*. Both are caused by motion in the medium. If the motion or velocity in the medium is generated by the application of an external force (e.g. by a pump, a blower, an agitator, etc.), heat transfer is said to occur by forced convection. But if the motion in the medium occurs as a result of density difference (which may be caused by a temperature difference), the concerned mode of heat transfer is called free convection. In this chapter we will confine our attention to forced convection. Free convection will be discussed in Chapter 5.

4.1 FORCED CONVECTION IN SYSTEMS OF SIMPLE GEOMETRIES

Here we identify a few simple flow situations for understanding the physical mechanism of forced convection heat transfer, without going into the mathematical analysis. Although the simplest way to visualize convective heat transfer to or from a surface is to assume the existence of a *stagnant fluid film* at the wall, *no* such film does exist in reality. It is true, however, that in the case of heat transfer from a surface, most of the resistance to heat transfer is offered by a narrow zone of fluid near the surface, called the boundary layer, but the fluid in this zone is never stagnant. The concept of boundary layer is of vital importance in understanding the flow and heat transfer characteristics near a wall.

4.1.1 Flow over a Flat Plate

Let us consider the flow of a fluid over an immersed wide flat plate [Fig. 4.1(a)] at zero angle of incidence (which means that the plate is oriented along the direction of flow of the bulk fluid).

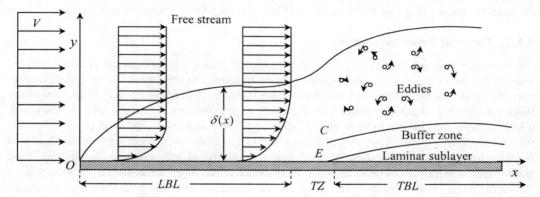

Fig. 4.1(a)   Boundary layer over a flat plate (at zero angle of incidence). (LBL: laminar boundary layer, TZ: transition zone, TBL: turbulent boundary layer.)

Fluid velocity at the surface of the plate is zero (because of no-slip condition), and gradually increases with distance from the plate. At a sufficiently large distance from the plate, the fluid velocity becomes equal to the 'free stream velocity', V. The region above the plate surface within which this change of velocity from zero to the free stream value occurs is called the *boundary layer,* or more specifically, the *velocity boundary layer* (also called the *momentum boundary layer* or the *hydrodynamic boundary layer*). The thickness of this region is called the *boundary layer thickness* and is denoted by δ. The boundary layer thickness increases with the distance x from the leading edge of the plate, i.e $\delta = \delta(x)$.

A theoretical analysis of boundary layer flow indicates that the transverse distance from the plate where the free stream velocity is attained is very large. As a result, the thickness of the boundary layer is rather arbitrarily defined as the distance from the plate at which a certain percentage, usually 99%, of the free stream velocity is attained. The equations of fluid motion in the boundary layer (see Section 11.3) can be solved in order to obtain the velocity profile in the boundary layer and to determine the boundary layer thickness.

The characteristics of boundary layer flow depend upon the Reynolds number defined as $\mathrm{Re}_x = \rho V x / \mu$, where x is the distance from the point O or the leading edge. Because x is a variable, Re_x is called the 'local Reynolds number'. The fluid motion in the boundary layer remains laminar up to a point on the plate where $10^5 < \mathrm{Re}_x < 5 \times 10^5$, which is the *critical Reynolds number* (it depends upon the roughness of the plate) for boundary layer flow. When Re_x increases beyond this value, transition to turbulent boundary layer begins. At $Re_x = 10^6$, the boundary layer becomes fully turbulent. However, these values should not be taken too strictly because they depend upon several factors including the surface roughness of the plate. The laminar, transition as well as the turbulent portions of the boundary layer are shown in Fig. 4.1(a). The velocity profile in the laminar region is also shown.

The turbulent boundary layer is visualized, on the basis of reported experimental measurements, as consisting of three zones—a laminar sublayer adjacent to the wall, a buffer zone and the turbulent zone. Change in the velocity from zero to V is virtually confined within the first two zones.

Mathematical analysis of boundary layer flow including solutions for the velocity distributions and the expressions for the drag force on the plate are available in standard books on fluid mechanics (e.g. Schlichting, 1968). This will be briefly discussed in Chapter 11.

4.1.2 Thermal Boundary Layer

Heat transfer from a hot plate to a flowing fluid occurs principally by convection. Similar to the velocity boundary layer, a thermal boundary layer is formed in the liquid that includes laminar, transition and turbulent zones in the sense of heat transport depending upon the variation of the Reynolds number along the plate. The thickness of the thermal boundary layer is, however, different from that of the hydrodynamic boundary layer, and considerably depends upon the thermal properties of the fluid. Just at the surface of the plate, the temperature of the fluid will be the same as the temperature of the plate (continuity of temperature). Again, the fluid temperature will be very close to the bulk temperature at the edge of the thermal boundary layer (i.e. at the edge of the thermal boundary layer, the difference from the bulk temperature remains within 1% of the total temperature drop across the boundary layer). Thus, it is not through the so-called stagnant film, but through the boundary layer that convective transfer of heat from a surface occurs.

Velocity distributions in a two-dimensional boundary layer [i.e. how the x- and the y-components of the velocity, u and v, depend upon the position (x, y)] govern the heat transfer coefficient in the boundary layer. A correlation that can be used to calculate the heat transfer coefficient (which is a function of x) will be described later. The thermal boundary layer over a flat plate is shown in Fig. 4.1(b). It may be noted that when the dimensionless group called the

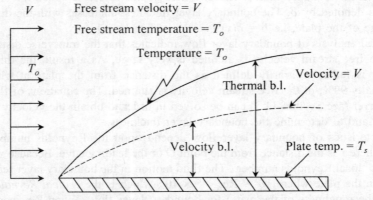

Fig. 4.1(b) Laminar velocity and temperature boundary layers over a heated plate.

Prandtl number, $\mathrm{Pr} = v/\alpha$ ($v = \mu/\rho$ is called the *momentum diffusivity*, and $\alpha = k/\rho c_p$ is called the *thermal diffusivity*) is less than unity, the momentum boundary layer remains within the thermal boundary layer. If $\mathrm{Pr} > 1$, the reverse becomes true.

4.1.3   Flow across a Cylinder

Transverse flow of a fluid across a single cylinder or a bank of cylinders is an important flow situation in industrial heat transfer. For example, a hot process stream flowing through a pipe may be cooled by blowing air across the pipe. Here too, a boundary layer forms over the cylinder as shown in Fig. 4.2(a). The point A is called the *stagnation point* because the fluid velocity is zero at this point and the pressure is maximum (the physical reason being conversion of velocity head

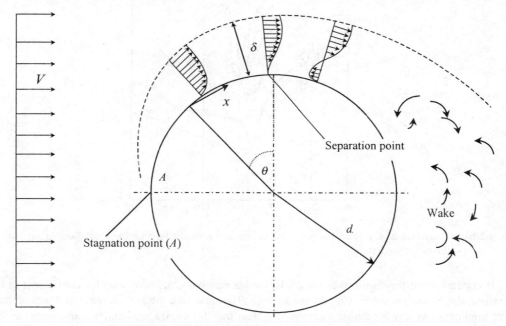

Fig. 4.2(a)   Velocity profile for cross-flow over a cylinder.

to pressure head). The distance of a point on the cylinder from the stagnation point is given by the linear measure x or the angular measure θ. As one moves away from the stagnation point along the periphery, the boundary layer decelerates and its thickness increases. The velocity profile within the boundary layer becomes gradually more and more flattener. Farther away from the stagnation point, the retarded layer is unable to overcome the static pressure and the boundary layer 'separates' from the cylinder surface. Downstream of this point of separation, reverse flow sets in near the surface but forward flow continues in the outer part of the boundary layer. On the rear side of the cylinder there are wakes, and the flow becomes turbulent.

The nature of the flow field in the boundary layer has profound influence on the heat transfer coefficient at the cylinder surface when the cylinder temperature is different from that of the fluid. The variation of the local heat transfer coefficient, h, in terms of the Nusselt number, $\mathrm{Nu} = hd/k$ vs. θ for air flowing across a heated cylinder at different Reynolds numbers is shown in Fig. 4.2(b).

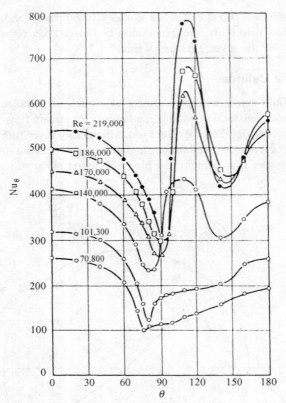

Fig. 4.2(b)   Variation of the local Nusselt number along the circumference for cross-flow of air over a heated cylinder.

It appears from the figure that for all Reynolds numbers, the heat transfer coefficient (and, therefore, the Nusselt number) first decreases with θ because of a gradual increase in the boundary layer thickness. At low Reynolds numbers, the heat transfer coefficient attains a minimum at the stagnation point and increases again because the eddies or wakes enhance heat transfer. But at higher Reynolds numbers each plot shows two minimum values of the heat transfer coefficient. The first minimum corresponds to the transition of the boundary layer from laminar to turbulent and the second one indicates the separation of the boundary layer from the cylinder surface.

For practical heat transfer calculations, an 'average heat transfer coefficient' rather than the local coefficient is used. We will now discuss the methods and the theoretical basis of development of correlations for heat transfer coefficients in forced convection.

Example 4.1   A steel sphere of diameter d and thermal conductivity k is being cooled in a flowing stream of air of bulk temperature T_o. The temperature distribution in the ball at a particular instant can be approximated as $T = a + br + cr^2$, where a, b and c are functions of time, and T is in K. What is the surface heat transfer coefficient? What is the unit of b?

SOLUTION   Because the ball is being cooled gradually, the problem is of unsteady state type. But at any instant the rate of heat flow from inside the sphere to its surface must be equal to the rate of heat loss to the ambient, as there cannot be any accumulation or depletion of heat at a surface.

$$\text{Heat flux to the surface of the sphere} = -k\left[\frac{\partial T}{\partial r}\right]_{r=d/2} = -k[b + 2cr]_{r=d/2} = -k(b + cd)$$

$$\text{Heat flux from the surface } (r = d/2) \text{ to the ambient} = h(T_{\text{surface}} - T_o)$$

$$= h(a + bd/2 + cd^2/4 - T_o)$$

Therefore,

$$-k(b + cd) = h(a + bd/2 + cd^2/4 - T_o)$$

The external heat transfer coefficient, $h = \dfrac{-4k(b + cd)}{4a + 2bd + cd^2 - 4T_o}$

The unit of b is K/m.

4.2 DIMENSIONAL ANALYSIS

Exact or approximate mathematical solutions to convective heat transfer problems in simple situations are often possible especially because of the great advancement in computers and computational techniques in recent times. But many of the practical situations encountered in industrial practice still defy our analytical and numerical tools, and we have to depend upon 'empirical correlations'. A large number of correlations have been developed over the years to relate the heat transfer coefficient (which itself is an empirical quantity for all practical purposes) with the relevant properties, parameters and variables of systems. Experience has taught us that it is, in general, most convenient to correlate the experimental data—not only in heat transfer but also in many other processes of science and engineering—by using a set of 'dimensionless groups'. A dimensionless group is composed of two or more variables, parameters or properties. The technique of dimensional analysis is a powerful tool to identify the dimensionless groups which are to be used to correlate the variables that influence a particular physical process. The first step is to identify the relevant variables, parameters or properties. For example, in the case of convective heat transfer to a fluid flowing through a circular pipe, we can say from our experimental findings and our understanding of the physical phenomenon that the heat transfer coefficient h depends upon the fluid velocity V, pipe diameter d, thermal conductivity of the fluid k, its specific heat c_p, density ρ, and viscosity μ. So, h, V, d, k, c_p, ρ, and μ are the variables and properties which we identify for the purpose of forming dimensionless groups. Dimensional analysis teaches us how to cluster two or more of these quantities. There are quite a few possible procedures (Taylor, 1974). We will discuss here the one based on the Buckingham Pi theorem.

4.2.1 Statement of Buckingham Pi Theorem

If the equation $\Phi(X_1, X_2, ..., X_n) = 0$ is the only relation among the quantities $X_1, X_2, ..., X_n$, and if it holds for any arbitrary choice of the units in which these quantities are measured, then the solution of this equation has the form $f(\Pi_1, \Pi_2, ..., \Pi_m) = 0$, where $\Pi_1, \Pi_2, ..., \Pi_m$ are independent dimensionless products of the Xs. Further, if p is the minimum number of primary quantities necessary to express the dimensions of Xs, then

$$n - m = p$$

4.2.2 Dimensionless Groups in Convective Heat Transfer to a Fluid Flowing through a Circular Pipe

We now apply the Buckingham Pi theorem to this particular problem of heat transfer in order to identify the dimensionless groups. The variables, parameters and properties which are important, are shown in Table 4.1. Our purpose is to correlate experimental data on heat transfer coefficient with the quantities cited in Table 4.1. The general functional form may be expressed as

$$\Phi \ (h, \ V, \ d, \ k, \ c_p, \ \rho, \ \mu) = 0 \tag{4.1}$$

Table 4.1   Important quantities in heat transfer in pipe flow

Quantity		Unit	Dimension
heat transfer coefficient,	h	W/m^2 °C	$Mt^{-3}T^{-1}$
average fluid velocity,	V	m/s	Lt^{-1}
pipe diameter,	d	m	L
thermal conductivity,	k	W/m °C	$MLt^{-3}T^{-1}$
specific heat,	c_p	W s/kg °C	$L^2t^{-2}T^{-1}$
density,	ρ	kg/m^3	ML^{-3}
viscosity,	μ	kg/m s	$ML^{-1}t^{-1}$

Four primary quantities [mass (M), time (t), length (L) and temperature (T)] are necessary to express the dimensions of the quantities in Eq. (4.1). Therefore, the number of dimensionless groups (m) will be $7 - 4 = 3$. We choose the following combinations to express Π_1, Π_2 and Π_3.

$$\Pi_1 = k^{a_{11}} \ \mu^{a_{12}} \ \rho^{a_{13}} \ d^{a_{14}} \ h^{a_{15}} \tag{4.2}$$

$$\Pi_2 = k^{a_{21}} \ \mu^{a_{22}} \ \rho^{a_{23}} \ d^{a_{24}} \ c_p^{a_{25}} \tag{4.3}$$

$$\Pi_3 = k^{a_{31}} \ \mu^{a_{32}} \ \rho^{a_{33}} \ d^{a_{34}} \ V^{a_{35}} \tag{4.4}$$

where a_{ij}s are unknown constants. We will take Eq. (4.2) first. The group Π_1 is dimensionless. Considering the dimensions of the various terms, we have

$$M^0 \, L^0 \, t^0 \, T^0 = (MLt^{-3}T^{-1})^{a_{11}} \ (ML^{-1}t^{-1})^{a_{12}} \ (ML^{-3})^{a_{13}} \ (L)^{a_{14}} \ (Mt^{-3}T^{-1})^{a_{15}}$$

Equating the powers of M, L, t and T on both sides, we get

$$a_{11} + a_{12} + a_{13} + a_{15} = 0 \tag{4.5a}$$

$$a_{11} - a_{12} - 3a_{13} + a_{14} = 0 \tag{4.5b}$$

$$-3a_{11} - a_{12} - 3a_{15} = 0 \tag{4.5c}$$

$$-a_{11} - a_{15} = 0 \tag{4.5d}$$

Now we have four equations in five unknowns. We solve them by expressing a_{11}, a_{12}, a_{13}, and a_{14} in terms of a_{15}.

$$a_{11} = -a_{15}; \qquad a_{12} = 0; \qquad a_{13} = 0; \qquad a_{14} = a_{15}$$

Thus

$$\Pi_1 = \left[\frac{hd}{k} \right]^{a_{15}}$$

Similarly, it can be shown from Eqs. (4.3) and (4.4) that

$$\Pi_2 = \left[\frac{c_p \mu}{k}\right]^{a_{25}} \quad \text{and} \quad \Pi_3 = \left[\frac{dV\rho}{\mu}\right]^{a_{35}}$$

Thus the seven variables in this problem can be related through the above three dimensionless groups. It is also possible to arrive at the same dimensionless groups without going through the formalism of the Pi theorem. For example, let us assume a power-law dependence of h on the remaining quantities in Table 4.1. Thus

$$h = C'\, k^{\beta_1}\, \mu^{\beta_2}\, \rho^{\beta_3}\, d^{\beta_4}\, c_p^{\beta_5}\, V^{\beta_6} \quad (C' \text{ and } \beta_i\text{s are constants})$$

Writing down the dimensions of the various quantities, we have

$$Mt^{-3}T^{-1} = C'(MLt^{-3}T^{-1})^{\beta_1}\, (ML^{-1}t^{-1})^{\beta_2}\, (ML^{-3})^{\beta_3}\, (L)^{\beta_4}\, (L^2t^{-2}T^{-1})^{\beta_5}\, (Lt^{-1})^{\beta_6}$$

Equating the powers of various dimensions, we get

$$\text{power of} \quad M: \quad \beta_1 + \beta_2 + \beta_3 \qquad\qquad\qquad = \quad 1$$
$$L: \quad \beta_1 - \beta_2 - 3\beta_3 + \beta_4 + 2\beta_5 + \beta_6 = \quad 0$$
$$t: \quad -3\beta_1 - \beta_2 - 2\beta_5 - \beta_6 \qquad\quad = -3$$
$$T: \quad -\beta_1 - \beta_5 \qquad\qquad\qquad\qquad = -1$$

There are four equations in six unknowns. Solving for any four unknowns (say β_1, β_2, β_3, and β_4), in terms of the remaining two (β_5 and β_6), we get

$$\beta_1 = 1 - \beta_5;\ \beta_2 = \beta_5 - \beta_6;\ \beta_3 = \beta_6;\ \beta_4 = \beta_6 - 1;\ \beta_5 = \beta_5;\ \beta_6 = \beta_6$$

Therefore,

$$h = C'(k)^{1-\beta_5}\, (\mu)^{\beta_5 - \beta_6}\, (\rho)^{\beta_6}\, (d)^{\beta_6 - 1}\, (c_p)^{\beta_5}\, (V)^{\beta_6}$$

On rearrangement,

$$\frac{hd}{k} = C'\left(\frac{c_p \mu}{k}\right)^{\beta_5} \left(\frac{dV\rho}{\mu}\right)^{\beta_6} \tag{4.6}$$

Experimental heat transfer data in turbulent pipe flow are mostly correlated in the form of Eq. (4.6). It should be noted, however, that dimensional analysis does not tell us how the dimensionless groups in a particular problem are related. In other words, it is silent about the possible functional form of the correlation. In Eq. (4.6), we arrived at a functional form of the dependence of hd/k (called the Nusselt number, Nu) on $c_p\mu/k$ (called the Prandtl number, Pr) and $dV\rho/\mu$ (the well known Reynolds number, Re) because we assumed a power-law type of dependence at the beginning.

4.3 DIMENSIONLESS GROUPS IN HEAT TRANSFER

Before we discuss the important heat transfer correlations available in the literature, it is pertinent to make a list and also to mention the physical significances of the dimensionless groups which are frequently used in forced convection heat transfer (Table 4.2).

Table 4.2   Important dimensionless groups in convective heat transfer

Name	Expression	Physical significance
Nusselt number, $\mathrm{Nu} = \dfrac{hl}{k}$		$\dfrac{\text{Wall temperature gradient}}{\text{Temperature gradient across the fluid in the pipe}}$
Reynolds number, $\mathrm{Re} = \dfrac{lV\rho}{\mu}$		$\dfrac{\text{Inertial force}}{\text{Viscous force}}$
Prandtl number, $\mathrm{Pr} = \dfrac{c_p\mu}{k} = \dfrac{\nu}{\alpha}$		$\dfrac{\text{Momentum diffusivity}}{\text{Thermal diffusivity}}$
Stanton number, $\mathrm{St} = \dfrac{h}{V\rho c_p}$		
$= \dfrac{(hl/k)}{(lV\rho/\mu)(c_p\mu/k)} = \dfrac{\mathrm{Nu}}{(\mathrm{Re})(\mathrm{Pr})}$		$\dfrac{\text{Rate of wall heat transfer by convection}}{\text{Rate of heat transfer by bulk flow}}$
Peclet number, $\mathrm{Pe} = \dfrac{\rho V l c_p}{k} = \dfrac{lV\rho}{\mu} \cdot \dfrac{c_p\mu}{k}$		$\dfrac{\text{Rate of heat transfer by bulk flow}}{\text{Rate of heat transfer by conduction}}$
$= \mathrm{Re} \cdot \mathrm{Pr}$		
Graetz number, $\mathrm{Gz} = \mathrm{Pe} \cdot d/L = \mathrm{Re} \cdot \mathrm{Pr} \cdot d/L$		Similar to Peclet number (used in connection with analysis of heat transfer in laminar flow in pipes)

The above table needs some more explanation. In these expressions (except the last one, Gz), l denotes the *characteristic length*. A characteristic length can be defined based on the geometry of the system. For example, in the case of flow, heat or mass transfer in a pipe, the pipe diameter is taken as the characteristic length. For flow past a flat plate, the distance from the leading edge is the characteristic length. So, in the case of pipe flow we use d as the characteristic length in place of l. In the expression for Graetz number, L is the pipe length over which heat transfer occurs.

The physical significance of Nusselt number given in Table 4.2 may be arrived at in the following way. Consider heating of a fluid while it flows through a pipe of diameter d. At a certain axial position, let the wall temperature be T_w and the bulk fluid temperature, T_b $(T_w > T_b)$. The heat flux at the wall is $q_r = k[dT/dr]_{r=d/2}$ (no negative sign is necessary, because temperature increases with r; dT/dr itself is positive. By the definition of heat transfer coefficient, we have

$$q_r = k\left[\frac{dT}{dr}\right]_{r=d/2} = h(T_w - T_b) = hd\frac{T_w - T_b}{d} = \frac{k\left(\dfrac{dT}{dr}\right)_{r=d/2}}{(T_w - T_b)/d}$$

$$= \frac{\text{Temperature gradient at the wall}}{\text{Temperature gradient across a distance equal to the pipe diameter}}$$

The Stanton number is defined as, $\mathrm{St} = h/V\rho\, c_p$. If we consider a temperature difference of ΔT across the wall and the bulk fluid, the *convective heat flux* becomes $h\Delta T$. Now, if a fluid flows at an average velocity V, the rate of energy transport by bulk flow or by the movement of the liquid per unit area of flow cross-section is = (mass flow rate)(specific heat)(temperature change) = $(V\rho)(c_p)(\Delta T)$.

The ratio of the above two quantities is

$$h\Delta T/(V\rho)(c_p)(\Delta T) = h/V\rho\, c_p = \mathrm{St}$$

Again, the *heat flux due to conduction* across a distance d under the same temperature driving force is $(k/d)\, \Delta T$. Therefore,

$$\frac{\text{Rate of heat transfer by bulk flow}}{\text{Rate of heat transfer by conduction}} = \frac{(V\rho)(c_p)(\Delta T)}{(k/d)(\Delta T)} = \frac{dV\rho c_p}{k} = \mathrm{Pe}$$

This is the physical significance of the *Peclet number*.

4.4 EXPERIMENTAL DETERMINATION OF THE HEAT TRANSFER COEFFICIENT

A correlation for the heat transfer coefficient can be developed only on the basis of a substantial amount of experimental heat transfer data for a particular system. For example, experimentally measured values of heat transfer coefficients for heating or cooling of various fluids in pipe flow can be correlated in the form of Eq. (4.6). Experiments should be performed with a large number of fluids so that a wide range of values of the physical properties (ρ, μ, c_p, k) is covered. In addition, pipes of different diameters should be used and the fluid velocity should be varied over a wide range. A simple experimental set-up for the determination of the heat transfer coefficient is shown in Fig. 4.3.

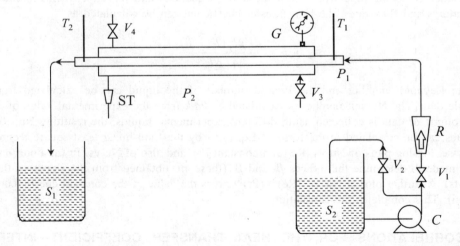

Fig. 4.3 An experimental set-up for the determination of heat transfer coefficient for pipe flow. (*C*: centrifugal pump; *G*: steam pressure gauge; P_1, P_2: inner and outer pipes; *R*: rotameter; S_1, S_2: liquid storage tank; ST: steam trap; T_1, T_2: thermometers or thermocouples; V_1, V_2, V_3, V_4: valves.)

The heat transfer element of the set-up consists of an assembly of two concentric pipes P_1 and P_2. The experimental liquid is stored at a constant temperature in the tank S_2. The centrifugal pump C forces the liquid through the inner pipe P_1, while saturated steam flows into the annulus between the pipes. Steam condenses in the annulus and transfers heat to the liquid flowing through the pipe P_1. The condensate is removed through the steam trap ST. A few valves are provided: V_1 and V_2, to regulate the liquid flow rate (the bypass line and the valve V_2 ensure better regulation of the flow rates); V_3, to regulate the steam pressure; V_4, to connect the system to a vent. Thermometers (or thermocouples), T_1 and T_2, measure the inlet and the outlet temperatures of the liquid. Steam pressure is measured by the gauge G; the steam temperature may be obtained from the steam table.

In a particular experiment, the temperatures, T_1 and T_2, and the liquid velocity V are measured after the steady state condition is attained. The values of the liquid properties (ρ, μ, c_p and k) are taken at the average temperature [i.e. $(T_1 + T_2)/2$]. If d is the pipe diameter and L the length of the heated section of the pipe, we have

$$\text{Rate of heat transfer to the liquid} = \frac{\pi}{4} d^2 V \rho c_p (T_2 - T_2) = (h)(\pi dL)(\Delta T_m)$$

Here πdL is the area of heat transfer and ΔT_m is the 'log mean temperature difference' (this will be explained in Section 4.7).

$$\Delta T_m = \frac{(T_w - T_1) - (T_w - T_2)}{\ln [(T_w - T_1)/(T_w - T_2)]} \qquad T_w \text{ is the temperature of the pipe wall.}$$

Because the heat transfer resistances offered by the pipe wall and by the condensing steam are very small, we can take the wall temperature to be the same as the steam temperature. Alternatively, the wall temperature can be measured directly by using a fine themocouple buried in or cemented to the pipe wall. From the above equations and the experimentally measured temperatures and flow rates, the heat transfer coefficient can be calculated as

$$h = \frac{dV \rho c_p (T_2 - T_1)}{4L \Delta T_m}$$

The Reynolds number and the Prandtl number of the liquid can be calculated from the available data. The Nusselt number is calculated as hd/k from the experimental value of h. If a large volume of data is collected using different experimental liquids, the resulting Nu, Re, and Pr values can be correlated in the form of Eq. (4.6) by the 'non-linear least square technique'. Otherwise, log-log plots of Nu vs. Re (at a constant Pr), and also of Nu vs. Pr (at a constant Re) can be made to determine the indices β_5 and β_6 (these are obtained from the slopes of the log-log plots). A further plot of Nu vs. $(\text{Re})^{\beta_5} (\text{Pr})^{\beta_6}$ gives the value of the constant coefficient C' in Eq. (4.6). This completes the correlation.

4.5 CORRELATIONS FOR THE HEAT TRANSFER COEFFICIENT—INTERNAL FLOWS

Systems of various shapes and different flow conditions are encountered in practical heat transfer calculations. The common shapes or geometries are cylindrical, spherical or flat (or planar). The most common flow problem involves flow through a circular pipe; the other important ones are flow through a non-circular duct (e.g. a rectangular channel), through a packed bed (e.g. of catalysts or of some inert material forming a regenerative bed), in an agitated vessel with internals, through the jacket over a reaction kettle or vessel, across a bank of tubes, etc. Again, the flow of a fluid in a system may be internal (e.g. flow through a pipe) or external (e.g. flow across the outer surface of a pipe). Here, we describe the more important correlations along with the ranges of applicability for the calculation of heat transfer coefficients. Some of these were developed even more than half a century ago, but are still widely used in engineering calculations. The important correlations for the case of internal flow are discussed below.

4.5.1 Laminar Flow through a Circular Pipe

Laminar flow (Re < 2100) may occur in industrial equipment for heating or cooling of a considerably viscous liquid or of a liquid or solution sensitive to shear stress.* Because it is possible to analytically solve the problem of heat transfer to or from a fluid in laminar flow through a pipe, a large volume of theoretical research work has been published in this area. Details are available in Shah and London (1978). For fully developed flow and a constant temperature of the pipe wall, the following correlation of Hausen (1943), although outdated, is recommended. This correlation incorporates a term called the 'thermal entry length'. (It is the distance from the point of entry to a pipe or channel over which the 'boundary layer approximations' are applicable; this will be discussed later.)

$$\text{Nu} = 3.66 + \frac{0.0668(d/L)(\text{Re})(\text{Pr})}{1 + 0.04[(d/L)(\text{Re})(\text{Pr})]^{2/3}} \qquad \left[\text{Nu} = \frac{\overline{h}\,d}{k}, \ \text{Re} = \frac{dV\rho}{\mu}\right] \qquad (4.7)$$

where
L is the pipe length
d is the pipe diameter
$\overline{h}$ is the average heat transfer coefficient over the pipe length L including the entry length.

A simpler correlation was suggested by Sieder and Tate (1936).

$$\text{Nu} = 1.86 \left[\frac{(\text{Re})(\text{Pr})}{L/d}\right]^{1/3} \left(\frac{\mu}{\mu_w}\right)^{0.14} \qquad (4.8)$$

The above correlation, Eq. (4.8), is applicable when

(a) $0.48 < \text{Pr} < 16{,}700$,

(b) the viscosity ratio is within the range, $0.0044 < \mu/\mu_w < 9.75$, where μ is the viscosity of the fluid at the bulk temperature, and μ_w is that at the wall temperature, T_w,

(c) $(\text{Pe } d/L) > 10^8$.

The fluid properties have to be known or evaluated at the mean bulk temperature of the fluid.

4.5.2 Turbulent Flow through a Circular Pipe

For 'fully developed' turbulent flow through a pipe, the following correlation suggested by Dittus and Boelter (1930) is used widely.

$$\text{Nu} = 0.023 \, (\text{Re})^{0.8} \, (\text{Pr})^n \qquad (4.9)$$

---

* Such a fluid, for example, the solution of a protein, may deteriorate if it is subjected to a high velocity or a high shear stress.

where

$n = 0.4$ for heating $(T_w > T)$

$n = 0.3$ for cooling $(T_w < T)$.

The conditions for applicability of this correlation are: (a) $0.7 \leq \Pr \leq 160$; (b) $d/L < 0.1$; and (c) $\text{Re} \geq 10,000$.

The above correlation, Eq. (4.9), is applicable for moderate values of the temperature difference betwen the wall and the bulk, $T_w - T$. The fluid properties are evaluated at the arithmetic mean of the bulk temperatures (i.e. the average of the inlet and the outlet temperatures of the fluid). The maximum error in the predicted value of the heat transfer coefficient does not exceed 25%.

When the temperature difference between the wall and the bulk is substantial, its effect on the fluid properties, particularly on viscosity, needs to be taken into account. For such cases, the Sieder-Tate equation (1936) is recommended.

$$\text{Nu} = 0.027 \, (\text{Re})^{0.8} \, (\text{Pr})^{0.33} \, (\mu/\mu_w)^{0.14} \tag{4.10}$$

Here μ/μ_w is the viscosity correction factor that has to be used when the wall viscosity at the wall temperature, μ_w, is substantially different from that, μ, at the bulk temperature. The conditions of applicability are: (a) $0.7 \leq \Pr \leq 16,700$; (b) $\text{Re} \geq 10,000$; and (c) $d/L \leq 0.1$. The bulk fluid properties have to be evaluated at the arithmetic mean bulk temperature.

Whitaker (1972) suggested the following equation which is based on the experimental data of a large number of researchers.

$$\text{Nu} = 0.015 \, (\text{Re})^{0.83} \, (\text{Pr})^{0.42} \, (\mu/\mu_w)^{0.14} \tag{4.11}$$

In the case of heat transfer to or from a liquid metal, flowing through a pipe, at a constant wall temperature, the correlation of Seban and Shimazaki (1951) may be used.

$$\text{Nu} = 5.0 + 0.025 \, (\text{Pe})^{0.8} \tag{4.12}$$

Equation (4.12) is applicable when $\text{Pe} > 100$ and $L/d > 60$. All the properties of the liquid metal should be evaluated at the mean bulk temperature.

4.5.3 Flow through a Non-circular Duct

Ducts of non-circular cross-sections, rectangular or square, are often used in the industrial and other applications. Flow through the annulus of a double-pipe heat exchanger, flow through the shell side of a shell-and-tube heat exchanger and flow of hot exhaust gases through rectangular or square ducts are a few common examples. The foregoing equations for estimation of the Nusselt number are generally applicable for heat transfer calculations for flow through non-circular ducts, but the *equivalent diameter* of the duct is to be used in the calculation of the Reynolds number. The equivalent diameter d_e is four times the *hydraulic radius*, r_H. The hydraulic radius is defined as

$$r_H = \frac{\text{Cross-sectional area of flow}}{\text{Wetted perimeter}}$$

and

$$d_e = 4r_H \tag{4.13a}$$

For example, in the case of flow through a rectangular duct of sides l_1 and l_2, the equivalent diameter is

$$d_e = \frac{4l_1l_2}{2(l_1 + l_2)}$$

It should be noted that the calculation of the wetted perimeter of a duct in the case of heat transfer calculation may be different from that used in the case of determination of pressure drop (Kern, 1950; Ludwig, 1983). To illustrate this point, we consider the flow through the annulus between two concentric tubes in a double-pipe heat exchanger (the construction of this kind of heat exchanger will be described later). Let the outside diameter of the inner pipe be d_1, and the inside diameter of the outer pipe be d_2. If our objective is to calculate the pressure drop for flow through this annulus, the wetted perimeter is $\pi(d_1 + d_2)$, because both the walls forming the annulus contribute to the frictional pressure drop. But if we consider heat transfer from a hot fluid flowing through the inner pipe to a cold fluid flowing through the annulus, *wetting* of the outer wall of the inner pipe only becomes relevant to the heat transfer coefficient in the annulus. So, here the wetted perimeter is simply πd_1. The equivalent diameters of the annulus in the respective cases are:

For pressure drop calculation,

$$d_{e,f} = 4\frac{(\pi/4)(d_2^2 - d_1^2)}{\pi(d_1 + d_2)} = d_2 - d_1 \tag{4.13b}$$

For heat transfer calculation,

$$d_{e,h} = 4\frac{(\pi/4)(d_2^2 - d_1^2)}{\pi d_1} = \frac{1}{d_1}(d_2^2 - d_1^2) \tag{4.13c}$$

4.6 CORRELATIONS FOR THE HEAT TRANSFER COEFFICIENT—EXTERNAL FLOWS

Convective heat transfer in external flows is encountered in many practical situations. Typical cases are flow over a flat surface, along or across a single pipe or a pipe bundle, over a sphere or a bed of spheres (e.g. convective heat transport from a wall, heating of air by a bundle of steam pipes, heat transfer in a packed bed of catalyst or that in a fluidized bed, etc.). Several important correlations that cover more common cases of external flows are discussed. below.

4.6.1 Flow over a Flat Plate

Heat transfer in flow over a flat plate occurs through the boundary layer formed on the plate. A physical description of the boundary layer including the laminar, transition and turbulent zones has been given in Section 4.1. Exact mathematical solutions for momentum and heat transfer in the laminar boundary layer and the analysis of transport in the turbulent boundary are available in the literature. These will be briefly described in Chapter 11. The following correlations are suggested for heat transfer calculations in the boundary layer flow.

For *heat transfer in laminar boundary layer flow,* the following correlation for the local Nusselt number can be obtained from an approximate solution of the boundary equations of momentum and energy. It depends upon the distance from the leading edge of the plate.

$$\mathrm{Nu}_x = 0.332\,(\mathrm{Re}_x)^{1/2}\,(\mathrm{Pr})^{1/3} \qquad (0.5 \le \mathrm{Pr} \le 50) \tag{4.14a}$$

where

Nu_x is the local Nusselt number $= x\, h_x/k$

Re_x is the local Reynolds number $= xV\rho/\mu$

Here x is the distance from the leading edge of the plate, and h_x is the local heat transfer coefficient. An average value of the heat transfer coefficient over a distance L may be obtained as

$$h_{\text{av}} = \frac{1}{L}\int_0^L h_x\, dx$$

or

$$\text{Nu}_L = 0.664\,(\text{Re}_L)^{1/2}\,(\text{Pr})^{1/3} \qquad (4.14\text{b})$$

For *heat transfer in turbulent boundary layer flow*, a simple correlation in common use is

$$\text{Nu}_x = 0.0296\,(\text{Re}_x)^{4/5}\,(\text{Pr})^{1/3}; \qquad 5\times10^5 < \text{Re}_x < 10^7, \quad 0.6 < \text{Pr} < 60 \qquad (4.15)$$

The correlation (4.15) was developed by using the experimental friction factor data for turbulent boundary layer flow past a flat plate in conjunction with the 'Colburn Analogy'.

Example 4.2   Air flows over a flat surface, 2 m in length oriented in the direction of flow and of sufficient breadth, maintained at 150°C. The pressure is 1 atm and the bulk air temperature is 30°C. If the air velocity is 12 m/s, determine (a) the local heat transfer coefficient as a function of longitudinal position, (b) the average heat transfer coefficient, and (c) the rate of heat loss from the surface.

SOLUTION   The local heat transfer coefficient, which varies with the distance from the leading edge, can be calculated by using a suitable correlation. The maximum Re has to be calculated before selecting the correlation to be used. The calculation is based on 1 m *breadth* of the plate.

Data: The relevant physical properties of air are taken at the mean 'film temperature', that is, at $(150 + 30)/2 = 90$°C. These are: $\rho = 0.962$ kg/m^3, $\mu = 2.131 \times 10^{-5}$ kg/m s; $k = 0.031$ W/m °C, $c_p = 1.01$ kJ/kg °C.

Length of the plate, $L = 2$ m; free stream velocity of air, $V = 12$ m/s

$$\text{Pr} = \frac{c_p\mu}{k} = \frac{(1010)(2.131 \times 10^{-5})}{0.031} = 0.694$$

Maximum Reynolds number of air, $\text{Re} = \dfrac{LV\rho}{\mu} = \dfrac{(2)(12)(0.962)}{2.131 \times 10^{-5}} = 1.083 \times 10^6$

Correlation to be used: The maximum Reynolds number at the end of the plate (at $x = 2$ m) calculated above indicates that over a portion of the plate near its end the flow is turbulent (transition to turbulent flow in the boundary layer over a flat plate occurs at a Reynolds number of about 5×10^5; see Section 4.1.1). Therefore, the heat transfer correlations for laminar and turbulent flows [Eqs. (4.14) and (4.15)] over the appropriate regions of the plate have to be used, and the 'local heat transfer coefficient' should be *averaged* over the respective regions.

Numerical calculations: We should first determine the distance from the leading edge (say L') where the laminar boundary layer flow ends and turbulence sets in. We put

$$5 \times 10^5 = \frac{L'V\rho}{\mu} = \frac{(L')(12)(0.962)}{2.131 \times 10^{-5}}$$

Therefore,

$$L' = 0.923 \text{ m}$$

We denote the local heat transfer coefficients by h' and h over the laminar and the turbulent regions, respectively, and x is the local distance from the leading edge.

For the laminar region we use Eq. (4.14a),

$$\frac{h'x}{k} = 0.332\left(\frac{xV\rho}{\mu}\right)^{1/2} (\text{Pr})^{1/3}$$

Putting the values of various quantities, we get

$$h' = \boxed{6.707x^{-1/2} \text{ W/m}^2 \text{ }^\circ\text{C}}; \qquad x \le L' (= 0.923 \text{ m})$$

(a) For the turbulent zone we use Eq. (4.15),

$$\frac{hx}{k} = 0.0296\left[\frac{xV\rho}{\mu}\right]^{4/5} (\text{Pr})^{1/3}$$

Putting the values of the different quantities, we get

$$h = \boxed{31.4\, x^{-1/5} \text{ W/m}^2 \text{ }^\circ\text{C}}; \qquad 0.923 < x \le L(= 2 \text{ m})$$

(b) The average heat transfer coefficient over the length L (= 2 m) of the plate is given by

$$h_{\text{av}} = \frac{1}{L}\left[\int_0^{L'} h'dx + \int_{L'}^{L} hdx\right]$$

$$= \frac{1}{2}\left[\int_0^{0.923} 6.707x^{-1/2}dx + \int_{0.923}^{2} 31.4x^{-1/5}dx\right]$$

$$= \frac{1}{2}\left\{(6.707)(2)\left[x^{1/2}\right]_0^{0.923} + (31.4)(5/4)\left[x^{4/5}\right]_{0.923}^{2}\right\}$$

$$= \boxed{22.2 \text{ W/m}^2 \text{ }^\circ\text{C}}$$

(c) The rate of heat loss from the surface (for 1 m breadth of the plate; area, A = 2 m $\times$ 1 m = 2 m^2)

$$Q = (h_{\text{av}})(\text{area})(\Delta T) = (22.2)(2)(150 - 30) = \boxed{5329 \text{ W}}$$

4.6.2 Flow across a Cylinder

Quite a few correlations for the average Nusselt number are available for the case of convective heat transfer in flow across a cylinder. For cross flow of a liquid over a cylinder, the heat transfer coefficient can be calculated by using:

(a) The correlation proposed by Fand (1965)

$$\text{Nu} = \frac{hd}{k} = \left[0.35 + 0.56(\text{Re})^{052}\right](\text{Pr})^{0.3}; \qquad \text{Re} = \frac{dV\rho}{\mu}; \quad 0.1 < \text{Re} < 10^5 \qquad (4.16)$$

(b) The correlations of Eckert and Drake (1972)

$$\text{Nu} = \left[0.43 + 0.50(\text{Re})^{05}\right](\text{Pr})^{0.38}\left(\text{Pr}/\text{Pr}_w\right)^{0.25}; \qquad 1 < \text{Re} < 10^3 \qquad (4.17)$$

$$\text{Nu} = 0.25(\text{Re})^{0.6}(\text{Pr})^{0.38}(\text{Pr}/\text{Pr}_w)^{0.25}; \qquad 10^3 < \text{Re} < 2 \times 10^5 \qquad (4.18)$$

where Pr_w is the Prandtl number at the wall temperature. These correlations are applicable to both gases and liquids.

(c) The correlation of Churchill and Burnstein (1977)

$$\text{Nu} = 0.3 + \frac{0.62(\text{Re})^{1/2}(\text{Pr})^{1/3}}{\left[1 + (0.4/\text{Pr})^{2/3}\right]^{1/4}}\left[1 + \left(\frac{\text{Re}}{2.82 \times 10^5}\right)^{5/8}\right]^{4/5} \qquad (4.19)$$

$$\text{for } 10^2 < \text{Re} < 10^7; \qquad (\text{Re})(\text{Pr}) > 0.2$$

The above correlation is based on the data for flow of air, water and liquid sodium. In the intermediate range of Reynolds number, however, the following correlation predicts the heat transfer coefficient more accurately.

$$\text{Nu} = 0.3 + \frac{0.62(\text{Re})^{1/2}(\text{Pr})^{1/3}}{\left[1 + (0.4/\text{Pr})^{2/3}\right]^{1/4}}\left[1 + \left(\frac{\text{Re}}{2.82 \times 10^5}\right)^{1/2}\right] \qquad (4.20)$$

$$\text{for } 2 \times 10^4 < \text{Re} < 4 \times 10^5; \qquad (\text{Re})(\text{Pr}) > 0.2$$

(d) Whitaker (1972) suggested the following correlation that involves a maximum error of $\pm 25\%$.

$$\text{Nu} = [0.40\,(\text{Re})^{1/2} + 0.06\,(\text{Re})^{2/3}]\,(\text{Pr})^{0.4}\,(\mu/\mu_w)^{0.25} \qquad (4.21)$$

$$\text{for } 40 < \text{Re} < 105, \qquad 0.65 < \text{Pr} < 300 \text{ and } 0.25 < \mu/\mu_w < 5.2$$

Example 4.3   A 0.724 mm diameter nichrome heating wire carries a current of 8.3 amperes. The wire has an electrical resistance of 2.625 ohm/m. Air flows across the heated wire at a velocity of 10 m/s. If the bulk air temperature is 27°C and the pressure is essentially atmospheric, what will be the temperature of the wire at steady state?

SOLUTION   At steady state, the rate of electrical heat generation should be equal to the rate of heat loss from the wire by forced convection. The temperature of the wire can be calculated from this heat balance. The heat transfer coefficient for forced convection at the surface of the wire can be calculated by using a suitable correlation for a cylinder under cross flow. The length of the wire is not needed for the calculation, and a 1 m section will be used as the basis. The wire temperature will be assumed to remain uniform.

Because the temperature of the wire is expected to be considerably higher than the bulk temperature of air, the physical properties of air should be taken at the mean temperature. But the temperature of the wire is yet unknown. Therefore, to start with, we take the values of the properties at the bulk air temperature. The relevant properties of air at the bulk temperature of 27°C (300 K) and 1 atm pressure are: $\rho = 1.1774$ kg/m^3; $c_p = 1.0057$ kJ/kg °C; $\mu = 1.983 \times 10^{-5}$ kg/m s; $k = 0.02624$ W/m °C; air velocity, $V = 10$ m/s.

Also, wire diameter, $d = 7.24 \times 10^{-4}$ m; surface area of 1 m wire = $(\pi)(7.24 \times 10^{-4})(1)$ = 2.274×10^{-3} m^2; current flowing, $I = 8.3$ amp; electrical resistance, $R = 2.625$ ohm.

Numerical calculations

$$\text{Re} = \frac{dV\rho}{\mu} = \frac{(7.24 \times 10^{-4})(10)(1.1774)}{1.983 \times 10^{-5}} = 430$$

$$\text{Pr} = \frac{c_p\mu}{k} = \frac{(1.0057 \times 1000)(1.983 \times 10^{-5})}{0.02624} = 0.76$$

We will use the Churchill–Burnstein correlation, Eq.(4.19), to calculate the heat transfer coefficient. Putting the values of Re and Pr in Eq. (4.19), we have

$$\text{Nu} = 0.3 + \frac{0.62(430)^{1/2}(0.76)^{1/3}}{\left[1 + (0.4/0.76)^{2/3}\right]^{1/4}} \left[1 + \left(\frac{430}{2.82 \times 10^5}\right)^{5/8}\right]^{4/5} = 10.8 = \frac{hd}{k}$$

Therefore,

$$h = 10.8 \left(\frac{k}{d}\right) = \frac{(10.8)(0.02624)}{7.24 \times 10^{-4}} = 391.4 \text{ W/m}^2 \text{ °C}$$

Rate of electrical heat generation in 1 m length of the wire

$$= I^2R = (8.3)^2(2.625) = 180.83 \text{ W } (= Q)$$

The heat generated is continuously removed by convection at steady state. That is,

$$Q = hA\Delta T$$

or

$$\Delta T = \frac{Q}{hA} = \frac{180.83}{(2.274 \times 10^{-3})(391.4)} = 203°C \ (= T_{\text{wire}} - T_{\text{air}})$$

Therefore, the steady state wire temperature = 203 + 27 = $\boxed{230°C}$

Once the wire temperature is calculated, we can take the values of the physical properties at the mean 'air-film' temperature [i.e. (230 + 27)/2 = 129°C] and repeat the calculations in order to get a more accurate estimate of the wire temperature.

The values of the properties of air at 129°C are: $\rho = 0.878$ kg/m^3; $c_p = 1.014$ kJ/kg °C; $\mu = 2.30 \times 10^{-5}$ kg/m s; $k = 0.0338$ W/m °C.

Therefore,

$$\text{Re} = \frac{(7.24 \times 10^{-4})(10)(0.878)}{2.30 \times 10^{-5}} = 276; \qquad \text{Pr} = \frac{(1014)(2.30 \times 10^{-5})}{0.0338} = 0.714$$

Putting the recalculated values of Re and Pr into the above correlation, we get

$$\text{Nu} = 8.47$$

and

$$h = \frac{(8.47)(0.0338)}{7.24 \times 10^{-4}} = 395 \text{ W/m}^2 \text{ °C}$$

$$\text{The revised } \Delta T = \frac{180.83}{395(2.274 \times 10^{-3})} = 201°C$$

The wire temperature, $T = 201 + 27 = 228°C$

Note: Although the recalculated value of the Reynolds number is considerably less than that calculated first, the change in the recalculated Nu is not large. Therefore, no appreciable improvement in accuracy is expected from a third trial.

4.6.3 Flow past a Sphere

The following important correlations are available for the case of heat transfer in flow over a single sphere:

(a) For flow of liquids past a sphere, Kramers (1946) proposed the following correlation:

$$\text{Nu (Pr)}^{-0.3} = 0.97 + 0.68 \text{ (Re)}^{0.5}; \qquad 1 < \text{Re} < 2000 \tag{4.22}$$

(b) Whitaker (1972) suggested the following correlation:

$$\text{Nu} = 2 + [0.4 \text{ (Re)}^{1/2} + 0.06 \text{ (Re)}^{2/3}] \text{ (Pr)}^{0.4} \ (\mu/\mu_w)^{0.25} \tag{4.23}$$

where μ is the viscosity of the fluid at the bulk temperature, and μ_w is that at the wall temperature. This correlation is applicable to both gases and liquids, and the error in the prediction remains within $\pm 30\%$. The number '2' arises out of the contribution of conduction only when there is no motion in the medium, i.e. when Re = 0 (see Problem 4.19). The Reynolds number is based on the diameter of the sphere.

Example 4.4 A ball of ice, 4 cm in diameter, at 0°C is suspended in a dry air stream at 25°C which is flowing at a velocity of 2 m/s. (a) What is the initial rate of melting of the ice? (b) How much time would be needed to melt away 50% of the ice? Assume that the shape of the ice-ball remains spherical all the time. Will the rate of melting be affected by the presence of moisture in air?

Given: heat of fusion of ice, $\lambda = 334$ kJ/kg; for air at 12.5°C, which is the mean air-film temperature, $\rho = 1.248$ kg/m³; $k = 0.026$ W/m °C; $\mu = 1.69 \times 10^{-5}$ kg/m s; $c_p = 1.005$ kJ/kg °C; density of ice, $\rho_{ice} = 920$ kg/m³.

SOLUTION The ice is at its melting point, and its temperature will remain constant at 0°C. The rate of melting will depend upon the air-film coefficient h. This quantity h depends upon the

Reynolds number, Re, and therefore on the diameter of the ice-ball. The heat transfer area (i.e. the surface area of the ball) also decreases with time. We assume that (a) the shape of the ball remains spherical at all time, (b) the water produced on melting drains out quickly, and the thickness of the water film on the ice-ball, and therefore the heat transfer resistance offered by it, is negligible, and (c) the heat transfer resistance of the air-film controls the rate of heat transfer.

Correlation to be used: We use the Whitaker correlation, Eq. (4.23), without viscosity correction (because the change in the viscosity over the temperature range of 0–25°C is small).

$$\text{Nu} = 2 + [0.4\,(\text{Re})^{1/2} + 0.06\,(\text{Re})^{2/3}]\,(\text{Pr})^{0.4}$$

Given: initial diameter of the ice-ball, $d_i = 4$ cm $= 0.04$ m; initial radius, $r_i = 0.02$ m; air velocity, $V = 2$ m/s.

(a) *Initial rate of melting of ice*

$$\text{Re} = \frac{d_i V \rho}{\mu} = \frac{(0.04)(2)(1.248)}{1.69 \times 10^{-5}} = 5908$$

$$\text{Pr} = \frac{c_p \mu}{k} = \frac{(1005)(1.69 \times 10^{-5})}{0.026} = 0.653$$

Therefore,

$$\text{Nu} = 2 + [0.4\,(5908)^{1/2} + (0.06)(5908)^{2/3}]\,(0.653)^{0.4} = 44.45$$

$$h = \text{Nu}\left(\frac{k}{d}\right) = 44.45\left(\frac{0.026}{0.04}\right) = 28.9 \text{ W/m}^2 \text{ °C}$$

Area of heat transfer = initial area of the sphere, $A_i = p(0.04)^2 = 5.026 \times 10^{-3}$ m^2

Initial rate of heat transfer,

$$Q_i = h\,A_i\,\Delta T = (28.9)(5.026 \times 10^{-3})(25 - 0) = 3.632 \text{ W} = 3.632 \text{ J/s}$$

Initial rate of melting of ice,

$$\frac{Q_i}{\lambda} = \frac{3.632 \text{ J/s}}{334 \text{ J/g}} = \boxed{0.0109 \text{ g/s}}$$

(b) *Time required for melting away 50% of the ice*

Let the radius of the ice-ball at any time t be r. Thus,

$$\text{Mass of ice in the ball} = \left(\frac{4}{3}\right)\pi\,r^3 \rho_{\text{ice}}$$

$$\text{Rate of melting = Rate of change of mass of the ball} = -\frac{d}{dt}\left(\frac{4}{3}\pi r^3\,\rho_{\text{ice}}\right) = -4\pi r^2\,\rho_{\text{ice}}\,\frac{dr}{dt}$$

$$\text{Rate of heat input required for melting} = -\lambda\left(4\pi r^2\,\rho_{\text{ice}}\,\frac{dr}{dt}\right)$$

$$\text{Area of the ice-ball at the instant } t = 4\pi r^2$$

If h be the heat transfer coefficient (which is a function of the radius, r), we can write the following heat balance equation

$$-\lambda\left(4\pi r^2\, \rho_{\text{ice}}\,\frac{dr}{dt}\right)= h\,(4\pi r^2)\,(\Delta T)$$

If t' be the time required for change of the radius of the ice-ball from r_i to r_f, integration of the above equation gives

$$-\int_{r=r_i}^{r_f}\frac{dr}{h} = \frac{\Delta T}{\lambda\rho_{\text{ice}}}\int_0^{t'} dt \qquad\qquad\text{(i)}$$

Now we express the instantaneous heat transfer coefficient h as a function of the radius r.

Reynolds number, $\text{Re} = \dfrac{2rV\rho}{\mu} = \dfrac{(2r)(2)(1.248)}{1.69 \times 10^{-5}} = (2.954 \times 10^5)\, r$

Nusselt number,

$$\text{Nu} = \frac{(h)(2r)}{k}$$

$$= 2 + [(0.4)(2.954 \times 10^5\ r)^{1/2} + (0.06)(2.954 \times 10^5\ r)^{2/3}](0.653)^{0.4}$$

or

$$h = \frac{0.026}{r} + 2.383\ (r^{-1/2} + 1.223\ r^{-1/3}) \qquad\qquad\text{(ii)}$$

Substituting (ii) in (i), we have

$$\int_{r=r_f}^{r_i}\frac{dr}{\dfrac{0.026}{r} + 2.383\left(r^{-1/2} + 1.223\,r^{-1/3}\right)} = \frac{(25-0)}{(920)(3.34 \times 10^5)}\,t' \qquad\qquad\text{(iii)}$$

The initial radius of the ice-ball, $r_i = 0.02$ m; the final radius (after 50% of the ice has melted away)

$$r_f = \left(\frac{r_i^3}{2}\right)^{1/3} = \left(\frac{(0.02)^3}{2}\right)^{1/3} = 0.01588 \text{ m}$$

Equation (iii) can be integrated numerically (by using Simpson's rule, for example). On integration, we get

$$1.355 \times 10^{-4} = 8.136 \times 10^{-8}\ t'$$

or

$$t' = \frac{1.355 \times 10^{-4}}{8.136 \times 10^{-8}} = 1665 \text{ s}$$

The time needed to melt away 50% of the ice in the ball = 1665 s = $\boxed{27 \text{ min } 45\text{ s}}$

If the air is not dry, the moisture of the air will diffuse through the air-film and condense on the surface of the ice-ball. The latent heat of condensation of moisture will significantly increase the rate of melting of ice. In such a case, the problem will involve 'simultaneous heat and mass transfer'.

4.6.4 Flow across a Bank of Tubes

Cross flow of a fluid, particularly of a gas like air, over a tube bank—either aligned or staggered—is important in many heat transfer applications and equipment. Typical applications are waste heat recovery from hot flue gases, flow of hot combustion gases across the tube bank in a water-tube boiler, air flow over tubes containing refrigerants in an air-cooler, etc. The cross-sections of aligned and staggered tube banks are shown in Figs. 4.4 and 4.5 with a few useful notations (S_T = transverse pitch, S_L = longitudinal pitch).

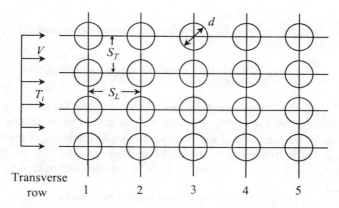

Fig. 4.4   Flow across a tube bank (aligned).

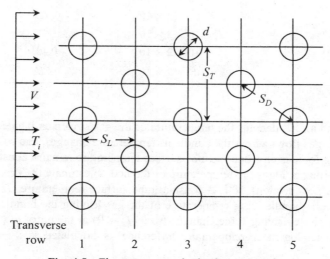

Fig. 4.5   Flow across a tube bank (staggered).

A good deal of research work has been done on heat transfer from a tube bank in cross flow. The following correlation for Nusselt number proposed by Zhukauskas (1972) is based on a large volume of experimental data.

$$\text{Nu} = \frac{\overline{h}\,d}{k} = C(\text{Re}_{d,\,max})^m \,(\text{Pr})^{0.36} \left(\text{Pr}/\text{Pr}_w\right)^{0.25} \tag{4.24}$$

where

$\overline{h}$ is the mean heat transfer coefficient
$\text{Re}_{d,\,max} = (d\rho V_{max})/\mu$
Pr is the bulk Prandtl number
Pr_w is the wall Prandtl number

and

$$V_{max} = \frac{S_T}{S_T - d}\,V \qquad \text{(for the aligned arrangement)} \tag{4.25}$$

$$= \frac{S_T}{2(S_D - d)}\,V \qquad \text{(for the staggered arrangement)} \tag{4.26}$$

However, if the V_{max} value calculated by using Eq. (4.26) for the staggered arrangement is larger than that calculated by Eq. (4.25), the former value should be used. The correlation (4.24) gives very good prediction when the number of tube rows in the bank, $N \geq 20$, $0.7 < \text{Pr} < 500$, and $10^3 < (\text{Re})_{d,\,max} < 2 \times 10^6$. However, it can be used even when $N < 20$. If $N = 4$, the error in prediction is around 25%. The constants C and m in the above correlation (4.24) are given in

Table 4.3 Constants C and m of correlation (4.24)

Configuration of bundle	$\text{Re}_{d,\,max}$	C	m
Aligned	10^3–(2×10^5)	0.27	0.63
Aligned	(2×10^5)–(2×10^6)	0.021	0.84
Staggered $[S_T/S_L < 2]$	10^3–(2×10^5)	$0.35(S_T/S_L)^{0.2}$	0.60
Staggered $[S_T/S_L > 2]$	10^3–(2×10^5)	0.40	0.60
Staggered	(2×10^5)–(2×10^6)	0.022	0.84

Table 4.3.

In Figs. 4.4 and 4.5, T_i denotes the temperature of the gas stream as it enters the tube bundle. However, when the gas flows across the bundle its temperature changes, and so does the gas phase temperature driving force even when the tube surface temperature remains constant. In such a case it is necessary to define a 'log mean temperature difference' to estimate the driving force (this will be discussed in the next section). If T_s = constant tube surface temperature, T_i = gas temperature at the inlet to the tube bundle, T_o = temperature of the gas leaving the bundle, then, for cooling of the gas, the gas phase driving force changes from $(T_i - T_s)$ at the entry to $(T_o - T_s)$ at the exit from the bundle. The 'log mean temperature difference' is calculated as

$$\Delta T_m = \frac{(T_i - T_s) - (T_o - T_s)}{\ln\left[(T_i - T_s)/(T_o - T_s)\right]} \tag{4.27}$$

While the inlet temperature T_i is known, the exit temperature T_o can be calculated by using Eq. (4.28). (Derivation of this equation is left as an exercise.)

$$\frac{T_o - T_s}{T_i - T_s} = \exp\left[-\frac{\pi dN\overline{h}}{\rho V N_T S_T c_p}\right] \tag{4.28}$$

where N is the total number of tubes and N_T is the number of tubes in a transverse row.

Charts and correlations for the calculation of friction factor and pressure drop for flow across a tube bundle are given in Zhukauskas (1972).

4.6.5 Heat Transfer Coefficient in a Packed and a Fluidized Bed

Heat transfer to or from a gas flowing through a packed bed of solid is important in various industrial applications. For example, in a gas-solid catalytic reactor heat generated by an exothermic chemical reaction inside the catalyst pellets is conducted through the pellets and then convected to the reaction mixture flowing through the catalyst bed. While drying a moist solid by passing a hot gas through it, heat transfer from the bulk gas to the surface of the solid occurs by forced convection through the packed bed of the material to be dried. A packed bed of inert solid is often used for the recovery of waste heat from a hot gas stream. The heated bed serves as a storage of thermal energy. The following heat transfer correlation for gas flow through a packed bed is useful.

$$\varepsilon \ (St) \ (Pr)^{2/3} = 2.06 \ (Re)_{d_p}^{-0.575}; \qquad Pr \approx 0.7, \quad 90 \leq (Re)_{d_p} \leq 4000 \tag{4.29a}$$

where

$St = Nu/(Re)(Pr)$ is the Stanton number

$(Re)_{d_p} = (d_p V_o \rho)/\mu$ is the particle Reynolds number

d_p is the diameter or the 'effective diameter' of a particle

V_o is the superficial fluid velocity (i.e. the fluid velocity based on the cross-section of the bed).

The bed porosity or void fraction is denoted by ε. Typically, ε varies from 0.3 to 0.5 (theoretically, $\varepsilon = 0.69$ for uniform-sized spheres, 0.71 for a bed of cubes and 0.79 for a bed of cylinders having length equal to the diameter). The group $(St)(Pr)^{2/3}$ is called the Colburn factor (see Section 4.8).

Heat transfer between the fluid and the solid in a fluidized bed is also very important in industrial applications such as fluidized bed dryers, reactors, etc. The following equation may be used for the estimation of heat transfer coefficient to or from particles in a fluidized bed.

$$\frac{h d_p}{k} = 2 + 0.6\left(\frac{d_p V_o \rho}{\mu}\right)^{1/2}\left(\frac{c_p \mu}{k}\right)^{1/3} \tag{4.29b}$$

Example 4.5 A stream of solid particles at 800°C is to be cooled to 550°C in a fluidized bed. Air is the fluidizing medium, and it has a mean temperature of 500°C in the bed. The pressure in the bed is maintained at 1.2 atm absolute and the superficial air velocity is 0.5 m/s. Calculate the average time of contact between the solid and the gas required for the desired cooling of the solid. Assume that the gas film heat transfer resistance controls.

The following data are given. Solid particles: average particle diameter, $d_p = 0.65$ mm; specific heat, $c_{ps} = 0.196$ kcal/kg °C; density, $\rho_s = 2550$ kg/m^3. Air at 550°C: $\rho = 0.545$ kg/m^3; $c_p = 0.263$ kcal/kg °C; $\mu = 3.6 \times 10^{-5}$ kg/m s; $k = 0.05$ kcal/h m °C.

SOLUTION   It is assumed that (i) the temperature inside a particle remains uniform at any time, and (ii) the gas film heat transfer resistance controls. Assumption (i) is quite reasonable if the solid has a small particle size, but its thermal conductivity is not too small.

If T_s is the temperature of a solid particle at any time t, T_g (= 500°C) the gas temperature and h the heat transfer coefficient between the gas and the solid particles, we can write the following heat balance equation.

$$-\frac{d}{dt}\left(\frac{1}{6}\pi d_p^3 \rho_s c_{ps} T_s\right) = \pi d_p^2 h (T_s - T_g)$$

or

$$-\frac{dT_s}{dt} = \frac{6h}{d_p \rho_s c_{ps}}(T_s - T_g)$$

At time $t = 0$, the temperature of the solid entering the bed, $T_{si} = 800°C$, and at time $t = t$, the temperature of the solid leaving the bed, $T_{so} = 550°C$. The above equation can now be integrated to the following form.

$$\ln \frac{T_{si} - T_g}{T_{so} - T_g} = \frac{6h}{d_p \rho_s c_{ps}} t \tag{i}$$

Calculation of the heat transfer coefficient, h

$$\text{Reynolds number, } \text{Re}_{d_p} = \frac{d_p V_o \rho}{\mu} = \frac{(6.5 \times 10^{-4})(0.5)(0.545)}{3.6 \times 10^{-5}} = 4.92$$

$$\text{Prandtl number, } \text{Pr} = \frac{c_p \mu}{k} = \frac{(0.263)(3.6 \times 10^{-5})(3600)}{0.05} = 0.69$$

From Eq. (4.29b), the heat transfer coefficient is

$$h = \left[\frac{0.05}{6.5 \times 10^{-4}}\right][2 + (0.6)(4.92)^{1/2}(0.69)^{1/3}] = 244 \text{ kcal/h m}^2 \text{ °C}$$

Putting the values of the various quantities in Eq. (i), the required contact time is

$$t = \frac{(6.5 \times 10^{-4})(2550)(0.196)}{(6)(244)} \ln \frac{800 - 500}{550 - 500} = 3.97 \times 10^{-4} \text{ h} = \boxed{1.43 \text{ s}}$$

4.7   HEAT TRANSFER WITH A VARIABLE DRIVING FORCE—COCURRENT AND COUNTERCURRENT OPERATIONS

The calculation of the rate of heat transfer becomes very simple if the overall heat transfer coefficient, the area of heat transfer and the temperature driving force all remain constant. But in many cases, one or more of these quantities may vary. Examples of heat transport through a variable area (in the case of radial transport through a cylinder or a sphere) or with a variable heat transfer coefficient (in the case of boundary layer heat transfer) have been discussed. However, variation of the temperature driving force with position in a heat transfer equipment or device is more common. In order to explain this, we will consider a very simple type of heat transfer device—the double-pipe heat exchanger.

The schematic of a double-pipe heat exchanger is shown in Fig. 4.6 (cocurrent) and Fig. 4.7 (countercurrent). The device consists of two concentric pipes welded at the ends and provided

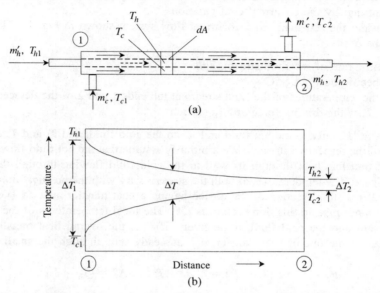

Fig. 4.6 (a) Double-pipe heat exchanger in parallel or cocurrent flow and (b) temperature distribution.

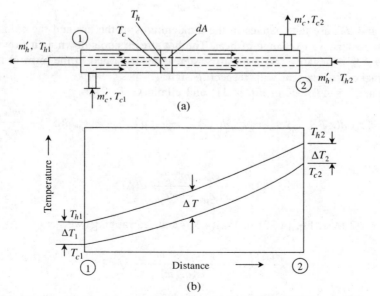

Fig. 4.7 (a) Double-pipe heat exchanger in countercurrent flow and (b) temperature distribution.

with flow nozzles so that a hot and a cold fluid can be brought in thermal contact in order to effect transfer of heat (see Chapter 8 for the details). One of the fluids flows through the pipe and the other through the annulus. If both the streams, the hot and the cold, flow in the same direction, the flow arrangement is called *parallel* or *cocurrent.* If the hot and the cold streams flow in the opposite directions, the arrangement is called *countercurrent* or *counterflow.* The countercurrent flow arrangement is more common because it allows exchange of more amount of heat between the streams than that by the cocurrent configuration.

Let us consider the schematic of a cocurrent flow system shown in Fig. 4.6. The following notations will be used:

m' is the flow rate of the fluid stream, kg/s
c_p is the specific heat, J/kg °C
T_1, T_2 are the temperatures of the fluid stream at the ends 1 and 2 of the device, respectively
$\Delta T = T_h - T_c$ = the driving force at any section.

The subscript 'h' refer to the hot fluid and 'c' to the cold fluid, and T_h and T_c are the 'local temperatures' of the respective streams. We arbitrarily assume that the hot fluid flows through the inner pipe and transfers heat through its wall to the cold fluid flowing through the annulus as shown in the figure. The temperatures of both the streams vary with the position along the device.

Now consider a 'thin' section of the device having a heat transfer area dA (i.e. the area of the wall of the inner pipe in this thin section is dA). The local temperatures of the fluids are T_h and T_c for the hot and the cold fluid, respectively. If U is the overall heat transfer coefficient (assumed constant), the rate of heat transfer, dQ, at steady state through the small area dA may be written as

$$dQ = U \, dA \, (T_h - T_c) = U \, dA \, \Delta T; \qquad \Delta T = (T_h - T_c) \tag{4.30}$$

Also, a heat balance over the thin section gives

$$dQ = -m'_h \, c_{ph} \, dT_h = m'_c \, c_{pc} \, dT_c \tag{4.31}$$

Here, dT_h and dT_c are the changes in the temperatures of the hot and the cold streams over the thin section because of exchange of heat. The hot stream cools down as it flows through the pipe and hence dT_h is negative. So a negative sign is incorporated in the term in the middle of Eq. (4.31) to make it consistent with respect to sign.

Next we put $T_h = \Delta T + T_c$ in Eq. (4.31) and eliminate T_c. Thus

$$dQ = -m'_h c_{ph} \, d(\Delta T + T_c) = -m'_h c_{ph} \, d(\Delta T) - m'_h c_{ph} \, dT_c = -m'_h c_{ph} \, d(\Delta T) - \frac{m'_h c_{ph}}{m'_c c_{pc}} \, dQ$$

Therefore,

$$dQ = -\frac{m'_h c_{ph} m'_c c_{pc}}{m'_h c_{ph} + m'_c c_{pc}} \, d(\Delta T) \tag{4.32}$$

Substituting for dQ from Eq. (4.32) in Eq. (4.30) and rearranging, we get

$$\frac{d(\Delta T)}{\Delta T} = -\frac{m'_h c_{ph} + m'_c c_{pc}}{m'_h c_{ph} m'_c c_{pc}} \, U \, dA \tag{4.33}$$

We now integrate Eq. (4.33) from end 1 to end 2 of the system.

At end 1, $A = 0$ and $\Delta T = T_{h1} - T_{c1} = \Delta T_1$ (say)

At end 2, $A = A$ and $\Delta T = T_{h2} - T_{c2} = \Delta T_2$ (say).

Therefore,

$$-\int_{\Delta T_1}^{\Delta T_2} \frac{d(\Delta T)}{\Delta T} = \frac{m'_h c_{ph} + m'_c c_{pc}}{m'_h c_{ph} \, m'_c c_{pc}} \, U \int_0^A dA$$

or

$$\ln\frac{\Delta T_1}{\Delta T_2} = \frac{m'_h c_{ph} + m'_c c_{pc}}{m'_h c_{ph} \, m'_c c_{pc}} \, UA \tag{4.34}$$

Now we write the heat balance equations for both the streams, the hot stream and the cold stream, over the heat exchange device. If Q is the total rate of heat transfer, we have

$$Q = m'_h c_{ph} \, (T_{h1} - T_{h2}) = m'_c c_{pc} \, (T_{c2} - T_{c1}) \tag{4.35}$$

or

$$\frac{Q}{m'_h c_{ph}} + \frac{Q}{m'_c c_{pc}} = (T_{h1} - T_{h2}) + (T_{c2} - T_{c1})$$

$$= (T_{h1} - T_{c1}) - (T_{h2} - T_{c2}) = \Delta T_1 - \Delta T_2$$

Therefore,

$$Q = \frac{m'_h c_{ph} m'_c c_{pc}}{m'_h c_{ph} + m'_c c_{pc}} \, (\Delta T_1 - \Delta T_2) \tag{4.36}$$

If we define ΔT_m as an appropriate *mean driving force* over the entire length of the device, we have

$$Q = U \, A \, \Delta T_m \tag{4.37}$$

From Eqs. (4.36) and (4.37), we get

$$UA \, \Delta T_m = \frac{m'_h c_{ph} m'_c c_{pc}}{m'_h c_{ph} + m'_c c_{pc}} \, (\Delta T_1 - \Delta T_2) \tag{4.38}$$

Comparing Eqs. (4.34) and (4.38), we get

$$\Delta T_m = \frac{\Delta T_1 - \Delta T_2}{\ln (\Delta T_1 / \Delta T_2)} \tag{4.39}$$

Equation (4.39) is one of the most important equations in heat transfer calculation. It shows that when the temperature driving force varies from one end of a heat exchange device to the other, the *log mean temperature difference*, ΔT_m, given above is the applicable mean driving force.

Next we will consider the case of heat exchange between a hot and a cold fluid in countercurrent flow as shown in Fig. 4.7. The notations used are very much the same as those used in the case of cocurrent flow.

The rate of heat transfer at steady state through the elementary area dA is

$$dQ = U \, dA \, (T_h - T_c) = U \, dA \, \Delta T \tag{4.40}$$

Also, if dT_h and dT_c are the temperature changes of the streams over this elementary area, we have

$$dQ = -m'_h c_{ph}\, dT_h = m'_c c_{ph}\, dT_c \tag{4.41}$$

The heat balance equations for the countercurrent flow situation are similar to those for the cocurrent flow. It is easy to verify that the final form of the mean driving force remains the same as Eq. (4.39) and the rate of heat transfer is then given by Eq. (4.37). The log mean driving force is abbreviated as LMTD. This is left as an exercise for the students.

Example 4.6 Lubricating oil used in the gearbox of a 14,000 rpm high speed blower is being recycled continuously through a double-pipe counterflow heat exchanger for cooling. The oil is to be cooled from 70°C to 40°C at the rate of 1000 kg/h using water entering at 28°C. The water temperature at the exit should not exceed 42°C. The specific heat of oil is 2.05 kJ/kg °C and that of water is 4.17 kJ/kg °C. Calculate the required rate of flow of water. If the heat exchange area is 3.0 m², calculate the overall heat transfer coefficient.

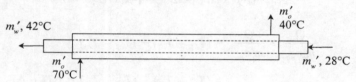

SOLUTION If m'_o and m'_w are the flow rates of oil and water, and c_{po} and c_{pw} are their specific heats, then a simple steady state energy balance over the heat exchanger gives:

$$\underset{\text{(heat lost by the oil)}}{m'_o\, c_{po}\, (70 - 40)} = \underset{\text{(heat gained by the water)}}{m'_w\, c_{pw}\, (42 - 28)}$$

or

$$(1000)(2.05)(30) = m'_w\, (4.17)(14)$$

or

$$m'_w = \boxed{1053 \text{ kg}/\text{h}} = \text{flow rate of water}$$

Heat duty: $Q = m'_o\, c_{po}\, (70 - 40)$ kJ/h $= \dfrac{(1000)(2.05)(30)}{3600} = 17.1$ kW

Temperature difference: hot end, $\Delta T_1 = 70 - 42 = 28°C$;   cold end, $\Delta T_2 = 40 - 28 = 12°C$.

Log mean temperature difference, LMTD $= \dfrac{\Delta T_1 - \Delta T_2}{\ln\,(\Delta T_1/\Delta T_2)} = \dfrac{28 - 12}{\ln\,(28/12)} = 18.9°C = (\Delta T)_m$

Now, $Q = UA\Delta T_m$. Given, $A = 3$ m². Therefore,

$$U = \frac{Q}{A\,\Delta T_m} = \frac{17,100 \text{ W}}{(3 \text{ m}^2)(18.9°C)} = \boxed{300 \text{ W}/\text{m}^2 \text{ °C}} = \text{overall heat transfer coefficient}$$

Example 4.7 Nitrogen gas is heated at a rate of 2000 kg/h before passing it through the trays in a tray drier in which an organic product is being dried. Medium pressure steam available from a waste heat boiler at a pressure of 5.7 bar (saturation temperature = 160°C) condenses within the tubes of a 'finned tube' heater. The gas flows outside the tubes. The heat duty is 38,700 kcal/h. The specific heat of nitrogen can be taken as 0.239 kcal/kg °C. The heat exchanger has an area of 10 m^2 and the overall heat transfer coefficient is estimated at 70 kcal/h m^2 °C. The fin efficiency is 63%. Calculate the inlet and the outlet temperatures of the gas.

SOLUTION   Heat balance of the gas, Q = (gas flow rate)(specific heat)$(T_o - T_i)$
or

$$38,700 = (2000)(0.239)(T_o - T_i)$$

or

$$T_o - T_i = 81°C \qquad (i)$$

where T_i and T_o are the inlet and the outlet temperatures of nitrogen.

Terminal driving forces (note that the steam side temperature remains constant at 160°C) are

hot end: $\Delta T_1 = 160 - T_o$;   cold end: $\Delta T_2 = 160 - T_i$.

Therefore,

$$\text{LMTD} = \frac{(160 - T_i) - (160 - T_o)}{\ln [(160 - T_i)/(160 - T_o)]} = \frac{T_o - T_i}{\ln [(160 - T_i)/(160 - T_o)]} = \Delta T_m$$

If η is the fin efficiency, the rate of heat transfer can be expressed as

$$Q = U A \eta \Delta T_m$$

or

$$\Delta T_m = \frac{Q}{UA\eta}$$

Given: Q = 38,700 kcal/h; A = 10 m^2; U = 70 kcal/h m^2 °C; η = 0.63.
Therefore,

$$\frac{T_o - T_i}{\ln [(160 - T_i)/(160 - T_o)]} = \frac{38,700}{(70)(10)(0.63)} = 87.8 \qquad (ii)$$

Solving the simultaneous Eqs. (i) and (ii), we get

$$T_i = \boxed{24°C} \quad \text{and} \quad T_o = \boxed{105°C}$$

Example 4.8   Hot water is flowing through a 3.5 cm (1-1/4″) schedule 40 steel pipe at a velocity of 1.8 m/s. The inlet temperature is 110°C, and the length of the pipe is 15 m. A 2 cm thick layer of insulation (k_c = 0.12 W/m °C) covers the pipe. The outside film coefficient is 10 W/m^2 °C, and the ambient temperature is 20°C. Calculate the drop in the temperature of the water over this section of the pipe.

SOLUTION   Here heat loss from the pipe occurs under a variable driving force that changes from one end of the pipe to the other. There are four heat transfer resistances in series—the tube-side resistance, the resistance owing to the pipe wall and that owing to the insulation, and the external resistance because of an air-film. The tube-side coefficient can be calculated by using the

Dittus-Boelter equation and then the overall resistance can be found out. The system is at steady state. Because the outlet temperature is not known (and, therefore, the mean water temperature is not known) the physical properties of water are taken at the inlet temperature of 110°C.

We will use the Dittus-Boelter equation [Eq. (4.9)] to calculate the tube-side (or the water-side) coefficient.

$$\text{Nu} = 0.023 \ (\text{Re})^{0.8} \ (\text{Pr})^{0.3} \qquad \text{[for cooling]}$$

Properties of water at 110°C: $\rho = 950$ kg/m³; $c_p = 4.23$ kJ/kg °C; $\mu = 2.55 \times 10^{-4}$ kg/m s; $k = 0.685$ W/m °C.

Actual i.d. of the pipe, $d_i = 3.5$ cm, $r_i = 0.0175$ m; wall thickness of 1-1/4 schedule 40 pipe = 0.0036 m; outer radius of the pipe, $r' = 0.0175 + 0.0036 = 0.0211$; outer radius of the insulation, $r_o = 0.0211$ m $+ 0.02$ m $= 0.0411$ m; pipe length $= 15$ m; water velocity, $V = 1.8$ m/s; $h_o = 10$ W/m² °C; thermal conductivity of the pipe wall (steel), $k_w = 43$ W/m °C; thermal conductivity of the insulation, $k_c = 0.12$ W/m °C.

The overall coefficient (inside area basis) is given by

$$\frac{1}{U_i} = \frac{1}{h_i} + \frac{r_i \ \ln (r'/r_i)}{k_w} + \frac{r_i \ \ln (r_o/r')}{k_c} + \frac{r_i}{r_o h_o}$$

Numerical calculations

$$\text{Re} = \frac{d_i V \rho}{\mu} = \frac{(0.035)(1.8)(950)}{2.55 \times 10^{-4}} = 234{,}700$$

$$\text{Pr} = \frac{c_p \mu}{k} = \frac{(4230)(2.55 \times 10^{-4})}{0.685} = 1.57$$

$$\text{Nu} = 0.023(234{,}700)^{0.8}(1.57)^{0.3} = 521.6$$

$$h_i = \frac{(\text{Nu})(k)}{d_i} = \frac{(521.6)(0.685)}{0.035} = 10{,}208$$

$$\frac{1}{U_i} = \frac{1}{10{,}208} + \frac{(0.0175) \ \ln (0.0211/0.0175)}{43} + \frac{(0.0175) \ \ln (0.0411/0.0211)}{0.12} +$$

$$\frac{0.0175}{(0.0411)(10)} = 0.14$$

or

$$U_i = 7.14 \ \text{W/m}^2 \ °C$$

Given: Inside area of a 15 m section of the pipe $= \pi d_i L = \pi(0.035)(15) = 1.649$ m² $=$ heat transfer area

Water flow rate $= \pi r_i^2 V \rho = \pi(0.0175)^2(1.8)(950) = 1.645$ kg/s $= m_w'$

Inlet water temperature, $T_i = 110$°C

Outlet temperature of water $= T_o$ (to be calculated)

Ambient temperature, $T_a = 20$°C

$$\text{LMTD} = \frac{(T_i - T_a) - (T_o - T_a)}{\ln (T_i - T_a)/(T_o - T_a)}$$

Rate of heat loss from water $= m'_w \, c_p \, (T_i - T_o)$

$$= (1.645)(4230)(T_i - T_o) = UA(\text{LMTD})$$

or

$$591(T_i - T_o) = \frac{T_i - T_o}{\ln [(T_i - T_a)/(T_o - T_a)]}$$

Putting the values of T_i and T_a and solving the above equation, we get

Outlet water temperature, $T_o = \boxed{109.8°C}$

The calculation of U_i shows that the resistances offered by the insulation and by the air-film virtually govern the overall heat transfer coefficient. The final result shows that the drop in the water temperature as it flows through the pipe is negligible.

Example 4.9 A 41 mm i.d. (1-1/2″ nominal bore) schedule 40 pipe carries water flowing at a rate of 1 kg/s. Water enters the pipe at 28°C and is heated by a stream of hot flue gas in cross-flow over the pipe. The gas velocity is 10 m/s. The arrangement essentially aims at recovering a part of the waste heat of the gas stream which has a bulk temperature of 250°C. The pressure is essentially atmospheric. The length of the pipe is 20 m. At what temperature does the water leave the pipe? The properties of the flue gas are about the same as those of the air.

SOLUTION Here heating occurs under a variable driving force which is maximum at the inlet to the pipe and decreases along the length because of a rise in the water temperature. As the exit water temperature is not known, the relevant properties of water are taken at the inlet temperature (28°C). The properties of the gas are taken at the mean of the inlet water temperature and the bulk gas temperature [i.e. (28 + 250)/2 = 139°C]. Once the outlet water temperature is calculated, the mean fluid temperatures can be calculated, the properties can be obtained at those mean temperatures and the heat transfer calculations may be redone to get more accurate results.

Data: For water at 28°C: $\rho = 996$ kg/m³; $c_p = 1.0$ kcal/kg °C; $\mu = 8.6 \times 10^{-4}$ kg/m s; $k_w = 0.528$ kcal/ h m °C.; flow rate = 1 kg/s = 3600 kg/h.
For the flue gas (same as air): $\rho = 0.891$ kg/m³; $c_p = 0.243$ kcal/kg °C; $\mu = 2.33 \times 10^{-5}$ kg/m s; thermal conductivity, $k_a = 0.0292$ kcal/h m °C; Pr = 0.69.
Inner diameter of the tube, $d_i = 41$ mm $= 4.1 \times 10^{-2}$ m

Water-side heat transfer coefficient

Cross-section of the pipe $= (\pi/4)d_i^2 = (\pi/4)(4.1 \times 10^{-2})^2 = 1.32 \times 10^{-3}$ m²

$$\text{Velocity of water, } V = \frac{1}{(996)(1.32 \times 10^{-3})} = 0.761 \text{ m/s}$$

$$\text{Re} = \frac{d_i V \rho}{\mu} = \frac{(0.041)(0.761)(996)}{8.6 \times 10^{-4}} = 3.61 \times 10^4$$

$$Pr = \frac{(4180)(8.6 \times 10^{-4})}{0.614} = 5.85$$

$$Nu = 0.023(3.61 \times 10^4)^{0.8} (5.85)^{0.4} = 206$$

[*Note:* The Whitaker equation, Eq. (4.11), gives Nu = 191 if we put $\mu = \mu_w$]
The water-side heat transfer coefficient,

$$h_i = 206\left(\frac{k_w}{d_i}\right) = 206\left(\frac{0.528}{0.041}\right) = 2653 \text{ kcal/h m}^2 \text{ °C}$$

Gas-side heat transfer coefficient

Wall thickness of 41 mm i.d. schedule 40 pipe = 3.7 mm
o.d. of the pipe, $d_o = 41 + 2(3.7)$ mm = 0.0484 m

$$Re = \frac{(0.0484)(10)(0.891)}{2.33 \times 10^{-5}} = 18,500$$

We will use Eq. (4.19) to calculate the heat transfer coefficient for cross flow of air over a cylinder.

$$Nu = \frac{h_o d_o}{k_a} = 0.3 + \frac{(0.62)(18,500)^{0.5}(0.69)^{1/3}}{[1 + (0.4/0.69)^{2/3}]^{1/4}}\left[1 + \left(\frac{18,500}{2.82 \times 10^5}\right)^{5/8}\right]^{4/5} = 75$$

or

$$h_o = 75\frac{k_a}{d_o} = \frac{(75)(0.0292)}{0.0484} = 45.2 \text{ kcal/ h m}^2 \text{ °C}$$

Overall heat transfer coefficient

We neglect the thermal resistance of the pipe wall. The overall coefficient based on the outside area is

$$\frac{1}{U_o} = \frac{r_o}{r_i h_i} + \frac{1}{h_o} = \frac{0.0484}{(0.041)(2653)} + \frac{1}{45.2}$$

or

$$U_o = 44.3 \text{ kcal/h m}^2 \text{ °C}$$

Heat balance

Flue gas temperature, $T_o = 250\text{°C}$

Inlet temperature of water, $T_1 = 28\text{°C}$

Outlet temperature of water, $T_2 = $ to be found out

Water flow rate, $m'_w = 1$ kg/s = 3600 kg/h

Outside area of 20 m pipe = $\pi(0.0484)(20) = 3.04 \text{ m}^2$

$$m'_w c_p(T_2 - T_1) = U_o A_o \Delta T_m$$

$$\Delta T_m = \text{LMTD} = \frac{(T_o - T_1) - (T_o - T_2)}{\ln\left[(T_o - T_1)/(T_o - T_2)\right]}$$

or

$$(3600)(1.0)(T_2 - 28) = (44.3)(3.04) \frac{T_2 - 28}{\ln\left[(250 - 28)/(250 - T_2)\right]}$$

The solution of the above equation gives the exit water temperature, $\boxed{T_2 = 36°C}$

As the increase in the water temperature is small, the mean gas-film temperature does not change appreciably. The recalculation of the coefficients is therefore not necessary.

Example 4.10 Hot engine oil has to be cooled from 110°C to 70°C in a counterflow double-pipe heat exchanger at a rate of 500 kg/h. The exchanger consists of a 25 mm (1 inch) 14 BWG inner tube and a 35 mm (1-1/4 inch) schedule 40 outer pipe. The oil flows through the annulus and the cooling water flows through the tube, entering at 29°C and leaving at 40°C. Calculate the length of the heat exchanger. The following data are available:

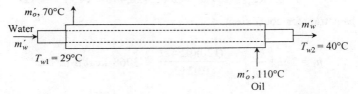

For oil: density, $\rho_o = 850$ kg/m³; specific heat, $c_{po} = 0.478$ kcal/kg °C; thermal conductivity, $k_o = 0.12$ kcal/h m °C

Viscosity of the oil, which is a strong function of temperature, is given as

$$\ln \mu_{\text{oil}} = [(5550/T) - 19] \text{ kg/m s} \quad (T \text{ in K}).$$

For water (at the mean liquid temperature): $k_w = 0.542$ kcal/h m °C; $\mu_{\text{water}} = 7.1 \times 10^{-4}$ kg/m s; $c_{pw} = 1$ kcal/kg °C

SOLUTION The required flow rate of water is not given. It has to be calculated by heat balance over the exchanger. The viscosity of the oil at the mean fluid temperature will be used for calculation of the Reynolds number. Also because the oil is quite viscous, viscosity correction of the oil-side coefficient will be necessary. Temperature dependence of other physical properties will be neglected. The thermal resistance of the tube wall will also be neglected.

Dimensions of the tube and of the pipe: inner tube: i.d. = d_{ti} = 2.12 cm; o.d. = d_{to} = 2.54 cm; outer pipe i.d. = d_{pi} = 3.50 cm (o.d. of the outer pipe is not required for the calculation).

Heat load of the exchanger,

$$= m_o' \, c_{po} \, (T_{o2} - T_{o1}) = (500)(0.478)(110 - 70) = 9560 \text{ kcal/h}$$

The required water flow rate can be calculated from heat balance,

$$m_w' \, c_{pw}(T_{w2} - T_{w1}) = 9560$$

or

$$m_w' = \frac{9560}{(1.0)(40 - 29)} = 869.1 \text{ kg/h} \ (= 2.42 \times 10^{-4} \text{ m}^3/\text{s})$$

Water-side heat transfer coefficient

Flow area of the tube $= \left(\dfrac{\pi}{4}\right)(0.0212)^2 = 3.53 \times 10^{-4} \text{ m}^2$

$$\text{Water velocity, } V_w = \frac{2.42 \times 10^{-4}}{3.53 \times 10^{-4}} = 0.686 \text{ m/s}$$

$$\text{Reynolds number, Re}_w = \frac{d_{ti} V_w \rho_w}{\mu_{\text{water}}} = \frac{(0.0212)(0.686)(1000)}{7.1 \times 10^{-4}} = 20,480$$

$$\text{Prandtl number of water, Pr}_w = \frac{(4170)(7.1 \times 10^{-4})}{0.63} = 4.7$$

$$\text{Nu}_w = 0.023(\text{Re}_w)^{0.8} \ (\text{Pr})^{0.4} = 0.023(20,480)^{0.8}(4.7)^{0.4} = 120$$

Water-side heat transfer coefficient,

$$h_i = \text{Nu}_w\left(\frac{k_w}{d_{ti}}\right) = \frac{(120)(0.542)}{0.0212} = 3068 \text{ kcal/h m}^2 \text{ °C}$$

Oil-side heat transfer coefficient

Flow area of the annulus $= \left(\dfrac{\pi}{4}\right)(d_{pi}^2 - d_{to}^2) = \left(\dfrac{\pi}{4}\right)[(0.035)^2 - (0.0254)^2] = 4.554 \times 10^{-4} \text{ m}^2$

$$\text{Velocity of the oil, } V_o = \frac{500}{(3600)(850)(4.554 \times 10^{-4})} = 0.359 \text{ m/s}$$

Equivalent diameter of the annulus (for heat transfer) is calculated by using Eq. (4.13b).

$$d_e = \frac{d_{pi}^2 - d_{to}^2}{d_{to}} = \frac{(0.035)^2 - (0.0254)^2}{0.0254} = 2.28 \times 10^{-2} \text{ m}$$

$$\text{Mean oil temperature} = \frac{110 + 70}{2} = 90°C = 363 \text{ K}$$

$$\ln \mu_{\text{oil}} = \left(\frac{5550}{363}\right) - 19 = -3.71; \qquad \mu_{\text{oil}} = 0.0246 \text{ kg/m s}$$

$$\text{Reynolds number, Re}_o = \frac{d_e V_o \rho_o}{\mu_{\text{oil}}} = \frac{(0.0228)(0.359)(850)}{0.0246} = 283 \ (\textit{it indicates laminar flow})$$

$$\text{Prandtl number of oil, Pr}_o = \frac{c_{po}\mu_{\text{oil}}}{k_o} = \frac{(0.478)(0.0246)(3600)}{0.12} = 353$$

Because the flow is laminar, the Sieder-Tate equation, Eq. (4.8), may be used for the calculation of the oil-side heat transfer coefficient. But this equation requires the tube length L which is not known (in fact, L is required to be calculated). Therefore, an 'iterative procedure' has to be used. We will start with a 'guess value' of the Nusselt number, calculate the overall heat transfer coefficient, the LMTD, the heat transfer area required and therefrom the tube length. Once a value of L is available, a more accurate value of the Nusselt number of the oil may be obtained by repeating the same procedure. But how to guess a value of the Nusselt number? If we look at the Hausen's equation, Eq. (4.7), for heat transfer coefficient in laminar flow, it appears that for a very long tube the Nusselt number becomes 3.66. Thus, 3.66 may be taken as the starting value or the 'first approximation' of Nu_o (the Nusselt number of oil).

$$Nu_o = 3.66; \quad \text{or} \quad h'_o = Nu_o \left(\frac{k_o}{d_e} \right) = 3.66 \left(\frac{0.12}{0.0228} \right) = 19.3 \text{ kcal/h m}^2 \text{ °C}$$

[h'_o is the first approximation to the true oil-side heat transfer coefficient h_o; h''_o is the second approximation, etc.]

Overall heat transfer coefficient (first approximation), $\dfrac{1}{U'_o} = \dfrac{1}{h'_o} + \dfrac{A_o}{A_i h_i}$

where

A_i = inside area of the tube = $\pi d_{ti} L$

A_o = outside area of the tube = $\pi d_{to} L$

Therefore,

$$\frac{1}{U'_o} = \frac{1}{19.3} + \frac{(\pi)(0.0254)L}{(\pi)(0.0212)L(3068)} = 0.0522$$

or

$$U'_o = 19.2 \text{ kcal/h m}^2 \text{ °C}$$

$$\text{LMTD} = \frac{\Delta T_2 - \Delta T_1}{\ln (\Delta T_2 / \Delta T_1)} = \frac{(110 - 40) - (70 - 29)}{\ln [(110 - 40)/(70 - 29)]} = 54.2 \text{°C}$$

Heat transfer area required A_o is given by

$$U'_o A_o \text{LMTD} = Q = 9560$$

or

$$A_o = \frac{9560}{(19.2)(54.2)} = 9.26 \text{ m}^2$$

$$\text{Tube length} = \frac{A_o}{\pi d_{to}} = \frac{9.26}{(\pi)(0.0254)} = 116 \text{ m}$$

Once the tube length L is known, we can have a better estimate of h_o from Eq. (4.8). However, because the wall temperature, and therefore the viscosity of the oil at the temperature of the wall, is not known, we will not use the viscosity correction term of Eq. (4.8) at this step.

$$Nu''_o = 1.86 \left[\frac{Re_o Pr_o}{L/d_e} \right]^{1/3} = 1.86 \left[\frac{(283)(353)}{116/0.0228} \right]^{1/3} = 5.02$$

or

$$h_o'' = 5.02 \frac{0.12}{0.0228} = 26.4 \text{ kcal/h m}^2 \, ^\circ\text{C}$$

Viscosity correction of the water-side coefficient is not required because the bulk and the wall temperatures are very close. However, the viscosity of the oil at the bulk temperature is expected to be much different from that at the wall temperature. To estimate the viscosities in the bulk and at the wall, the corresponding temperatures should be known. However, both these temperatures vary with distance along the tube. So we have to calculate the average bulk temperature and the average wall temperatures of the oil.

The average bulk temperature of the oil $= \dfrac{110 + 70}{2} = 90^\circ\text{C} = 363 \text{ K}$

The average temperature of the cooling fluid (water) $= \dfrac{29 + 40}{2} = 34.5^\circ\text{C}$

The average wall temperature can be determined by equating the heat transfer rates for oil and water sides on the basis of unit length of the tube. That is,

$$h_o'' (\pi d_{to})(1)(T_{\text{oil}} - T_{\text{wall}}) = h_i(\pi d_{ti})(1)(T_{\text{wall}} - T_{\text{water}})$$

or

$$(26.4)(0.0254)(90 - T_{\text{wall}}) = (3068)(0.0212)(T_{\text{wall}} - 34.5)$$

or

$$T_{\text{wall}} = 35^\circ\text{C} = 308 \text{ K}$$

Viscosity of the oil at the average wall temperature is

$$\ln \mu_{\text{wall}} = \frac{5550}{308} - 19 = -0.981; \quad \mu_{\text{wall}} = 0.375 \text{ kg/m s}$$

Now we recalculate the Nusselt number from Eq. (4.8) considering the viscosity correction term.

$$\text{Nu}_o'' = 1.86 \left[\frac{\text{Re}_o \text{Pr}_o}{L/d_e} \right]^{1/3} \left[\frac{\mu}{\mu_{\text{wall}}} \right]^{0.14} = 1.86 \left[\frac{(283)(353)}{116/0.0228} \right]^{1/3} \left[\frac{0.0246}{0.375} \right]^{0.14} = 3.43$$

Therefore,

$$h_o'' = 3.43 \frac{0.12}{0.0228} = 18.1 \text{ kcal/h m}^2 \, ^\circ\text{C}$$

$$\frac{1}{U_o''} = \frac{1}{h_o''} + \frac{A_o}{A_i h_i} = \frac{1}{18.1} + \frac{0.0254}{(0.0212)(3068)} = 0.0556$$

or

$$U_o'' = 18.1 \text{ kcal/h m}^2 \, ^\circ\text{C} \text{ (as water-side resistance is very small)}$$

and

$$A_o = \frac{Q}{U_o'' \text{LMTD}} = \frac{9560}{(18.1)(54.2)} = 9.75 \text{ m}^2$$

Therefore,

$$\text{Tube length, } L = \frac{A_o}{\pi d_{to}} = \frac{9.75}{\pi (0.0254)} = \boxed{122 \text{ m}}$$

The recalculated tube length is reasonably close to the value obtained before. No further iteration is necessary. (In this example, the oil offers most of the heat transfer resistance. As the tube length is large, a serpentine configuration of the exchanger may be suitable.)

Example 4.11 A large volume of exhaust gas available at 260°C is used to heat process water. The gas flows across a tube bank and water flows through the tubes. The outside diameter of the tubes is 6.0 cm. The tube bank has 15 transverse rows and 14 longitudinal rows. The tubes are spaced 15 cm centre-to-centre in an equilateral triangular arrangement. The hot gas approaches the tube bank at a velocity of 16 m/s. If the tube surface temperature is 70°C, calculate the rate of heat transfer to water per metre length of the tube bank.

SOLUTION We will use the notations of Section 4.6.4 and Fig. 4.5. The tubes are spaced in an equilateral triangular arrangement. So,

$$S_T = S_D = 15 \text{ cm} = 0.15 \text{ m}$$

$$S_L = \frac{\sqrt{3}}{2} S_D = 0.13 \text{ m}$$

Given: $V = 16$ m/s; $d = 0.06$ m; properties of the exhaust (taken to be the same as those of air at 260°C): v (momentum diffusivity) $= 4.43 \times 10^{-5}$ m^2/s; $\rho = 0.73$ kg/m^3; $c_p = 0.248$ kcal/kg °C; $k = 0.0375$ kcal/h m °C

Prandtl number of bulk air, $\text{Pr} = \dfrac{(0.248)(4.43 \times 10^{-5})(3600)(0.73)}{0.0375} = 0.62$

Prandtl number of the air at the wall temperature (70°C), $\text{Pr}_w = 0.70$
Number of tubes in a transverse row, $N_T = 15$
The total number of tubes, $N = 15 \times 14 = 210$
By Eq. (4.25), we have

$$V_{\text{max}} = \frac{S_T}{S_T - d} V = \frac{0.15}{0.15 - 0.06} (16) = 26.7 \text{ m/s}$$

By Eq. (4.26), we have

$$V_{\text{max}} = \frac{S_T}{2(S_D - d)} V = \frac{(0.15)(16)}{2(0.15 - 0.06)} = 13.35 \text{ m/s}$$

We select the larger value, $V_{\text{max}} = 26.7$ m/s.
Therefore,

$$\text{Re}_{d, \text{max}} = \frac{d V_{\text{max}}}{v} = \frac{(0.06)(26.7)}{4.43 \times 10^{-5}} = 36,163$$

Equation (4.24) will be used to calculate the average heat transfer coefficient of the gas phase. The constants in the equation are taken from Table 4.3.

The pitch ratio, $\dfrac{S_T}{S_L} = \dfrac{0.15}{0.13} = 1.154 < 2$ (Note: in this problem, $S_T = S_D$)

From Table 4.3, we have

$$m = 0.6, \quad C = 0.35\left(\frac{S_T}{S_L}\right)^{0.2} = 0.36$$

Therefore,

$$\frac{\bar{h}d}{k} = C\left(\mathrm{Re}_{d,\,max}\right)^m (\mathrm{Pr})^{0.36}\left(\frac{\mathrm{Pr}}{\mathrm{Pr}_w}\right)^{0.25}$$

$$= (0.36)(36,163)^{0.6}\,(0.62)^{0.36}\left(\frac{0.62}{0.7}\right)^{0.25} = 160$$

or

$$\bar{h} = (160)\left(\frac{k}{d}\right) = (160)\left(\frac{0.0375}{0.06}\right) = 100 \text{ kcal/h m}^2\,°\text{C}$$

If T_o is the exit temperature of the flue gas, then by Eq. (4.28), we have

$$(T_o - T_s) = (T_i - T_s)\exp\left[-\frac{\pi d N\bar{h}}{\rho V N_T S_T c_p}\right]$$

$$= (260 - 70)\exp\left[-\frac{\pi(0.06)(210)(100)}{(0.73)(16)(3600)(15)(0.15)(0.248)}\right] = 161°\text{C}$$

LMTD [by Eq. (4.27)], $\Delta T_m = \dfrac{(T_i - T_s) - (T_o - T_s)}{\ln\left[(T_i - T_s)/(T_o - T_s)\right]} = \dfrac{(260 - 70) - 161}{\ln\left[(260 - 70)/161\right]} = 163°\text{C}$

For 1 m length of the tube in the tube bank,

Heat transfer area, $A = \pi d L N = \pi(0.06)(1)(210) = 39.6 \text{ m}^2$

Rate of heat transfer,

$$Q = \bar{h}\,(\text{area})(\Delta T_m) = (100)(39.6)(163) = \boxed{6.45 \times 10^5 \text{ kcal/h}}$$

4.8 MOMENTUM AND HEAT TRANSFER ANALOGIES

It is well known that the basic laws of transport of momentum and heat are expressible in similar forms. Consider, for example, a fluid in laminar motion through a circular pipe (Fig. 4.8). The wall of the pipe is maintained at a higher temperature and the fluid gets heated as it flows through the pipe. Also, radial transport of momentum in the pipe occurs from a faster-moving layer to a slower-moving layer according to the Newton's law of viscosity.

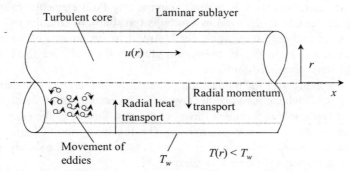

Fig. 4.8 Radial transport of momentum and heat in turbulent pipe flow.

$$\tau = -\mu \left(\frac{du_x}{dr} \right) \tag{4.42}$$

where τ is the shear stress (also called the momentum flux), μ is the viscosity, and $u(r)$ is the radial distribution of velocity in the pipe (u is the axial velocity and is a function of the radial position; the x-axis is the axis line of the pipe). Equation (4.42) may be rewritten as

$$\tau = -\frac{\mu}{\rho} \frac{d}{dr} (\rho u) = -v \frac{d}{dr} (\rho u) \tag{4.43}$$

The quantity ρu [(kg m/s)/m^3] is the volumetric concentration of x-momentum (or the momentum in the axial direction), $v = (\mu/\rho)$ is the momentum diffusivity, also known as kinematic viscosity. Equation (4.43) physically means that

Momentum flux = (Momentum diffusivity) (Gradient of the concentration of momentum)

$$\tag{4.44}$$

Now we consider the case of heat transfer to the fluid from the wall. The radial heat flux at the wall is given by the Fourier's law

$$q = k \frac{dT}{dr} \tag{4.45}$$

Because T increases with r, the negative sign is not used on the right-hand side. The equation can be rewritten as

$$q = \frac{k}{\rho c_p} \frac{d}{dr} (\rho c_p T) = \alpha \frac{d}{dr} (\rho c_p T) \tag{4.46}$$

Here α is the thermal diffusivity, and $\rho c_p T$ (kJ/m^3) is the volumetric concentration of heat energy. Therefore,

Heat flux = (Thermal diffusivity)(Gradient of the concentration of heat energy)   (4.47)

The flux Eqs. (4.43) and (4.46), and their physical representations given in Eqs. (4.44) and (4.47) show the similarity of the basic laws of momentum and heat transport. The diffusivities of momentum and heat, i.e. v and α, have identical units, m^2/s.

However, the transports of momentum or heat (or mass) in a turbulent medium are not governed by the above simple laws. Randomly moving tiny fluid elements, called *eddies*, act as the carriers of momentum, heat (or mass) in a turbulent medium. In fact, the movement of eddies or 'eddy exchange' is primarily responsible for transport in the regions of a medium where the intensity of turbulence is high. But in the region close to the wall, the fluid motion is almost laminar, and diffusional transport dominates.

A simple way of expressing the momentum or heat flux in a turbulent medium was originally put forward by Reynolds who argued that the laws of diffusional transport as given by Eqs. (4.43) and (4.46), are still applicable in turbulent flow, but the contribution of the eddy exchange should be incorporated in terms of separate parameters. The resulting modified transport 'laws' in a turbulent medium are given below where ε_M and ε_H stand for the 'eddy diffusivities' of momentum and heat respectively (see also Chapter 11).

$$\text{Turbulent transport of momentum,} \quad \tau = -(v + \varepsilon_M)\frac{d}{dr}(\rho u) \tag{4.48}$$

$$\text{Turbulent transport of heat,} \quad q = (\alpha + \varepsilon_H)\frac{d}{dr}(\rho c_p T) \tag{4.49}$$

[It should be remembered that if the transport of heat occurs from the fluid to the wall (i.e. cooling of the fluid), the right-hand side of Eq. (4.49) should include a negative sign because the radial temperature gradient is negative.]

Equation (4.48), written for $r = R$ (i.e. at the wall), gives the wall shear stress. Thus, we have

$$-(v + \varepsilon_M)\left[\frac{d(\rho u)}{dr}\right]_{r=R} = \tau_w = \frac{1}{2}f\rho V^2 \tag{4.50}$$

or

$$-\left[\frac{d\hat{u}}{dr}\right]_{r=R} = \frac{\tau_w}{\rho V(v + \varepsilon_M)} = \frac{fV}{2(v + \varepsilon_M)} \tag{4.51}$$

The RHS of Eq. (4.50) follows from the definition of Fanning's friction factor f, the wall shear stress τ_w, and the 'mean fluid velocity' V in the pipe.

Equation (4.51) gives the *dimensionless velocity gradient* at the wall,

$$\hat{u} = u/V = \text{dimensionless velocity}$$

Similarly, Eq. (4.49) can be written at $r = R$ for the wall heat flux and for the wall temperature profile. Thus,

$$(\alpha + \varepsilon_H)\left[\frac{d(\rho c_p T)}{dr}\right]_{r=R} = q_w = h(T_w - T_m) \tag{4.52}$$

That is,

$$\left[\frac{d\hat{T}}{dr}\right]_{r=R} = -\frac{h}{\rho c_p(\alpha + \varepsilon_H)}; \quad \hat{T} = \frac{T}{T_w - T_m} \tag{4.53}$$

where
q_w is the wall heat flux
h is the wall heat transfer coefficient
$\hat{T}$ is the dimensionless temperature
T_m is the mean fluid temperature.

The analogy between momentum and heat transfer in pipe flow was first quantified by Reynolds on the specific assumptions that:

(i) the gradients of the dimensionless velocity and the dimensionless temperature at the wall are equal, and

(ii) $(v + \varepsilon_M) = (\alpha + \varepsilon_H)$

Then , from Eqs. (4.51) and (4.53), we have

$$\frac{fV}{2(v + \varepsilon_M)} = \frac{h}{\rho c_p(\alpha + \varepsilon_H)} \tag{4.54}$$

Putting $(v + \varepsilon_M) = (\alpha + \varepsilon_H)$ and rearranging, we get

$$\frac{hd/k}{(c_p\mu/k)(dV\rho/\mu)} = \frac{f}{2}$$

or

$$\frac{Nu}{(Re)(Pr)} = St = \frac{h}{\rho V c_p} = \frac{f}{2} \tag{4.55}$$

Equation (4.55) is called the *Reynolds analogy* and can be used to determine the heat transfer coefficient if the friction factor f is known.

Prandtl, in 1910, provided a more realistic picture of turbulent transport by assuming that momentum and heat transfer occur through eddy exchange or eddy transport in a 'turbulent core' and through diffusive transport in the 'laminar sub-layer' near the wall. With this assumption and by using the 'universal velocity profile' (see Chapter 11) near the wall, Prandtl developed the following equation relating the Stanton number with the Fanning's friction factor.

$$St = \frac{f/2}{1 + 5\sqrt{f/2}\,(Pr - 1)} \tag{4.56}$$

The above equation is called the *Prandtl analogy*. It reduces to the Reynolds analogy in the case of $Pr = 1$.

Chilton and Colburn (1934) observed that experimental heat transfer data could be better correlated by replacing $1 + 5\,(f/2)^{1/2}\,(Pr - 1)$ by $(Pr)^{2/3}$ in Eq. (4.56). That is,

$$St = \frac{f/2}{(Pr)^{2/3}}$$

or

$$\frac{Nu}{(Re)(Pr)^{1/3}} = \frac{f}{2} \tag{4.57}$$

The LHS of Eq. (4.57) is the well known Colburn j-factor (j_H). That is,

$$\frac{Nu}{(Re)(Pr)^{1/3}} = j_H = \frac{f}{2} \tag{4.58}$$

Equation (4.58) is called the *Chilton-Colburn analogy*. Using the well known correlation for the friction factor, $f = 0.046$ $(Re)^{-0.2}$ for pipe flow, the j-factor is given by

$$j_H = 0.023 \ (Re)^{-0.2} \qquad (4.59)$$

Quite a few other analogies (e.g. those proposed by Martinelli, Diessler, Lyon, etc.) are described in the literature.

Example 4.12 Aniline is a tonnage organic chemical used as a raw material for dye intermediates, rubber chemicals and for isocyanates (isocyanates are the starting materials for polyurethanes, a very important commercial polymer). The modern method of manufacture of aniline is vapour phase catalytic reduction of nitrobenzene by hydrogen.

The reaction gas mixture leaving the catalytic reactor in an aniline plant is condensed in a shell-and-tube heat exchanger. Condensation occurs in the shell-side while cooling water flows through the tubes. The tubes are 3 m long and 25 mm o.d. with 14 BWG wall. Water flows at a rate of 0.057 m^3 per minute per tube. Water enters at 32°C. The tube wall temperature may be assumed to remain constant at 80°C. Calculate the rise in temperature of water as it flows through the tubes. The heat transfer coefficient may be estimated from the Dittus-Boelter equation as well as from the heat transfer analogies described above, taking the physical properties of water at the mean fluid temperature.

SOLUTION   Inside diameter of 25 mm o.d. 14 BWG tube, $d_i = 21.2$ mm

$$\text{Water velocity,} \quad V = \left(\frac{0.057}{60}\right) \frac{1}{(\pi/4)(0.0212)^2} = 2.7 \text{ m/s}$$

The physical properties of water should be taken at the mean liquid temperature. The outlet temperature of water is not yet known. So, the calculation will be done on the basis of the liquid properties at the inlet condition. Once the outlet temperature is known, a revised calculation will be done to get more accurate results.

For water at 32°C (i.e. the inlet condition),

density, $\rho = 995$ kg/m^3;   viscosity, $\mu = 7.65 \times 10^{-4}$ kg/ms;

thermal conductivity, $k = 0.623$ W/m °C;   specific heat, $c_p = 4.17$ kJ/kg °C

$$\text{Reynolds number, Re} = \frac{(0.0212)(2.7)(995)}{7.65 \times 10^{-4}} = 7.44 \times 10^4$$

$$\text{Prandtl number, Pr} = \frac{(4170)(7.65 \times 10^{-4})}{0.623} = 5.12$$

Use the Dittus-Boelter equation to calculate the heat transfer coefficient

$$\text{Nu} = (0.023) \ (7.44 \times 10^4)^{0.8} \ (5.12)^{0.4} = 349$$

Let us now calculate the Nusselt number by using the three available analogies. In order to use an analogy, we need the friction factor. The friction factor is calculated by using the following well known equation.

$$f = 0.0014 + 0.125(Re)^{-0.32} = 0.0014 + (0.125)(7.44 \times 10^4)^{-0.32} = 0.00485$$

Reynolds analogy [Eq. (4.55)]

$$St = \frac{Nu}{(Re)(Pr)} = \frac{f}{2}$$

or

$$Nu = (Re)(Pr)\left(\frac{f}{2}\right) = (7.44 \times 10^4)(5.12)(0.00485/2) = 924$$

Prandtl analogy [Eq. (4.56)]

$$St = \frac{f/2}{1 + 5(Pr - 1)\sqrt{f/2}} = \frac{0.00485/2}{1 + (5)(5.12 - 1)\sqrt{0.00485/2}} = 0.001204$$

i.e.,

$$Nu = (St)(Re)(Pr) = (0.001204)(7.44 \times 10^4)(5.12) = 458.7$$

Colburn analogy [Eq. (4.58)]

$$\frac{Nu}{(Re)(Pr)^{1/3}} = j_H = \frac{f}{2}$$

or

$$Nu = (Re)(Pr)^{1/3}\left(\frac{f}{2}\right) = (7.44 \times 10^4)(5.12)^{1/3}(0.00485/2) = 311$$

It appears that Reynolds analogy predicts a very high value of Nu. Prandtl analogy also predicts a significantly higher Nu compared to the Dittus-Boelter equation or the Colburn analogy. We will do further calculations using the Colburn analogy. The calculations will be similar if any other analogy is used.

$$h = (Nu)\left(\frac{k}{d}\right) = (311)\left(\frac{0.623}{0.0212}\right) = 9140 \text{ W/m}^2 \, ^\circ C$$

Given: Inlet water temperature, $T_i = 32^\circ C$; outlet water temperature $= T_o$ (to be determined); wall temperature, $T_w = 80^\circ C$; water flow rate, $m'_w = 0.057$ m^3/min $= 0.95$ kg/s

$$Q = m'_w c_p (T_o - T_i) = h A \text{ (LMTD)} \tag{i}$$

$$LMTD = \frac{T_o - T_i}{\ln[(T_w - T_i)/(T_w - T_o)]}$$

Considering a single tube of the heat exchanger,

Area of heat transfer, $A = \pi d_i L = \pi(0.0212)(3) = 0.2$ m^2

From Eq. (i),

$$(0.95)(4170)(T_o - 32) = (9140)(0.2) \frac{T_o - 32}{\ln[(80 - 32)/(80 - T_o)]}$$

or

$$T_o = 50^\circ C$$

Now we repeat the calculation taking the physical properties of water at the mean liquid temperature, i.e. $(50 + 32)/2 = 41^\circ C$.

At this temperature (41°C), $\rho = 991$ kg/m^3; $\mu = 6.2 \times 10^{-4}$ kg/ms; $k = 0.623$ W/m °C; $c_p = 4.17$ kJ/kg °C

$$\text{Reynolds number, Re} = \frac{(0.0212)(2.7)(991)}{6.2 \times 10^{-4}} = 9.15 \times 10^4$$

$$\text{Prandtl number, Pr} = \frac{(4170)(6.2 \times 10^{-4})}{0.623} = 4.11$$

$$f = 0.0014 + 0.125(\text{Re})^{-0.32} = 0.0014 + (0.125)(9.15 \times 10^4)^{-0.32} = 0.00463$$

From the Colburn analogy, we have

$$\text{Nu} = (\text{Re})(\text{Pr})^{1/3}\left(\frac{f}{2}\right) = (9.15 \times 10^4)(4.11)^{1/3}\left(\frac{0.00463}{2}\right) = 339$$

$$h = (\text{Nu})\left(\frac{k}{d}\right) = (339)\left(\frac{0.623}{0.0212}\right) = 9962 \text{ W/m}^2 \text{ °C}$$

If T_o' is the outlet temperature, then

$$(0.95)(4170)(T_o' - 32) = (9962)(0.2)\frac{T_o' - 32}{\ln\left[(80 - 32)/(80 - T_o')\right]}$$

or

$$T_o' = 51°C$$

So the outlet temperature of water for one pass through the tubes = $\boxed{51°C}$

SHORT QUESTIONS

1. Distinguish between free and forced convection heat transfer.

2. Explain the terms: (i) hydrodynamic boundary layer, (ii) local Reynolds number, (iii) boundary layer thickness, and (iv) thermal boundary layer.

3. Under what condition does the thermal boundary layer remain within the hydrodynamic boundary layer?

4. Define local and average heat transfer coefficients in boundary layer flow over a flat plate.

5. What is stagnation point? How does boundary layer separation occur?

6. What do you mean by 'dimensional analysis'? State Buckingham's Pi-theorem.

7. Name four important dimensionless groups in heat transfer. What are their physical significances?

8. What is hydraulic diameter? What is the hydraulic diameter of a duct of equilateral triangular cross-section of side a?

9. Why is flow across a tube bank important in engineering applications?

10. Make a sketch of a double-pipe heat exchanger. What are the modes of flow in such an exchanger? Which flow mode ensures the best heat recovery?

11. How is LMTD defined? How does its magnitude compare with the arithmetic average temperature?

12. Differentiate between Reynolds analogy and Prandtl analogy. What is the significance of the Colburn j-factor?

13. Give typical values of heat transfer coefficients for turbulent pipe flow of (i) water, (ii) air, and (iii) an organic stream, say kerosene. What is the nature of dependence of the heat transfer coefficient h on (i) fluid velocity, (ii) thermal conductivity, and (iii) heat capacity?

PROBLEMS

4.1 Warm water is required at the rate of 500 kg/h for washing a filter cake, and it is decided to use a 25 mm steam-heated tube for the purpose. The tube wall is maintained at 130°C by condensing steam on the outside surface. Calculate the length of the tube required to heat the water from 30° to 50°C at the required rate. Use the Dittus-Boelter equation to calculate the heat transfer coefficient. The i.d. of the tube is 21.2 mm.

   Given: $\mu = 6.82 \times 10^{-4}$ kg/m s, $k = 0.63$ W/m °C, $c_p = 4.174$ kJ/kg °C. Also, calculate the heat transfer coefficient using the Whitaker correlation [Eq. (4.11)] and compare with the value obtained above. Neglect the resistance of the tube wall.

   [*Hint:* $\Delta T_m = \dfrac{(130 - 30) - (130 - 50)}{\ln\left[(130 - 30)/(130 - 50)\right]}$; $Q = (500)(4.174)(50 - 30)$ kJ/h; $Q = h_i A \Delta T_m$
   Calculate h and A, and determine the length of the tube.]

4.2 Water at 32°C enters a 1-1/2 inch schedule 40 pipe (o.d. = 4.82 cm, i.d. = 4.1 cm) at the rate of 800 kg/h. It is heated by a stream of flue gas in cross-flow over the pipe at 20 m/s. If the bulk gas temperature is 250°C, calculate the local heat flux at a position in the pipe where the water temperature is 40°C. What is the wall temperature at this location? What length of pipe is required to heat the water to 60°C? Assume that the properties of flue gas are the same as those of air. The thermal resistance of the pipe wall may be neglected.

4.3 Air at 70°C and essentially atmospheric pressure is flowing through a 15 cm × 10 cm rectangular duct at a velocity of 12 m/s. If the wall temperature of the duct is 32°C, and the length of the duct is 50 m, calculate the drop in the air temperature.

4.4 An aqueous solution of methanol (30% methanol) is required to be heated from 28° to 60°C at the rate of 2000 kg/h before it is pumped to the distillation column for purification. Five tubes of 1.25 cm o.d. (0.5 inch, 14 BWG) are required to be used in parallel. The tube wall temperature is maintained constant at 125°C. The actual i.d. of the tubes is 0.83 cm. What length of the tubes is necessary?

   Data supplied: for methanol, $k = 0.156$ kcal/h m °C, $\rho = 800$ kg/m^3, $c_p = 0.60$ kcal/kg °C, $\mu = 4.207 \times 10^{-4}$ kg/m s (all at the average fluid temperature of 44°C). Calculate the thermal conductivity of the mixture by using the Filippov equation (see Example 3.5). The viscosity of the liquid mixture can be estimated by using the approximate equation,

$\ln \mu_m = x_1 \ln \mu_1 + x_2 \ln \mu_2$, where x_1 and x_2 are mole fractions of the components 1 and 2. Heat capacity of the mixture, $c_{p,\ m} = x_1 c_{p1} + x_2 c_{p2}$ (for details of estimation of properties of pure substances or of mixtures, see Reid et al., 1988).

4.5 Combustion air to an oil-fired furnace is preheated by countercurrent heat exchange with hot flue gas in a device consisting of two 10 m long concentric pipes. The flue flows through the inner pipe, 0.3 m diameter. The inside diameter of the outer pipe is 0.5 m. The inner pipe has a thin wall. The flue gas enters the pipe at 300°C at a rate of 1500 kg/h; 1450 kg/h air at 31°C is sucked in through the annulus by an I.D. (induced draft) fan. At what temperature does the air leave this heat exchanger assembly? Assume that the flue gas has properties like air.

4.6 A 5 m long nichrome heating wire, having 0.045 cm diameter and 6.66 ohm/m electrical resistance, is connected to a 220 V power supply. Air at a velocity of 3 m/s is in cross-flow to the wire. If the bulk air temperature is 25°C, what is the steady state temperature of the wire?

4.7 Polystyrene beads are often manufactured by suspension polymerization of styrene monomer in a stirred reactor. An efficient temperature control of the reactor is necessary to get a good quality product. If a batch reactor is used, the reaction is carried out over different temperature ranges for pre-determined periods of time.

The temperature in a batch reactor is required to be maintained at 80°C for a certain period of time. The rate of heat generation is 4300 kcal/h. The reactor has a cooling coil made of 25.4 mm diameter thin-walled tube (i.d. = 21.2 mm). The coil diameter is 50 cm, and the cooling water enters the coil at 25°C at the rate of 800 kg/h. The outer surface heat transfer coefficient is 300 kcal/h m² °C. The tube-side coefficient can be calculated by using the Dittus-Boelter equation. Calculate (a) the exit water temperature, and (b) the number of turns required for the cooling coil.

4.8 A narrow metal tube of 12.5 mm o.d. and 1.2 mm wall thickness (18 BWG), having an electrical heating wire wound on it, is used to heat water. The wire diameter is 0.91 mm and its resistance is 165 ohm/km. The tube is 5 m long. The heating wire, 40 m long, is connected to 220 V power supply. There is a 0.5 cm thick insulation (k = 0.08 W/m °C) on the tube and the outer air-film coefficient is 22 W/m² °C. Water flows through the tube at 0.5 m/s entering at 24°C. The arrangement essentially provides 'constant heat flux condition'. Calculate the average tube-side heat transfer coefficient. What is the steady state temperature of the tube and the percentage of heat loss to the ambient which is at 28°C?

4.9 An absorption oil is frequently used for the recovery of hydrocarbon vapour in a gas absorption tower. The oil, loaded with the recovered solute, leaves the tower and enters a steam stripping unit in which the hydrocarbons are removed from the oil. The regenerated oil is cooled and recycled back to the absorption tower. In such an application, cooling of the hot oil by direct contact with water is often done in industrial practice. The oil is sparged at the bottom of a cylindrical column. The droplets of the hot oil move up through water that enters the column at the top and leaves at the bottom. The arrangement provides countercurrent direct contact between the hot and the cold streams (see Fig. 4.9).

In a particular unit, hot oil enters the column at 120°C. The flow rate of water is sufficiently high and the water temperature may be assumed to remain constant at an

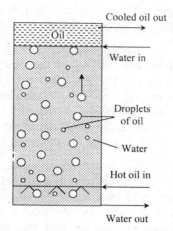

Cooled oil out

Water in

Droplets of oil

Water

Hot oil in

Water out

Fig. 4.9 A gas absorption tower (Problem 4.9).

average value of 30°C. The average drop size is 6 mm, and the velocity of rise is 5 cm/s. The exit oil temperature should not be above 35°C. Using an estimated 'drop-side heat transfer coefficient' of 35 kcal/h m² °C, calculate the height of the column required.

   Given: for the oil, $\rho = 900$ kg/m³, $c_p = 0.478$ kcal/kg °C. Use a suitable correlation (for flow past a sphere) to estimate the water-side heat transfer coefficient.

   [*Hint:* Calculate the drop Reynolds number, and calculate the water-side coefficient by using a suitable correlation for heat transfer for flow past a sphere [Eq (4.22) or (4.23) may be used]. Calculate the overall heat transfer coefficient U. Assume that the temperature inside the droplet remains uniform (i.e. it is not a function of the position inside the drop; 'lumped parameter model', see Problem 3.1). Write down the heat balance equation $-mc_p(dT/dt) = UA(T - T_w)$, where m is the mass of a drop of area A, c_p is its specific heat, T is the temperature of the drop at time t, and T_w is the temperature of water (assumed constant). Integrate this equation from $T = 120$°C to $T = 35$°C, and calculate the time t. The height of the column is droplet velocity times t.]

4.10 Spray towers are sometimes used for cooling process water. In one such tower, the water droplets have an average diameter of 1.5 mm. The relative velocity between the falling drops and the upflowing air is 0.9 m/s. In a certain section of the tower the air temperature is 23°C, and the water temperature is 42°C. Calculate the convective heat transfer coefficient using the equations given before and also the Ranz and Marshal equation, $Nu = 2.0 + 0.60(Re)^{0.5} (Pr)^{0.33}$

4.11 Water at 25°C flows over a wide hot plate at 1.5 m/s free stream velocity at zero angle of incidence. The plate is 1 m long in the direction of flow. (a) Determine the local and average heat transfer coefficients. (b) Calculate the heat supply necessary to maintain the plate at 70°C. Heat flow occurs from one side of the plate only.

4.12 Air at 1 atm and 28°C is flowing parallel to a flat plate at a velocity of 20 m/s. If the plate length is 2 m and its temperature 125°C, calculate the average heat transfer coefficient and the total rate of heat transfer from the plate (the critical Re = 5×10^5).

4.13 Air at 20°C flows over a block of ice, 1 m × 1 m free surface area, at a velocity of 8 m/s. If the ice is at its melting point, calculate the rate of melting of ice.

4.14 An oil ($k = 0.138$ W/m °C) flowing through a 25 mm radius pipe exchanges heat with an external medium. At a particular section of the pipe, the radial velocity and the temperature distribution in the oil are: $V = 0.5 - 800r^2$; $T = 60 + 1.728 \times 10^5 r^2 - 3.072 \times 10^6 r^3$, where V is in m/s, r in m, and T in °C. (a) Determine whether the oil is in laminar or in turbulent flow. (b) Is the oil being heated or cooled? (c) Calculate the average velocity of the oil. (d) Calculate the 'cup-mixing temperature' of the oil at the given section. (e) Find the local (i) wall heat flux, (ii) heat transfer coefficient, and (iii) Nusselt number.

4.15 The exhaust gas from a furnace flows across a tube bundle at an undisturbed velocity of 15 m/s. The bundle consists of seven transverse rows and six longitudinal rows of 38 mm tubes (1.5 inch, 14 BWG; o.d. = 38 mm; i.d. = 34 mm), and carries process water for heating. If the gas temperature is 260°C and the tube wall temperature 70°C, calculate (a) the gas-side film coefficient, and (b) the heat flux to water. The exhaust gas has thermophysical properties like air.

4.16 Polyethylene terephthalate is a polymer of great commercial importance having a variety of uses from fabrics to moulded articles (e.g. PET bottles). A stream of waste ethylene glycol is recovered from the reaction products in a PET plant by distillation in a column and the hot stream of glycol at 100°C has to be cooled down to 45°C, before pumping to the storage tank, at the rate of 1000 kg/h by countercurrent heat exchange in a double-pipe heat exchanger using cooling water. The cooling water enters at 25°C and leaves at 35°C. Glycol flows through the inner pipe, 35 mm i.d. and 42 mm o.d. (1.25 inch, schedule 40 IPS), and the outer water pipe has an i.d. of 78 mm (3 inch, schedule 40 IPS). Calculate the heat transfer area required. The following data for ethylene glycol are available: $\rho = 1075$ kg/m^3, $c_p = 0.611$ kcal/kg °C, $k = 0.223$ kcal/h m°C, viscosity, $\ln \mu = A + B/T$ (μ is the viscosity in cP; $A = -7.811$, $B = 3.143 \times 10^3$, T is in K). Viscosity correction of the estimated heat transfer coefficient is necessary.

4.17 Mercury purified by distillation is cooled by passing it through a double-pipe heat exchanger [inner tube: 13.5 mm i.d., 2.75 mm wall thickness (3/4 inch, 12 BWG); outer pipe: 41 mm i.d. (1-1/2 inch, schedule 40 IPS)]. The inner tube of the exchanger carries mercury flowing at a rate of 1500 kg/h, entering at 250°C and leaving at 40°C. Cooling water enters the outer pipe at 26°C and leaves at 40°C. Calculate the required length of the tube. The following physical property values of mercury at the average temperature may be used: $\rho = 13,200$ kg/m^3, $v = 8.5 \times 10^{-8}$ m^2/s, $c_p = 0.14$ kJ/kg °C, $k = 11.5$ W/m °C.

4.18 A small oil-fired boiler consumes 200 kg of fuel oil per hour. The oil is stored in a tank at the ambient temperature of 28°C, and must be heated to 90°C so that it can be atomized satisfactorily in the oil burner. Saturated steam at 140°C is used in the jacket of a double-pipe heat exchanger, while oil flows through the inner pipe of 25 mm i.d. (1 inch, schedule 40 IPS). Calculate the length of the pipe required. The following properties of fuel oil may be used: $\rho = 978$ kg/m^3; $c_p = 1.88$ kJ/kg °C; $k = 0.112$ W/m °C; $\ln \mu = -11.846 + 5221/T$, μ in cP, and T in K. The thermal resistance of the wall as well as that of condensing steam may be neglected. Viscosity correction of the oil-side heat transfer coefficient is required.

4.19 Consider steady state heat transfer (by conduction) from a sphere of diameter d, having a constant surface temperature T_s, to a large volume of 'stagnant' fluid. The temperature of the fluid is T_∞ at a point far away from the sphere. Show that the Nusselt number is Nu = 2.

[*Hint:* Make a shell balance in the fluid and solve it to get the steady state temperature distribution $T(r)$ in the fluid phase. Determine $h = -k[dT/dr]_{r = d/2}$ and obtain the value of Nu.]

4.20 Derive Eq. (4.28).

[*Hint:* Considering a distance dx in the longitudinal direction and a gas flow area $S_T l$ (l = length of the tube), the energy balance is given by $-S_T \, lV \, c_p dT = (dx/S_L)\pi \, d \, l \, \bar{h} \, (T - T_S)$, where dx/S_L is the number of tubes that lie within the distance l. Integrate this equation and use the condition $x = 0$, $T = T_i$. For the longitudinal dimension $x = L$ (say) of the bundle, the number of longitudinal rows = $L/S_L = N/N_T$. At $x = L$, the gas leaves the bundle at a temperature T_o. This will lead to Eq. (4.28).]

4.21 In the final stage of manufacture of polyvinyl chloride (PVC) beads, moisture present in the washed beads is removed by drying in a rotary drier. Hot air flows through the drier countercurrent to the beads when the water adhering to the beads vaporizes out. In a particular PVC plant, the drying air required for the purpose is heated in cross-flow across a tube bundle. Air approaches at a velocity of 7 m/s and a temperature of 25°C. The tube bundle consists of 49 tubes (1-1/4 inch, 16 BWG) arranged in a 6 cm square pitch in equal number of longitudinal and square rows. The tube bundle has an effective length of 1 m. The tube wall temperature is maintained constant at 130°C by condensing low pressure steam within the tubes. Calculate the rate of heat transfer to the air.

4.22 An air pre-heater in the form of a tube bundle is used to pre-heat air by heat exchange with hot flue gas. The feed air flows through the tubes and the flue is in cross-flow across the bundle. There are ten transverse rows with twelve 1 inch o.d. tubes in each row. The tubes are in staggered arrangement with a transverse pitch $S_T = 2.5$ cm, and a longitudinal pitch $S_L = 3$ cm. The upstream velocity of the flue gas is 5 m/s and the temperature is 250°C. The average tube surface temperature is 120°C. Calculate (a) the drop in the flue gas temperature as it leaves the bundle, and (b) the rate of heat transfer to air.

4.23 Water is flowing through a 1 inch schedule 40 pipe at a rate of 160 ft³ per hour. The pipe is 10 ft long and the pipe wall has a constant temperature of 215°F. If the inlet temperature of water is 85°F, calculate the exit water temperature by using (i) Reynolds analogy, (ii) Prandtl analogy, and (iii) Colburn analogy. Neglect entrance effects.

REFERENCES AND FURTHER READING

Churchill, S.W. and M. Burnstein, "A correlating equation for forced convection from gases and liquids to a circular cylinder in cross-flow", *J. Heat Transfer*, **99** (1977), 300.

Dittus, F.W. and L.M.K. Boelter, University, California (Berkeley), Pub. Eng., Vol. 2 (1930), 443.

Eckert, E.R.G. and R.M. Drake, *Analysis of Heat and Mass Transfer*, McGraw-Hill, New York, 1972.

Fand, R.M., "Heat transfer by forced convection to a cylinder in cross-flow", *Intern. J. Heat Mass Transfer,* **8** (1965), 995.

Hausen, H., *Beih. Verfahrenstech.,* **4** (1943), 91.

Kern, D.Q., *Process Heat Transfer*, McGraw-Hill, New York, 1950.

Knudsen, J.D. and D.L. Katz, *Fluid Dynamics and Heat Transfer*, McGraw-Hill, New York, 1958.

Kramers, H., "Heat transfer from spheres to flowing media", *Physica*, **12** (1946), 61.

Ludwig, E., *Applied Process Design for Chemical and Petrochemical Plants*, Vol III, Gulf Publ. Co., Houston, Texas, 1983.

Reid, R.C., J.M. Prausnitz, and B.E. Poling, *The Properties of Gases and Liquids*, 4th ed., McGraw-Hill, New York, 1988.

Schlichting, H., *Boundary Layer Theory*, McGraw-Hill, New York, 1968.

Seban, R.A. and T.T. Shimazaki, "Heat transfer to a fluid flowing turbulently in a smooth pipe with walls at constant temperature", *Trans. ASME*, **83** (1951), 803.

Shah, R.K. and A.L. London, *Laminar Flow Convection in Ducts*, Academic Press, New York, 1978.

Sieder, E.W. and C.E. Tate, "Heat transfer and pressure drop of liquids in tubes", *Ind. Eng. Chem.*, **28** (1936), 1429.

Taylor, E.S., *Dimensional Analysis for Engineers,* Clarendon Press, Oxford, 1974.

Whitaker, S., "Forced convection heat transfer correlations for flow in pipes, past flat plates, single cylinders, single spheres, and flow in packed beds and tube bundles", *AIChE J.*, **18** (1972), 361.

Zhukauskas, A., "Heat transfer from tubes in cross-flow", *Adv. Heat Transfer*, **8**, 1972.

5

Free Convection

Free convection or natural convection pertains to the motion of a fluid caused by density differences during heating or cooling of the fluid. A common demonstration is the heating of a solution in a beaker in the laboratory. The liquid at the bottom of the beaker gets heated up and its density decreases. The liquid above is at a lower temperature and, therefore, has a higher density. Thus a buoyancy force comes into play; the lighter liquid at the bottom moves up, and the heavier liquid at the top flows down creating free convection currents. The upflowing hot liquid transfers heat from the bottom region to the top region. The phenomenon is called *free convection* or *natural convection,* because no mechanical device or externally imposed force is used to create motion in the fluid.*

There are numerous examples of free convection heat transfer in industrial operations. Process equipment and accessories at an elevated temperature (or at a sub-atmospheric temperature) are insulated to avoid heat loss (or gain). The *skin temperature* of the insulation will be more than the ambient temperature. This causes free convection currents on the surface of the insulation and consequent heat loss. Some of the examples are heat losses from steam pipes, reactors vessels, furnaces, etc.

5.1 QUALITATIVE DESCRIPTION OF FREE CONVECTION FLOWS

The buoyancy force that causes convective flow is often called *body force* and is gravitational in nature. This force, i.e. buoyant force comes into play when the fluid is heated at the bottom (or cooled at the top). Such a system is *hydrodynamically unstable* for all practical purposes. When the fluid is heated at the top, for example, by confining the liquid in the region between two horizontal parallel surfaces, the system is *stable*. No convection current sets in because the buoyancy force simply does not exist. This situation is shown in Fig. 5.1. Free convection flow

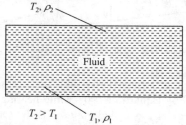

Fig. 5.1 A fluid confined between two parallel surfaces, the upper surface is at a higher temperature.

---

* Natural convection current is also induced when a solute kept suspended in a liquid undergoes dissolution creating a solution of density higher than that of the liquid below. This convection current causes 'natural convection mass transfer'.

at or over a heated surface—for example, a wall, a wire or a cylinder—occurs in the form of a *plume*. The plume width varies with position as shown in Fig. 5.2.

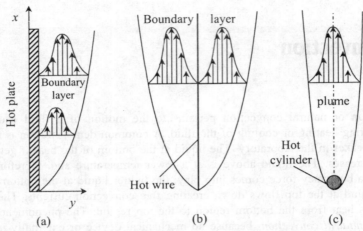

Fig. 5.2   Free convection boundary layer: (a) along a heated plate, (b) along a hot vertical wire, and (c) plume over a hot horizontal cylinder.

However, when a fluid is enclosed between two horizontal surfaces (at a distance L apart, say), the lower surface being hot and the upper surface cold, heat transfer occurs practically by pure conduction when the Grashof number Gr_L, (this is a dimensionless group characterizing free convection; see Section 5.2) is less than about 1700. Above this value of the Grashof number, free convection begins and a pattern of hexagonal prismatic cells is formed as shown in Fig. 5.3. At the middle of each prism, the warmer fluid ascends; the colder fluid descends along the 'walls'

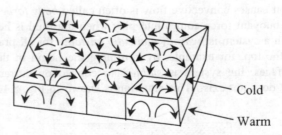

Fig. 5.3   Benard cells in a layer of liquid heated from below (schematic).

of the hexagonal prisms. The top surface looks like the cross-section of a honeycomb. This phenomenon was first observed by Benard in 1901 while heating a thin layer of spermaceti from below. These free convection cells are called *Benard cells*. Other common situations of free convection heat transfer are from a horizontal (or inclined) plate, hot or cold, facing upwards or downwards. Free convection in a vertical or inclined enclosure made up of two parallel, cylindrical or spherical surfaces is also important from the practical angle. The natural convection flow patterns in a few common geometries are qualitatively shown in Fig. 5.4.

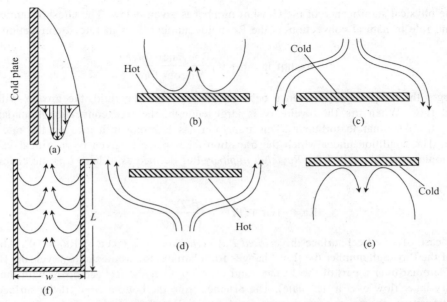

Fig. 5.4 Free convection flow patterns: (a) on a cold vertical surface, (b) on a hot surface facing up, (c) on a cold surface facing up, (d) on a hot surface facing down, (e) on a cold surface facing down, and (f) on a vertical enclosure with cold parallel surfaces.

5.2 HEAT TRANSFER CORRELATIONS FOR FREE CONVECTION

Analytical solutions for velocity and temperature distributions as well as for the heat transfer rate in free convective motion are available for a few simple situations (see Chapter 11). These solutions are based on the *similarity principle* (see, for example, Schlichting, 1968) and boundary layer approximations. However, in many practical cases we have to depend on empirical correlations. Free convection heat transfer correlations involve the Nusselt number, the Prandtl number and the Grashof number. The expression for the Grashof number Gr_L is given below (the details of how the Grashof number arises in the analysis of free convection heat transfer are not given here).

$$Gr_L = \frac{g\beta(T_s - T_o)L^3}{\nu^2} \tag{5.1}$$

where

$$\beta = -\frac{1}{\rho}\left(\frac{\partial\rho}{\partial T}\right)_P = \text{the coefficient of volumetric expansion}$$

g is the acceleration due to gravity
L is a characteristic length
T_o is the ambient fluid temperature
T_s is the surface temperature
ρ is the fluid density
P is the total pressure.

The physical significance of the Grashof number is given below. The Grashof number plays the same role in natural convection as the Reynolds number does in forced convection.

$$\text{Grashof number} = \frac{\text{Bouyancy force}}{\text{Viscous force}}$$

The larger the temperature difference between the wall and the fluid, the greater will be the induced flow. When the fluid velocity is large enough, the free convection boundary layer changes from laminar to turbulent. This transition has a strong influence on the rate of heat transfer. The condition under which the transition takes place is given by a critical value of a dimensionless number called the *Rayleigh number,* Ra_L defined as (where L is the characteristic length)

$$Ra_L = (Gr_L)\,(Pr) = \frac{g\beta(T_s - T_o)L^3}{\nu\alpha} \tag{5.2}$$

For the case of a vertical surface the *critical Rayleigh number* is taken as $Ra_L \approx 10^9$. Above this value of the Rayleigh number the flow changes from laminar to turbulent. However, the flow may remain laminar over a part of the surface, and turbulent over the rest (compare with the case of boundary layer flow over a flat plate). The distance from the bottom edge (if the surface is hot) where transition occurs is given by the local Rayleigh number, $Ra_x = 10^9$. A few important correlations for free convection heat transfer from surfaces of common geometries are discussed below. These correlations are usually of the form $Nu = f(Ra, Pr)$, and the relevant physical properties are taken at the mean film temperature.

5.2.1   Free Convection from a Flat Surface

The rate of free convection heat transfer from a flat surface or plate greatly depends upon the orientation of the plate—whether it is vertical, inclined, or horizontal. Again, a horizontal plate may be hot or cold, and the hot or the cold surface may face upwards or downwards. A single correlation does not cover all these situations.

(i) For a vertical plate, hot or cold, the following equation suggested by Churchill and Chu (1975) may be used for the entire range of Rayleigh number.

$$\overline{Nu}_L = \frac{\overline{h}L}{k} = \left[0.825 + \frac{0.387\,(Ra_L)^{1/6}}{\left\{1 + (0.492/Pr)^{9/16}\right\}^{8/27}} \right]^2 \tag{5.3}$$

However, if the free convective flow over the plate is laminar (i.e. $Ra_L < 10^9$), the following equation gives a better prediction.

$$\overline{Nu}_L = \frac{\overline{h}L}{k} = 0.68 + \frac{0.67\,(Ra_L)^{1/4}}{\left\{1 + (0.492/Pr)^{9/16}\right\}^{4/9}} \tag{5.4}$$

These equations are also applicable for an inclined plate if the angle of inclination with the vertical, $\theta < 60°$. In this chapter, in all the correlations, $\overline{h}$ denotes the average heat transfer coefficient and $\overline{Nu}$ the average Nusselt number.

Free convection heat transfer from a flat surface is of much practical importance because heat transfer from furnace walls and many other equipment occurs by this mechanism.

Example 5.1 A furnace with a steel door, having an inner lining of an insulating material, is at a temperature of 65°C. The door, 1.5 m high and 1 m wide, loses heat to an ambient at 25°C. Calculate the rate of heat loss from the door at steady state.

SOLUTION The average free convection heat transfer coefficient $\bar{h}$ can be calculated using Eq. (5.3). The Rayleigh number will be calculated first in order to check the flow regime.

The average air-film temperature, $T_f = (65 + 25)/2 = 45°C = 318$ K

The relevant properties of the air at this temperature are: Prandtl number, Pr = 0.695; $v = 1.85 \times 10^{-5}$ m²/s; $\beta = 1/T_f = 1/318 = 3.145 \times 10^{-3}$ K^{-1}; $k = 0.028$ W/m °C.

Other relevant data are: door height, $L = 1.5$ m; temperature drop, $T_s - T_o = 65 - 25 = 40°C$; $g = 9.8$ m/s².

$$\text{Gr}_L = \frac{g\beta(T_s - T_o)L^3}{v^2} = \frac{(9.8)(3.145 \times 10^{-3})(40)(1.5)^3}{(1.85 \times 10^{-5})^2} = 1.216 \times 10^{10}$$

The Rayleigh number, $\text{Ra}_L = (\text{Gr}_L)\,(\text{Pr}) = (1.216 \times 10^{10})(0.695) = 8.45 \times 10^9$

From Eq. (5.3), $\overline{\text{Nu}}_L = \dfrac{\bar{h}L}{k} = \left[0.825 + \dfrac{0.387(\text{Ra}_L)^{1/6}}{\left\{1 + (0.492/\text{Pr})^{9/16}\right\}^{8/27}} \right]^2$

$$= \left[0.825 + \frac{0.387(8.45 \times 10^9)^{1/6}}{\left\{1 + (0.492/0.695)^{9/16}\right\}^{8/27}} \right]^2$$

$$= 238.4$$

The average heat transfer coefficient,

$$\bar{h} = (238.4)\left(\frac{k}{L}\right) = \frac{238.4 \times 0.028}{1.5} = 4.45 \text{ W/m}^2 \text{ °C}$$

Door area = (1.5)(1.0) = 1.5 m²

Temperature driving force, $\Delta T = 65 - 25 = 40°C$

Rate of heat loss = $\bar{h}\,A\,\Delta T = (4.45)(1.5)(40) = \boxed{267 \text{ W}}$

(ii) Different flow situations arise when heat transfer by natural convection occurs from a horizontal plate depending upon whether the plate is hot or cold, or whether the heat transfer surface faces up or down. The nature of flow in such cases has been shown qualitatively before. A characteristic length in all these cases is defined as below.

$$L = \frac{\text{Area of the plate}}{\text{Perimeter of the plate}}$$

A few useful correlations are given below:

(a) Heat transfer *from the upper surface* of a flat plate:

$$\overline{\text{Nu}} = \frac{\overline{h}L}{k} = 0.54 \, (\text{Ra}_L)^{1/4}; \qquad 10^4 \le \text{Ra}_L \le 10^7 \tag{5.5}$$

(b) Heat transfer *to the lower surface* of a cold plate:

$$\overline{\text{Nu}} = \frac{\overline{h}L}{k} = 0.15 \, (\text{Ra}_L)^{1/3}; \qquad 10^7 \le \text{Ra}_L \le 10^{11} \tag{5.6}$$

(c) Heat transfer *from the lower surface* of a hot plate or *to the upper surface* of a cold plate:

$$\overline{\text{Nu}} = \frac{\overline{h}L}{k} = 0.27 \, (\text{Ra}_L)^{1/4}; \qquad 10^5 \le \text{Ra}_L \le 10^{10} \tag{5.7}$$

Example 5.2   A thin metal plate 1 m × 1 m is placed on a rooftop. It receives radiant heat from the sun directly at the rate of 170 W/m². If heat transfer from the plate to the ambient occurs purely by free convection, calculate the steady state temperature of the plate. Assume that there is no heat loss from the bottom of the plate. The ambient temperature is 25°C.

SOLUTION   At steady state, the rate of radiant heat input to the plate must be equal to the rate of heat loss to the ambient at 25°C by free convection. Thus the problem reduces to: What should be the plate temperature in order that the free convection heat flux is 170 W/m²? We need to calculate the Grashof number in order to calculate the heat transfer coefficient, and we must know the plate temperature in order to calculate the Grashof number. But the plate temperature is what we are to find out. So the problem has to be solved by trial, starting with a guess value of the plate temperature.

Let us assume a plate temperature, $T_s = 60°C$.
The mean film temperature = $(60 + 25)/2 = 42.5°C = 315.5$ K
Given: $T_s - T_o = 60 - 25 = 35°C$; Properties of the air at the mean film temperature: $\beta = 1/315.5 = 3.17 \times 10^{-3}$ K⁻¹; $v = 1.85 \times 10^{-5}$ m²/s; Pr = 0.7; $k = 0.028$ W/m °C, and $g = 9.8$ m/s².
For free convection from a horizontal plate, the characteristic length L is taken as

$$L = \frac{\text{Plate area}}{\text{Perimeter}} = \frac{1 \, \text{m}^2}{4 \, \text{m}} = 0.25 \, \text{m}$$

$\text{Ra}_L = (\text{Gr}_L) \, (\text{Pr})$

$$= \frac{g\beta(T_s - T_o)L^3}{v^2} \, (\text{Pr}) = \frac{(9.8)(3.17 \times 10^{-3})(35)(0.25)^3}{(1.85 \times 10^{-5})^2} \, (0.70) = 3.475 \times 10^7$$

We will use Eq. (5.5) for calculating the Nusselt number though the value of Ra_L is a little out of the range.

$$\overline{Nu}_L = \frac{\overline{h}L}{k} = 0.54\,(Ra_L)^{1/4} = (0.54)(3.475 \times 10^7)^{1/4} = 41.46$$

or

$$\overline{h} = \frac{(41.46)(k)}{L} = \frac{(41.46)(0.028)}{0.25} = 4.64 \text{ W}/\text{m}^2 \text{ °C}$$

Rate of heat transfer from the plate $= (4.64)(1.0)(T_s - 25) = 170 \text{ W/m}^2$

Therefore,

$$\text{Plate temperature, } T_s = \boxed{61.6°\text{C}}$$

The calculated temperature is very close to the initial guess and therefore no further trial is necessary.

5.2.2 Free Convection from a Cylinder

(a) For heat transfer from or to a *horizontal cylinder*, Morgan (1975) recommended a correlation of the following form

$$\overline{Nu} = \frac{\overline{h}d}{k} = C(Ra_d)^n \tag{5.8}$$

where d is the diameter of the cylinder. The constant C and the exponent n may be obtained from Table 5.1 for different ranges of the Rayleigh number.

Table 5.1   Values of C and n for Eq. (5.8)

Ra_d	C	n
10^{-10} – 10^{-2}	0.675	0.058
10^{-2} – 10^2	1.02	0.148
10^2 – 10^4	0.85	0.188
10^4 – 10^7	0.48	0.250
10^7 – 10^{12}	0.125	0.333

The following correlation by Churchill and Chu (1975) may also be used for free convection heat transfer calculation from a horizontal cylinder.

$$\overline{Nu} = \frac{\overline{h}d}{k} = \left[0.60 + \frac{0.387(Ra_d)^{1/6}}{\left\{1 + (0.559/Pr)^{9/16}\right\}^{8/27}} \right]^2 ; \qquad 10^{-5} < Ra_d < 10^{12} \tag{5.9}$$

(b) In the case of a *vertical cylinder*, the equations for a vertical flat plate are applicable if the thickness of the free convection boundary layer is much smaller than the cylinder diameter. The cylinder may be considered to be equivalent to a vertical flat plate of breadth equal to the circumference of the cylinder. The criterion for this is quantitatively given by

$$\frac{d}{L} \geq \frac{35}{(Gr_L)^{1/4}} \tag{5.10}$$

Here L is the height of the vertical cylinder. However, if the cylinder diameter is small and the above condition is not fulfilled, the numerical results of Minkowycz and Sparrow (1974) may be used.

Example 5.3   A long horizontal cylindrical carbon steel rod, 2.54 cm in diameter and 40 cm long, at 80°C cools down by free convection heat transfer to an ambient at 30°C. Calculate the time required for cooling of the rod down to 35°C. Assume that the temperature of the rod remains uniform at any instant ('lumped parameter model'; see Problem 3.1). The following simple correlation (applicable for free convection heat transfer from a horizontal cylinder in air) may be used: $h = 1.32 \ (\Delta T/d)^{0.25} \ \text{W/m}^2 \ °C$, where ΔT (°C) is the temperature difference between the surface and the ambient, and d (in m) is the diameter of the cylinder. For carbon steel, $\rho = 7800 \ \text{kg/m}^3$, $c_p = 0.473 \ \text{kJ/kg °C}$.

SOLUTION   Here the heat transfer process occurs at unsteady state. If m is the mass of the cylinder, c_p its specific heat, A the area of heat transfer, and T_o the ambient fluid temperature, the differential equation for variation of the temperature of the body is given by

$$-mc_p \frac{dT}{dt} = hA(T - T_o)$$

or

$$-\frac{d\hat{T}}{dt} = \frac{hA}{mc_p}\hat{T} \quad \text{where } \hat{T} = T - T_o$$

Given: diameter of the cylinder, $d = 0.0254$ m;   length, $l = 0.4$ m;   mass $= (\pi/4)d^2l\rho = (\pi/4)(0.0254)^2(0.4)(7800) = 1.581$ kg

If we neglect the area of the flat surfaces of the cylinder, the area of heat transfer,

$$A = \pi \, d \, l = \pi \ (0.0254)(0.4) = 0.0319 \ \text{m}^2$$

Because, $T - T_o$ is the instantaneous temperature difference between the cylinder and the ambient,

$$\hat{T} = T - T_o = \Delta T$$

Heat transfer coefficient, $h = 1.32(\hat{T}/d)^{0.25} = 1.32(\hat{T}/0.0254)^{0.25} = 3.306(\hat{T})^{0.25} \ \text{W/m}^2 \ °C$
Therefore,

$$-\frac{d\hat{T}}{dt} = \frac{(3.306)(0.0319)}{(1.581)(473)}(\hat{T})^{0.25}\hat{T}$$

or

$$-\frac{d\hat{T}}{(\hat{T})^{5/4}} = 1.410 \times 10^{-4} \ dt$$

Integrating, we have

$$4(\hat{T})^{-1/4} = 1.410 \times 10^{-4} \ t + C_1, \quad \text{where } C_1 \text{ is an integration constant.}$$

Using the initial condition, $t = 0$, $\hat{T} = 80 - 30 = 50$ °C, we get

$$C_1 = 4(50)^{-1/4} = 1.504$$

and

$$(\hat{T})^{-1/4} = 0.376 + 0.352 \times 10^{-4} \ t$$

When the temperature drops down to 35°C, $\hat{T} = 35 - 30 = 5$°C. Therefore,

$$(5)^{-1/4} = 0.376 + 0.352 \times 10^{-4}\, t$$

or

$$t = 8316 \text{ s} = 2.31 \text{ h}$$

So the required time for cooling is $\boxed{2.31\,\text{h}}$.

Example 5.4 A horizontal steam pipe, 78 mm i.d. and 89 mm o.d., carries saturated steam at 15 kg/cm^2 gauge pressure. It is lagged only with a 2 cm thick layer of pre-formed mineral fibre of thermal conductivity 0.05 W/m °C which is exposed to ambient air at 25°C. Calculate the rate of heat loss by free convection per metre length of the pipe.

SOLUTION Here, heat loss occurs from the steam pipe in the presence of two resistances in series—one of the insulation and the other of the free convection boundary layer outside the insulation. As the skin temperature of the insulation is not known, the free convection film coefficient cannot be calculated directly (because the coefficient itself depends upon the difference between the skin temperature and the ambient temperature). However, the rate of heat transfer through the insulation must be equal to the rate of heat transfer from the surface of the insulation by free convection at *steady state*. The problem has to be solved by trial and error starting with an assumed or guess value of the skin temperature of the insulation (or, alternatively, we may proceed by assuming a value of the free convection heat transfer coefficient). If the heat transfer resistance offered by the pipe wall is neglected, the temperature at the inner surface of the insulation (say T_i) is equal to the steam temperature. The temperature of the saturated steam at 15 kg/cm^2 (or 15.72 bar absolute pressure) is 473.5 K (from steam table).

Let us assume an insulation skin temperature, $T_s = 50$°C. In order to calculate the heat transfer rate through the insulation, we use the following data: pipe length, $l = 1$ m; thermal conductivity of the insulation, $k_c = 0.05$ W/m °C; $T_i = 473.5$ K $= 200.5$°C; $T_s = 50$°C (assumed); outer radius of the insulation, $r_s = (8.9/2)$ cm $+ 2$ cm $= 0.0645$ m; inner radius of the insulation, $r_i = 8.9/2$ cm $= 0.0445$ m.

The rate of heat transfer through the insulation per metre length of the pipe is

$$Q = \frac{2\pi L k_c (T_i - T_s)}{\ln (r_s / r_i)} = \frac{2\pi (1)(0.05)(200.5 - 50)}{\ln (0.0645 / 0.0445)} = 127.4 \text{ W}$$

Now we calculate the convection film coefficient. The properties of the air taken at the mean film temperature, i.e. $T_f = (50 + 25)/2 = 37.5$°C $(= 310.5$ K), are: $v = 1.76 \times 10^{-5}$ m^2/s; $\beta = (1/310.5) = 3.22 \times 10^{-3}$ K^{-1}; $k = 0.027$ W/m °C; Pr $= 0.705$.

Also, the o.d. of the insulated pipe, $d_s = 2r_s = 0.129$ m; $g = 9.8$ m^2/s; ambient temperature, $T_o = 25$°C.

Grashof number, $\mathrm{Gr}_d = \dfrac{g\beta(T_s - T_o)d_s^3}{v^2} = \dfrac{(9.8)(3.22 \times 10^{-3})(50 - 25)(0.129)^3}{(1.76 \times 10^{-5})^2} = 5.47 \times 10^6$

Rayleigh number, $\mathrm{Ra}_d = (\mathrm{Gr}_d)\,(\mathrm{Pr}) = (5.47 \times 10^6)\,(0.705) = 3.85 \times 10^6$

Using Eq. (5.9), we have

$$\overline{\text{Nu}} = \frac{\overline{h}\,d_s}{k} = \left[0.60 + \frac{0.387(3.85 \times 10^6)^{1/6}}{\left\{1 + (0.559/0.705)^{9/16}\right\}^{8/27}}\right]^2 = 21.34$$

or

$$\overline{h} = (21.34)\left(\frac{k}{d_s}\right) = \frac{21.34 \times 0.027}{0.129} = 4.47 \text{ W/m}^2 \text{ °C}$$

The rate of heat loss, $Q = (\pi\, d_s l)(\overline{h})(T_s - T_o)$

Using the value of Q calculated before, we get

$$127.4 = \pi(0.129)(1)(4.47)(T_s - 25)$$

or

$$T_s = 95.3°C$$

This is the estimated value of the skin temperature of the insulation. It is much higher than our initial guess of 50°C. Let us calculate the whole thing again with a revised guess value of $T_s = 70°C$.

The rate of heat loss through the insulation is

$$Q = \frac{2\pi l k_c(T_i - T_s)}{\ln\,(r_s/r_i)} = \frac{2\pi(1.0)(0.05)(200.5 - 70)}{\ln\,(0.0645/0.0445)} = 110.4 \text{ W}$$

At the average air-film temperature, $T_f = (25 + 70)/2 = 47.5°C \ (= 320.5 \text{ K})$, the properties of air are: $\beta = (1/320.5) = 3.12 \times 10^{-3}$, $v = 1.8 \times 10^{-5} \text{ m}^2/\text{s}$; Pr = 0.703; $k = 0.0275 \text{ W/m °C}$.

$$\text{Ra}_d = (\text{Gr}_d)(\text{Pr}) = \frac{g\beta(T_s - T_a)d^3}{v^2}\,\text{Pr} = \frac{(9.8)(3.12 \times 10^{-3})(70 - 25)(0.129)^3}{(1.8 \times 10^{-5})^2}\,(0.703) = 6.4 \times 10^6$$

Using Eq. (5.9), we get

$$\frac{\overline{h}\,d_s}{k} = \left[0.60 + \frac{(0.387)(6.4 \times 10^6)^{1/6}}{\left\{1 + (0.559/0.703)^{9/16}\right\}^{8/27}}\right] = 24.97$$

or

$$\overline{h} = (24.97)(0.0275/0.129) = 5.32 \text{ W/m}^2 \text{ °C}$$

and

$$Q = 110.4 = \pi dl\overline{h}\,(T_s - T_o) = \pi(0.129)(1)(5.32)(T_s - 25)$$

or

$$T_s \doteq 76.4°C$$

The recalculated skin temperature is again a little higher than the second guess temperature of 70°C. We now assume a skin temperature of 74°C, and calculate the whole thing once again. It can be shown that the calculated temperature is $T_s = 73.9°C$, which is very close to the assumed value. Taking $T_s = 74°C$, the rate of heat loss through the insulation per metre length of the pipe is

$$Q = \frac{(2\pi)(1)(0.05)(200.5 - 74)}{\ln(0.0645/0.0445)} = \boxed{107\ \text{W}}$$

Note: The calculated insulation skin temperature, 74°C, is too high to be acceptable. In fact, the temperature is high because the insulation thickness (50 mm) is too low. The use of a larger insulation thickness (75 to 100 mm) will solve the problem.

Example 5.5 The sulphur dioxide converter in a sulphuric acid plant is a tall cylindrical vessel housing four catalyst chambers for stage-wise conversion to sulphur trioxide with inter-stage cooling of the gas mixture. A 100 tpd (tons per day) plant has a converter 10.5 m in height and 4 m in diameter. It is insulated with a layer of mineral wool ($k_i = 0.0602$ kcal/ h m °C). Although the temperature in the reactor varies from one stage to another, an average reactor wall temperature of 460°C may be assumed. If heat loss from the insulated reactor occurs only by free convection, what thickness of insulation should be used so that the insulation skin temperature does not exceed 65°C? The ambient temperature may be taken as 30°C.

SOLUTION Heat is transferred from the reactor wall to the ambient through two resistances in series—those of the insulation and of the outer air-film. Because the insulation skin temperature is given, it is possible to calculate the free convection heat transfer coefficient and therefore the rate of heat transfer. The reactor is a vertical cylinder, but here the Nusselt number correlation for a vertical plate may be used if the criterion (5.10) is satisfied. The outside area of the insulated vessel is not known. Hence we have to start with an assumed value of the insulation thickness.

We will first check the validity of the criterion, Eq.(5.10), by calculating the Grashof number. The mean air-film temperature is

$$T_f = 65 + 30/2 = 47.5°C = 320.5\ \text{K}$$

The values of the relevant quantities are: $\beta = 1/320.5 = 3.12 \times 10^{-3}$ K^{-1}; $v = 1.84 \times 10^{-5}$ m^2/s; $T_s = 65°C$; $T_o = 30°C$; $L = 10.5$ m; $g = 9.8$ m/s^2; uninsulated vessel diameter, $d_i = 4$ m; for air, Prandtl number, Pr = 0.705, and $k = 0.0241$ kcal/h m °C. Now

$$\text{Gr}_L = \frac{g\beta(T_s - T_o)L^3}{v^2} = \frac{(9.8)(3.12 \times 10^{-3})(65 - 30)(10.5)^3}{(1.84 \times 10^{-5})^2} = 3.66 \times 10^{12}$$

$$\frac{d_i}{L} = \frac{4\ \text{m}}{10.5\ \text{m}} = 0.38$$

and

$$\frac{35}{(\text{Gr}_L)^{1/4}} = \frac{35}{(3.66 \times 10^{12})^{1/4}} = 0.025$$

Therefore,

$$\frac{d_i}{L} > \frac{35}{(\text{Gr}_L)^{1/4}}$$

Because the criterion (5.10) is satisfied, we will use the correlation for heat transfer for a vertical flat plate, Eq. (5.3).

$$\overline{Nu} = \frac{\overline{h}l}{k} = \left[0.825 + \frac{0.387 \, (Ra_L)^{1/6}}{\left\{ 1 + (0.496/Pr)^{9/16} \right\}^{8/27}} \right]^2$$

$$Ra_L = (Gr_L) \, (Pr) = (3.66 \times 10^{12})(0.705)$$

Therefore,

$$\frac{\overline{h}L}{k} = 1504$$

or

$$\overline{h} = (1504)\left(\frac{0.0241}{10.5}\right) = 3.45 \text{ kcal/h m}^2 \text{ °C}$$

If w is the required thickness of the insulation, the outer diameter of the insulated vessel is $(4 + 2w)$ metres. Because the calculated insulation thickness is expected to be much smaller than the vessel diameter, it is reasonable to use the 'arithmetic mean area' of the insulation for the purpose of calculation of the rate of heat transfer through the insulation.

Average diameter of the insulation $= \dfrac{4 + (4 + 2w)}{2} = 4 + w$

Average heat transfer area of the insulation $= \pi(4 + w)L = \pi(10.5)(4 + w)$

Rate of heat transfer through the insulation,

$$Q_i = \frac{\pi(4 + w)(10.5)(0.0602)(460 - 65)}{w}$$

Rate of heat transfer from the outer surface of the insulation by free convection,

$$Q_c = \overline{h} \, \pi(4 + 2w)L(65 - 30) = (3.45)(\pi)(4 + 2w)(10.5)(35)$$

Putting $Q_i = Q_c$ at *steady state*, we get

$$w = \boxed{0.188 \text{ m}} = \text{ the required insulation thickness}$$

Note: In actual practice, insulation of a SO_2 converter of this size is commonly done by about a 200 mm thick layer of mineral wool. The thickness calculated in this example is very close to that used in an operating plant.

5.2.3 Free Convection from a Sphere

The following correlation suggested by Churchill (1983) may be used.

$$\overline{Nu} = \frac{\overline{h}d}{k} = 2 + \frac{0.589 \, (Ra_d)^{1/4}}{\left\{ 1 + (0.469/Pr)^{9/16} \right\}^{4/9}}; \qquad Pr \geq 0.7; \qquad Ra_d \leq 10^{11} \qquad (5.11)$$

5.2.4 Free Convection in an Enclosure

Another important class of problems consists of convection in enclosed spaces. A lot of theoretical and experimental work in this area was reviewed by Ostrach (1972). Correlations for heat transfer in a horizontal, inclined or vertical rectangular enclosure, in the space between concentric and eccentric cylinders and spheres and also in enclosures of various other geometries are available (Churchill, 1983). Here we present the correlations of McGregor and Emery (1969) for free convection heat transfer in a *vertical rectangular* cavity (that has heated or cooled vertical surfaces and adiabatic horizontal surfaces).

$$\overline{Nu}_w = 0.42(Ra_w)^{0.25} \, (Pr)^{0.012} \left(\frac{L}{w}\right)^{-0.3} \tag{5.12}$$

$$\text{for} \quad 10 < \frac{L}{w} < 40; \quad 1 < Pr < 2 \times 10^4; \quad 10^4 < Ra_w < 10^7$$

$$\overline{Nu}_w = 0.046 \, (Ra_w)^{1/3} \tag{5.13}$$

$$\text{for} \quad 1 < \frac{L}{w} < 40; \quad 1 < Pr < 20; \quad 10^6 < Ra_w < 10^9$$

Here, w/L is called the *aspect ratio* of an enclosure [see Fig. 5.4(f)]. Both Nusselt number and Rayleigh number are based on the width of the enclosure, w (i.e. w is taken as the characteristic length).

A simple correlation for estimation of the heat transfer rate in the enclosure formed by two concentric horizontal cylinders was suggested by Raithby and Hollands (1975). They introduced an 'effective thermal conductivity' of the medium in the enclosure.

$$\frac{k_{\text{eff}}}{k} = 0.386 \left[\frac{Pr}{0.861 + Pr}\right]^{0.25} \left(Ra_{c,e}^*\right)^{0.25}; \quad 10^2 < Ra_{c,e}^* < 10^7 \tag{5.14}$$

where $Ra_{c,e}^*$ is a modified Rayleigh number given by

$$Ra_{c,e}^* = \frac{[\ln (d_o/d_i)]^4}{L^3 (d_i^{-0.6} + d_o^{-0.6})^5} \, Ra_L \tag{5.15}$$

where
d_i is the outer diameter of the inner tube

d_o is the inner diameter of the outer tube

$L = d_o - d_i$ = the characteristic length of the annular enclosure

The rate of heat flow by natural convection per unit length is the same as that through the annular cylindrical region having an effective thermal conductivity k_{eff}. That is,

$$q = \frac{2\pi k_{\text{eff}} (T_i - T_o)}{\ln (d_o/d_i)} \tag{5.16}$$

where T_i and T_o are the temperatures of the inner and the outer cylindrical walls, respectively.

A few more correlations applicable to natural convection in enclosures of other geometries are available in Holman (1992) and Incropera and DeWitt (1996).

Example 5.6 A refrigerator is placed near the partition wall of a room such that there is only a 4 cm gap in between. The surface of the refrigerator facing the wall is 1.6 m in height and 0.8 m in breadth. If the temperature of the surface is 22°C and the wall temperature is 30°C, calculate the rate of heat gain by the cooler surface.

SOLUTION This is a case of free convection heat transfer in a rectangular vertical enclosure (we assume that the partition wall and a surface of the refrigerator form an enclosure). Equation (5.12) or (5.13) can be used to calculate the Nusselt number depending upon the values of L/w, Pr, and Ra_w. Here, we have

Height of the enclosure, $L = 1.60$ m; width, $w = 0.04$ m; and $L/w = 1.6$ m/0.04 m $= 40$.
Mean air temperature $= (22 + 30)/2 = 26°C$
Prandtl number, Pr $= 0.7$
For air at 26°C: $\beta = 1/299$ K^{-1}; $v = 1.684 \times 10^{-5}$ m^2/s; $k = 0.026$ W/m °C; thermal diffusivity, $\alpha = 2.21 \times 10^{-5}$ m^2/s

Temperatures of the 'walls' of the enclosure: $T_s = 30°C$; $T_i = 22°C$; and $g = 9.8$ m/s^2. Therefore,

$$Ra_w = \frac{g\beta(T_s - T_i)w^3}{v\alpha} = \frac{(9.8)(1/299)(30 - 22)(0.04)^3}{(1.684 \times 10^{-5})(2.21 \times 10^{-5})} = 4.51 \times 10^4$$

Equation (5.12) is used for the calculation of the Nusselt number. Thus, we have

$$\overline{Nu}_w = 0.42(Ra_w)^{0.25}(Pr)^{0.012}\left(\frac{L}{w}\right)^{-0.3}$$

$$= 0.42(4.51 \times 10^4)^{0.25}(0.7)^{0.012}(40)^{-0.3}$$

$$= 2.02$$

Therefore,

$$\overline{h} = \overline{Nu}_w\left(\frac{k}{w}\right)$$

$$= (2.02)\left(\frac{0.026}{0.04}\right) = 1.313 \text{ W/m}^2 \text{ °C}$$

Rate of heat transfer to the refrigerator through this wall $= \overline{h}(\text{area})(T_s - T_i)$

$$= (1.313)(1.6 \times 0.8)(30 - 22) = \boxed{13.5 \text{ W}}$$

5.3 COMBINED FREE AND FORCED CONVECTION

There are practical situations in which the contributions of both free and forced convection should be taken into account for heat transfer calculations. The following conditions determine the regimes of free convection, forced convection, and combined free and forced convection (or mixed convection).

$$\frac{Gr_L}{Re_L^2} \ll 1, \textit{forced convection regime} \text{ (negligible free convection)} \tag{5.17}$$

$$\frac{Gr_L}{Re_L^2} \gg 1, \textit{free convection regime} \text{ (negligible forced convection)} \tag{5.18}$$

$$\frac{Gr_L}{Re_L^2} \sim 1, \textit{mixed convection regime} \text{ (both free and forced convection are important) (5.19)}$$

If heat transfer occurs in the mixed convection regime, the following equation may be used to calculate the Nusselt number.

$$Nu^m = \left(Nu_{forced}\right)^m \pm \left(Nu_{free}\right)^m \tag{5.20}$$

where Nu_{forced} and Nu_{free} are Nusselt numbers for forced and free convection, respectively. A value of $m = 3$ is usually recommended. A positive or a negative sign is taken depending upon whether the free convection flow occurs in the same or the opposite direction of forced convection flow.

Example 5.7   A horizontal section of an uninsulated pipe, 60 mm o.d., carrying warm water runs below an exhaust fan in a factory shed. The fan creates a suction that causes a mild upward cross-flow velocity of 0.3 m/s on the pipe. If the pipe wall temperature is 60°C, calculate the rate of heat loss by the combined free and forced convection per metre length of the pipe. The ambient air temperature is 30°C.

SOLUTION   Mean air-film temperature $= \dfrac{60 + 30}{2} = 45°C = 318$ K

For air at 318 K, we have

$\rho = 1.105$ kg/m³; $c_p = 0.24$ kcal/kg °C; $\mu = 1.95 \times 10^{-5}$ kg/m s;   Pr $= 0.7$;
$v = \mu/\rho = 1.85 \times 10^{-5}$ m²/s; $k = 0.0241$ kcal/h m °C; $\beta = 1/318 = 3.145 \times 10^{-3}$ K⁻¹; and cross flow air velocity, $V = 0.3$ m/s.

Surface temperature of the pipe, $T_s = 60°C$

Bulk temperature of air, $T_o = 30°C$

Diameter of the pipe, $d = 60$ mm $= 0.06$ m

Calculation of the free convection Nusselt number

$$Ra_d = (Gr_d)(Pr) = \frac{g\beta(T_s - T_o)d^3}{v^2}(Pr)$$

$$= \frac{(9.8)(3.145 \times 10^{-3})(60 - 30)(0.06)^3}{(1.85 \times 10^{-5})^2}(0.7)$$

$$= 4.08 \times 10^5$$

Using Eq. (5.9), we get

$$\overline{Nu}_{free} = \left[0.60 + \frac{(0.387)(4.08 \times 10^5)^{1/6}}{\left\{ 1 + (0.559/0.7)^{9/16} \right\}^{8/27}} \right]^2 = 11.31$$

Calculation of the forced convection Nusselt number

Using the Churchill and Burnstein correlation, Eq. (4.19), we get

$$Re = \frac{dV}{v} = \frac{(0.06)(0.3)}{1.85 \times 10^{-5}} = 972$$

$$\overline{Nu}_{forced} = 0.3 + \frac{(0.62)(972)^{1/2}(0.7)^{1/3}}{\left[1 + (0.4/0.7)^{2/3} \right]^{1/4}} \left[1 + \left(\frac{972}{2.82 \times 10^5} \right) \right]^{4/5} = 15.8$$

We use Eq. (5.20) to calculate the Nusselt number for mixed convection [the positive sign should be used in the expression on the RHS of Eq. (5.20), because the air flows upwards and free and forced convection processes are augmenting each other].

$$\overline{Nu}^3 = \left(\overline{Nu}_{forced} \right)^3 + \left(\overline{Nu}_{free} \right)^3 = (15.8)^3 + (11.31)^3 = 3944 + 1447 = 5391$$

or

$$\overline{Nu} = 17.5 = \frac{hd}{k}$$

The combined heat transfer coefficient, therefore, is

$$h = \frac{(17.5)(0.0241)}{0.06} = 7.03 \text{ kcal/h m}^2 \text{ °C}$$

The rate of heat loss per metre length of the pipe

$$= h \, A \, \Delta T = (7.03)(\pi)(0.06)(1)(60 - 30) = \boxed{39.7 \text{ kcal/h}}$$

SHORT QUESTIONS

1. A layer of fluid enclosed between two horizontal plates is heated from above. Is the system unstable?

2. What is a plume? Make a sketch of the plume along a cold vertical metal wire placed in a warm environment.

3. What are Benard cells?

4. How do you define Grashof number? What is its forced convection analogue? State its physical significance.

5. What is the physical significance of Rayleigh number? At what value of the Rayleigh number does transition of a boundary layer from laminar to turbulent occur at a heated vertical surface?

6. Give the approximate range of values of the free convection heat transfer coefficient at a vertical hot surface in contact with (i) water, (ii) air.

7. In a case of combined free and forced convection heat transfer, mention the conditions under which the role of free convection is (i) significant, (ii) moderate, and (iii) controlling.

8. How do you calculate the Nusselt number in a mixed convection regime?

9. What is meant by the aspect ratio of an enclosure?

10. A liquid stored in an insulated storage tank has to be maintained at 12°C. It is decided to keep it cool by passing chilled water at 4°C through a coil. Where should the coil be placed? Just below the top surface of the liquid, or at the bottom of the tank?

11. Under what condition can a cylindrical surface be approximated to a flat surface?

PROBLEMS

5.1 A square furnace of outer dimensions 3 m × 3 m and 2 m in height has a skin temperature of 60°C. Calculate the total rate of heat loss from the four vertical walls and the top flat surface by free convection to an ambient at 27°C. Hence compare the average natural convection heat flux from a vertical surface with that from a hot horizontal surface facing upwards.

5.2 A laboratory water bath has an immersed horizontal heating element, 2.54 cm in diameter and 30 cm in length, with a power input of 500 W. If the bulk water temperature is 38°C and heat transfer occurs by free convection only, calculate the surface temperature of the heater.

5.3 A 40 mm i.d. horizontal pipe (1.5 inch schedule 40 IPS) of wall thickness 3.7 mm carries low pressure steam at 130°C. What is the rate of free convection heat loss per metre of the bare pipe? If the pipe is now insulated with a 3 cm thick layer of fibre glass (k = 0.043 W/m °C), calculate the percentage reduction in heat loss. The thermal resistance of the pipe wall is negligible. The ambient temperature is 30°C.

5.4 A 30 cm × 30 cm horizontal square duct made of sheet metal carries warm air. The duct is uninsulated and is 10 m long. Air at 70°C enters the duct at a velocity of 3 m/s. If heat loss from the outer surface occurs by free convection only, and if the ambient temperature is 27°C, calculate the drop in the air temperature at the exit. (Note that the top surface of the channel faces upwards, the bottom surface faces downwards, and the other two surfaces are vertical. Assume that the free convection flow at one surface does not affect that at another. The inside forced convection heat transfer coefficient is also to be calculated.)

5.5 A 4 cm diameter steel ball (ρ = 7800 kg/m³, c_p = 473 J/kg °C) at 90°C is suspended in still air at 25°C. If the ball cools down by free convection heat loss only, calculate the time required for the temperature of the ball to drop down to 35°C.

5.6 A room has a nearly horizontal flat roof of thin GI (galvanized iron) sheet. On a sunny day the roof receives 700 W/m² of radiation. Assuming that 70% of this energy is transferred to the ambient medium at 32°C by free convection from both upper and lower surfaces of the roof, calculate the steady state temperature of the roof.

5.7 A horizontal steam pipe, 52.5 mm i.d. and 60.5 mm o.d. (2 inch schedule 40 IPS), carries saturated steam at 138°C. The pipe is covered with a 25 mm thick layer of a preformed insulation ($k = 0.17$ W/m °C). The insulated pipe, 40 m long, runs through an ambient at 21°C. If the steam at the inlet is saturated, calculate the "quality" of the steam at the other end of the pipe section. The average linear velocity of steam in the pipe is 20 m/s. The necessary data for steam may be taken from the steam table.

5.8 Considerable dissipation of energy occurs in a transformer used in an electrical substation. Cooling of a transformer is done by keeping it immersed in oil (this also provides electrical insulation). A small transformer is kept immersed in a cylindrical shell, 0.9 m in diameter and 1.5 m in height. If the energy loss is 2 kW, calculate the wall temperature of the shell. The ambient air is at 28°C.

5.9 A 100 m³ cylindrical flat roof storage tank (height/diameter = 1.2) resting on a concrete base is used to store coal tar in a low temperature coal carbonization plant. The tar has to be maintained warm so that it does not solidify and can be pumped to the tar distillation unit. A temperature of not less than 65°C has to be maintained in the tank for this purpose.

Heating by immersed steam coils placed near the bottom of the tank is the most convenient method of maintaining the temperature of a molten liquid. On the basis of some design calculations, a process engineer recommends 3.0 m² area of the horizontal steam coil (tube dia = 47 mm) which is to be supplied with saturated steam at 5 kg/cm² gauge. Do you think that the design will work? If not, what additional area of the steam coil should be provided? The tank is insulated with a 5 cm layer of glass fibre blanket ($k = 0.135$ W/m °C). The ambient temperature is 20°C, and heat loss from the tank surfaces occurs by free convection (heat loss from the bottom surface may be neglected). The following properties of the tar may be used for calculation: $\rho = 1080$ kg/m³; $\beta = 7 \times 10^{-4}$ /°C; $\nu = 60$ stoke; $k = 0.2$ W/m °C; $c_p = 2.1$ kJ/kg °C.

5.10 A 500 W immersion heater has the heating element in the form of a horizontal circular loop that ends up in two parallel arms as shown in the Fig. 5.5. The diameter of the element is 1.0 cm and the total length of the element is 28 cm. The heater is placed in a bath having

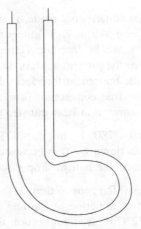

Fig. 5.5 Heater element (Problem 5.10).

water at 24°C, with the arms vertical, and then switched on. Find the surface temperature of the loop. Also calculate the surface temperature of the loop if the heater is placed in air at the same temperature (24°C) and in the same orientation.

5.11 The contents of a stirred tank reactor, 1.5 m o.d. and 2.5 m tall, have to be maintained at 150°C by circulating hot oil through a coil in the reactor. The tank rests on the first floor of a factory shed and is supported by three brackets in such a way that half of the tank remains above the floor as shown in the Fig. 5.6. A layer of glass fibre insulation has to be applied on the tank so that the skin temperature does not exceed 40°C when the ambient temperature is 30°C. Calculate the required thickness of the insulation. State any simplifying assumption you make.

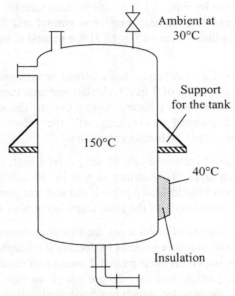

Fig. 5.6 Insulated tank (Problem 5.11).

Note: The tank has two nozzles, a manhole and other fittings. However, for heat loss calculation it will be sufficient to consider that the top and bottom surfaces are flat. The effects due to the fittings may be neglected.

5.12 Water at 4°C enters a horizontal carbon steel pipe, 102 mm i.d. and 114 mm o.d., lagged with a 30 mm thick layer of polyurethane foam insulation ($k = 0.019$ W/m °C). The velocity of water in the pipe is 1.5 m/s. If heat loss from the insulated pipe occurs by free convection to an ambient at –20°C, calculate the length of the pipe section at the end of which formation of ice crystals will start.

5.13 A double-walled thermos box is used for storing foodstuff. One vertical face of the box measures 1 m × 1 m, and the gap between the walls is 3 cm. If the inner wall is at 4°C and the outer wall has a temperature of 25°C, calculate the rate of heat gain by the box through this wall by free convection. What is the equivalent thermal conductivity of the gap? If the gap is filled with cork having thermal conductivity of 0.045 W/m °C, what would be the rate of heat gain?

5.14 A cylindrical solar collector, 0.2 m o.d., having an absorbing outer surface, is protected by a thin, transparent cylindrical casing thus forming an assembly of two concentric cylinders. The confined space contains nitrogen gas at 1 atm pressure. If the wall temperature of the solar collector (i.e. the temperature of the inner cylinder) is 80°C, and the width of the enclosure (i.e. the gap between the cylinders) is 0.07 m, calculate the rate of heat loss from the outer wall of the assembly to the ambient at 30°C.

[*Hint:* Assume a value of the outer wall temperature at steady state. Calculate the natural convection heat transfer coefficient at the outer surface. Calculate the effective thermal conductivity and the thermal resistance of the annular space using Eqs. (5.14) – (5.16). Then follow the procedure of Example 5.4.]

5.15 Develop the criteria given by Eqs. (5.17)–(5.19) for heat transfer from a vertical flat surface. Use the available correlations for heat transfer in natural and forced convection.

[*Hint:* Use the simplified form of Eq. (5.3) for natural convection, and Eq. (4.14) for forced convection.]

5.16 A vertical furnace door, 1.2 m in height, has a surface temperature of 120°C over which air is flowing upwards at a velocity of 2 m/s. Calculate the total rate of heat loss from the door and determine the contribution of natural convection to the rate of heat loss. At what (i) upward and (ii) downward air velocity will the effect of natural convection be insignificant compared to that of forced convection?

5.17 A horizontal steam pipe (2 inch schedule 40 IPS) carries high pressure steam at 230°C. It is lagged with a 3 inch layer of insulation ($k = 0.16$ W/m °C). Gentle wind at 1 m/s is blowing in cross flow over the insulated pipe. If the ambient temperature is 32°C, calculate the rate of heat loss per foot length of the pipe, considering both free and forced convection.

5.18 In the process of manufacture of toilet soap, the raw soap containing about 13% moisture is extruded to about 5 mm diameter noodles by forcing it through a 'plodder'. The extrudate is cut into small pieces which are then passed through two closely spaced rolls in order to disintegrate any coarse particle that may be present in the raw soap.

In a particular plant, the plodder, which is a horizontal cylindrical equipment maintained at 5°C by a cooling jacket, is being operated without any insulation on it. As a result it absorbs a lot of heat from the ambient medium, thus increasing the refrigeration load and the power bill. It is now planned to apply polyurethane foam insulation to the plodder. But, before a final decision is taken, it is desired to calculate the optimum insulation thickness and the payback period of the investment on insulation. Determine the payback period based on the following data and information at 18% rate of interest.

Heat absorption by the bare surface of the plodder occurs by combined natural and forced convection as well as by condensation of moisture of air on it. The plodder has an essentially cylindrical surface, 2 m long and 0.5 m diameter. It is estimated that a nearby exhaust fan in the factory shed causes a cross-flow air velocity of 0.75 m/s over the plodder. It is further estimated that condensation of moisture on the surface of the plodder releases heat at the rate of 300 kcal/h m². The ambient temperature may be taken as 33°C on the average. The cost of insulation (inclusive of labour and supervision) is Rs. 16,000/m³ of the material. The refrigeration unit used for cooling the plodder has an overall efficiency of 80%, and the power tariff is Rs. 3.50 per kWh.

REFERENCES AND FURTHER READING

Churchill, S.W. and H.H.S. Chu, "Correlating equations for laminar and turbulent free convection from a horizontal cylinder", *Intern. J. Heat Mass Transfer,* **18** (1975), 1323.

Churchill, S.W., "Free Convection in Layers and Enclosures," in E.U. Schlunder (Ed.), *Heat Exchanger Design Handbook*, Hemisphere Publ. Co., New York, 1983.

Gebhart, B., *Heat Transfer*, 2nd ed., McGraw-Hill, New York, 1970.

Holman, J.P., *Heat Transfer*, 7th ed., McGraw-Hill, New York, 1992.

Incropera, F.P. and D.P. DeWitt, *Introduction to Heat Transfer*, 2nd ed., John Wiley, New York, 1996.

McGregor, P.K. and A.P. Emery, "Free convection through vertical plane layers: Moderate and high Prandtl number fluids", *J. Heat Transfer*, **91** (1969), 391.

Minkowycz, W.J. and E.M. Sparrow, "Local nonsimilar solutions for natural convection on a vertical cylinder", *J. Heat Transfer*, **96** (1974), 178.

Morgan, V.T., "The Overall Convective Heat Transfer from Smooth Circular Cylinders", in T.F. Irvin, and J.P. Hartnett (Eds.), *Advances in Heat Transfer*, Vol. 11, Academic Press, New York, 1975.

Ostrach, S., "Natural Convection in Enclosures", in T.F. Irvin and J.P. Hartnett, (Eds.), *Advances in Heat Transfer*, Vol. 8, Academic Press, New York, 1972.

Raithby, G.D. and K.G.T. Hollands: ibid, Vol. 11 (1975), 265–315.

6
Boiling and Condensation

Many heat transfer activities in industrial practice involve a change of phase like solid to liquid or liquid to solid, and liquid to vapour or vapour to liquid. The examples are numerous. Melting of a solid is common to many polymer processing operations, where the polymer beads are melted before moulding. Freezing of water in an ice plant or conversion of drops of urea-melt to particles in a urea prilling tower involves a change of phase from liquid to solid. A change of phase from the liquid to the vapour or from the vapour to the liquid is encountered much more frequently. For example, a distillation column is provided with a reboiler at the bottom and a condenser at the top. In many chemical processes the feed liquid is vaporized before it enters the reactor (for example, liquid ammonia is vaporized before it is mixed with air and fed to the ammonia oxidation reactor in a nitric acid plant). Vaporization of water to produce high pressure steam, and condensation of steam leaving the turbine are the two major heat transfer operations in a thermal power plant.

It is important to study the fundamentals of the phenomena of boiling and condensation prior to learning the design methodologies of related heat transfer equipment. The rate of heat transfer during boiling or condensation is generally much faster than the rate of conduction or convection heat transfer. This is because in conduction or convection, transport of heat occurs from a surface (or an interface) to the bulk of a medium. But in boiling or condensation, the heat transfer phenomenon is often limited to a very narrow region near the surface. Change of phase, for example, from liquid to vapour or from vapour to liquid, occurs within this region. Nevertheless, convection plays a major role in heat transfer during boiling or condensation.

6.1 THE BOILING PHENOMENON

When some water is taken in a beaker and placed on a heater, heat transfer to the water from the hot bottom of the beaker occurs by free convection at the beginning. When the water gets considerably heated, evaporation from the free surface becomes visible. As heating proceeds, the water eventually reaches its boiling point and the formation of small vapour bubbles starts. However, at this stage most of the tiny bubbles which are formed at the bottom surface collapse before reaching the top. This means that although the water at the bottom is at its boiling point, the temperature of the upper layers is still below the boiling point. The cooling of the upper layers occurs by evaporation heat loss at the free surface. However, this situation does not continue for long, and soon the liquid starts boiling. Vapour bubbles are generated at the bottom, grow in size, and rise up vigorously through the liquid. This is a simple and brief picture of the different stages that appear during heating of a liquid from the cold to the boiling condition. As a matter of fact, the temperature of a boiling liquid remains above its true boiling point, that is, it has a little *superheat*. The amount of energy a liquid element possesses because of the superheat is released when a vapour bubble in contact with the element grows in size and draws the heat of vaporization from the *superheated* liquid element.

The course of heating of a liquid first leading to evaporation and then to boiling can be followed quantitatively if the temperature of the hot surface, and consequently the heat flux, is varied systematically over a wide range. There are quite a few classical works on the boiling phenomenon (e.g. Nukiyama, 1934; Cichelli and Bonilla, 1945; Farber and Scorah, 1948; Bromley, 1950). An electrically heated surface, for example, a wire was used by many workers because of the ease with which the rate of heat input and the temperature can be controlled.

The experimental arrangement including the results of Farber and Scorah (1948) is described here briefly. A schematic of their experimental set-up is shown in Fig. 6.1. An electrically heated wire immersed in the liquid supplies the heat required for boiling. The rate of heat input is

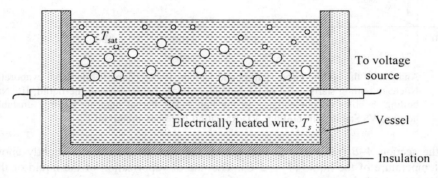

Fig. 6.1   Schematic of a set-up for experimental study of boiling heat transfer.

calculated from the measured values of the applied voltage and the current. This measurement also gives the resistance of the wire and hence the wire temperature T_s. The surface heat flux q_s can be calculated from the rate of electrical heat input and the area of the wire. A set of temperature-flux data can be generated by changing the applied voltage over a reasonably wide range.

In a boiling liquid, the temperature of the hot surface must be higher than the boiling point of the liquid, i.e. $T_s > T_{sat}$, where T_{sat} is the temperature of the saturated liquid or the boiling point of the liquid. The temperature driving force for the boiling liquid which is also called the 'excess temperature' is defined as follows.

$$\text{Excess temperature, } T_e = T_s - T_{sat} \qquad (6.1)$$

A plot of the logarithm of the heat transfer coefficient at the wire surface (h_b) vs. the excess temperature (T_e) typically looks like Fig. 6.2(a). The modes or regimes of boiling over different ranges of excess temperature are identified in this figure. It is to be noted that these results are characteristic of *pool boiling*. Pool boiling refers to boiling of a quiescent liquid (that is, a liquid pool) in which the motion is caused by free convection and by the formation, growth, detachment and rise of the bubbles. Pool boiling of the liquid occurs in the experimental arrangement shown in Fig. 6.1. There is another kind of boiling, as distinguished from pool boiling, called *forced convection boiling* in which motion in the boiling medium is caused by an external means (like a pump) in addition to the factors involved in pool boiling.

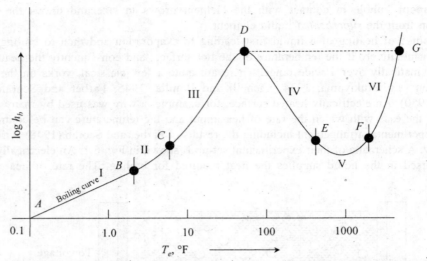

Fig. 6.2(a)   Analysis of the boiling curve (Farber and Scorah, 1948). [Zone I: Interfacial evaporation; II: Nucleate boiling starts, most of the bubbles collapse before leaving the liquid; III: Nucleate boiling, bubbles do not collapse in the liquid; IV: Partial nucleate boiling, unstable film boiling; V: Stable film boiling; VI: Radiation becomes important.]

Over the section A–B (regime I) in Fig. 6.2(a), the wire temperature is slightly above the saturation temperature of the liquid ($T_e \leq 2°C$). The rate of vaporization of the liquid or the rate of formation of vapour bubbles is pretty small. Motion in the liquid medium is caused principally by free convection. The hot liquid vaporizes only at the free top surface. This regime is called the *interfacial evaporation regime*.

The section B–D is characterized by *nucleate boiling*. Here the excess temperature varies from $T_e \approx 2°C$ at B to $T_e \approx 35°C$ at D. The nucleate boiling regime may be subdivided into two smaller regimes. Over the section B–C ($T_e \approx 2$ to $6°C$), representing regime II, isolated bubbles are formed at the surface of the wire, but most of them collapse before reaching the free surface of the liquid. Because bubble formation at a reasonable rate starts at B, this point indicates the onset of nucleate boiling. Over the remaining section, C–D, which is regime III, a large number of vapour bubbles are generated that vigorously rise through the liquid and escape at the free surface. Near the point D, the vapour bubbles rise as jets or columns and then form bigger bubble slugs. Breakage and coalescence of vapour bubbles also occur because of the intense motion generated in the liquid.

When the excess temperature exceeds that at the point D, the mode of boiling gradually changes from nucleate to *film boiling*. Here the bubbles are formed in so large numbers that they coalesce right on the heating surface to form a film of vapour. This greatly reduces the heat transfer coefficient. The film boiling regime may be subdivided into regimes IV, V, and VI. In regime IV (section D–E), the formation of the vapour film on the wire surface is not complete, and the film is discontinuous. The film may even disappear momentarily and then reappear. This regime represents *partial nucleate boiling* or *transition boiling*. Beyond the point E, starts the regime V (section E–F) where *stable film boiling* occurs and a continuous vapour film always blankets the heating surface. The surface heat flux decreases significantly although the excess temperature may be as high as a few hundred degrees. After the point F, the wire temperature increases so much that radiative heat transfer becomes important. This is characteristic of regime VI (section F–G). If the excess temperature T_e, is too high, the heating wire may even melt. This, in fact, occurred in the boiling heat transfer experiments of Nukiyama (1934) who used a nichrome heating wire that was burnt out at a high temperature. But Farber and Scorah (1948) used a platinum wire and could reach a higher temperature and heat flux without burn-out of the wire. The melting or burn-out of the heat transfer surface is sometimes called the *boiling crisis*. In some cases, a direct shift from the point D to the point G may occur as the temperature increases. The excess temperature and the heat flux in some electrical heating devices or in a nuclear reactor may be so large that it may lead to boiling crisis. This possibility should be taken into account at the design stage. The boiling phenomena in the different regimes are also shown in Fig. 6.2(b).

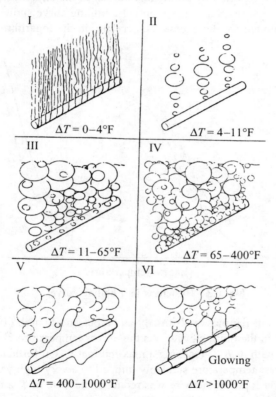

Fig. 6.2(b)   Schematic of the boiling phenomenon in different boiling regimes (after Farber and Scorah, 1948).

It is pertinent at this point to mention the so-called *Leidenfrost phenomenon*. When water droplets fall on a sufficiently hot plate, the droplets bounce up and down a few times before they disappear by vaporization. This phenomenon is common in the kitchen where it occurs when a little water falls on a hot oily pan. If some liquid air or liquid nitrogen spills on the floor of a laboratory, the droplets move so briskly that they appear to be dancing before they disappear.

The above phenomenon was first observed by Leidenfrost in 1756. The mechanism is related to film boiling. When a water droplet lands on a hot surface, a film of steam immediately forms between the droplet and the hot surface. Because the thickness of this vapour film increases very quickly by further vaporization of the liquid, the droplet experiences an upward thrust and it bounces. It falls back on the surface and bounces again by the same mechanism and reduces in size at every touch with the surface. The droplet disappears eventually. So the Leidenfrost phenomenon is caused by film boiling and occurs for evaporating droplets on a hot surface corresponding to regime V of the boiling curve.

6.2   HYSTERESIS IN THE BOILING CURVE

This phenomenon was first reported by Nukiyama (1934) in his experiments on pool boiling of water using a hot nichrome wire. Nukiyama reported a boiling curve shown in Fig. 6.3 in which the surface heat flux is plotted vs. the excess temperature on the logarithmic scale. Although he

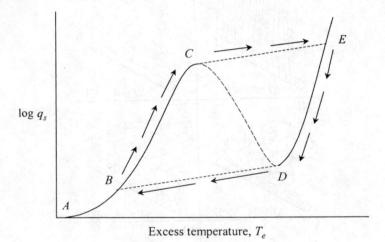

Fig. 6.3   Hysteresis in a boiling curve.

did not identify the different regimes as distinctly as Farber and Scorah (1948) did, he observed a hysteresis phenomenon in the boiling curve. As the excess temperature T_e increased, the surface heat flux q_s followed the path $A–B–C$ reaching a maximum at C. On further increase in the power input to the wire, the excess temperature suddenly jumped to a very high value at E following the path $C–E$. When the power input to the wire was reduced, the drop in T_e and the flux q_s followed the path $E–D–B–A$. Had there been no jump in the temperature after the point C during heating, the boiling curve should have been like $A–B–C–D–E$, although $C–D$ represents an unstable regime. The point C in Fig. 6.3 corresponds to the maximum heat flux or the *critical heat flux*. The Leidenfrost phenomenon of an evaporating droplet also occurs at the point C.

6.3   THE MECHANISM OF NUCLEATE BOILING

Nucleate boiling proceeds through the formation of bubble nuclei in a superheated liquid (as it has been stated earlier, a boiling liquid always has a degree of superheat), growth of the bubbles and their escape from the liquid. There are quite a few questions associated with this phenomenon. Does a bubble form in the bulk of the superheated liquid or on the hot surface with which the liquid is in contact? How to determine the rate of bubble growth? After it is formed on the hot surface, when does the bubble get detached from it? Not all of these questions will be addressed here. Our discussion will rather be confined to the phenomenon of formation of a bubble nucleus.

Formation of a tiny bubble in a liquid is called *nucleation*. A superheated liquid is said to be in a *metastable condition,* i.e. it cannot exist in this condition indefinitely and any disturbance in the system tends to bring it to the thermodynamic equilibrium state. A bubble itself in a superheated liquid may be considered to be a source of disturbance. If a bubble nucleus is formed in the bulk of a superheated liquid, it is called *homogeneous nucleation*. On the other hand, if a nucleus is formed on a hot surface or on a solid particle suspended in the superheated liquid, we call it *heterogeneous nucleation*.

The condition of mechanical equilibrium at a liquid–vapour (or a liquid–gas) interface is given by the Gibbs equation,

$$p_v - p_l = \sigma \left(\frac{1}{r_1} + \frac{1}{r_2} \right) \tag{6.2}$$

where p_v and p_l are the pressures exerted by the vapour and the liquid phases at the interface, σ is the surface tension of the liquid, and r_1 and r_2 are the principal radii of curvature of the interface. If the interface is spherical (i.e. if we consider a spherical vapour bubble in a liquid), $r_1 = r_2 = r$ (say), then Eq. (6.2) reduces to

$$p_v - p_l = \frac{2\sigma}{r} \tag{6.3}$$

It is clear from the above equation that the pressure inside the bubble is higher than that in the neighbouring liquid. The pressure difference, $p_v - p_l$, can be related to the liquid superheat by using the Clausius–Clapeyron equation assuming that the vapour phase behaves like an ideal gas. Thus, we have

$$\frac{d \ln p}{dT} = \frac{L_v}{RT^2} \tag{6.4}$$

where L_v is the molar heat of vaporization and R the gas constant. Let us assume that T_l is the temperature of the superheated liquid, p_l is the pressure in the superheated liquid, and T_{sat} is the corresponding boiling point of the liquid. The vapour in the bubble is in thermal equilibrium with the liquid, i.e. the vapour temperature is also T_l. As the vapour is saturated, the pressure in the bubble p_v is equal to the vapour pressure of the liquid at the temperature T_l. Let us integrate Eq. (6.4) within the appropriate limits. Thus, we have

$$\int_{p_l}^{p_v} d \ln p = \frac{L_v}{R} \int_{T_{sat}}^{T_l} \frac{dT}{T^2}$$

or

$$\ln \frac{p_v}{p_l} = \frac{L_v}{R} \left(\frac{1}{T_{sat}} - \frac{1}{T_l} \right)$$

Because

$$\ln \frac{p_v}{p_l} = \ln \left(1 + \frac{p_v - p_l}{p_l} \right) \approx \frac{p_v - p_l}{p_l}$$

Therefore,

$$\frac{p_v - p_l}{p_l} = \frac{L_v}{R}\left(\frac{T_l - T_{sat}}{T_l T_{sat}}\right)$$

Substituting for $p_v - p_l$ from Eq. (6.3), we get

$$T_l - T_{sat} = \frac{2\sigma}{r\,p_l}\frac{RT_l T_{sat}}{L_v} \approx \frac{2\sigma}{r\,p_l}\frac{RT_{sat}^2}{L_v} \qquad (6.5)$$

The radius of a vapour bubble in mechanical equilibrium in a superheated liquid can be calculated from the above equation (Rohsenow, 1966). For example, if we consider a pool of water with 10°C superheat at atmospheric pressure, we have

$T_l = 110°C$, $T_{sat} = 100°C$, $p_l = 1.013$ bar, $\sigma = 0.063$ N/m, $L_v = 4.07 \times 10^4$ kJ/kg-mol, $R = 8.314$ kJ/K kg-mol.

Therefore, we have from Eq. (6.5)

$$T_l - T_{sat} = 10°C \approx \frac{(2)(0.063)}{r(1.013 \times 10^5)}\frac{(8.314)(373)^2}{4.07 \times 10^4}$$

or

$$r = 3.55 \times 10^{-6}\text{ m}$$

What happens if a bubble nucleus of a radius smaller than above, say 10^{-6} m, is generated in the liquid by some way? This nucleus cannot remain in mechanical equilibrium if the superheat is only 10°C. It is sure to collapse as per the theory. Equation (6.5) thus explains why some tiny bubbles collapse in the liquid if the degree of superheat is insufficient.

Equation (6.5) also tells us that a high degree of superheat is necessary for the generation of a tiny nucleus in the bulk liquid. However, our experience shows that a liquid boils vigorously even at a superheat of several degrees. The reason is that bubble nuclei are formed by heterogeneous nucleation at the minute cavities and crevices present at the heating surface. This is more favoured on a rough surface than it is on a smooth one. Bumping of boiling water in a glass beaker occurs because it has a smooth surface. This does not happen when water boils in an aluminium pan. Hans and Griffith (1965) developed an expression for the critical size of a nucleation site (i.e. a cavity or a crevice on the surface) that will produce a bubble nucleus at a given degree of superheat. Theoretical and experimental studies on bubble growth and detachment have been reported by a large number of researchers. The different stages of nucleation and bubble growth at a cavity are shown schematically in Fig. 6.4.

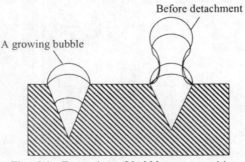

Fig. 6.4 Formation of bubbles over cavities.

Example 6.1   Consider nucleate pool boiling of ethyl acetate at 1 atm pressure. (a) If the superheat in the liquid is 5°C, calculate the diameter of a cavity on the boiling surface which produces a bubble nucleus that does not collapse after release from the surface. (b) If a cavity has a radius of 1.0 μm, what degree of superheat is necessary so that a bubble nucleus grows in size after detachment from the cavity.

SOLUTION   The radius of a bubble nucleus in mechanical equilibrium in a superheated liquid can be determined from Eq. (6.5). This is our working equation for this problem.

$$T_l - T_{sat} = \frac{2\sigma}{r p_l} \frac{R T_{sat}^2}{L_v}$$

We assume that the radius of a bubble nucleus is equal to the radius of the cavity at the moment of detachment.

(a) Boiling point of ethyl acetate at 1 atm pressure is, T_{sat} = 350 K (77°C). The following physical data are used in the calculation [for the correlations used for estimation of physical properties, see Reid et al. (1988)]. Liquid temperature, T_l = 350 + 5 = 355 K (because the liquid has 5°C superheat).

The vapour pressure of ethyl acetate at this temperature can be calculated by using the Antoine equation

$$\log p_l = 4.22658 - \frac{1244.95}{\theta + 217.88}$$

where p_l is the vapour pressure in bar, and θ the temperature in °C.

At θ = 77 + 5 = 82°C, we get p_l = 1.189 bar.

The surface tension of the liquid is given by

$$\sigma = 26.29 - 0.1161\theta \text{ dyne/cm}$$

At θ = 82°C, we get

$$\sigma = 16.8 \text{ dyne/cm} = 1.68 \times 10^{-2} \text{ N/m}$$

Molar heat of vaporization, L_v = 33,605 kJ/kg-mol = 3.3605×10^7 J/kg-mol
The universal gas constant, R = 0.08314 m³ bar/kg-mol K
Substituting the values of the various quantities in the working equation, we get

$$(355 - 350) \text{ K} = \frac{(2)(1.68 \times 10^{-2} \text{ N/m})(0.08314 \text{ m}^3 \text{ bar/kg-mol K})(350 \times 350 \text{ K}^2)}{(r\text{m})(1.189 \text{ bar})(3.3605 \times 10^7 \text{ J/kg-mol})}$$

or

$$r = 1.713 \times 10^{-6} \text{ m} = \boxed{1.713 \text{ μm}}$$

Therefore, a bubble nucleus larger than 1.713 μm that has been detached from a cavity will not collapse in the liquid.

(b) In this case, r = 1 μm (given) = 10^{-6} m. It is required to calculate T_l. At the temperature T_l K = $(T_l - 273)$°C, the vapour pressure and surface tension of ethyl acetate are given by

$$\log p_l = 4.22658 - \frac{1244.95}{T_l - 273 + 217.88} \text{ bar}$$

$$\sigma = 0.02629 - 1.161 \times 10^{-4} \, (T_l - 273), \text{ N/m}$$

The values of T_{sat}, L_v, and R are as in part (a) above. Putting all these quantities and expressions in the working equation, [Eq. (6.5)], we get a transcendental equation in the unknown T_l. This equation, given below, can be solved by any suitable technique (for example, the Newton-Rhapson method).

$$T_l - 350 = \frac{(2)[0.02629 - 1.161 \times 10^{-4}(T_l - 273)](0.08314)(350)^2}{(10^{-6})(3.3605 \times 10^7)[10^{4.2266 - (1244.95)/(T_l - 273 + 217.88)}]}$$

The solution is

$$T_l = 358 \text{ K or } 85°C; \quad \text{superheat of the liquid} = 85 - 77 = \boxed{8°C}$$

Therefore, if the liquid has a superheat of 8°C, a bubble nucleus detached from a 1 μm radius cavity will not collapse. It will rather grow in size after detachment.

6.4 CORRELATIONS FOR POOL BOILING HEAT TRANSFER

While the preceding discussion targetted the physical mechanism of boiling phenomenon, it does not help us in calculating the rate of heat transfer in a boiling liquid. The second aspect is more important in process calculations and in the design of liquid boiling equipment. A large number of empirical and semi-empirical correlations available in the literature can be used for this purpose. The more important and practically useful correlations are given below.

6.4.1 Nucleate Boiling

Liquid boiling equipment in the process industry is designed to operate in the nucleate boiling regime because it is in this regime that heat transfer occurs more effectively. The equipment configuration is mostly of the shell-and-tube type (see Chapter 8). The liquid boils either in the shell or in the tube, and the heating medium is steam or any other hot fluid. Therefore, correlations on boiling in or outside a horizontal or vertical tube are most important for the purpose of design. Also, because the rate of boiling, and for that matter the rate of heat transfer, depends upon quite a few major factors like the characteristics of the boiling surface, physical properties of the liquid (ρ, c_p, k, μ, σ, etc.) pressure, surface temperature (T_s), etc., no single correlation can be satisfactory for all situations. Further, it is difficult to quantitatively take into account the first factor (i.e. surface characteristics, which include the size distribution of cavities on the surface). Even if this factor is included in a correlation, it cannot be used unless the surface characteristics of the material used in the boiler are known.

Rohsenow (1952) proposed the following correlation for heat flux in nucleate pool boiling based on his experimental data on boiling of quite a few liquids.

$$q_s = \mu_l L_v \left[\frac{g(\rho_l - \rho_v)}{\sigma} \right]^{1/2} \left[\frac{c_{pl} T_e}{C_{sf} L_v (\text{Pr}_l)^n} \right]^3 \tag{6.6}$$

where
  μ_l is the liquid viscosity (Pa s)
  L_v is the enthalpy of vaporization of the liquid (J/kg)

g is the acceleration due to gravity (m/s^2)

ρ is the density (kg/m^3)

σ is the surface tension (N/m)

c_p is the specific heat (J/kg °C)

T_e is the excess temperature = $T_s - T_{sat}$ (°C)

T_s is the temperature of the boiling surface (K)

Pr$_l$ is the liquid Prandtl number

q_s is the heat flux (W/m^2)

the subscripts mean: l = liquid, and v = vapour

The values of the constant C_{sf} which depend upon the surface characteristics, are listed in Table 6.1 for a few materials. The value of the exponent n is: $n = 1$ for water, and $n = 1.7$ for most organic liquids. The liquid properties should be taken at the saturation temperature T_{sat}. Error in the prediction is pretty high, and may be even 50% in some cases.

Table 6.1 Values of C_{sf} in Eq. (6.6) for several fluid–surface combinations

Fluid–surface	C_{sf}	n
Water–stainless steel		
mechanically polished	0.013	1.0
ground and polished	0.006	1.0
chemically etched	0.013	1.0
Water–brass	0.006	1.0
Water–copper	0.013	1.0
Benzene–chromium	0.010	1.7
n-Pentane–chromium	0.015	1.0
Ethyl alcohol–chromium	0.0027	1.7
iso-Propyl alcohol–copper	0.0025	1.0
Carbon tetrachloride–copper	0.013	1.0
Aqueous K$_2$CO$_3$ (35%)–copper	0.0054	1.0

Another useful correlation for the heat flux proposed by Mostinski has been cited by Kakac (1981). A conservative form of his correlation is

$$h_b = 0.00341 \, (P_c)^{2.3} \, (T_e)^{2.33} \, (P_r)^{0.566} \text{ W/m}^2 \text{ °C} \tag{6.7}$$

where

P_c is the critical pressure in bar

P_r is the reduced pressure

$T_e = T_s - T_{sat}$ in °C

6.4.2 Critical Heat Flux

The estimation of critical heat flux is very important in the design of a boiling equipment. If the temperature of the heating fluid is much higher than the boiling of the liquid, film boiling of the liquid may occur. So the critical heat flux corresponding to the point C in Fig. 6.3 is calculated first, and the temperature of the heating fluid may be selected accordingly so as to maintain boiling in the nucleate regime. We will cite here two important correlations for the critical heat flux.

(i) A simple correlation for the critical heat flux q_{max} was developed by Lienhard et al. (1973) by fitting their experimental data to a theoretical equation suggested by Zuber (1958).

$$q_{max} = 0.149 \, L_v \rho_v \left[\frac{\sigma g(\rho_l - \rho_v)}{\rho_v^2} \right]^{1/4} \quad \text{W/m}^2 \tag{6.8}$$

where the terms have significances and units as in Eq. (6.6).

(ii) Peak boiling heat flux on a horizontal cylinder can be estimated by using the following correlation of Sun and Lienhard (1970).

$$q_{max} = q'_{max} \left[0.89 + 2.27 \exp(-3.44\sqrt{r'} \right] \quad \text{for } r' > 0.15 \tag{6.9}$$

$$r' = \text{dimensionless radius} = r \left[\frac{g(\rho_l - \rho_v)}{\sigma} \right]^{1/2} \tag{6.10}$$

where q'_{max}, the peak heat flux on an infinite horizontal plate, is given by

$$q'_{max} = 0.131 L_v \sqrt{\rho_v} \left[\sigma g(\rho_l - \rho_v) \right]^{1/4} \quad \text{W/m}^2 \tag{6.11}$$

6.4.3　Stable Film Boiling

The arm $E-F$ of the boiling curve, Fig. 6.2(a), represents stable film boiling, sometimes called 'film pool boiling'. The correlation of Bromley (1950) can be used to calculate the boiling heat transfer coefficient in this regime.

$$h_b = 0.62 \left[\frac{k_v^3 \rho_v(\rho_l - \rho_v)g(L_v + 0.4 \, c_{pv} T_e)}{d \, \mu_v \, T_e} \right]^{1/4} \quad \text{W/m}^2 \, ^\circ\text{C} \tag{6.12}$$

where the subscript v denotes the vapour phase, and d is the tube diameter. The above equation is applicable to film boiling outside horizontal tubes. However, this equation can be used for a vertical plate after replacing d by L (plate height) and the constant 0.62 by 0.7.

If the surface temperature is high enough to make the contribution of radiative heat transfer important, the 'total heat transfer' coefficient may be calculated by using the following equation.

$$h = h_b(h_b/h)^{1/3} + h_r \tag{6.13}$$

where h_r is the radiation heat transfer coefficient, and h_b is given by Eq. (6.12).

Example 6.2　Calculate the rate of boiling of water in a 0.35 m diameter stainless steel pan at 1 atm pressure if the bottom of the pan is maintained at 115°C.

SOLUTION　It is expected that water will boil in the nucleate pool boiling regime at an excess temperature $T_e = 115 - 100 = 15°C$. We will first calculate the rate of boiling using the Rohsenow's correlation, [Eq. (6.6)], and then check the result by comparing with the critical heat flux. The various physical properties involved are taken at the saturation temperature (i.e. 100°C) under the given conditions.

For water at 100°C: $\mu_l = 2.79 \times 10^{-4}$ N s/m^2; $c_{pl} = 4.22$ kJ/kg °C; $\rho_l = 958$ kg/m^3; $L_v = 2257$ kJ/kg; $\sigma = 0.059$ N/m; $\mathrm{Pr}_l = 1.76$.

For saturated steam: $\rho_v = 0.5955$ kg/m^3.

For the stainless steel pan, take $C_{sf} = 0.013$, and $n = 1$ (Table 6.1). Also, $g = 9.8$ m^2/s.

Putting these values in Eq. (6.6), we have

$$q_s = (2.79 \times 10^{-4})(2257)\left[\frac{(9.8)(958 - 0.5955)}{0.059}\right]^{1/2}\left[\frac{(4.22)(15)}{(0.013)(2257)(1.76)}\right]^3$$

$$= 462.5 \text{ kW/m}^2$$

The rate of boiling $= \dfrac{462.5 \text{ kW/m}^2}{2257 \text{ kJ/kg}} = 0.205$ kg/m^2 s

Area of the pan $= (\pi/4)(0.35)^2 = 0.0962$ m^2

The total rate of boiling $= (0.205)(0.0962) = 0.01972$ kg/s $= \boxed{71 \text{ kg/h}}$

Let us now calculate the critical heat flux using the Lienhard's correlation, [Eq. (6.8)].

$$q_{max} = 0.149 \, L_v \rho_v \left[\frac{\sigma g(\rho_l - \rho_v)}{\rho_v^2}\right]^{1/4}$$

$$= (0.149)(2257)(0.5955)\left[\frac{(0.059)(9.8)(958 - 0.5955)}{(0.5955)^2}\right]^{1/4} = 1259 \text{ kW/m}^2$$

The actual heat flux is well below the critical heat flux value. So it is reasonable to assume that nucleate pool boiling of water will occur under the given conditions. The boiling curve [see Fig. 6.2(b)] also indicates that pool boiling will occur at an excess temperature, $T_e = 15$°C.

It will be interesting to calculate the boiling heat flux using the simple correlation of Mostinski [Eq. (6.7)] and to compare it with the value predicted by the Rohsenow's equation.

$$h_b = 0.00341(P_c)^{2.3}(T_e)^{2.33}(P_r)^{0.566}$$

The given pressure $= 1$ atm $= 1.013$ bar. For water, $P_c = 221.2$ bar, and the reduced pressure, $P_r = 1.013/221.2 = 0.00458$.

Therefore,

$$h_b = (0.00341)(221.2)^{2.3}(15)^{2.33}(0.00458)^{0.566} = 22 \text{ kW/m}^2 \text{ °C}$$

$$\text{Heat flux, } q_s = h_b (T_s - T_{sat}) = h_b T_e = (22)(15) = 330 \text{ kW/m}^2$$

This value compares reasonably well with the value of 462.5 kW/m^2 predicted by Rohsenow's equation.

Example 6.3 Formaldehyde is one of the raw materials for the manufacture of a variety of resins like PF (phenol-formaldehyde), UF (urea-formaldehyde), MF (melamine-formaldehyde), etc. The technical method of production of formaldehyde is catalytic air oxidation of methanol (finely divided silver is one of the catalysts used).

In a formaldehyde plant, a hot stream of flue gas is used to vaporize methanol in a horizontal shell-and-tube vaporizer at a gauge pressure of 0.8 bar so that a part of the waste heat may be recovered. The vaporized methanol is then mixed with air and fed to the reactor. In the vaporizer, the hot gas flows through the tubes and the methanol boils in the shell. The tubes are 1-1/4 inch, 12 BWG (AISI 316L), and the surface temperature is 200°C. Calculate the rate of boiling.

SOLUTION   The problem does not indicate the boiling regime; it does not provide any physical property data. We will follow the following steps in order to solve the problem.

(i) Calculate the boiling point of methanol at 0.8 bar gauge (1.813 bar absolute) pressure. The Antoine equation with the constants taken from the literature may be used.

(ii) Calculate the temperature excess, T_e.

(iii) If the experimental boiling curve for a particular liquid is not available, prediction of the boiling regime (nucleate or film) becomes difficult. As we do not have the boiling curve for methanol, we do the following.

(iv) Calculate (a) the heat flux assuming nucleate pool boiling and (b) the critical flux. If the critical heat flux is larger, then the assumption of nucleate boiling is reasonable and no further calculation is necessary.

(v) If the calculated critical heat flux appears to be less than that obtained by assuming nucleate boiling, obviously the system is not operating in the pool boiling regime. Therefore, a suitable boiling correlation should be used to calculate the heat flux and the boiling rate.

(vi) Physical property data for many common fluids are available in Reid et al. (1988).
  [If the data necessary for a particular problem are not available, these can be estimated by using the available correlations. Reid et al. (1988) provide an excellent documentation and assessment of correlations for the estimation of physical properties of pure fluids and mixtures. The use of a few such correlations is illustrated in this example.]

Estimation of liquid and vapour properties necessary for the given problem

To calculate the boiling point of methanol at 1.813 bar, we use the Antoine equation

$$\ln p_v = A - \frac{B}{T} = 12.5673 - \frac{4234.6}{T}$$

where p_v is the vapour pressure in bar and T is the absolute temperature.
  Putting $p_v = 1.813$ bar, we get

$$T = 353.7 \text{ K} = \text{saturation temperature, } T_{sat.}$$

Tube wall temperature = 200°C = 473 K

Temperature excess, $T_e = 473 - 353.7 = 119.3°C$

$$\text{Mean vapour temperature} = \frac{473 + 353.7}{2} = 413.3 \text{ K}$$

The relevant liquid properties (required for the calculation of pool boiling heat flux) are evaluated at the saturation temperature. The vapour properties (required for the calculation of the film boiling heat flux) are estimated at the mean vapour temperature.

Liquid properties (ρ_l, c_{pl}, σ, μ, Pr, L_v) *at 353.7 K*

(a) Saturation liquid density (ρ_l): Use the modified Rackett technique (Reid et al., p. 67).

Molar volume of methanol at the boiling point, $V_s = \dfrac{RT_c}{P_c}(Z_{RA})^{1 + (1 - T_r)^{2/7}}$ (i)

For methanol: T_c = critical temperature = 512.6 K; T_r = reduced temperature = T/T_c = 353.7/512.6 = 0.69; Z_{RA} = 0.29056 − 0.08775w, w = acentric factor = 0.556; R = the universal gas constant = 0.08314 m^3 bar/g-mol K; P_c = critical pressure = 80.9 bar.

Substituting for the different quantities in Eq (i), we get

Molar volume, V_s = 0.0461 m^3/kg-mol

Molecular weight of methanol = 32; Saturated liquid density, $\rho_l = \dfrac{32}{0.0461}$ = 694 kg/m^3

(b) Liquid heat capacity (c_{pl}): Use the group contribution method of Missenard (Reid et al., p. 139).

Temperature, K	Contribution of		c_{pl} (J/g-mol °C)
	− CH$_3$ group	−OH group	
348	45.8	61.7	107.5
373	48.3	71.1	119.4

By linear interpolation at 353.7 K, we have

$c_{pl} = 107.5 + (119.4 - 107.5)\dfrac{353.7 - 348}{373 - 348}$ = 110.2 kJ/kg-mol °C = 3.444 kJ/kg °C

(c) Surface tension (σ) at 353.7°C: Data available (Reid et al., p. 635) are

Temp., K: 313   333

σ (dyne/cm):   20.96   19.4

By linear extrapolation, at 353.7°C, σ = 17.8 dyne/cm = 1.78 × 10^{-2} N/m

(d) Liquid viscosity (μ_l): If the viscosity of a liquid is available at one temperature, that at another temperature can be calculated using the following equation (Reid et al., p. 439).

$$\mu^{-0.2661} = (\mu')^{-0.2661} + \frac{T - T'}{233}$$

where μ and μ' are the viscosities (in cP) at temperatures T and T', respectively.

Viscosity of methanol at 25°C (298 K) is μ' = 0.55 cP (from Reid et al.). Therefore, at 353.7 K, we get

$$\mu^{-0.2661} = (0.55)^{-0.2661} + \frac{353.7 - 298}{233}$$

or
$$\mu = 0.274 \text{ cP}$$
That is,
$$\text{Liquid viscosity, } \mu_l = 0.274 \text{ cP} = 2.74 \times 10^{-4} \text{ kg/m s}$$

(e) Prandtl number: μ and c_{pl} have been obtained before. The thermal conductivity of methanol is calculated by using the following equation (Reid et al., Table 10.5, p. 547).

$$k_l = a + bT + cT^2 \text{ W/m °C}; \quad a = 0.3225, \quad b = -4.785 \times 10^{-4}, c = 1.168 \times 10^{-7}$$

At $T = 353.7$ K, we have

$$k_l = 0.168 \text{ W/m °C}; \quad \text{and Pr}_l = \frac{c_{pl}\mu}{k_l} = \frac{(3444)(2.74 \times 10^{-4})}{0.168} = 5.6$$

(f) Heat of vaporization (L_v): The heat of vaporization of methanol at its normal boiling point (337.7 K) is 1100 kJ/kg. Because the boiling temperature of methanol at the given pressure is not much different from its normal boiling point, we use the above value of the heat of vaporization, i.e. $L_v = 1100$ kJ/kg.

Properties of methanol vapour (k_v, ρ_v, c_{pv} and μ_v) *at the mean vapour temperature*, 413.3 K

(a) Density of the vapour (ρ_v): We assume ideal gas behaviour (which is reasonable in consideration of the given pressure), and put

$$P = 1.813 \text{ bar}, T = 413.3 \text{ K}, R = 0.08314 \text{ m}^3 \text{ bar/kg-mol K}$$

Molar volume, $V = \dfrac{RT}{P} = 18.95 \text{ m}^3\text{/kg-mol}$

Density of the vapour, $\rho_v = \dfrac{32}{18.95} = 1.688 \text{ kg/m}^3$ (mol. wt. = 32)

(b) Thermal conductivity of the vapour (k_v): We use the equation,

$$k_v = a + bT + cT^2 + dT^3 \text{ W/m °C} \quad \text{(Reid et al., Table 10.3, p. 516)}$$

$$a = -7.797 \times 10^{-3}, b = 4.167 \times 10^{-5}, c = 1.214 \times 10^{-7}, d = -5.184 \times 10^{-11}$$

Putting $T = 413.3$ K, we get
$$k_v = 0.0265 \text{ W/m °C}$$

(c) Heat capacity of the vapour (c_{pv}): We use the equation,

$$c_{pv} = a + bT + cT^2 + dT^3 \text{ kJ/kg-mol °C}$$

where
$$a = 21.15, \quad b = 7.092 \times 10^{-2}, \quad c = 2.589 \times 10^{-5}, d = -2.852 \times 10^{-8} \text{ (Reid et al., p. 617).}$$

At $T = 413.3$ K, $c_{pv} = 52.86$ kJ/kg-mol K = 1652 J/kg K

(d) Viscosity of vapour (μ_v): The following data are available (Reid et al.)

Temperature, °C :	67	127
Viscosity, μP :	112	132

We use linear extrapolation (for simplicity, and because the temperature range involved is not large) to calculate the viscosity at 413.3 K (140.3°C). Thus, $\mu_v = 1.364 \times 10^{-5}$ kg/m s.

The necessary physical properties of the liquid and of the vapour having been estimated, we now proceed to calculate the heat flux.

If nucleate pool boiling occurs (it does not appear likely here in view of the large excess temperature of 119.3°C), the Rohsenow's equation, Eq. (6.6), may be used. Take $C_{sf} = 0.0027$ and $n = 1.7$ (see Table 6.1; in the absence of specific data for the methanol–stainless steel combination, we use the data for ethanol–stainless steel). Putting the values of the various quantities in Eq. (6.6), we get

$$q = (2.74 \times 10^{-4})(1100)\left[\frac{(9.8)(694 - 1.688)}{1.78 \times 10^{-2}}\right]^{1/2}\left[\frac{(3.444)(119.3)}{(0.0027)(1100)(5.6)^{1.7}}\right]^3 \text{ kW/m}^2$$

$$= 75.3 \text{ MW/m}^2$$

To calculate the critical heat flux, we use the Sun and Lienhard correlation [Eqs. (6.9)–(6.11)].

$$q'_{max} = 0.131 L_v \sqrt{\rho_v}\left[\sigma g(\rho_l - \rho_v)\right]^{1/4}$$

$$= (0.131)(1100)(1.688)^{1/2}[(1.78 \times 10^{-2})(9.8)(694 - 1.688)]^{1/4} = 620.5 \text{ kW/m}^2$$

$$r' = r\left[\frac{g(\rho_l - \rho_v)}{\sigma}\right]^{1/2}$$

$$= (0.016)\left[\frac{(9.8)(694 - 1.688)}{1.78 \times 10^{-2}}\right]^{1/2} = 9.88$$

(r = outer radius of 1-1/4 inch 12 BWG tube = 0.016 m)

$$q_{max} = q'_{max}\left[0.89 + 2.27 \exp(-3.44\sqrt{r'})\right]$$

$$= (620.5)[0.89 + 2.27 \exp\{(-3.44)(9.88)^{1/2}\}] = 552.2 \text{ kW/m}^2$$

It appears that the calculated heat flux for nucleate pool boiling exceeds the critical heat flux which is not acceptable. So the assumption of pool boiling is not correct. There will be film boiling of methanol instead. The film boiling heat transfer coefficient can be calculated using the Bromley's correlation, Eq. (6.12). That is,

$$h_b = 0.62\left[\frac{k_v^3 \rho_v(\rho_l - \rho_v)g(L_v + 0.4 c_{pv}T_e)}{d\mu_v T_e}\right]^{1/4} \text{ W/m}^2 \text{ °C}$$

$$= 0.62\left[\frac{(0.0265)^3(1.688)(694 - 1.688)(9.8)\{(11 \times 10^5) + (0.4)(1652)(119.3)\}}{(0.032)(1.364 \times 10^{-5})(119.3)}\right]^{1/4}$$

$$= 163.4 \text{ W/m}^2 \text{ °C (Note that the heat transfer coefficient value is pretty low.)}$$

The heat flux, $q_b = (h_b)(T_e) = (163.4)(119.3) = 19.49 \text{ kW/m}^2$

The boiling rate $= \dfrac{19.49 \text{ kW/m}^2}{1100 \text{ kJ/kg}} = 0.01772 \text{ kg/m}^2 \text{ s} = \boxed{63.79 \text{ kg/m}^2 \text{ h}}$

6.5 FORCED CONVECTION BOILING

In pool boiling, the motion in the liquid is created by the rising vapour bubbles and also by free convection. Forced convection boiling, on the other hand, is associated with flow of the liquid along with the vapour bubbles driven by an externally imposed pressure gradient, or is caused by some kind of natural circulation. Six flow regimes or zones that can be identified in forced convection boiling in a vertical tube are shown in Fig. 6.5. Two-phase upflow (i.e. upflow of the liquid with vapour bubbles dispersed in it) occurs in most part of the tube.

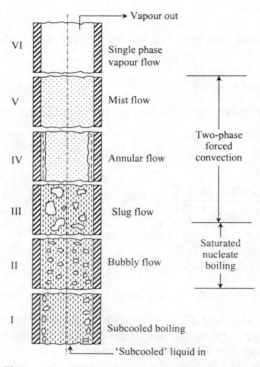

Fig. 6.5　Flow regimes in forced convection boiling in a vertical tube.

　　In zone I of Fig. 6.5, the liquid is still below its boiling point and *subcooled boiling* occurs at the wall. Most of the tiny bubbles formed at the wall collapse in the bulk liquid (see Section 6.3). In zone II, the liquid has sufficient superheat to nurture the vapour bubbles; the superheated liquid flows up along with the bubbles (*bubbly flow*). Because of growth and coalescence, the bubbles gradually become large in size and vapour slugs are formed in zone III (*slug flow*). In zone IV (*annular flow*), the liquid adheres to the wall as a film surrounding the inner vapour core. The inner core also contains tiny suspended liquid particles carried over from the zone below (these liquid particles are formed during vigorous bubble breakage). Over zone V, there is very little liquid on the wall but there are minute liquid droplets in the flowing vapour (*mist flow*). There is nothing else than vapour in zone VI. The rates of heat transfer are very high in zones II, III, and IV, but very low over zones V and VI where heat transfer occurs predominantly through the vapour.

Extensive research work has been carried out on forced convection boiling heat transfer, and excellent reviews are available. Here we shall cite the correlation by Chen (1966) that covers a considerably wide range of flow rates and regimes—both saturated nucleate boiling, and two-phase forced convection boiling. Chen considered the heat transfer coefficient to consist of two parts—a nucleate boiling part (h_b), and a forced convection boiling part (h_c). Thus,

$$h = h_b + h_c \tag{6.14}$$

$$h_c = 0.023 \, (\text{Re}_l)^{0.8} \, (\text{Pr}_l)^{0.4}\left(\frac{k_l}{d}\right) F \quad \text{W/m}^2 \text{ K}, \quad \text{Re}_l = \frac{G(1-w)d}{\mu_l} \tag{6.15}$$

$$h_b = 1.218 \times 10^{-3} \, \frac{(k_l)^{0.79}(c_{pl})^{0.45}(\rho_l)^{0.49}}{(\sigma)^{0.5}(\mu_l)^{0.29}(L_v)^{0.24}(\rho_v)^{0.24}}(T_e)^{0.24}\,(\Delta p_{\text{sat}})^{0.75} S \quad \text{W/m}^2 \text{ K} \tag{6.16}$$

$$\Delta p_{\text{sat}} = \frac{T_e L_v}{T_{\text{sat}} v_{lv}}; \qquad T_e = T_s - T_{\text{sat}}; \qquad v_{lv} = v_v - v_l \tag{6.17}$$

$$S = \left[1 + 0.12\,(\overline{\text{Re}})^{1.14}\right]^{-1.0}; \qquad \text{for } \overline{\text{Re}} < 32.5 \tag{6.18}$$

$$= \left[1 + 0.42\,(\overline{\text{Re}})^{0.78}\right]^{-1.0}; \qquad \text{for } 32.5 < \overline{\text{Re}} < 70 \tag{6.19}$$

$$= 0.1; \qquad \text{for } \overline{\text{Re}} > 70 \tag{6.20}$$

$$\overline{\text{Re}} = 10^{-4} \, \text{Re}_l \, F^{1.25} \tag{6.21}$$

$$F = 1.0 \quad \text{for } \frac{1}{\chi} < 0.10; \quad F = 2.35\left(\frac{1}{\chi} + 0.231\right)^{0.736} \quad \text{for } \frac{1}{\chi} > 0.10 \tag{6.22}$$

$$\frac{1}{\chi} = \left[\frac{w}{1-w}\right]^{0.9}\left[\frac{\rho_l}{\rho_v}\right]^{0.5}\left[\frac{\mu_v}{\mu_l}\right]^{0.1} \tag{6.23}$$

where

G is the fluid mass velocity, kg/m^2 s; w is the mass fraction vapour (or 'quality') in the fluid;
$v_{lv} = v_v - v_l$, m^3/kg (v = specific volume)
subscripts l and v denote the liquid and the vapour phase, respectively
T_s is the surface temperature
Δp_{sat} (N/m^2) is the difference of vapour pressure between the superheated liquid (at surface temperature T_s) and the saturated liquid (at temperature T_{sat})
d is the inner diameter of the tube
σ is the surface tension of the liquid (N/m) at the bulk temperature
other terms have significances as stated before and are in SI units.

In Eq. (6.15), Re$_l$ is the Reynolds number based on the liquid flow rate only (the mass of vapour is subtracted from the total flow). The factor S in Eq. (6.16) is called the *boiling suppression factor*. The factor F in Eq. (6.21) modifies the 'liquid only Reynolds number', Re$_l$, to a 'two-phase' Reynolds number.

In the liquid preheating region, if any, the Dittus-Boelter equation, Eq. (4.9), may be used to calculate the heat transfer coefficient.

Example 6.4   A mixture of benzene and toluene, taken a cut from a reformate distillation column in a refinery, has to be separated further in another distillation column. A reboiler is required to partially boil out the bottom liquid of the column. The vapour generated flows up through the column. The bottom liquid contains a little benzene, but for the purpose of heat transfer calculations the physical properties of the liquid can be assumed to be the same as those of toluene. So far as the boiling calculations are concerned, we formulate the problem as follows.

Toluene, as a saturated liquid, enters the bottom of a 14 BWG 25 mm vertical tube at 0.3 bar gauge pressure at the rate of 200 kg/h, 30% of which is vaporized per pass. The tube wall temperature is maintained at 160°C (433 K) by saturated steam condensing outside the tube at 4.7 bar (gauge). It is desired to calculate the required length of the tube. The following data under the given conditions are supplied.

Boiling point of toluene at the given pressure (0.3 bar gauge), $T_{sat} = 120°C$ (393 K), heat of vaporization, $L_v = 3.63 \times 10^5$ J/kg.

Other properties of the liquid: $k_l = 0.112$ W/m °C, $c_{pl} = 1968$ J/kg °C, $\rho_l = 753$ kg/m^3, $\sigma = 1.66 \times 10^{-2}$ N/m, $\mu_l = 2.31 \times 10^{-4}$ kg/m s.

For toluene vapour: $\rho_v = 3.7$ kg/m^3, viscosity of toluene vapour, $\mu_v = 10^{-5}$ kg/m s.

SOLUTION   Here forced convection boiling occurs in the tube. We can use the correlations of Chen [Eqs. (6.14)–(6.23)].

The boiling heat transfer coefficient under forced convection depends upon the 'quality' of the vapour-liquid mixture, i.e. the mass fraction of vapour w in the mixture. However, the nature of such dependence is not very simple. Here the quality, w, varies from 0 to 0.3. We adopt a stepwise calculation procedure. In each step (starting from $w = 0$) we increase the value of w by 0.05. Heat transfer coefficients and the area are calculated for each step taking the mean value of w in the interval. The total area required can then be obtained by adding the areas required for the individual steps. Given below are the detailed calculations for the interval $w = 0.1$ to $w = 0.15$, the mean value being $w = 0.125$.

Calculation of h_c and h_b for $w = 0.125$

$$d = \text{i.d. of 14 BWG 25 mm tube} = 0.0211 \text{ m}$$

$$\text{Mass flow rate, } G = \frac{200}{(3600)(\pi d^2/4)} = 159 \text{ kg/m}^2 \text{ s}$$

$$\text{Re}_l = \frac{G(1-w)d}{\mu_l} = \frac{(159)(1 - 0.125)(0.0211)}{2.31 \times 10^{-4}} = 1.271 \times 10^4$$

$$\text{Pr}_l = \frac{c_{pl}\mu_l}{k_l} = \frac{(1968)(2.31 \times 10^{-4})}{0.112} = 4.06$$

From Eq. (6.23), we have

$$\frac{1}{\chi} = \left[\frac{w}{1-w}\right]^{0.9}\left[\frac{\rho_l}{\rho_v}\right]^{0.5}\left[\frac{\mu_v}{\mu_l}\right]^{0.1} = \left[\frac{0.125}{0.875}\right]^{0.9}\left[\frac{753}{3.7}\right]^{0.5}\left[\frac{10^{-5}}{2.31 \times 10^{-4}}\right]^{0.1} = 1.81$$

From Eq. (6.22), $F = 2.35 \left[\dfrac{1}{\chi} + 0.231\right]^{0.736} = 3.973$

From Eq. (6.21), $\overline{Re} = 10^{-4}\,Re_l\,F^{1.25} = (10^{-4})(1.271 \times 10^4)(3.973)^{1.25} = 7.13$

From Eq. (6.18), $S = \left[1 + 0.12\,(\overline{Re})^{1.14}\right]^{-1.0} = 0.47$

From Eq. (6.15), the forced convection heat transfer coefficient,

$$h_c = (0.023)(1.271 \times 10^4)^{0.8}\,(4.06)^{0.4}\left(\dfrac{0.112}{0.0211}\right)(3.973) = 1630 \text{ W/m}^2 \text{ °C}$$

From Eq. (6.16), the boiling heat transfer coefficient can be calculated.

Excess temperature, $T_e = 160 - 120 = 40°C$

$$v_{lv} = \dfrac{1}{\rho_v} - \dfrac{1}{\rho_l}; \qquad \Delta p_{sat} = \dfrac{T_e L_v}{T_{sat} v_{lv}} = \dfrac{(40)(3.63 \times 10^5)}{(393)\left(\dfrac{1}{3.7} - \dfrac{1}{753}\right)} = 1.374 \times 10^5 \text{ N/m}^2$$

Therefore,

$$h_b = 1.218 \times 10^{-3}\,\dfrac{(k_l)^{0.79}\,(c_{pl})^{0.45}(\rho_l)^{0.49}}{(\sigma)^{0.5}(\mu_l)^{0.29}(L_v)^{0.24}(\rho_v)^{0.24}}\,(T_e)^{0.24}\,(\Delta p_{sat})^{0.75}\,S$$

$$= (1.218 \times 10^{-3})\dfrac{(0.112)^{0.79}(1968)^{0.45}(753)^{0.49}(40)^{0.24}(1.374 \times 10^5)^{0.75}}{(1.66 \times 10^{-2})^{0.5}(2.31 \times 10^{-4})^{0.29}\,(3.63 \times 10^5)^{0.24}(3.7)^{0.24}}\,(0.47)$$

$$= 4077 \text{ W/m}^2 \text{ °C}$$

Total heat transfer coefficient, $h = h_c + h_b = 1630 + 4077 = 5707 \text{ W/m}^2 \text{ °C}$

Calculation of the required heat transfer area

Rate of vaporization in order to change the quality w by 5%

$$= (200)(0.05) = 10 \text{ kg/h} = 2.778 \times 10^{-3} \text{ kg/s}$$

Heat load, $Q = (2.778 \times 10^{-3})(L_v) = (2.778 \times 10^{-3})(3.63 \times 10^5) = 1008 \text{ W}$

If A is the area of heat transfer, $Q = h\,A\,\Delta T = (5707)(A)(160 - 120)$

or

$$A = 0.004416 \text{ m}^2$$

If the required length of the tube is l,

$$l = \dfrac{A}{\pi d} = \dfrac{0.004416}{\pi(0.0211)}$$

$$= 0.0666 \text{ m} = 6.66 \text{ cm}$$

The calculations have to be done for all the intervals of w considered. The results are shown in Table 6.2 (the calculations are not shown and are left as an exercise).

Table 6.2   Example 6.4

Step size w	Average w	h_c (W/m^2 °C)	h_b (W/m^2 °C)	$h = h_c + h_b$ (W/m^2 °C)	Q (W)	Tube length (m)
0 – 0.05	0.025	736	6364	7100	1008	0.0535
0.05 – 0.10	0.075	1235	6872	6162	1008	0.0617
0.10 – 0.15	0.125	1630	4077	5707	1008	0.0666
0.15 – 0.20	0.175	1970	3504	5474	1008	0.0695
0.20 – 0.25	0.225	2273	3088	5361	1008	0.0709
0.25 – 0.30	0.275	2546	2776	5322	1008	0.0714

Total tube length = $\boxed{0.393 \text{ m}}$

Comments.   It is seen that the total heat transfer coefficient is very high, and a rather small length of the tube (0.393 m) offers the necessary heat transfer area. But does it happen in actual practice? No. In reality, the overall heat transfer coefficient gets drastically reduced by the 'dirt factor' (see Chapter 8). So a much larger tube length will be necessary in practice to achieve the desired boiling rate. For boiling of organics, an overall heat transfer coefficient in the range 500–1000 W/m^2 °C (100–200 Btu/h ft^2 °F) is used in the design of reboilers (Ludwig, 1983; Walas, 1990; also see Chapter 8).

6.6   THE CONDENSATION PHENOMENON

Condensation means the change of phase from the vapour to the liquid. If the temperature of a vapour is reduced below its saturation temperature, condensation occurs. A liquid at its boiling point is called a *saturated liquid* and the vapour in equilibrium with this liquid is called *saturated vapour*. A liquid or vapour above the saturation temperature is called superheated. Like boiling, condensation is also an important heat transfer operation. A distillation column is always equipped with a condenser to condense the vapour coming out of the top of the column. Products coming out of a reactor are often obtained as a vapour, which is then required to be condensed for recovery. In an evaporator or a vaporizer, steam is frequently used as the source of heat. The steam condenses on one side of the heat transfer surface to supply the necessary heat of vaporization to the liquid boiling on the other side. In a steam power plant, huge condensers are used for condensation of the low pressure steam leaving the turbine.

If a mixture of a vapour and a gas is cooled, the vapour condenses to form minute droplets suspended in the carrier gas. This is called *homogeneous condensation*. In contrast, if a vapour or a gas–vapour mixture comes in contact with a cool surface, condensation occurs on the surface. This is called *surface condensation*. If the condensate wets the surface, it flows down as a continuous film. This is called *film condensation*. However, if the surface is not wetted by the condensate, drops (instead of a continuous film) appear on the surface, which grow in size and then trickle down the surface. This is called *dropwise condensation*. This occurs when the surface is contaminated or the condensate liquid does not have any 'affinity' for the surface. These two modes of condensation are shown in Fig. 6.6.

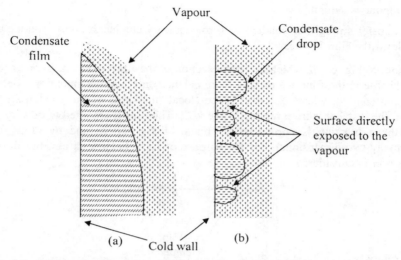

Fig. 6.6 Modes of condensation: (a) film condensation and (b) dropwise condensation.

In film condensation, the latent heat is transferred through the liquid film (which blankets the surface on which condensation occurs) and then conducted through the wall to the cooling fluid on the other side of the wall. The condensate film offers considerable heat transfer resistance. In dropwise condensation, on the other hand, a part of the surface is covered by liquid drops and the rest is directly exposed to the vapour. As a result, heat transfer coefficient for film condensation is considerably lower than that for dropwise condensation. A vapour, that normally condenses as a film on a particular surface, can be made to condense in the dropwise mode by applying a suitable coating on the surface. But this technique has not met with much success in practice. In dropwise condensation, a drop nucleus first forms on the surface and then grows in size. The approximate *practical* ranges of film-coefficient for condensation are: condensation of steam (dropwise)—5,500–16,000 W/m² °C (1000–3000 Btu/h ft² °F); organic vapour (film condensation)—1000–3000 W/m² °C (150–550 Btu/h ft² °F). These values may be much *higher* for *very clean* surfaces—around 5500 W/m² °C (1000 Btu/h ft² °F) for film condensation of organics, and 55,000–450,000 W/m² °C (10,000–80,000 Btu/h ft² °F) for dropwise condensation of steam.

Extensive experimental and theoretical research work has been done with a view to fundamental understanding of the phenomenon of condensation and also with a view to developing equations and correlations for the calculation of the heat transfer rate in condensation. A few simple approaches and results are described below.

6.7 FILM CONDENSATION ON A VERTICAL SURFACE

A theoretical analysis of laminar film condensation of a vapour on a vertical plate was given by Nusselt in 1916. This analysis is based on the following major assumptions:

(i) The condensate film is in 'locally' fully developed laminar flow with zero interfacial shear and constant liquid properties.

(ii) The vapour is saturated.

(iii) Heat transfer through the condensate film occurs by conduction only, and the temperature profile in the film is linear.

Let us refer to Fig. 6.7 in which a vertical section of the film is shown. In order to determine the velocity profile in the film, we make a force balance on a liquid element or 'control volume' of sides dx and $(\delta - y)$, where $\delta = \delta(x)$ is the 'local' thickness of the condensate film which depends upon the distance x from the top of the wall. The film is assumed to be very wide in the z-direction (the z-direction is normal to the plane of the paper), and therefore the problem becomes virtually two-dimensional. For the purpose of force balance, a unit breadth of the film in the z-direction is considered.

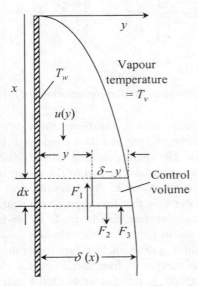

Fig. 6.7 Control volume in a condensate film in laminar flow.

The three forces, F_1, F_2, and F_3, representing the shear force, gravity force and the buoyancy force, acting on the control volume are also shown in Fig. 6.7. By a force balance, we have

$$F_1 = F_2 - F_3$$

or

$$\mu \frac{du}{dy} dx = \rho_l g(\delta - y)dx - \rho_v g(\delta - y)dx$$

Integrating, we get, $u = \dfrac{(\rho_l - \rho_v)g}{\mu}\left(y\delta - \dfrac{y^2}{2}\right) + C'$; $\quad C'$ is the integration constant.

Using the boundary condition $u = 0$ at $y = 0$ (i.e. 'no-slip condition' at the wall), we have $C' = 0$. Therefore,

$$u = \frac{(\rho_l - \rho_v)g}{\mu}\left(y\delta - \frac{y^2}{2}\right) \tag{6.24}$$

Equation (6.24) gives the velocity profile in the freely flowing film. The profile is half-parabolic. The rate of flow of the condensate (per unit breadth of the film) at any location x is obtained by integration of this equation.*

$$m' = \int_0^\delta u(dy.1)\rho_l = \int_0^\delta \rho_l \frac{(\rho_l - \rho_v)g}{\mu}\left(y\delta - \frac{y^2}{2}\right)dy = \frac{\rho_l(\rho_l - \rho_v)g\delta^3}{3\mu} \tag{6.25}$$

Here, as stated before, δ is the 'local thickness' of the condensate film.

As x increases downwards, the film thickness also increases because of condensation of more and more vapour. The rate of condensation on an 'elementary surface' of size $dx \times 1$ exposed to the vapour can be obtained from the rate of heat transfer through this area. The temperature profile in the condensate film is assumed to be linear and we may, therefore, write

$$\text{The rate of heat transfer} = k_l \frac{T_v - T_w}{\delta}(dx)(1)$$

Here T_v and T_w are the vapour and the wall temperature, respectively.** The change in the rate of condensate flow over a distance dx is

$$dm' = k_l \frac{T_v - T_w}{\delta L_v}dx$$

or

$$\frac{dm'}{dx} = k_l \frac{T_v - T_w}{\delta L_v}$$

Substituting for m' from Eq. (6.25), differentiating with respect to x and rearranging, we get

$$\frac{\rho_l(\rho_l - \rho_v)g}{\mu_l}\delta^3 \, d\delta = k_l \frac{T_v - T_w}{L_v}dx$$

Integrating the above equation and using the appropriate boundary condition (at $x = 0$, $\delta = 0$) it is easy to get an expression for the local thickness of the condensate film. That is,

$$\delta(x) = \left[\frac{4\mu_l k_l x(T_v - T_w)}{gL_v \rho_l(\rho_l - \rho_v)}\right]^{1/4} \tag{6.26}$$

If h is the heat transfer coefficient for the condensate film, heat flux through the film at any location x is

$$h = \frac{k_l}{\delta}$$

Using the expression for $\delta = \delta(x)$ given by Eq. (6.26), we get

---

* Because the breadth of the film (i.e. its dimension in the z-direction) is unity, the elementary area of flow normal to the x-direction (i.e. the direction of flow) is $1.dy = dy$.

** The vapour temperature is considered uniform throughout, and the temperature at the vapour-condensate interface is T_v. The temperature drop across the condensate film is $T_v - T_w$.

$$h = k_l \left[\frac{4\mu_l k_l x(T_v - T_w)}{gL_v\rho_l(\rho_l - \rho_v)} \right]^{-1/4} = \left[\frac{gL_v\rho_l(\rho_l - \rho_v)k_l^3}{4\mu_l x(T_v - T_w)} \right]^{1/4} \tag{6.27}$$

The local Nusselt number is

$$\mathrm{Nu}_x = \frac{hx}{k_l} = \left[\frac{gL_v\rho_l(\rho_l - \rho_v)x^3}{4\mu_l k_l(T_v - T_w)} \right]^{1/4} \tag{6.28}$$

The average heat transfer coefficient over a length L is

$$\bar{h} = \frac{1}{L}\int_0^L h\,dx = \frac{1}{L}\left[\frac{gL_v\rho_l(\rho_l - \rho_v)k_l^3}{4\mu_l(T_v - T_w)} \right]^{1/4} \left(\frac{4}{3}\right)L^{3/4}$$

$$= 0.943\left[\frac{gL_v\rho_l(\rho_l - \rho_v)k_l^3}{4\mu_l L(T_v - T_w)} \right]^{1/4} \tag{6.29}$$

While using Eq. (6.29) the liquid properties should be taken at the mean film temperature, i.e. at $(T_v + T_w)/2$. If condensation occurs on an inclined surface, g in the above equation should be replaced by $g\cos\theta$, where θ is the inclination of the surface to the vertical.

Nusselt's analysis of film condensation on a vertical surface has several limitations. These are: (i) it is grossly approximate to assume that the flow in the condensate film is one-dimensional and that the velocity profile has a half-parabolic form at each location along the plate, (ii) non-consideration of convection effects and assumption of a linear temperature distribution in the film are far from adequate. A more refined theoretical treatment of the problem was given by Sparrow and Gregg (1959) on the basis of two-dimensional boundary layer approximation for flow and temperature fields. It is, however, interesting to note that, despite its limitations, the Nusselt's equation (6.28) predicts the condensation heat transfer coefficient on a flat vertical (or inclined) surface with an error of only a few per cent in many situations of practical importance.

The accuracy of the Nusselt's equation can be improved by incorporating a modification suggested by Rohsenow et al. (1985). At any position on the plate, the liquid-film temperature changes from T_v at the free surface to T_w at the plate surface. This means that some amount of sensible heat is also removed in addition to the latent heat of condensation. To take this into account, Rohsenow et al. (1985) suggested the use of a modified latent heat term $L'_v = L_v(1 + 0.68\mathrm{Ja})$ in place of the heat of vaporization, L_v. The quantity Ja is called the 'Jacob number', Ja $= c_{pl}(T_v - T_w)/L_v$.

6.8 TURBULENT FILM CONDENSATION

If the rate of condensation is low or the height of the condensing surface is small, the thickness of the condensate film remains small and the flow remains laminar. The free surface of a film in laminar flow appears smooth and ripple-free. The nature of flow is determined by the *film Reynolds number*, Re_f. The local average liquid velocity in the film can be obtained from Eq. (6.24).

$$\bar{u} = \frac{1}{\delta}\int_0^\delta u\,dy = \frac{1}{\delta}\int_0^\delta \frac{(\rho_l - \rho_v)g}{\mu_l}\left(y\delta - \frac{y^2}{2}\right)dy = \frac{(\rho_l - \rho_v)g\delta^2}{3\mu_l}$$

If the 'hydraulic diameter'* of the condensate film (on the basis of unit breadth of the film) is taken as the characteristic length, the film Reynolds number can be defined as follows:

$$\text{Re}_f = \frac{4\delta\bar{u}\rho_l}{\mu_l} = \frac{4\rho_l(\rho_l - \rho_v)g\delta^3}{3\mu_l^2} = \frac{4m'}{\mu_l} \tag{6.30}$$

where m' is the rate of condensation per unit breadth of the plate. It is easy to visualize that the film thickness is small near the top of the plate and increases towards the bottom. So the film Reynolds number also increases with distance along the plate. It has been experimentally found that when $\text{Re}_f \leq 30$, the film remains *laminar* and the free surface of the film remains wave-free. If $30 \leq \text{Re}_f \leq 1600$, *waves and ripples* appear on the surface, although the flow is still laminar. If the film Reynolds number is more than 1600, the film becomes *wavy and turbulent*. The following correlations can be used to calculate the average condensation heat transfer coefficient over three ranges of the film Reynolds number, Re_f (Rohsenow et al., 1985), expressed on the basis of a length L of the condensate film.

$$\text{Nu}' = \frac{\bar{h}}{k_l}\left[\frac{\mu_l^2}{\rho_l(\rho_l - \rho_v)g}\right]^{1/3} = 1.47\,(\text{Re}_f)^{-1/3}; \qquad \text{Re}_f \leq 30 \quad \text{(laminar)} \tag{6.31}$$

$$= \frac{\text{Re}_f}{1.08(\text{Re}_f)^{1.22} - 5.2}; \qquad 30 \leq \text{Re}_f \leq 1600 \quad \text{(wavy)} \tag{6.32}$$

$$= \frac{\text{Re}_f}{8750 + 58\left[(\text{Re}_f)^{0.75} - 253\right]\text{Pr}_l^{-0.5}}; \qquad \text{Re}_f \geq 1600 \quad \text{(turbulent)} \tag{6.33}$$

Equation (6.31) defines a *modified Nusselt number*, Nu', which is also called the *condensation number*; $\bar{h}$ is the average heat transfer coefficient.

The above correlations for condensation on a vertical plate are also applicable to condensation inside or outside vertical tubes if the tube diameter is not too small. Equation (6.31) can be deduced from Eq. (6.29) if Re_f is taken at the bottom of a plate of length L.

Example 6.5 Saturated propane vapour at 32°C condenses on a vertical surface, 0.3 m high, at a pressure of 11.2 atm. The surface is maintained at 25°C. Calculate the rate of condensation of propane.

Data supplied: At 25°C, density of liquid propane, $\rho_l = 483$ kg/m³; viscosity, $\mu_l = 0.091$ cP; thermal conductivity, $k_l = 0.09$ W/m K; heat of vaporization, $L_v = 326$ kJ/kg; specific heat, $c_{pl} = 2.61$ kJ/kg K; density of saturated propane vapour at 32°C and 11.2 atm pressure, $\rho_v = 24.7$ kg/m³.

---

* In this case, the hydraulic diameter $= \dfrac{(4)(\text{cross section})}{\text{wetted perimeter}} = \dfrac{4(\delta)(1)}{1} = 4\delta$ (see Section 4.5.3).

SOLUTION   The regime of flow of the condensate (i.e. whether it is laminar, or wavy or turbulent) is not known. In fact, over a part of the surface the flow is expected to be laminar, and the length of this part is to be determined first. If the length of the plate is larger than this, the lower part of the film will be wavy. For a sufficiently long plate, there may even be a turbulent region. Unit breadth of the plate is considered in the calculations.

First we calculate the average condensation film coefficient using Eq. (6.29) and replacing L_v by L'_v. Thus

$$\bar{h} = 0.943 \left[\frac{gL'_v \rho_l (\rho_l - \rho_v) k_l^3}{4\mu_l L(T_v - T_w)} \right]^{1/4}$$

where

$$T_v - T_w = 32 - 25 = 7°C$$

$$L'_v = L_v (1 + 0.68Ja) = 326 + (0.68)(2.61)(7) = 338.4 \text{ kJ/kg}$$

Therefore,

$$\bar{h} = 0.943 \left[\frac{(9.8)(338.4 \times 10^3)(483)(483 - 24.7)(0.09)^3}{L(9.1 \times 10^{-5})(7)} \right]^{1/4} = 902.6 \, (L)^{-1/4}$$

For unit breadth of the plate, the rate of heat transfer,

$$Q = \bar{h}(L \times 1)(T_v - T_w) = 902.6 \, (L)^{-1/4}(L)(7) = 6318.2(L)^{3/4} \text{ W}$$

The rate of condensation, $m' = \dfrac{Q}{L'_v} = \dfrac{6318.2(L)^{3/4}}{338.4 \times 10^3} \text{ kg/s} = 1.867 \times 10^{-2}(L)^{3/4} \text{ kg/s}$

The flow remains laminar till the film Reynolds number becomes 30. Using this value of the film Reynolds number, the corresponding distance L from the top can be calculated.

$$\text{Re}_f = \frac{4m'}{\mu_l} = \left(\frac{4}{9.1 \times 10^{-5}} \right)(1.867 \times 10^{-2})(L)^{3/4} = 30$$

or

$$L = 0.012 \text{ m} \quad \text{and} \quad m' = 6.77 \times 10^{-4} \text{ kg/s}$$

Therefore, the condensate film flow remains laminar only over a length of 1.2 cm from the top. The film becomes wavy after this distance.

In order to calculate the rate of condensation over the remaining length, we assume the flow to be wavy. This assumption will, however, be checked for its validity. We use Eq. (6.32).

$$\text{Nu}' = \frac{\bar{h}}{k_l} \left[\frac{\mu_l^2}{\rho_l(\rho_l - \rho_v)g} \right]^{1/3} = \frac{\text{Re}_f}{1.08(\text{Re}_f)^{1.22} - 5.2} \tag{i}$$

The length of the plate over which the flow is wavy is, $L' = 0.3 - 0.012 = 0.288 \text{ m}$

The corresponding area of condensation, $A = (0.288 \text{ m})(1 \text{ m}) = 0.288 \text{ m}^2$

Let the rate of condensation over the total length of the plate be

$$m'' = [m']_{\text{laminar}} + [m']_{\text{wavy}}$$

or

$$m'' = 6.77 \times 10^{-4} + \frac{\overline{h}A(T_v - T_w)}{L_v'} = 6.77 \times 10^{-4} + \frac{\overline{h}(0.288)(7)}{338.4 \times 10^3}$$

$$= 6.77 \times 10^{-4} + 5.96 \times 10^{-6} \, (\overline{h})$$

$$\text{Re}_f = \frac{4m''}{\mu_l} = \frac{4}{9.1 \times 10^{-5}} \left[6.77 \times 10^{-4} + 5.96 \times 10^{-6}(\overline{h}) \right] = 29.76 + 0.262 \, (\overline{h})$$

From Eq. (i) above, we have

$$\frac{\overline{h}}{0.09} \left[\frac{(9.1 \times 10^{-5})^2}{(483)(483 - 24.7)(9.8)} \right]^{1/3} = \frac{29.76 + 0.262 \, (\overline{h})}{(1.08)[29.76 + 0.262(\overline{h})]^{1.22} - 5.2}$$

or

$$\overline{h} = 1425 \text{ W/m}^2 \text{ °C}$$

$$\text{Re}_f = 29.76 + (0.262)(\overline{h}) = 29.76 + (0.262)(1425) = 403$$

This calculated Reynolds number at the bottom edge of the plate remains within the wavy flow regime ($30 \leq \text{Re}_f \leq 1600$). Therefore, the use of the wavy flow equation is justified.

Total rate of condensation, $m'' = 6.77 \times 10^{-4} + 5.96 \times 10^{-6} \, (\overline{h})$ kg/s = $\boxed{33.01 \text{ kg/h}}$

It may be noted that, although the film remains purely laminar over only 0.012 m (i.e. 4% of the plate length) this accounts for about 7.4% (check it!) of the total rate of condensation. Why is it so? In the top region, the film thickness is small and the heat transfer resistance is also small. Hence, the rate of condensation is relatively high in the upper or laminar region.

6.9 CONDENSATION OUTSIDE A HORIZONTAL TUBE OR A TUBE BANK

Condensation outside a vertical tier of horizontal tubes occurs in many condensation units. The possible physical pictures of the condensation phenomenon over a tube bank are shown in Fig. 6.8. If there is a single horizontal tube (or even a row of horizontal tubes), condensate flows as a film along the cylindrical surface [Fig. 6.8(a)]. If there is another tube below, the condensate film flows down from the bottom edge of the upper tube to the upper edge of the lower tube [Fig. 6.8(b)]. This goes on from an upper tube to the lower of the tube tier, and the thickness of the condensate film grows. It may also happen that the condensate from the bottom edge of the upper tube drips down to the lower tube [Fig. 6.8(c)]. The case of film condensation on a sphere is shown in Fig. 6.8(d). Given below are the useful correlations for the average heat transfer coefficient in the above situations. These are based on the Nusselt's analysis.

Condensation on a single horizontal tube,

$$\overline{h} = 0.728 \left[\frac{g\rho_l(\rho_l - \rho_v)k_l^3 L_v}{d\mu_l(T_v - T_w)} \right]^{1/4} \tag{6.34}$$

Condensation on a vertical tier of N horizontal tubes,

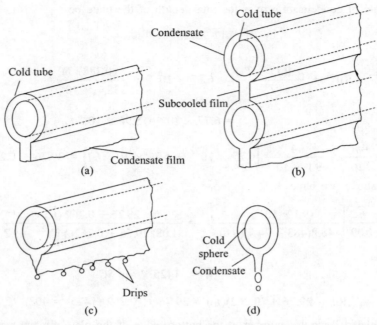

Fig. 6.8   Film condensation on curved surfaces: (a) single horizontal tube, (b) vertical tube bank (continuous film), (c) vertical tube bank (dripping), and (d) sphere.

$$\bar{h} = 0.728 \left[\frac{g\rho_l(\rho_l - \rho_v)k_l^3 L_v}{Nd\mu_l(T_v - T_w)} \right]^{1/4} \tag{6.35}$$

Condensation on a sphere,

$$\bar{h} = 0.815 \left[\frac{g\rho_l(\rho_l - \rho_v)k_l^3 L_v}{d\mu_l(T_v - T_w)} \right]^{1/4} \tag{6.36}$$

The various terms have significances as described before. Chen (1961) suggested the following modified form of Eq. (6.35) taking into account the condensation occurring on the subcooled film between two adjacent tubes [see Fig. 6.8(b)].

$$\bar{h} = 0.728 \left[1 + 0.2\frac{c_{pl}(T_v - T_w)}{L_v}(N - 1) \right] \left[\frac{g\rho_l(\rho_l - \rho_v)k_l^3 L_v'}{Nd\mu_l(T_v - T_w)} \right]^{1/4} \tag{6.37}$$

Equation (6.37) has been found to agree well with experimental results. All the other equations above *underpredict* the heat transfer coefficient by around 20%. (This means that the heat transfer coefficient calculated by using any of the above equations is about 20% less than the actual value.) Thus, so far as condenser design is concerned, the estimates are conservative and safe. Rohsenow et al. (1985) suggested that in Eqs. (6.35) and (6.36), L_v should be replaced by L_v' (see Section 6.7 last para) for improved accuracy.

Example 6.6 Trichloroethylene (TCE) is an important solvent (it is a carcinogen) for various applications (the major uses are for cleaning and degreasing). A manufacturer of porous battery separators uses it to leach out a non-volatile oil dispersed in the moulded PVC sheets in order to generate pores. TCE is recovered by stripping it out of the solution containing the oil followed by condensation.

Condensation of the vapour leaving the stripper is done in a horizontal condenser at essentially the atmospheric pressure. The vapour, which is of virtually pure TCE, condenses on the outside of 25.4 mm o.d. tubes having a surface temperature of 25°C.

Calculate the rate of condensation of TCE (a) on a single horizontal tube, 0.7 m long, and (b) in a condenser that has 36 tubes of the same diameter and length, but arranged in a 38 mm square pitch.

The following data for TCE are available:
normal boiling point = 360.4 K (87.4°C)
heat of vaporization = 320.8 kJ/kg
specific heat = 1.105 kJ/kg °C
liquid viscosity = 0.45 cP
thermal conductivity of the liquid = 0.1064 W/m °C
liquid density = 1375 kg/m^3
density of the vapour = 4.44 kg/m^3

The liquid properties are taken at the mean film temperature [i.e. $(25 + 87.4)/2 = 56.2$°C].

SOLUTION (a) We will use Eq. (6.34) in order to calculate the heat transfer coefficient for condensation outside a single horizontal tube.

Given: $d = 0.0254$ m, $T_v - T_w = 87.4 - 25 = 62.4$°C, $\mu_l = 4.5 \times 10^{-4}$ kg/m s.

We replace L_v by

$$L'_v = L_v + 0.68c_{pl}(T_v - T_w)$$
$$= 320.8 + (0.68)(1.105)(62.4)$$
$$= 3.677 \times 10^5 \text{ J/kg}$$

Putting the values of the various quantities in Eq. (6.34), we get

$$\bar{h} = (0.728)\left[\frac{(9.8)(1375)(1375 - 4.44)(0.1064)^3(3.677 \times 10^5)}{(0.0254)(4.5 \times 10^{-4})(62.4)}\right]^{1/4} = 1340 \text{ W/m}^2 \text{ °C}$$

Area of the tube = $\pi d L = (\pi)(0.0254)(0.7) = 0.0559$ m^2

Rate of heat transfer, $Q = (1340)(0.0559)(62.4) = 4670$ W

Rate of condensation = $m' = \dfrac{Q}{L'_v} = \dfrac{4670}{3.677 \times 10^5} = 1.27 \times 10^{-2}$ kg/s = $\boxed{45.7 \text{ kg/h}}$

(b) There are 6 tubes in a vertical tier, and six such tiers in total in the multi-tube condenser. We will use Eq. (6.35) to calculate the heat transfer coefficient. Putting the values of the various quantities and putting $N = 6$, we get

$$\bar{h} = (0.728)\left[\frac{(9.8)(1375)(1375 - 4.44)(0.1064)^3(3.677 \times 10^5)}{(6)(0.0254)(4.5 \times 10^{-4})(62.4)}\right]^{1/4} = 856 \text{ W/m}^2 \text{ °C}$$

Area of the tubes (36 tubes in total) = $(36)(\pi)(0.0254)(0.7) = 2.012 \text{ m}^2$

The rate of heat transfer, $Q = (856)(2.012)(62.4) = 107.5 \text{ kW}$

The rate of condensation, $m' = \dfrac{Q}{L'_v} = \dfrac{107.52}{367.7} = 0.2924 \text{ kg/s} = \boxed{1053 \text{ kg/h}}$

It will be interesting to calculate the condensation heat transfer coefficient for the tier of tubes using the Chen's correlation [Eq. (6.37)]. It uses a correction term for Eq. (6.34).

$$\left[1 + 0.2\frac{c_{pl}(T_v - T_w)}{L'_v}(N - 1)\right] = 1 + \frac{(0.2)(1.105)(62.4)(6 - 1)}{367.7} = 1.187$$

Thus, there will be a 18.7% increase in the calculated rate of heat transfer and, therefore, in the rate of condensation, according to the Chen's correlation.

6.10 CONDENSATION INSIDE A HORIZONTAL TUBE

This is another situation of much practical interest. Condensation of refrigerants mostly occurs inside horizontal or vertical tubes. When a vapour condenses in a horizontal tube, the condensate film flows around the periphery to form a bottom layer of liquid that moves towards the exit end of the tube (Fig. 6.9) in a kind of *open channel flow*. If the tube length is short or the rate of condensation is low (i.e. if the thickness of the flowing condensate layer at the bottom is small), the correlation suggested by Chato (1962) is useful.

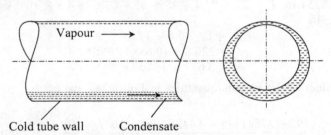

Fig. 6.9 Film condensation in a horizontal tube.

$$\bar{h} = 0.555\left[\frac{g\rho_l(\rho_l - \rho_v)k_l^3 L''_v}{d_i\mu_l(T_v - T_w)}\right]^{1/4} \tag{6.38}$$

where $L''_v = L_v[1 + (3/8)\text{Ja}]$, and the vapour Reynolds number is below 35,000.

If the rate of condensation is high or the tube is long, the correlations of Ackers et al. (1958) may be used (the error involved may sometimes be as high as 50%).

$$\frac{\bar{h}d_i}{k} = 0.026\left(\text{Re}_m\right)^{0.8}\left(\text{Pr}_l\right)^{1/3}$$

$$\text{Re}_m = \frac{d}{\mu_l}\left[G_l + G_v\left(\frac{\rho_l}{\rho_v}\right)^{1/2}\right] \tag{6.39}$$

where G_l and G_v are the liquid and vapour mass velocities (kg/m^2 s) calculated on the basis of the cross-section (= $\pi d_i^2/4$) of the tube. These correlations are useful when the vapour Reynolds number is over 20,000, and the liquid Reynolds number is over 5000.

Example 6.7 Saturated vapour of benzene at 1 atm pressure enters a horizontal 16 BWG 3/4 inch tube (o.d. = 19 mm, i.d. = 16 mm) at a mass velocity of 20 kg/m^2 s. A coolant flows outside the tube, and the wall temperature is maintained at 55°C. What fraction of the vapour will condense if the tube is 0.5 m long? The following data for benzene are available at the mean film temperature, i.e. (80 + 55)/2 = 67.5°C: ρ_l = 815 kg/m^3, μ_l = 0.381 cP, c_{pl} = 1.94 kJ/kg °C, k_l = 0.13 W/m °C, L_v = 391 kJ/kg. Normal boiling point of benzene is 80°C (353 K); for saturated benzene vapour at 1 atm pressure, ρ_v = 2.7 kg/m^3, μ_v = 89 µP).

SOLUTION The mass flow rate of benzene, G_v = 20 kg/m^2 s; tube diameter, d_i = 0.016 m.

The vapour Reynolds number, $\mathrm{Re}_v = \dfrac{d_i G_v}{\mu_v} = \dfrac{(0.016)(20)}{8.9 \times 10^{-6}} = 36{,}000$

We will use the correlation of Chato, Eq. (6.38), for calculation of the heat transfer coefficient.

$$\bar{h} = 0.555 \left[\frac{g\rho_l(\rho_l - \rho_v)k_l^3 L_v''}{d_i \mu(T_v - T_w)} \right]^{1/4}$$

where

$$L_v'' = L_v + (3/8)\, c_{pl}(T_v - T_w) = 391 + (3/8)(1.94)(80 - 55) = 409.2 \text{ kJ/kg}$$

Therefore,

$$\bar{h} = 0.555 \left[\frac{(9.8)(815)(815 - 2.7)(0.13)^3 \times (409.2 \times 10^3)}{(0.016)(3.81 \times 10^{-4})(80 - 55)} \right]^{1/4} = 1378 \text{ W/m}^2\text{ °C}$$

Area of the tube available for heat transfer = $\pi d_i\, l$ = $(\pi)(0.016)(0.5)$ = 0.0251 m^2
The rate of heat transfer, Q = (1378)(0.0251)(80 − 55) = 864.3 W
The rate of condensation of benzene,

$$m' = \frac{Q}{L_v''} = \frac{864.3 \text{ W}}{409.2 \text{ kJ/kg}} = 2.112 \times 10^{-3} \text{ kg/s} = 7.603 \text{ kg/h}$$

The rate of input of benzene vapour = $(20 \text{ kg/m}^2 \text{ s})\left(\dfrac{\pi}{4}\right)[(0.016)^2 \text{ m}^2]$ = 4.02×10^{-3} kg/s

Fraction of the input vapour condensed = $\dfrac{2.112 \times 10^{-3}}{4.02 \times 10^{-3}}$ = $\boxed{52.5\%}$

6.11 EFFECT OF NON-CONDENSABLE GASES

The rate of vapour condensation is significantly reduced in the presence of non-condensable gases. In such a situation, the vapour molecules reach the condensate-vapour interface by diffusion through the vapour-gas mixture. This creates a mass transfer or diffusional resistance besides the heat transfer resistance for condensation. The partial pressure of vapour at the interface is lower than that in the bulk. Therefore, the actual rate of condensation will be less than what is expected on the basis of partial pressure of the vapour in the bulk medium. Accurate calculation of the rate of condensation requires the solution of the concerned coupled heat and mass transfer problem. The solution methodology of such equations will not be discussed here. It is, however, pertinent to refer to the classical work of Colburn and Hougen (1934) and a later modification by Votta (1964).

6.12 DROPWISE CONDENSATION

In dropwise condensation, the vapour condenses on tiny condensate nuclei on the surface. These nuclei grow in size to form drops. The bigger drops run down the surface and further condensation occurs on the nuclei. The nuclei originate in cracks and pits of the surface. The condensate drops typically occupy about 90% of the surface area.

As stated before, the condensation heat transfer coefficient is usually very high for dropwise condensation, and contributes very little towards the total heat transfer resistance. In most practical situations this contribution can be safely neglected. The details of the mechanism of dropwise condensation and prediction of the heat transfer coefficient are available in the literature [see, for example, Rohsenow et al. (1985)]. Dropwise condensation of steam is most common, and the presence of non-condensable gases drastically reduces the rate of condensation.

It is possible to ensure dropwise condensation by treating the surface with a substance that has a strong affinity for the particular surface but repels water. Such a substance is called the *promoter* of dropwise condensation. Silicone, higher organic acids (like octanoic acid) etc. have been reported to be good promoters.

SHORT QUESTIONS

1. Explain the terms: superheated liquid, superheated vapour, quality of a vapour, subcooled boiling, nucleate boiling, film boiling, pool boiling, transition boiling, homogeneous and heterogeneous nucleation.

2. What are the different boiling regimes? Define critical heat flux and boiling crisis. What is the range of excess temperature over which nucleate pool boiling of an organic liquid occurs?

3. When does radiation play a role in boiling heat transfer? What is Leidenfrost phenomenon?

4. Discuss the occurrence of hysteresis in a boiling curve.

5. What is forced convection boiling? Is it common in industrial boilers and vaporizers?

6. Give the criterion for the collapse of a vapour bubble in a liquid. Can a bubble collapse in a superheated liquid?

7. What are the different flow regimes in forced convection boiling of a liquid in a vertical tube? In which regime is the boiling heat transfer coefficient the largest?

8. Which boiling regime is preferred in industrial boilers and vaporizers?

9. Explain the terms: homogeneous condensation, surface condensation, filmwise condensation, dropwise condensation, and condensation promoters. Give the approximate ranges of values of heat transfer coefficients for filmwise and dropwise condensation on different types of surfaces.

10. What are the basic assumptions of Nusselt's theory of film condensation? What is the nature of temperature distribution in the condensate film according to this theory?

11. Define Jacob number and condensation number.

12. How does the average condensation heat transfer coefficient on an inclined plate depend upon the length of the plate?

13. How is the film Reynolds number defined? At what Reynolds number does the flow in a film turn turbulent?

14. Consider a vertical tier of three horizontal tubes on which condensation of a vapour is taking place. On which of the tubes is the rate of condensation maximum? On which is it minimum? Give explanations.

15. How does the presence of non-condensable gases affect the rate of condensation of a vapour?

16. Will the rate of condensation on a surface be more at 'zero gravity'? Explain. How will the rate of condensation be affected by enhanced gravity?

17. At what kind of site on a surface does the nucleus of a bubble (in boiling) or a drop (in condensation) originate?

PROBLEMS

6.1 Water is boiling in a pan at atmospheric pressure. If the water has a 6°C superheat, calculate the pressure inside a vapour bubble of 5 mm diameter.

6.2 Cyclohexane boiling in a vaporizer at 1 atm pressure has 7°C superheat. What is the maximum size of a cavity on the boiling surface for which the bubble nucleus detached from the cavity does not collapse? If there is a cavity of 8 micron diameter, what should be the superheat in the liquid so that a bubble nucleus leaving this cavity does not collapse?

6.3 In a flat-bottom copper kettle, 0.75 ft in diameter, water boils at atmospheric pressure at a rate of 40 lb/h. What is the temperature of the bottom surface of the kettle?

6.4 Isopropyl alcohol is boiling at 1.2 bar pressure in a copper pan. Low pressure steam is condensing in an outer jacket in order to maintain a pan surface temperature of 110°C. Calculate the rate of boiling using both Rohsenow's equation and Mostinski's equation. What is the critical boiling heat flux for isopropyl alcohol? What should be the boiling surface temperature to yield the critical heat flux?

6.5 A chrome-plated electric immersion heater is used to boil water at 1 atm pressure. The surface temperature of the heater is 120°C. Calculate the rate of boiling of water and power supply to the heater if the surface area of the cylindrical heating element is 4.2×10^{-5} m^2.

6.6 Large quantities of propylene are used for the manufacture of polypropylene (polypropylene has a variety of uses, for example, in making pipes, moulded articles, films, coatings, etc.).

Separation of propylene from propane by distillation is an important step in its manufacture. In a particular plant, the distillation column receives vapour from a horizontal kettle-type reboiler. The reboiler liquid has a low concentration of propylene, and has properties essentially the same as those of propane. The pressure in the reboiler is 18 bar. Two heating fluids are available: (i) low pressure steam at 1.5 bar gauge, and (ii) a hot waste gas at 200°C. In the former case, the temperature of the boiling surface is estimated to be 90°C, and in the case of heating by the hot waste gas the same can be taken as 70°C. These temperatures are estimated after taking into consideration other heat transfer resistances present. What are the modes of boiling of the liquid in each case? Calculate the heat fluxes and the vapour generation rate in each case.

6.7 A 25 mm diameter heating element having a surface temperature of 300°C is immersed in water in a vessel. The ambient pressure is atmospheric. Calculate the film boiling heat flux using Bromley's equation.

6.8 Derive Bromley's equation given by Eq. (6.12).

6.9 Saturated ethyl benzene at 1.1 bar gauge pressure enters the bottom of a 1.25 inch 14 BWG tube at a rate of 250 kg/h in which 35% of the liquid is vaporized per pass. The tube is vertical and its surface temperature is maintained at 170°C by steam condensing outside it. Using Chen's correlations, calculate the length of the tube required to achieve vaporization.

6.10 Saturated vapour of ethanol at 1 atm pressure condenses on a vertical plate maintained at 60°C by cooling water. (a) Calculate the length of the plate over which the condensate film remains laminar. What is the thickness of the film at the end of the laminar zone? (b) If the plate is 0.5 m in length, calculate the distance over which the film becomes turbulent. Determine the average heat transfer coefficient and the rate of condensation in this zone. (c) Calculate the average heat transfer coefficient for the entire plate.

6.11 Saturated steam condenses at 1 atm pressure on a vertical wall, 1 m high, maintained at a uniform temperature of 78°C. If film condensation occurs, calculate the following at a location 0.5 m below the top of the plate: (a) the film thickness, (b) the average velocity of the condensate, and (c) the local heat transfer coefficient. Also, calculate the average heat transfer coefficient over the total length of the plate.

6.12 Saturated n-butane at 4 bar absolute pressure condenses on a plane surface oriented at an angle of 30° with the vertical. The surface temperature of the plate is maintained at 40°C. (a) Calculate the local heat flux, the local rate of condensation and the local film thickness at a position 0.15 m from the top (the distance is taken along the plate). (b) Calculate the total rate of condensation if the plate is 0.4 m long. (c) By what per cent does the rate of condensation increase if the plate length is increased by 50%?

6.13 Derive Eq. (6.31) from Eq. (6.29).

6.14 Extend the Nusselt analysis to the case of condensation on the surface of a right-circular cone having wall temperature T_w and conical angle θ and find an expression for the average heat transfer coefficient.

6.15 If condensation of a saturated vapour occurs on the outside of a tube under 'zero gravity' condition, a condensate film will form on the surface but will not run off. Show that in such a condition if a pseudo steady state approximation is done, the heat transfer coefficient is given by

$$h = \frac{k_l}{R \ln (R_\delta - R)}$$

where R is the radius of the tube, and R_δ is the outer radius of the film at time t and is given by

$$\frac{1}{2} + \left[\frac{R_\delta}{R}\right]^2 \left[\ln \frac{R_\delta}{R} - \frac{1}{2}\right] = \frac{2k_l}{\rho_l L_v} \frac{T_w - T_v}{R^2} t$$

6.16 Calculate the rate of condensation of saturated ethanol at 1 atm pressure outside a 25 mm diameter vertical tube, 0.5 m long, given, $T_w = 40°C$.

6.17 Derive Eq. (6.34) assuming laminar flow of the condensate film along the curved surface of a horizontal cylinder.

6.18 Calculate the rate of condensation of saturated acetone at 2 bar total pressure on a tier of six horizontal tubes, given, $T_w = 40°C$.

6.19 What should be the surface temperature of a 10 cm diameter sphere so that the rate of condensation of saturated methanol at 1 atm pressure is 15 kg/h?

6.20 Saturated vapour of ethylene chloride at 1 atm pressure is passed through a 16 BWG 25 mm tube at a rate of 25 kg/h. The tube wall is maintained at 75°C. (a) What fraction of the vapour will condense if a 0.7 m section of tube is used? (b) What length of the tube is required to condense 80% of the vapour?

REFERENCES AND FURTHER READING

Ackers, W.W., H.A. Deans, and O.K. Crosser, "Condensing heat transfer within horizontal tubes", *Chem. Engg. Prog. Symp. Ser.*, **55**, No. 29 (1958), 171.

Bromley, L.A., "Heat transfer in stable film boiling", *Chem. Eng. Prog.*, **46** (1950), 221.

Chato, J.C., "Laminar condensation inside horizontal and inclined tubes", *J. Am. Soc. Refrig. Air Cond. Engrs.*, **4** (Feb 1962), 52.

Chen, M.M., "An analytical study of laminar film condensation, Part 2. Single and multiple horizontal tubes", *J. Heat Transfer*, **83** (1961), 55.

Chen, J.C., "Correlations for boiling heat transfer to saturated liquids in convective flow", *Ind. Eng. Chem. Proc. Des. Dev.*, **5** (1966), 322.

Cichelli, M.T. and C. F. Bonilla, "Heat transfer to liquids boiling under pressure", *Trans A. I. Ch. E.*, **41** (1945), 755.

Colburn, A.P. and O.A. Hougen, "Design of cooler condensers for mixtures of vapors with noncondensing gas", *Ind. Eng. Chem.*, **26** (1934), 1178.

Collier, J.G., *Convective Boiling and Condensation*, McGraw-Hill, New York, 1972.

Farber, E.A. and E.L. Scorah, "Heat transfer to water boiling under pressure", *Trans. ASME*, **70** (1948), 369.

Ginoux, J.N., *Two-phase Flow and Heat Transfer*, Hemisphere Publication, New York, 1978.

Hans, C.Y. and P. Griffith, "The mechanism of heat transfer in nucleate pool boiling, Part I. Bubble initiation, growth and departure", *Intern. J. Heat Mass Transfer*, **8** (1965), 887.

Kakac, S., A.E. Burgles, and F. Mayinger (Eds.), *Heat Exchangers*, Hemisphere Publication, New York, 1981.

Lienhard, J.H., V.K. Dhir, and D.M. Riherd, "Peak pool boiling heat flux measurements on finite horizontal flat plates", *J. Heat Transfer*, **95** (1973), 477–482.

Ludwig, E., *Applied Process Design for Chemical and Petrochemical Plants*, Gulf Publication, Houston, Texas, 1983.

Malek, R.G., "Predict nucleate boiling transfer rates", *Hydrocarbon Proc.*, February 1973, 89–92.

Nukiyama, S., "The maximum and minimum values of heat transmitted from metal to boiling water under atmospheric pressure", *J. Jap. Soc. Mech. Engrs.*, **37** (1934), 367.

Reid, R.C., J.M. Prausnitz, and B.E. Poling, *The Properties of Gases and Liquids*, 4th ed., McGraw-Hill, New York, 1988.

Rohsenow, W.M., "A method of correlating heat transfer data for surface boiling liquids", *Trans. ASME*, **74** (1952), 969.

Rohsenow, W.M., "Nucleation and boiling heat transfer", *Ind. Eng. Chem.*, **58**, No. 1 (1966), 41.

Rohsenow, W.M., J.P. Hartnett, and E.N. Ganik (Eds.), *Handbook of Heat Transfer Fundamentals*, 2nd ed., McGraw-Hill, New York, 1985.

Sparrow, E.M. and J.L. Gregg, "A boundary layer treatment of laminar film condensation", *J. Heat Transfer*, **82** (1959), 13.

Sun, K.H. and J.H. Lienhard, "The peak boiling heat transfer on horizontal cylinders", *Inter. J. Heat Mass Transfer*, **13** (1970), 1425.

Votta, F., "Condensing from vapour-gas mixtures", *Chem. Eng.*, **71** (June 8, 1964), 223.

Zuber, N., "On the stability of boiling heat transfer", *Trans. ASME*, **80** (1958), 711.

7

Radiation Heat Transfer

Heat transfer by conduction or convection from one point in a medium to another occurs in the presence of a temperature difference. Heat transfer from a body by radiation, to the contrary, *does not need* a temperature driving force or a medium. However, *radiation heat exchange* between two bodies at different temperatures always results in a *net transfer* of heat energy from the body at a higher temperature to the other at a lower temperature. Heat transfer by radiation plays an important role in many heating or cooling operations and equipment such as combustion of fossil fuels, operation of a furnace, thermal cracking, the tube stills in petroleum refineries, different types of kilns, etc. Radiation heat loss from a process equipment becomes significant when its temperature is considerably different from that of the ambient. A basic understanding of the process of radiation is, therefore, necessary for thermal calculations in a variety of physical situations.

Thermal radiation is emitted by a body in the form of electromagnetic waves. All bodies around us emit radiation, the intensity of which depends upon the temperature and a few other characteristics of the body. A body at absolute zero temperature does not emit any radiation. Emission of radiation usually occurs from the surface of a body. Radiation propagates at the speed of light, $c = 2.998 \times 10^8$ m/s. A radiation or an electromagnetic wave is characterized by its wavelength λ or its frequency v related by

$$c = v\lambda \tag{7.1}$$

The wavelength λ is usually expressed in Å ($= 10^{-10}$ m), or nanometre, nm ($= 10^{-9}$ m), or μm (one micrometre or micron $= 10^{-6}$ m). Emission of radiation from a body is not continuous, but occurs only in the form of discrete quanta. Each quantum has an energy of

$$E = hv \tag{7.2}$$

where the Planck's constant, $h = 6.6256 \times 10^{-34}$ Joule-second(J s). Emission of radiation from a body is related to the oscillation or transition of electrons in the molecules (or atoms). The electromagnetic spectrum (i.e. the distribution of electromagnetic waves according to wavelengths) comprises extremely short wavelength (or high frequency) gamma rays and X-rays on one end to long wavelength microwaves and radio waves on the other. The intermediate range is occupied by the ultraviolet (UV), visible, and thermal radiation. The complete electromagnetic spectrum is shown in Fig. 7.1. The names of the different bands of wavelengths of the spectrum are also given in Table 7.1. It follows from Eqs. (7.1) and (7.2) that the energy of an electromagnetic wave varies inversely with its wavelength. Long wavelength radiation has lower values of energy.

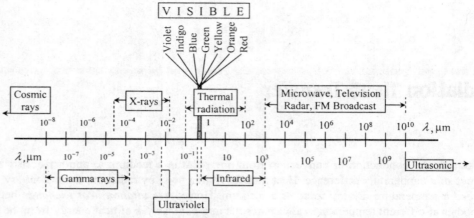

Fig. 7.1   The complete electromagnetic spectrum.

Table 7.1   Common electromagnetic waves and their wavelength bands

Type	Wavelength band (μm)
Cosmic rays	Up to 4×10^{-7}
Gamma rays	(4×10^{-7})–(1.4×10^{-4})
X-rays	(1×10^{-5})–(2×10^{-2})
Ultraviolet rays	(1×10^{-2})–(3.9×10^{-1})
Visible light	(3.9×10^{-1})–(7.8×10^{-1})
Infrared rays	(7.8×10^{-1})–(1×10^{3})
Thermal radiation	(1×10^{-1})–(1×10^{2})
Microwave, radar, TV, radio waves	(1×10^{3})–(2×10^{10})

7.1   BASIC CONCEPTS OF RADIATION FROM A SURFACE

Thermal radiation incident on a body tends to increase its temperature. However, depending upon the nature of the material constituting the body and its surface characteristics, the incident radiation may be absorbed, reflected, or transmitted, partly or fully. For example, light falling on an opaque body is mostly absorbed, and only a small part may be reflected. Absorption of radiation in such a body occurs in an extremely thin layer of the medium near the surface causing a rise in the local temperature. Conduction of heat from the surface to the interior of the body follows, tending to raise the temperature of the entire body. When thermal radiation falls on a polished surface, it is partly reflected as in the case of visible light. The law of reflection of light is valid for thermal radiation too. If a body is permeable to the incident radiation, a part of it is transmitted through the body. For example, visible radiation incident on a glass plate is mostly reflected and transmitted, and a small fraction may be absorbed. In the case of a gaseous medium, a part of the radiation may be absorbed, and the rest transmitted. When light falls on a metal plate, a part of it is absorbed and the rest is reflected, but there is no transmission. The fraction of incident radiation absorbed by a body is called *absorptivity* (α), the fraction reflected is *reflectivity* (ρ), and the fraction transmitted through the body is the *transmissivity* (τ). These fractions should add up to unity. That is,

$$\alpha + \rho + \tau = 1 \tag{7.3}$$

This is shown in Fig. 7.2(a). Most solid bodies are impermeable to thermal radiation.

Absorption and reflection of light on a surface are responsible for our perception of colour. It is well known that the colour of a body does not result from emission (in fact, a body above absolute zero emits radiation irrespective of whether it receives any radiation or not). When radiation of the visible range is incident on a 'real' body, light of certain wavelengths is absorbed and the rest is reflected. The colour of the body is determined by the wavelength of the radiation it reflects. For example, a red car appears so because its surface reflects only the 'red component' of light and absorbs the rest. A surface which absorbs light of all wavelengths in the visible range is *black*, whereas a surface that reflects light of all wavelengths in this range and does not absorb any preferentially, is called *white*. Thus for an opaque black surface, $\rho = 0$ and $\tau = 0$, i.e. $\alpha = 1$.

For an opaque white surface, $\alpha = 0$ and $\tau = 0$, i.e. $\rho = 1$.

Reflection from a surface may be of two types—*specular* and *diffuse*. If the angle of incidence of radiation is equal to the angle of reflection (i.e. if the law of reflection is applicable), then the reflection is called specular [or mirror-like; see Fig. 7.2(b)]. But if the incident radiation is reflected in all directions uniformly, the reflection is called diffuse [Fig. 7.2(c)]. Real surfaces have reflective properties in between these two extremes of totally specular and totally diffuse natures. Similarly, transmission of radiation through a body may be specular or diffuse. When the

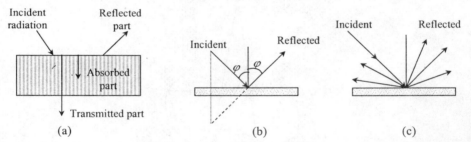

Fig. 7.2 (a) Absorption, reflection and transmission of incident radiation, (b) specular reflection, and (c) diffuse reflection.

incident radiation is transmitted (partly or wholly) straight through a body, the transmission is specular. But if the radiation is scattered while passing through the body, and emerges from the other side of it with random spatial orientation, the transmission is called diffuse.

Another important property of the surface of a substance is its ability to emit radiation. *Emission* of radiation should not be confused with *reflection*. Whereas reflection may occur only when a surface receives radiation, emission of radiation always occurs if the temperature of the surface is above absolute zero. Like reflection, emission of radiation depends upon the surface characteristics. *Emissivity* (denoted by ε) of a surface is a measure of how good it is as an emitter. This term is quantitatively defined in Section 7.1.5.

7.1.1 Blackbody Radiation

The fundamental laws and relations for radiation have been developed for surfaces on the basis of the concept of a blackbody. A *blackbody* is a surface that has the following properties:

(i) A blackbody completely absorbs the incident radiation irrespective of its wavelength. A blackbody is 'black' because it does not reflect any radiation.

(ii) A blackbody is a perfect emitter. No other surface can emit more radiation than a blackbody provided they are at the same temperature. The emissivity of a blackbody is taken as $\varepsilon = 1$.

(iii) Emission from a blackbody occurs in all possible directions. A blackbody is a *perfectly diffuse emitter*.

Thus, a blackbody is an ideal absorber ($\alpha = 1$) and also an ideal emitter ($\varepsilon = 1$) of radiation. It serves as a standard with which a real surface can be compared in order to determine how good the surface is as an absorber or as an emitter. Furthermore, whether a surface is a blackbody or not is not always determined by its 'blackness' to the human eye. For example, milk appears white to the eye, but is essentially 'black' with respect to the long wavelength thermal radiations which are completely absorbed by it. Similarly, snow and ice are also practically 'black' to these radiations. If radiation enters a cavity in a body through a small aperture then it may so happen that the incident radiation undergoes repeated reflections on the wall of the cavity (Fig. 7.3), and is eventually absorbed completely. Thus, a cavity approximates the blackbody behaviour.

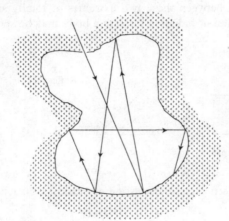

Fig. 7.3 A cavity approaching blackbody behaviour—repeated reflections at the wall.

7.1.2 Planck's Law

At a given temperature, a surface emits radiation of different wavelengths—theoretically varying from zero to infinity. For the same range of wavelength, the surface radiates more energy as the temperature increases. In 1900, Max Planck, with the help of his quantum theory, derived the following equation for *monochromatic emisssive power* or *spectral emissive power* $E_{b\lambda}$ of a blackbody as a function of wavelength of radiation (the subscript b means black, and λ refers to the wavelength of the monochromatic radiation). The term monochromatic refers to a radiation consisting of electromagnetic waves of a single wavelength. We define the monochromatic emissive power $E_{b\lambda}$ as the amount of radiant energy emitted by a surface per unit area, per unit time and per unit wavelength. Its unit is W/m² µm.

$$E_{b\lambda} = \frac{2\pi hc^2 \lambda^{-5}}{\exp\,(hc/\lambda kT) - 1} = \frac{K_1 \lambda^{-5}}{\exp\,(K_2/\lambda T) - 1} \tag{7.4}$$

where

$E_{b\lambda}$ is the monochromatic emissive power of a blackbody

λ is the wavelength of the monochromatic radiation emitted

T is the absolute temperature of the blackbody

h is the Planck's constant

k_B is the Boltzmann constant

c is the velocity of light

$K_1 = 2\pi c^2 h$

$$K_2 = \frac{hc}{k}$$

Putting $c = 2.998 \times 10^8$ m/s, $h = 6.6256 \times 10^{-34}$ J s, $k_B = 1.3805 \times 10^{-23}$ J/K, we get

$$K_1 = 3.743 \times 10^8 \ \mathrm{W}(\mu\mathrm{m})^4/\mathrm{m}^2 \ \text{and} \ K_2 = 1.4387 \times 10^4 \ \mu\mathrm{m} \ \mathrm{K}$$

Equation (7.4) is known as the Planck's law or Planck's distribution. The Planck's distribution is a one-parameter distribution function; the parameter is the absolute temperature T. For a given temperature T, $E_{b\lambda} \, d\lambda$ is the radiant energy flux leaving the surface within the range of wavelength λ to $\lambda + d\lambda$. A plot of the Planck's distribution can be easily prepared; it is shown in Fig. 7.4. The plot says quite a few things. At any temperature, a blackbody emits radiation covering a range of wavelengths. Any of the curves corresponding to a particular temperature has a maximum (at $\lambda = \lambda_{\max}$, say), and the energy of the emitted radiation is rather concentrated in

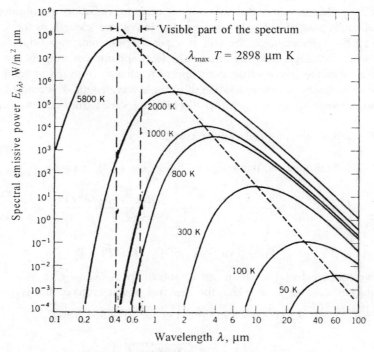

Fig. 7.4 Spectral emissive power of a blackbody.

a region around the wavelength λ_{max} (see Example 7.1). The peak of the curve is shifted towards the shorter wavelength side as the temperature increases. The value of λ_{max} decreases as the temperature of the blackbody increases. This means that a blackbody at a higher temperature emits more of short wavelength radiation. For example, the sun (which may be considered as a blackbody at a surface temperature of about 5800 K) emits maximum amount of radiant energy around the visible range. But the earth with an average surface temperature of around 300 K emits radiant energy in the infrared range.

7.1.3 Wein's Displacement Law

It can be deduced from Eq. (7.4) that the wavelength λ_{max}, corresponding to the peak of the λ vs. $E_{b\lambda}$ plot (Fig. 7.4), is inversely proportional to the temperature of the blackbody. Or, in other words,

$$\lambda_{max}T = \text{constant} = 2898 \ \mu m \ K \tag{7.5}$$

Equation (7.5) is called the Wein's displacement law. The derivation of this law from the Planck's distribution is shown in Example 7.1. However, Wein had propounded his law in 1893, before Planck's quantum theory was proposed. In the context of this law it is interesting to note how the gradual change in the 'colour' of a body occurs as it is heated. At a low temperature, the body emits predominantly infrared radiation that cannot be detected by the eye. As the temperature of the body increases, it emits more and more short wavelength radiation. Only when the body is hot enough, it emits sufficient amount of radiation in the visible range, and its colour because of its temperature becomes perceptible. The visible range of the spectrum covers radiation of 0.39–0.78 μm (390–780 nm) wavelength. While being heated, the colour of the body, at some point of time, appears dark red. This indicates that the body is emitting radiation in considerable amount of wavelength equal to that of red light (that has the lowest wavelength in the visible range). At gradually increasing temperatures, the colour appears bright red, then bright yellow, and then white. In the *white-hot* stage, the body emits sufficient radiation covering the entire visible range. However, never does the colour of the body appear to change from red to orange to yellow, etc. because at no stage the body emits radiation of a single wavelength. The emitted radiation always covers a band or range of wavelength that theoretically varies from zero to infinity.

Example 7.1 Derive Wein's displacement law.

SOLUTION To determine the maximum of $E_{b\lambda}$ in Eq. (7.4), we put

$$\frac{dE_{b\lambda}}{d\lambda} = \frac{K_1(-5)\lambda^{-6}}{\exp(K_2/\lambda T) - 1} - \frac{K_1\lambda^{-5}(-K_2/\lambda^2 T)}{\exp[(K_2/\lambda T) - 1]^2} = 0$$

or

$$5[1 - \exp(-K_2/\lambda T)] - (K_2/\lambda T) = 0$$

The above transcendental equation can be solved for $K_2/\lambda T$ to give $K_2/\lambda T = 4.965$. Therefore, the wavelength at which the emissive power is maximum is given by

$$\lambda_{max}T = K_2/4.965 = 1.4387 \times 10^4/4.965 \ \mu m \ K$$

or

$$\lambda_{max}T = \boxed{2898 \ \mu m \ K}$$

7.1.4 The Stefan-Boltzmann Law

The Planck's distribution, Eq. (7.4), gives the monochromatic emissive power of a blackbody. It is rather simple to determine the *total emissive power* of a blackbody over the entire spectrum by integration of $E_{b\lambda}$. If the total emissive power (also called the *hemispherical* total emissive power) of a blackbody is denoted by E_b (omitting the other subscript, λ), then

$$E_b = \int_0^\infty E_{b\lambda}\, d\lambda = \int_0^\infty \frac{K_1 \lambda^{-5}}{\exp(K_2/\lambda T) - 1} d\lambda = \sigma T^4 \tag{7.6}$$

Evaluation of the above integral is shown in Example 7.2. Here $\sigma\,[= (\pi^4/15)(K_1 K_2^4)]$ is called the Stefan-Boltzmann constant.

Putting the values of K_1 and K_2 as obtained earlier, we get

$$\sigma = 5.669 \times 10^{-8} \text{ W/m}^2 \text{ K}^4$$

Equation (7.6) is called the Stefan-Boltzmann law which allows us to calculate the total amount of radiation of all wavelengths emitted in all possible directions by a blackbody. The value of the Stefan-Boltzmann constant σ has been determined experimentally by a number of researchers. The accepted experimental value

$$\sigma = 5.729 \times 10^{-8} \text{ W/m}^2 \text{ K}^4$$

is also used in radiation calculations.

Starting with the Planck's distribution, Eq. (7.4), it is easy to calculate the fraction of the total energy emission from a blackbody in a given range of wavelength. If $F_{b,\,(0-\lambda)}$ is the fraction of the energy emitted by a blackbody as radiation of wavelength in the range $0-\lambda$, then

$$F_{b,\,(0-\lambda)} = \frac{\int_0^\lambda E_{b\lambda}\, d\lambda}{\int_0^\infty E_{b\lambda}\, d\lambda} = \frac{1}{\sigma T^4} \int_0^\lambda \frac{K_1 \lambda^{-5}}{\exp(K_2/\lambda T) - 1}\, d\lambda$$

$$= \frac{K_1}{\sigma} \int_0^{\lambda T} \frac{(\lambda T)^{-5}}{\exp(K_2/\lambda T) - 1}\, d(\lambda T) \tag{7.7}$$

As shown above, the integral can be conveniently expressed as a function of λT rather than λ. A list of the values of this integral against λT is given in Table 7.2. Similarly, the fraction of radiant energy corresponding to the wavelength range $\lambda_1 - \lambda_2$ is given by

Table 7.2 Planck's radiation functions

λT (μm K)	$F_{b,\,(0-\lambda)}$	λT (μm K)	$F_{b,\,(0-\lambda)}$
1222.2	0.0025	5333.3	0.6731
1333.3	0.0053	5666.7	0.7076
1666.7	0.0254	6000.0	0.7383
2000.0	0.0667	6333.3	0.7643
2222.2	0.1051	7000.0	0.8081
2555.0	0.1734	7333.3	0.8262
3000.0	0.2733	7666.7	0.8421
3333.3	0.3474	8333.3	0.8688
3666.7	0.4171	9444.4	0.9017
4000.0	0.4809	10555.6	0.9247
4333.3	0.5381	11666.7	0.9411
4666.7	0.5890	15000.0	0.9689
5000.0	0.6337	33333.3	0.9963

$$F_{b,(\lambda_1-\lambda_2)} = \frac{K_1}{\sigma} \int_{\lambda_1 T}^{\lambda_2 T} \frac{(\lambda T)^{-5}}{\exp(K_2/\lambda T) - 1} \, d(\lambda T) = F_{b,(0-\lambda_2)} - F_{b,(0-\lambda_1)} \qquad (7.8)$$

The fractions of the radiant energy in the wavelength ranges $(0-\lambda_1)$ and $(\lambda_1-\lambda_2)$ are shown by the shaded areas in Fig. 7.5.

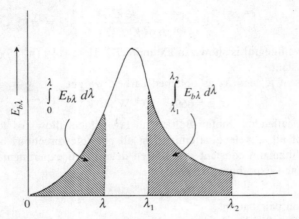

Fig. 7.5 Energy of radiation over a range of wavelength.

Example 7.2 Starting with the Planck's distribution, Eq. (7.4), obtain the total emissive power of a blackbody as given by Eq. (7.6).

SOLUTION Since $E_b = \int_{\lambda=0}^{\infty} E_{b\lambda} \, d\lambda$

$$= -\int_{\lambda=0}^{\infty} \frac{K_1 T^4}{K_2^4} \left[\frac{K_2}{\lambda T}\right]^3 \frac{1}{\exp(K_2/\lambda T) - 1} \, d(K_2/\lambda T)$$

putting $\xi = K_2/\lambda T$, and noting that $\lambda = 0 \Rightarrow \xi = \infty$, and $\lambda = \infty \Rightarrow \xi = 0$, we have

$$E_b = \frac{K_1 T^4}{K_2^4} \int_{\xi=0}^{\infty} \frac{\xi^3 \, d\xi}{e^\xi(1 - e^{-\xi})}$$

The integral on the RHS

$$I = \int_{\xi=0}^{\infty} \frac{\xi^3 \, d\xi}{e^\xi(1 - e^{-\xi})} = \int_0^{\infty} \xi^3 e^{-\xi}(1 + e^{-\xi} + e^{-2\xi} + \cdots) \, d\xi$$

$$= \int_0^{\infty} \xi^3 (e^{-\xi} + e^{-2\xi} + e^{-3\xi} + \cdots) \, d\xi = \sum_{n=1}^{\infty} \int_0^{\infty} \xi^3 e^{-n\xi} \, d\xi$$

Putting $n\xi = \zeta$, we get

$$I = \sum_{n=1}^{\infty} \frac{1}{n^4} \int_0^{\infty} \zeta^3 e^{-\zeta} \, d\zeta$$

The integral can now be obtained in terms of the Gamma function*

$$I = \sum_{n=1}^{\infty} \frac{1}{n^4}(3!) = \sum_{n=1}^{\infty} \frac{6}{n^4} = \frac{\pi^4}{15}$$

Therefore,

$$E_b = \frac{\pi^4}{15} \frac{K_1 T^4}{K_2^4} = \boxed{\sigma T^4}$$

Example 7.3 The sun may be considered to be a blackbody with a surface temperature of 5780 K. Calculate (a) the fraction of solar radiation that falls in the visible range, (b) the fraction that occurs on the left of the visible range ($\lambda < 0.4$ μm), (c) the fraction that occurs on the right of the visible range ($\lambda > 0.7$ μm), (d) the wavelength and frequency of maximum spectral emissive power, (e) the maximum spectral emissive power, and (f) the hemispherical total emissive power.

SOLUTION Here the sun is considered as a blackbody. Hence its emissive power, $E_{b\lambda}$, is given by the Planck's equation [Eq. (7.4)].

(a) The fraction of solar radiation that corresponds to the range of wavelength λ_1–λ_2 is given by Eq. (7.8).

$$F_{b,(\lambda_1-\lambda_2)} = \frac{1}{\sigma T^4} \int_{\lambda_1}^{\lambda_2} E_{b\lambda} \, d\lambda = F_{b,(0-\lambda_2)} - F_{b,(0-\lambda_1)}$$

The visible range of the spectrum covers λ from 0.4 to 0.7 μm. So,

$$\lambda_1 = 0.4, \ \lambda_1 T = (0.4)(5780) = 2312 \text{ μm K}$$

$$\lambda_2 = 0.7, \ \lambda_2 T = (0.7)((5780) = 4046 \text{ μm K}$$

From Table 7.2, $F_{b,(0-\lambda_1)} = 0.1229$ and $F_{b,(0-\lambda_2)} = 0.4889$.
Therefore, the fraction of radiation lying within the wavelength range 0.4–0.7 μm is

$$= 0.4889 - 0.1229 = \boxed{0.366}$$

(b) and (c) The fraction of radiation that occurs within 0–0.4 μm $= F_{b,(0-\lambda_1)} = \boxed{0.1229}$
The fraction that occurs above 0.7 μm

$$= F_{b,(\lambda_2-\infty)} = 1 - F_{b,(0-\lambda_2)} = 1 - 0.4889 = \boxed{0.5111}$$

(d) The wavelength corresponding to the maximum spectral emissive power can be obtained from the Wein's displacement law,

$$\lambda_{\max} T = 2898 \text{ μm K. Here } T = 5780 \text{ K}$$

Therefore,

$$\lambda_{\max} = \frac{2898}{5780} = \boxed{0.5014 \text{ μm}}; \quad \text{this lies in the visible range}$$

---

*The Gamma function is defined as $\Gamma(p) = \int_0^{\infty} e^{-x} x^{p-1} dx = (p-1)!$

The corresponding frequency is given by

$$v = c/\lambda_{max} = \frac{2.998 \times 10^8 \text{ m/s}}{0.5014 \times 10^{-6} \text{ m}} = \boxed{5.98 \times 10^{14} \text{ s}^{-1}}$$

(e) Since $\lambda_{max} = 0.5014$ μm, and $\lambda_{max}T = 2898$ μm K, putting these values and also the values of K_1 and K_2 in Eq. (7.4), we get

$$E_{b, \lambda_{max}} = \frac{2\pi(6.6256 \times 10^{-34})(2.998 \times 10^8)^2(0.5014 \times 10^{-6})^{-5}}{\exp\left[\dfrac{(6.6256 \times 10^{-34})(2.998 \times 10^8)}{(1.3805 \times 10^{-23})(2898 \times 10^{-6})}\right] - 1} = \boxed{8.297 \times 10^{13} \text{ W/m}^2 \text{ m}}$$

(f) The hemispherical total emissive power is given by

$$E_b = \sigma T^4 = (5.669 \times 10^{-8})(5780)^4 \text{ W/m}^2 = \boxed{6.327 \times 10^7 \text{ W/m}^2}$$

Example 7.4 A small blackbody has a total hemispherical emissive power of 4 kW/m². Determine its surface temperature and the wavelength of emission maximum. In which range of the spectrum does this wavelength fall?

SOLUTION If T is the surface temperature of the blackbody, its total emissive power is given by Eq. (7.6). Putting $E_b = 4000$ W/m², we get

$$E_b = 4000 = 5.669 \times 10^{-8} T^4$$

or

$$T = \boxed{515 \text{ K}}$$

The wavelength of emission maximum is given by Wein's law, $\lambda_{max}T = 2898$ μm K. That is,

$$\lambda_{max} = \frac{2898}{515} = \boxed{5.63 \text{ μm}}$$

From Fig. 7.1, it is seen that this wavelength falls in the *infrared* region of the spectrum.

7.1.5 Kirchhoff's Law

Let us consider a large enclosure of surface temperature T_s. The enclosure virtually behaves like a blackbody (compare with Fig. 7.3). It is assumed that the radiant heat flux incident on any surface in the enclosure is q. There are several bodies in the enclosure which are sufficiently small so that the radiation effect on the surface of the enclosure is small. Let the surface area of one of the bodies (call it the first body) be A_1. If E_1 denotes the emissive power of the body, we have

the rate at which body-1 emits radiant energy = $E_1 A_1$

the rate at which body-1 receives radiant energy = $\alpha_1 q A_1$

where α_1 is the absorptivity (which is the fraction of incident radiation absorbed) of the body. Therefore, if body-1 is in thermal equilibrium with the enclosure (i.e. if it is also at the temperature T_s), then the rate of emission of radiation must be equal to the rate of absorption.

Therefore,

$$E_1 A_1 = \alpha_1 q A_1$$

or

$$E_1 = \alpha_1 q \tag{7.9}$$

Now if body-1 is replaced by a blackbody, we should put $E_1 = E_b$, and $\alpha_1 = \alpha_b = 1$ (because the absorptivity of a blackbody is unity).
Therefore,

$$E_b = 1 \cdot q = q \tag{7.10}$$

Dividing Eq. (7.9) by Eq. (7.10), we have

$$\frac{E_1}{E_b} = \frac{\alpha_1 q}{q} = \alpha_1 \tag{7.11}$$

Let us now define emissivity ε of a body or surface as the ratio of its emissive power to that of a blackbody. So for body-1, we have

$$\varepsilon_1 = \frac{E_1}{E_b}$$

Therefore from Eq. (7.11), we get

$$\varepsilon_1 = \alpha_1 \tag{7.12}$$

Similar relation is also valid for any other body in the enclosure. So removing the subscript in Eq. (7.12), we get

$$\varepsilon = \alpha \tag{7.13}$$

Equation (7.13) states that *the emissivity of a body, which is in thermal equilibrium with its surroundings, is equal to its absorptivity*. This is Kirchhoff's law established by him in 1860. However, it should be remembered that Kirchhoff's law is strictly applicable only when the source temperature of the irradiation is equal to the temperature of the irradiated surface. But, as a matter of fact, the absorptivity of most real surfaces is relatively insensitive to temperature and wavelength. So, for practical purposes, it is customary to assume that the emissivity and the absorptivity of a surface are equal even when it is not in thermal equilibrium with its surroundings. This assumption leads to the concept of a *gray body* for which the emissivity is independent of the wavelength of radiation. Emissivities of several real surfaces are given in Table 7.3.

7.1.6 Gray Body

A blackbody is an ideal substance or surface whose emissivity and absorptivity are both equal to unity. Any real substance emits (or absorbs) less radiation than a blackbody at the same temperature. Also the emissivity and the absorptivity of a real substance may vary with its temperature or the wavelength of the radiation emitted or absorbed. A gray body is defined as a substance whose emissivity and absorptivity are independent of wavelength. Thus a gray body is also an ideal body, but its ε and α values are both less than unity. Figure 7.6 shows qualitatively the nature of dependence of monochromatic emissive power of a blackbody, a gray body, and a real surface upon the wavelength of radiation.

For a real surface, in general, the emissivity depends upon the wavelength of radiation. Correspondingly, we define the *spectral emissivity* $\varepsilon_\lambda(\lambda)$ as the ratio of intensity of radiation emitted by a body at wavelength λ to that of the radiation emitted by a blackbody at the same temperature and wavelength.

Table 7.3   Emissivities of selected surfaces

Surface	Emissivity, ε
Aluminium (highly polished)	0.039–0.057 (500–850 K)
Commercial sheet	0.09 (373 K)
Asbestos board	0.96 (300 K)
Brass, polished	0.1 (300–600 K)
rolled plate	0.06 (293 K)
Brick, building	0.45 (1273 K)
fireclay	0.75 (1273 K)
magnesite refractory	0.38 (1273 K)
Chromium, polished	0.08 (310 K); 0.36 (1370 K)
Iron, oxidized cast iron	0.64 (470 K); 0.78 (870 K)
oxidized steel	0.79 (470–870 K)
rolled steel sheet	0.66 (293 K)
Lampblack	0.78–0.84
Ordinary glass	0.94 (293 K)
Paints, aluminium paints, and laquers, 10% Al	0.52 (373 K)
black or white laquer	0.8–0.95
Plaster	0.91 (283–360 K)
Rubber, hard glassy	0.94 (295 K)
Silver, pure and polished	0.02 (500 K); 0.032 (900 K)
Stainless steel, polished	0.074 (373 K)
type 301B	0.54 (500 K); 0.63 (1200 K)
type 316A	0.57 (500 K); 0.66 (1200 K)

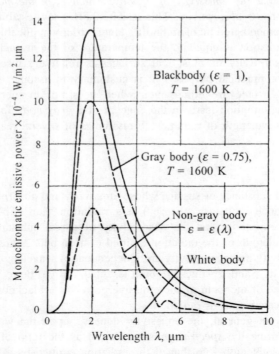

Fig. 7.6   Monochromatic emissive power of a blackbody, a gray body and a non-gray body.

Example 7.5   The spectral emissivity of a non-gray surface at 1500 K can be approximated by a step function of wavelength as shown in Fig. 7.7. Calculate (a) the total (hemispherical) emissive power, and (b) the total (hemispherical) emissivity.

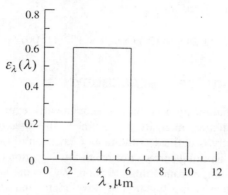

Fig. 7.7   Example 7.5.

SOLUTION   (a) At a wavelength λ, the spectral emissivity of the given surface is $\varepsilon_\lambda(\lambda)$ and the spectral emissive power of a blackbody $E_{\lambda b}$ is given by Eq. (7.4). Therefore, the spectral emissive power of the given non-gray surface = $\varepsilon_\lambda(\lambda)E_{b\lambda}$.

$$\text{The total emissive power, } E = \int_0^\infty \varepsilon_\lambda(\lambda)E_{b\lambda}\,d\lambda$$

The nature of dependence of $\varepsilon_\lambda(\lambda)$ on λ is given in Fig. 7.7. Accordingly, we split the above integral into four sub-integrals.

$$E = \int_0^2 \varepsilon_\lambda(\lambda)E_{b\lambda}\,d\lambda + \int_2^6 \varepsilon_\lambda(\lambda)E_{b\lambda}\,d\lambda + \int_6^{10}\varepsilon_\lambda(\lambda)E_{b\lambda}\,d\lambda + \int_{10}^\infty \varepsilon_\lambda(\lambda)E_{b\lambda}\,d\lambda$$

From Fig. 7.7, we have

$$\varepsilon_\lambda(\lambda) = 0.2 \qquad 0 < \lambda < 2$$
$$\varepsilon_\lambda(\lambda) = 0.6 \qquad 2 < \lambda < 6$$
$$\varepsilon_\lambda(\lambda) = 0.1 \qquad 6 < \lambda < 10$$
$$\varepsilon_\lambda(\lambda) = 0 \qquad\quad \lambda > 10$$

or

$$E = (0.2)\int_0^2 E_{b\lambda}\,d\lambda + (0.6)\int_2^6 E_{b\lambda}\,d\lambda + (0.1)\int_6^{10}E_{b\lambda}\,d\lambda + (0)\int_{10}^\infty E_{b\lambda}\,d\lambda$$

The above integrals may be evaluated by using Table 7.2 for the temperature of the body, $T = 1500$ K. For example, if $\lambda = 2$ μm, $T = 1500$ K, then $\lambda T = 3000$ μm K.

From Table 7.2, $F_{(0-2)} = 0.2733$.

Similarly, $F_{(2-6)} = 0.89 - 0.2733 = 0.6167$; $F_{(6-10)} = 0.9689 - 0.89 = 0.0789$

Therefore,

$$E = [(0.2)F_{(0-2)} + (0.6)F_{(2-6)} + (0.1)F_{(6-10)}]\, E_b$$

$$= [(0.2)(0.2733) + (0.6)(0.6167) + (0.1)(0.0789)]\, \sigma T^4$$

$$= (0.4325)(5.669 \times 10^{-8})(1500)^4 = \boxed{1.241 \times 10^5 \text{ W/m}^2}$$

(b) If ε is the total hemispherical emissivity of the surface, then we can write

$$E = \varepsilon \, \sigma \, T^4$$

Using the value of E calculated above, we get

$$\varepsilon = E/\sigma T^4 = (1.241 \times 10^5)/(5.669 \times 10^{-8})(1500)^4 = \boxed{0.4324}$$

7.2   RADIATION INTENSITY OF A BLACKBODY

The rate of emission of radiation by a surface is conveniently expressed in terms of *radiation intensity*. Two types of radiation intensity are defined—the *spectral intensity*, and the *total intensity*. The spectral intensity is defined as the rate of emission of radiant energy in a particular direction per unit area of the emitting surface normal to this direction (i.e. projected area), per unit wavelength λ, per unit solid angle around this direction. We denote it by $I_{b\lambda}$, where the subscript λ means that the spectral intensity is a function of wavelength, and the subscript b implies black. It has the unit of $W/(m^2)(\mu m)(sr)$, where 'sr' is steradian, the unit of solid angle. The spherical coordinate system and a solid angle are shown in Fig. 7.8.

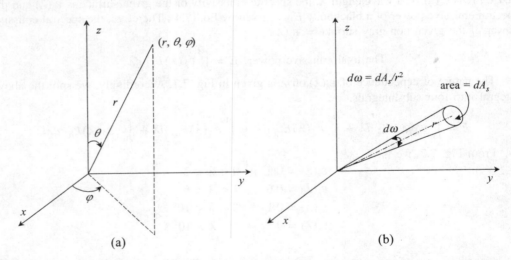

Fig. 7.8   (a) The spherical coordinate system and (b) the spherical angle or the solid angle $d\omega$ corresponding to the area dA_s.

The quantity 'spectral intensity' $I_{b\lambda}$, may be explained with the help of Figs. 7.8 and 7.9. Consider an elementary 'black' area dA that emits radiation and is surrounded by a hemisphere of arbitrary radius r and corresponding surface area $2\pi r^2$. The hemisphere is oriented in such a way that the elementary emitting surface dA superimposes on the base of the hemisphere and the midpoint of dA coincides with the centre of the hemisphere. The hemisphere is assumed to behave like a blackbody. Consider another elementary area dA_s on the surface of the hemisphere located at the cone angle θ and the polar angle φ, (also called the *azimuthal angle*) in the spherical coordinate system (Fig. 7.9). Now, the projection of dA normal to the the r-direction is $dA \cos \theta$

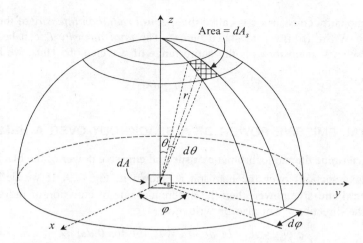

Fig. 7.9 Emission from a small surface in a hemispherical enclosure.

[θ is the angle between the r- and the z-directions (see Fig. 7.8(a)]. Therefore, the differential rate of radiant energy emission from the surface around a wavelength λ should be proportional to: (i) the projected area of the surface element normal to the r-direction, $dA \cos \theta$; (ii) the solid angle $d\omega$ subtended by the elementary receiving surface dA_s of the hemisphere (see Fig. 7.10), and the small interval of wavelength $d\lambda$. Hence,

$$d^3 Q_{b\lambda} \propto dA \cos \theta \, d\omega d\lambda \qquad (7.14)$$

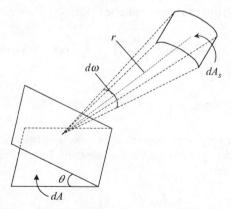

Fig. 7.10 The solid angle $d\omega$ at the centre of dA (projected area of dA normal to the radial direction $= dA \cos \theta$).

where $d^3 Q_{b\lambda}$ is the rate of radiant energy emission from the small area dA in the direction r covering a solid angle $d\omega$ for the interval of wavelength $d\lambda$. Note that d^3 implies that it is a third order differential term, proportional to the product of three first order terms dA, $d\omega$, and $d\lambda$.

Introducing a proportionality constant $I_{b\lambda}$ and putting $d\omega = dA_s/r^2$ [see Fig 7.8(b)], we get

$$d^3 Q_{b\lambda} = \frac{I_{b\lambda} \, dA \, \cos \theta \, dA_s \, d\lambda}{r^2} \qquad (7.15)$$

The proportionality constant $I_{b\lambda}$ is called the *spectral radiation intensity* of the blackbody dA and has the unit $W/(m^2)(\mu m)(sr)$, as stated before. The *total intensity* I_b can be determined by integrating the spectral intensity over the entire range of wavelength. Thus, we have

$$I_b = \int_0^\infty I_{b\lambda}\, d\lambda \tag{7.16}$$

7.3 SPECTRAL EMISSIVE POWER OF A BLACKBODY OVER A HEMISPHERE

We shall now determine the monochromatic or spectral emissive power of the blackbody (referred to in the previous section) over the hemisphere shown in Fig. 7.9. If we define $E'_{b\lambda}$ as the directional spectral emissive power, then the differential rate of emission of radiant energy from the surface in the direction r can also be written as

$$d^3 Q_{b\lambda} = E'_{b\lambda} dA\, d\omega\, d\lambda = I_{b\lambda}\, dA\, \cos\theta\, d\omega\, d\lambda$$

In the spherical polar coordinate system, the elementary or differential area dA_s on the surface of the sphere is given by

$$dA_s = r^2 \sin\theta\, d\theta\, d\varphi$$

i.e.

$$d\omega = dA_s/r^2 = \sin\theta\, d\theta\, d\varphi \tag{7.17}$$

Therefore,

$$E'_{b\lambda}\, d\omega = I_{b\lambda} \cos\theta \sin\theta\, d\theta\, d\varphi \tag{7.18}$$

The hemispherical spectral emissive power $E_{b\lambda}$ can be obtained by integrating the directional spectral emissive power $E'_{b\lambda}$ over the hemispherical surface. Hence, we have

$$E_{b\lambda} = \int_\omega E'_{b\lambda}\, d\omega = \int_{\varphi=0}^{2\pi} \int_{\theta=0}^{\pi/2} I_{b\lambda} \cos\theta \sin\theta\, d\theta\, d\varphi$$

$$= I_{b\lambda} \int_{\varphi=0}^{2\pi} \left[-\frac{1}{4} \cos 2\theta \right]_0^{\pi/2} d\varphi = \pi I_{b\lambda} \tag{7.19a}$$

Equation (7.19a) embodies a very important result, i.e. *the hemispherical spectral emissive power of a blackbody is π times the spectral radiation intensity*. Integrating over the entire range of wavelength, we get

$$\int_0^\infty E_{b\lambda}\, d\lambda = \pi \int_0^\infty I_{b\lambda}\, d\lambda$$

or

$$E_b = \pi I_b \tag{7.19b}$$

7.4 RADIATIVE HEAT EXCHANGE BETWEEN SURFACES—THE VIEW FACTOR

Calculation of heat exchange between two or more surfaces is of great practical importance. Heat transfer by radiation between neighbouring surfaces is pretty common in industrial practice. For example, a furnace wall or door loses heat by radiation, in addition to other modes of heat loss, to the surroundings. A steam pipe loses heat to its ambient. A natural question is: what are the factors that determine the rate of heat exchange between two bodies? The factors are: the temperatures of the individual surfaces, their emissivities (which is unity for a 'black' surface), and *how well one surface can see the other.* Let us attempt to express the last factor quantitatively in terms of a 'view factor' assuming that the intervening medium between the surfaces does not absorb any radiation.

A 'diffuse' radiating surface emits radiation in all directions. Hence, all the radiation emitted may not fall on another surface in the neighbourhood. The fraction of the total radiant energy that is emitted by the surface i and is intercepted by the surface j is called the *view factor* or *configuration factor* or *shape factor*. We denote the view factor by F_{ij}, where the subscript i refers to the emitting surface and j the absorbing or intercepting surface. It is to be noted that the radiation intercepted by the surface j will be fully absorbed if j is a blackbody. Otherwise, it will be partly absorbed and the rest will be reflected and/or transmitted. Thus, the view factor can be expressed as

$$\text{View factor, } F_{ij} = \frac{\text{The fraction of the total radiation emitted by } i \text{ and intercepted by } j}{\text{Total radiation emitted by } i}$$

In order to develop a mathematical expression for view factor, we consider two surfaces A_i and A_j of arbitrary geometry and orientation. Two small surface elements of A_i and A_j are denoted by dA_i and dA_j and are located a distance r apart. The distance segment r subtends an angle θ_i with the normal to the surface element dA_i, and an angle θ_j with the normal to the surface element dA_j (Fig. 7.11).

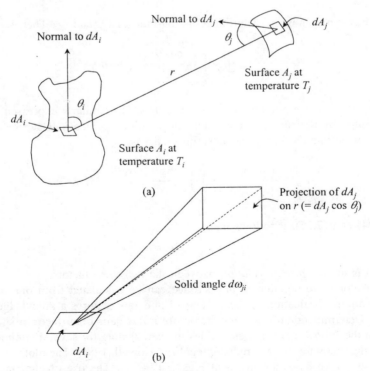

Fig. 7.11 (a) Surface elements dA_i and dA_j and (b) the solid angle $d\omega_{ij}$ subtended by dA_j at any point of dA_i.

If I_{bi} is the 'total intensity of radiation' emitted by the surface A_i, then the amount of radiation emitted by the small surface element dA_i and intercepted by another small surface element dA_j (both of which are assumed to be blackbodies) is given by [from Eq. (7.15)]

$$d^2Q_{ij} = I_{bi}\, dA_i \cos\theta_i\, d\omega_{ij} \qquad (7.20)$$

But the solid angle, $d\omega_{ij} = (dA_j \cos\theta_j)/r^2$ [see Fig. 7.11(b)]. Therefore, we have

$$d^2Q_{ij} = \frac{I_{bi}\, dA_i\, dA_j \cos\theta_i \cos\theta_j}{r^2} \qquad (7.21)$$

The notation d^2Q_{ij} indicates that it is a second order differential term involving the product of two first order differential terms dA_i and dA_j. Integration of Eq. (7.21) over the surfaces A_i and A_j will give the amount of radiation emitted by the surface A_i and intercepted by the surface A_j.

$$Q_{ij} = \iint_{A_i A_j} \frac{I_{bi}\, dA_i\, dA_j \cos\theta_i \cos\theta_j}{r^2} = I_{bi} \iint_{A_i A_j} \frac{\cos\theta_i \cos\theta_j}{r^2} dA_i\, dA_j \qquad (7.22)$$

The total amount of radiation Q_i emitted by the surface A_i is obtained from its hemispherical emissive power [see Eq. (7.19b)].

$$Q_i = A_i E_{bi} = A_i \pi I_{bi} \qquad (7.23)$$

The view factor may now be determined as the ratio of Q_{ij} and Q_i. That is,

$$F_{ij} = \frac{Q_{ij}}{Q_i} = \frac{1}{\pi A_i} \iint_{A_i A_j} \frac{\cos\theta_i \cos\theta_j}{r^2}\, dA_i\, dA_j \qquad (7.24)$$

The view factor F_{ji} corresponding to the surface A_j can be similarly expressed as the fraction of radiation emitted by the surface j and intercepted by the surface i. It can be obtained from Eq. (7.24) simply by interchanging the subscripts. That is,

$$F_{ji} = \frac{Q_{ji}}{Q_j} = \frac{1}{\pi A_j} \iint_{A_j A_i} \frac{\cos\theta_j \cos\theta_i}{r^2}\, dA_j\, dA_i \qquad (7.25)$$

From Eqs. (7.24) and (7.25), it follows that,

$$A_i F_{ij} = A_j F_{ji} \qquad (7.26)$$

Equation (7.26) is the *reciprocity relation* between the two view factors.

The view factor corresponding to any surface can be calculated from one or more of the above three equations. Evaluation of the concerned integrals requires a knowledge of analytical solid geometry. Determination of the view factors for a few pairs of surfaces of simple geometry are discussed in the following examples. Tables of view factors for several common geometries are available in the literature (for example, Spiegel and Howell, 1981). The plots of radiation view factors for a few simple cases are given in Figs. 7.12–7.14. The use of algebraic equations is preferable to plots or charts for better accuracy.

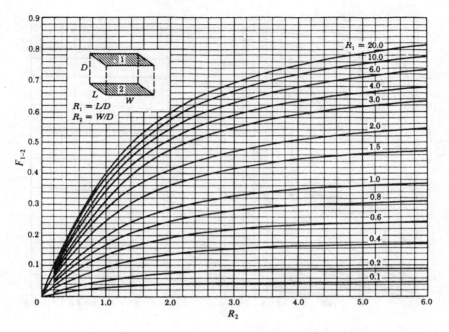

Fig. 7.12 The view factor for parallel, directly opposed rectangles.

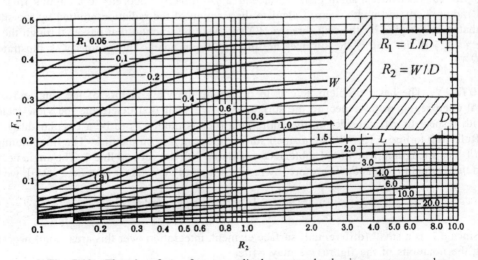

Fig. 7.13 The view factor for perpendicular rectangles having a common edge.

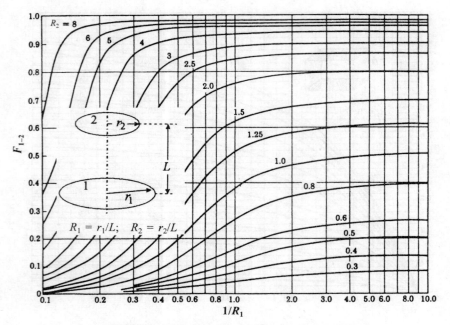

Fig. 7.14 The view factor for parallel coaxial disks.

Example 7.6 Consider a small plane surface area dA_1 directly opposed* to a circular ring A_2 of inner radius 5 cm and width 3 cm. Calculate the fraction of the radiation emitted by the surface dA_1 that is intercepted by the ring (area $= A_2'$) and also the fraction that passes through the hole (area $= A_2''$) in the ring if the surfaces are placed 20 cm apart. The geometry is illustrated in Fig. 7.15.

SOLUTION The basic problem here is to determine the view factors of the ring and of the central open region with respect to the surface dA_1. We shall first determine the view factor of A_2 with respect to dA_1 considering the entire area (i.e. assuming A_2 to be a circular disk).

Referring to Fig. 7.15, we take a small surface element dA_2 on the disk. The line joining dA_1 and dA_2 has a length R. The line R subtends an angle θ_1 with the vertical line L. The angle between R and the normal to dA_2 is θ_2. Therefore, by the definition of view factor [see Eq. (7.24)], we have

$$F_{dA_1 \to A_2} = \frac{1}{\pi dA_1} \int_{A_2} \frac{\cos \theta_1 \cos \theta_2 \, dA_1 \, dA_2}{R^2}$$

Since dA_1 is a small (differential) surface element, integration over this area is not necessary. Using the notations of the figure, we may write

$$R^2 = L^2 + r^2; \qquad \cos \theta_1 = \cos \theta_2 = \frac{L}{\sqrt{L^2 + r^2}}; \qquad dA_2 = r \, dr \, d\phi$$

---

*Directly opposed means that the surfaces are parallel and the line joining the centres of the surfaces is normal to both the surfaces.

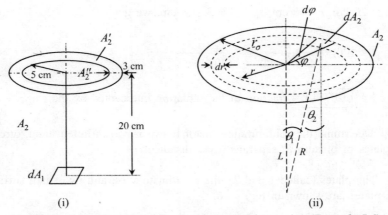

Fig. 7.15 A small surface area directly opposed to a ring (Example 7.6).

Therefore, from Fig. 7.15(ii)

$$F_{dA_1 \to A_2} = \frac{1}{\pi} \int_{\phi=0}^{2\pi} \int_{r=0}^{r_o} \frac{1}{L^2 + r^2} \frac{L}{\sqrt{L^2 + r^2}} \frac{L}{\sqrt{L^2 + r^2}} \, r \, dr \, d\phi$$

$$= \frac{L^2 2\pi}{\pi} \int_0^{r_o} \frac{r \, dr}{(L^2 + r^2)^2} = L^2 \left[\frac{-1}{(L^2 + r^2)} \right]_{r=0}^{r_o}$$

$$= L^2 \left[\frac{1}{L^2} - \frac{1}{L^2 + r_o^2} \right] = \frac{r_o^2}{L^2 + r_o^2}$$

If we put $r_o = 5$ cm and $L = 20$ cm, the view factor of the hole (area A_2'') in the ring can be calculated from the above relation. Therefore,

$$F_{dA_1 \to A_2''} = \frac{5^2}{20^2 + 5^2} = \boxed{0.0588}$$

= fraction of the radiation that passes through the hole.

Now, we can use the following relation for the ring.

$$F_{dA_1 \to A_2'} = \frac{1}{\pi} \int_{\phi=0}^{2\pi} \int_{r=r_i}^{r_o} \frac{L^2 r \, dr \, d\phi}{(L^2 + r^2)^2}$$

$$= L^2 \left[-\frac{1}{L^2 + r^2} \right]_{r=r_i}^{r_o}$$

$$= L^2 \left[\frac{1}{L^2 + r_i^2} - \frac{1}{L^2 + r_o^2} \right]$$

Given: $L = 20$ cm, $r_i = 5$ cm, $r_o = 5 + 3 = 8$ cm, we get

$$F_{dA_1 \to A_2'} = (20)^2 \left[\frac{1}{(20)^2 + (5)^2} - \frac{1}{(20)^2 + (8)^2} \right]$$

$$= \boxed{0.0791} = \text{fraction of the radiation intercepted by the ring.}$$

Example 7.7 Determine the configuration factor between two infinitely long, directly opposed parallel flat plates of breadth B, separated by a distance H.

SOLUTION The plates (called 1 and 2), the coordinate axes, and two small surface elements, (one on each plate), are shown in Fig. 7.16.

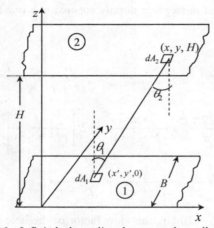

Fig. 7.16 Infinitely long directly opposed parallel flat plates.

A small element dA_1 on the surface–1 has coordinates $(x', y', 0)$, and another small surface element dA_2 on the surface–2 has coordinates (x, y, H). The line dA_1–dA_2 subtends an angle θ_1 with the normal to the area element dA_1 and an angle θ_2 with the normal to the area element dA_2. So $\cos\theta_1$ is the direction cosine of dA_1–dA_2 with the z-axis.

By definition, the configuration factor is [see Eq. (7.24)]

$$F_{1 \to 2} = \frac{1}{\pi A_1} \iint_{A_1 A_2} \frac{\cos\theta_1 \cos\theta_2 \, dA_1 \, dA_2}{r^2}$$

From Fig. 7.16, we have

$$\cos\theta_1 = H/r; \qquad r^2 = (x - x')^2 + (y - y')^2 + H^2; \qquad \cos\theta_1 = \cos\theta_2$$

Area elements are: $dA_1 = dx'dy'$ and $dA_2 = dxdy$. Hence, we obtain

$$F_{1 \to 2} = \frac{1}{\pi A_1} \int_{A_1} dA_1 \int_{y=0}^{B} \int_{x=-\infty}^{\infty} \frac{H^2 dxdy}{\left[(x - x')^2 + (y - y')^2 + H^2 \right]^2}$$

$$= \frac{H^2}{\pi A_1} \int_{A_1} dA_1 \int_{y=0}^{B} dy \int_{x=-\infty}^{\infty} \frac{dx}{\left[\xi^2 + (x - x')^2\right]^2} ; \qquad \xi^2 = H^2 + (y - y')^2$$

$$= \frac{H^2}{\pi A_1} \int_{A_1} dA_1 \int_{y=0}^{B} dy \left[\frac{x - x'}{2\xi^2\{\xi^2 + (x - x')^2\}} + \frac{1}{2\xi^3} \tan^{-1}\left(\frac{x - x'}{\xi}\right)\right]_{x=-\infty}^{\infty}$$

$$= \frac{H^2}{A_1} \int_{A_1} dA_1 \int_{y=0}^{B} \frac{dy}{2\xi^3} = \frac{H^2}{2 A_1} \int_{A_1} dA_1 \int_{y=0}^{B} \frac{dy}{[H^2 + (y - y')^2]^{3/2}}$$

$$= \frac{H^2}{2 A_1} \int_{A_1} dA_1 \left[\frac{y - y'}{H^2\{H^2 + (y - y')^2\}^{1/2}}\right]_{y=0}^{B} ; \qquad \because \int \frac{dx}{(a^2 + x^2)^{3/2}} = \frac{x}{a^2(a^2 + x^2)^{1/2}}$$

$$= \frac{1}{2 A_1} \int_{A_1} dA_1 \left[\frac{B - y'}{\{H^2 + (y' - B)^2\}^{1/2}} + \frac{y'}{\{H^2 + (y')^2\}^{1/2}}\right]$$

Now we put $dA_1 = dx'\, dy'$, make the coordinates x', y' current, and then integrate. Because the plate extends from $x = -\infty$ to $x = \infty$, the plate area may be written as

$$\lim_{L \to \infty} (2L)(B)$$

Therefore,

$$F_{1\to 2} = \frac{1}{2} \lim_{L \to \infty} \frac{1}{(2L)(B)} \int_{x=-L}^{L} \int_{y=0}^{B} \left[\frac{y}{(H^2 + y^2)^{1/2}} - \frac{y - B}{\{H^2 + (y - B)^2\}^{1/2}}\right] dx\, dy$$

$$= \frac{1}{2B} \left[(H^2 + y^2)^{1/2} - \{H^2 + (y - B)^2\}^{1/2}\right]_{y=0}^{B}$$

$$= (1 + H^2/B^2)^{1/2} - H/B$$

Hence,

$$\boxed{F_{1\to 2} = \left[1 + \frac{H^2}{B^2}\right]^{1/2} - \frac{H}{B}}$$ is the desired configuration factor.

7.5 VIEW FACTOR ALGEBRA

Determination of view factor by using Eq. (7.24) often becomes difficult for bodies having complex geometries. Sometimes the problem can be tackled by making use of the definition of view factor and the reciprocity relation, Eq. (7.26). *View factor algebra* is a methodology to determine the view factor for a pair of surfaces from known view factors of surfaces of other geometries and orientations. Let us explain it by a simple example of radiant heat exchange between two bodies of areas A_1 and A_2 (Fig. 7.17). The view factor is F_{12}. The area A_2 is the combination of two sub-areas A_3 and A_4. So, the process may be considered as energy exchange

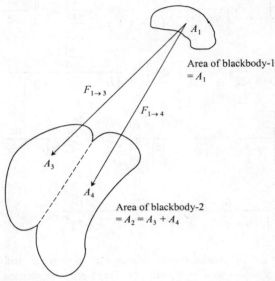

Fig. 7.17   Energy exchange between the areas A_1 and A_2 (the area A_2 is divided into areas A_3 and A_4 by the dotted line).

between the surfaces A_1 and A_3 with a view factor F_{13}, and that between A_1 and A_4 with a view factor F_{14}. Because

$$A_2 = A_3 + A_4$$

the fraction of radiant energy emitted by A_1 and intercepted by A_2 becomes equal to the sum of the fractions of the same intercepted by A_3 and A_4 individually. Hence,

$$F_{12} = F_{13} + F_{14} \qquad (7.27)$$

Again, by using the reciprocity relation for the surface pairs (1, 4), we have

$$A_1 F_{14} = A_4 F_{41} \qquad (7.28)$$

From Eqs. (7.27) and (7.28), we have

$$F_{41} = \frac{A_1}{A_4} F_{14} = \frac{A_1}{A_4}\left(F_{12} - F_{13}\right) \qquad (7.29)$$

Thus, the unknown view factor F_{41} can be calculated if the view factors F_{12} and F_{13} and the areas A_1 and A_4 are known.

In the case of a blackbody enclosure (Fig. 7.18) having N number of surfaces or walls, the radiation emitted by any surface, for example, surface-1, may be incident on any of the surfaces *including itself* (because, if a surface is concave, it can see 'itself'). So,

$$F_{11} + F_{12} + F_{13} + \ldots + F_{1N} = 1 \qquad (7.30)$$

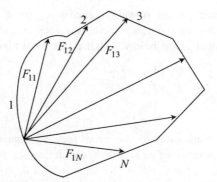

Fig. 7.18 A blackbody enclosure having N surfaces.

For any arbitrary surface i of the enclosure, we have

$$\sum_{j=1}^{N} F_{ij} = 1 \tag{7.31}$$

It is easy to understand that, unlike a concave surface, the radiation emitted by a flat or a convex surface can never hit itself. In other words, a flat or a convex surface cannot see 'itself'. So, if p is such a surface,

$$F_{pp} = 0$$

Let us now consider a simple example of radiation exchange in an enclosure formed by two concentric spherical surfaces (Fig. 7.19). The outer surface has an area A_2 and the inner one has an area A_1. The inner surface cannot see itself and all the radiation it emits is intercepted by the

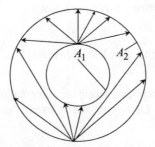

Fig. 7.19 Radiation exchange in an enclosure formed by two concentric black spherical surfaces.

area A_2 of the outer sphere. So, $F_{12} = 1$. But the outer spherical surface, being concave inwards, can see itself partly ('partly', because its view is partially obstructed by the surface A_1). Using the reciprocity relation, Eq. (7.26), we have

$$A_1 F_{12} = A_2 F_{21}$$

or

$$F_{21} = \frac{A_1}{A_2} F_{12} = \frac{A_1}{A_2}$$

Now we shall show that for an N-walled enclosure, some of the view factors may be determined from a knowledge of the rest. There is a total of N^2 view factors which may be represented in the matrix form given below. What is the total number of equations or relations

$$\begin{bmatrix} F_{11} & F_{12} & \cdots & F_{1N} \\ F_{21} & F_{22} & \cdots & F_{2N} \\ F_{N1} & F_{N2} & \cdots & F_{NN} \end{bmatrix}$$

at our disposal for the determination of the view factors? Equation (7.31), when written for $i = 1, 2, \ldots, N$, provides N number of equations. Considering the reciprocity relation, Eq. (7.26), for the N-walled enclosures, we have $N(N-1)/2$ equations in the view factors [How? For a given i, we have $N-1$ values of j. Now i may be varied from 1 to N. So there are $N(N-1)$ equations. But each equation occurs twice. Therefore, the actual number of equations is $N(N-1)/2$]. Thus the total number of equations available is

$$N + \frac{N(N-1)}{2} = \frac{N(N+1)}{2}$$

which is the number of unknown view factors that can be determined.

The rest of the view factors

$$N^2 - \frac{N(N+1)}{2} = \frac{N(N-1)}{2}$$

must be known. For example, if we have a 4-surface enclosure:

The total number of view factors involved $= N^2 = 4^2 = 16$

The number of view factors that should be known $= \dfrac{N(N-1)}{2} = \dfrac{(4)(3)}{2} = 6$

The number of view factors that can be determined $= \dfrac{N(N+1)}{2} = \dfrac{(4)(5)}{2} = 10$

Example 7.8   Consider an enclosure consisting of a hemisphere of diameter D and a flat surface of the same diameter (Fig. 7.20). Determine the relevant view factors.

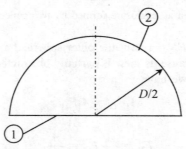

Fig. 7.20   Enclosure consisting of a hemisphere and a flat surface (Example 7.8).

SOLUTION We have here the hemispherical surface-2, and the flat surface-1. Because there are two surfaces in total, the number of view factors is $2^2 = 4$. These are: F_{11}, F_{12}, F_{21}, and F_{22}.

The flat surface cannot 'see itself'. So, $F_{11} = \boxed{0}$

By Eq. (7.31), $F_{11} + F_{12} = 1$. Therefore, $F_{12} = \boxed{1}$

By the reciprocity relation, Eq. (7.26), we have

$$A_1 F_{12} = A_2 F_{21}$$

Putting $A_1 = \dfrac{\pi}{4} D^2$, $\quad A_2 = \dfrac{\pi}{2} D^2$ and $\quad F_{12} = 1$,

$$F_{21} = \frac{\pi D^2 / 4}{\pi D^2 / 2} = \boxed{0.5}$$

Using Eq. (7.31) once again, we have

$$F_{21} + F_{22} = 1$$

Therefore,

$$F_{22} = 1 - F_{21} = 1 - 0.5 = \boxed{0.5}$$

Example 7.9 Consider an enclosure formed by closing one end of a cylinder (diameter = D, height = H) by a flat surface and the other end by a hemispherical dome. Determine the view factors of all the surfaces of the enclosure if the height is twice the diameter.

SOLUTION The enclosure is shown in Fig. 7.21. Let us mark the three surfaces as—1: the flat surface; 2: the cylindrical surface; and 3: the hemispherical surface. There will be $N^2 = 3^2 = 9$ view factors in total.

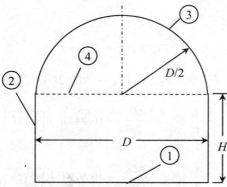

Fig. 7.21 Cylinder enclosed by flat surface at one end and hemispherical dome at the other end (Example 7.9).

Number of view factors that should be known = $N(N-1)/2 = 3$ (see Section 7.5). Using the results of Example 7.8, $F_{11} = 0$, $F_{33} = 0.5$

It may be noted that the fraction of radiation emitted by '1' and intercepted by the hemispherical end '3' is the same as that intercepted by the imaginary flat end represented by the dotted line '4'. This fraction can be obtained from the following equation. [For the notations R_1 and R_2, see Fig. 7.14. The view factor can be obtained directly from Fig. 7.14 too.]

$$F_{14} = \frac{1}{2}\left[X - \sqrt{X^2 - 4(R_2/R_1)^2}\right]$$

where

$$R_1 = \frac{D/2}{H} = 0.25; \qquad R_2 = \frac{D/2}{H} = 0.25; \qquad X = 1 + [(1 + R_2^2)/R_1^2] = 18 \ (\because H = 2D)$$

Therefore,

$$F_{14} = \frac{1}{2}\left[18 - \sqrt{18^2 - 4}\right] = \boxed{0.056} = F_{13}$$

We write Eq. (7.31) for the surface '1'.

$$F_{11} + F_{12} + F_{13} = 1$$

But $F_{11} = 0$ and $F_{13} = 0.056$. Therefore,

$$F_{12} = 1 - 0.056 = \boxed{0.944}$$

Now we consider the surface '3'.

$$F_{31} + F_{32} + F_{33} = 1$$

But $F_{33} = \boxed{0.5}$ and $A_3 F_{31} = A_1 F_{13}$ $\quad$ or $\quad$ $F_{31} = [(\pi D^2/4)/(\pi D^2/2)](0.056) = \boxed{0.028}$

Therefore,

$$0.028 + F_{32} + 0.5 = 1$$

or

$$F_{32} = \boxed{0.472}$$

For the surface '2', we have

$$F_{21} + F_{22} + F_{23} = 1$$

But

$$A_2 F_{21} = A_1 F_{12}, \qquad A_2 = (\pi D)(2D) = 2\pi D^2$$

$$F_{21} = \left(\frac{A_1}{A_2}\right) F_{12} = \frac{\pi D^2/4}{2\pi D^2} \ (0.944) = \boxed{0.118}$$

$$F_{23} = \left(\frac{A_3}{A_2}\right) F_{32} = \frac{\pi D^2/2}{2\pi D^2} \ (0.472) = \boxed{0.118}$$

Therefore,

$$0.118 + F_{22} + 0.118 = 1$$

or

$$F_{22} = \boxed{0.764}$$

The view factor matrix is, therefore, given by

$$\begin{bmatrix} F_{11} & F_{12} & F_{13} \\ F_{21} & F_{22} & F_{23} \\ F_{31} & F_{32} & F_{33} \end{bmatrix} = \begin{bmatrix} 0 & 0.944 & 0.056 \\ 0.118 & 0.764 & 0.118 \\ 0.028 & 0.472 & 0.5 \end{bmatrix}$$

Example 7.10 An enclosure consists of two black surfaces formed by cutting off a 0.3 m diameter spherical shell by a plane at a distance 0.1 m from the centre and closing it by a black circular disk (see Fig. 7.22). Determine the view factors of the surfaces.

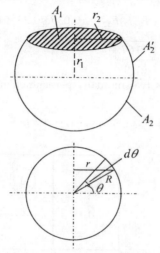

Fig. 7.22 Spherical shell cut by a plane (Example 7.10).

SOLUTION Here we have two black surfaces of determinable areas. The surface A_1 is flat and cannot see itself. The radiation it emits is completely intercepted by surface A_2. So, by the definition of view factor, $F_{12} = \boxed{1}$

We shall use the reciprocity relation, [Eq. (7.26)], $A_1 F_{12} = A_2 F_{21}$, to calculate F_{21}.

Calculation of area A_2. We have

$$A_2 = 2\pi R^2 + A_2'$$

where A_2' is the area of the hemisphere, less the part cut off by the surface A_1.

From Fig. 7.22, an elementary area over the spherical surface at an angular position θ is

$$dA_2' = 2\pi r R \, d\theta$$

where $\cos \theta = r/R$; or $-\sin \theta \, d\theta = (1/R)dr$.

Therefore,

$$A_2' = -\int_{r=R}^{r_2} (2\pi R) \frac{r \, dr}{\sqrt{R^2 - r^2}} = 2\pi R \sqrt{R^2 - r_2^2} \qquad \left(\because d\theta = -\frac{dr}{R \sin \theta} = -\frac{dr}{\sqrt{R^2 - r^2}} \right)$$

Given: $R = 0.15$ m (radius of the sphere), distance of the disk from the centre $= 0.1$ m $= r_1$. Therefore,

$$r_2 = \sqrt{R^2 - r_1^2} = 0.1118 \text{ m}$$

and

$$\text{Area } A_1 = \pi r_2^2 = \pi(0.1118)^2 = 0.0393 \text{ m}^2$$

$$A_2 = 2\pi R^2 + 2\pi R\sqrt{R^2 - r_2^2} = 0.2356 \text{ m}^2$$

Thus,

$$F_{21} = (A_1/A_2)F_{12} = (0.0393/0.2356)(1)$$

$$= \boxed{0.167}$$

Example 7.11 Figure 7.23 shows two mutually perpendicular rectangles. Prove that the view factor F_{13} is given by the relation

$$2A_1F_{13} = A_{(1,2)}F_{(1,2)(3,4)} - A_1F_{14} - A_2F_{23}$$

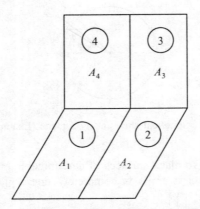

Fig. 7.23 Two mutually perpendicular rectangles (Example 7.11).

The notation $F_{(1,2)(3,4)}$ is the view factor giving the fraction of radiation emitted by surfaces 1 and 2 together and intercepted by surfaces 3 and 4 together and $A_{(1,2)}$ denotes the total area of surfaces 1 and 2.

SOLUTION Using Eq. (7.27) and the reciprocity relation Eq. (7.26), we have

$$F_{(1,2)(3,4)} = F_{(1,2)3} + F_{(1,2)4} = \frac{A_3F_{3(1,2)}}{A_{(1,2)}} + \frac{A_4F_{4(1,2)}}{A_{(1,2)}}$$

$$= \frac{A_3}{A_{(1,2)}}(F_{31} + F_{32}) + \frac{A_4}{A_{(1,2)}}(F_{41} + F_{42})$$

Therefore,

$$A_{(1,2)}F_{(1,2)(3,4)} = A_3F_{31} + A_3F_{32} + A_4F_{41} + A_4F_{42} \tag{i}$$

It can be proved that $\qquad A_1F_{13} = A_2F_{24}$

Also, we have the relations

$$A_3F_{31} = A_1F_{13}; \quad A_4F_{41} = A_1F_{14}; \quad A_3F_{32} = A_2F_{23}; \quad A_2F_{24} = A_4F_{42}$$

Putting these relations in Eq. (i) above, we get

$$A_{(1,2)}F_{(1,2)(3,4)} = A_1F_{13} + A_2F_{23} + A_1F_{14} + A_1F_{13}$$

Therefore,

$$2A_1F_{13} = A_{(1,2)}F_{(1,2)(3,4)} - A_1F_{14} - A_2F_{23}$$

Once F_{13} is known, F_{24} or F_{42} can be determined in this surface configuration.

7.6　RATE OF RADIATION EXCHANGE BETWEEN BLACKBODIES

The definition of view factor can be used to calculate the rate of exchange of radiation between two blackbodies of any geometry and orientation. Consider again two blackbodies 1 and 2 of surface areas A_1 and A_2, temperatures T_1 and T_2, and total emissive powers E_{b1} and E_{b2}, respectively. Then,

the rate at which radiation emitted by body-1 is intercepted and absorbed by body-2

$$= Q_{12} = A_1E_{b1}F_{12}$$

the rate at which radiation emitted by body-2 is intercepted and absorbed by body-1

$$= Q_{21} = A_2E_{b2}F_{21}$$

Therefore, the net rate of radiation exchange between the bodies 1 and 2

$$= [Q_{12}]_{\text{net}} = A_1E_{b1}F_{12} - A_2E_{b2}F_{21}$$

Putting $E_{b1} = \sigma T_1^4$, $E_{b2} = \sigma T_2^4$, and using the reciprocity relation, $A_1F_{12} = A_2F_{21}$, we get

$$[Q_{12}]_{\text{net}} = A_1F_{12}\,\sigma\left(T_1^4 - T_2^4\right) \tag{7.32}$$

What does $[Q_{12}]_{\text{net}}$ mean physically? It means the *net rate of gain of radiation* by body-2 because of its radiative interaction with body-1.

The result given by Eq. (7.32) can be easily extended to radiative exchange among the N surfaces of the enclosure of Fig. 7.18. The net rate of radiant energy loss by the surface i because of its radiative interaction with all the surfaces including itself is given by

$$[Q_i]_{\text{net}} = \sum_{j=1}^{N} A_iF_{ij}\sigma\left(T_i^4 - T_j^4\right) \tag{7.33}$$

It is to be noted that if the surfaces of the enclosure are at thermal equilibrium, we have

$$T_i = T_j, \quad \text{and} \quad [Q_i]_{\text{net}} = 0$$

Example 7.12　A carbon steel sphere, 0.3 m in diameter and at 800 K, cools down by radiative heat loss to an ambient at 30°C. If other modes of heat loss are neglected, and if the ball and the ambient are assumed to be black, calculate the time required for the ball to cool down to 70°C. The density and specific heat of carbon steel are 7801 kg/m³ and 0.473 kJ/kg °C, respectively.

SOLUTION If the temperature of the sphere is T_1 K at any time t, the net rate of heat loss by the sphere is

$$Q_{12} = A_1 F_{12}\sigma\left(T_1^4 - T_2^4\right)$$

We use the following notations:
The sphere: surface-1; the ambient: surface-2
F_{12} = view factor = 1 (because the entire radiation emitted by '1' is intercepted by the surroundings)
A_1 is the area of the sphere.
T_2 is the ambient temperature.
By unsteady state heat balance, we get

$$-\frac{d}{dt}(mc_p T_1) = Q_{12}$$

or

$$-\frac{dT_1}{dt} = \frac{A_1 F_{12}\sigma(T_1^4 - T_2^4)}{mc_p}$$

The values of various quantities are:
R = radius of the sphere = 0.15 m
$A_1 = 4\pi R^2 = 0.2827$ m²
$m = (4/3)\pi R^3 \rho = 110.3$ kg
$c_p = 473$ J/kg °C
$F_{12} = 1$

Therefore,

$$-\frac{dT_1}{dt} = \frac{(0.2827)(1)(5.669 \times 10^{-8})(T_1^4 - T_2^4)}{(110.3)(473)}$$

$$= 3.072 \times 10^{-13}(T_1^4 - T_2^4)$$

Integrating, we have

$$\int \frac{dT_1}{T_1^4 - T_2^4} = -3.072 \times 10^{-13}t + C \quad (C = \text{integration constant})$$

$$\text{LHS} = \frac{1}{2T_2^2}\int\left(\frac{1}{T_1^2 - T_2^2} - \frac{1}{T_1^2 + T_2^2}\right)dT_1 = \frac{1}{4T_2^3}\ln\frac{T_1 - T_2}{T_1 + T_2} - \frac{1}{2T_2^3}\tan^{-1}\frac{T_1}{T_2}$$

Hence,

$$\frac{1}{2}\ln\frac{T_1 - T_2}{T_1 + T_2} - \tan^{-1}\frac{T_1}{T_2} = -2T_2^3(3.072 \times 10^{-13}\,t) + C \tag{i}$$

Given: ambient temperature, $T_2 = 30°C = 303$ K; and at $t = 0$, temperature of the sphere, $T_1 = 800$ K.

Therefore,

$$C = \frac{1}{2} \ln \frac{800 - 303}{800 + 303} - \tan^{-1} \frac{800}{303} = -1.607$$

Putting the values of C and $T_2 (= 303$ K) in Eq. (i), we get

$$\frac{1}{2} \ln \frac{T_1 - T_2}{T_1 + T_2} - \tan^{-1} \frac{T_1}{T_2} = -1.709 \times 10^{-5} t - 1.607 \qquad \text{(ii)}$$

The above equation gives the temperature history of the ball as it cools down. If the final temperature is 70°C (343 K), the corresponding time t can be calculated as follows:

$$\frac{1}{2} \ln \frac{343 - 303}{343 + 303} - \tan^{-1} \frac{343}{303} = -1.709 \times 10^{-5} t - 1.607$$

Therefore,

$$t = 36930 \text{ s} = \boxed{10.2 \text{ h}}$$

Note: In reality, the sphere will cool down much faster because of the convective heat loss.

7.7 EXCHANGE OF RADIATION BETWEEN DIFFUSE GRAY SURFACES

In the preceding section, we discussed the method of calculation of the rate of exchange of radiation in a black enclosure, that is, an enclosure having walls that behave like black bodies. It has also been mentioned before that a blackbody has an ideal surface with respect to absorption or emission of radiation (for a blackbody, absorptivity = emissivity = 1). A real surface does not behave like a blackbody, although it may sometimes be a close approximation to a blackbody. The calculation of the radiation exchange rate for a blackbody requires its temperature, relevant view factor, and surface area. The situation is not so simple for *real* bodies or *non-black* bodies. A real surface cannot absorb all the radiation incident on it (a part of it may even be transmitted if the transmittance is not zero). Thus, for a non-black body we can generally say that the reflectivity ρ is not zero, and the total quantity of radiation that leaves a non-black surface is the combination of emitted radiation and reflected radiation. So, more information is needed for the calculation of radiation exchange in a non-black environment and the equations for such calculations have to be developed.

Before we proceed to analyse radiation exchange between non-black bodies, we shall discuss a little more about their characteristics and learn a few more terms. We have already defined a gray body or surface (which is also an idealization). A gray body or surface is sometimes called *diffuse gray*. A surface is called diffuse gray if its spectral emissivity ε_λ and the absorptivity α_λ are independent of the angle of incidence or the angle of emission. (Note that spectral emissivity and absorptivity are defined in the same way as emissivity or absorptivity except that they pertain to the wavelength λ.) In fact, a surface is called diffuse if it emits (or absorbs) radiation equally in all directions without any directional preference (see Fig. 7.24).

As a matter of fact, for a real surface, the spectral emissivity and absorptivity may depend upon the direction of emitted radiation or the incident radiation as the case may be. Also, reflection from a real surface may be secular or diffuse (see Section 7.2). However, the analysis will be quite complicated if all these characteristics are taken into consideration. Because many

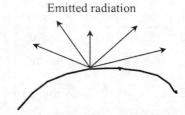

Fig. 7.24 A diffuse gray surface.

real surfaces have characteristics close to those of diffuse gray surfaces, the equations for radiation exchange that will be presented here can be used in many practical situations with satisfactory accuracy.

We will introduce here two new terms: *irradiation*, denoted by G, and *radiosity*, denoted by J. These are defined as under [see Fig. 7.25(a)].

Irradiation, G = total radiation (from all sources) that hits a surface per unit area per unit time (W/m²).

Radiosity, J = total radiation that leaves a surface (by way of emission or reflection) per unit area per unit time (W/m²).

Since a blackbody does not reflect any radiation ($\rho = 0$), its radiosity is equal to its emissive power (i.e. $J = E_b$).

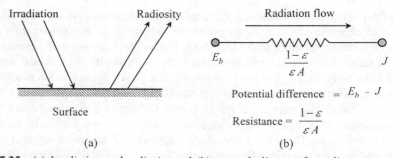

Fig. 7.25 (a) Irradiation and radiosity and (b) network diagram for radiant energy loss.

7.7.1 Radiation Exchange in a Gray Enclosure

Radiosity of a diffuse gray body can be related to the irradiation it receives in the following way. If E_b is the total emissive power of a blackbody at the same temperature (that is, at the temperature of the diffuse gray body under consideration), then, we have:

$$\text{Emissive power of the gray body} = \varepsilon E_b$$

$$\text{Radiation reflected by the gray body} = \rho\, G$$

where ε is its emissivity, and ρ its reflectivity. Therefore, the total rate at which radiant energy leaves the surface is

$$J = \varepsilon E_b + \rho\, G \tag{7.34}$$

If the transmissivity of the body is zero ($\tau = 0$), we have $\alpha + \rho = 0$ [from Eq. (7.3)]. Also, by Kirchhoff's law, $\alpha = \varepsilon$.

Therefore,

$$\rho = 1 - \alpha = 1 - \varepsilon$$

Putting this result in Eq. (7.34), we get

$$J = \varepsilon E_b + (1 - \varepsilon)G \qquad (7.35)$$

If A is the surface area of the body, the net rate of loss of radiant energy from unit area of the surface is

$$\left(\frac{Q}{A}\right)_{net} = J - G = \varepsilon E_b + (1 - \varepsilon)G - G = \varepsilon E_b - \varepsilon G$$

But from Eq. (7.35), we have

$$G = \frac{J - \varepsilon E_b}{1 - \varepsilon}$$

Therefore,

$$\left(\frac{Q}{A}\right)_{net} = \varepsilon E_b - \frac{\varepsilon(J - \varepsilon E_b)}{1 - \varepsilon} = \frac{\varepsilon(E_b - J)}{1 - \varepsilon}$$

Hence,

$$Q_{net} = \frac{E_b - J}{(1 - \varepsilon)/\varepsilon A} \qquad (7.36)$$

We may consider the RHS of Eq. (7.36) as the ratio of a potential difference and a surface resistance for radiation heat transfer [compare with Eq. (2.8)]. The radiant heat loss can, therefore, be *simulated* by an electric current flow network diagram as shown in Fig. 7.25(b).

Now let us consider radiation heat exchange in an enclosure as shown in Fig. 7.18, assuming the walls to be diffuse gray. It is easy to understand that the view factors and the emissivities of the surfaces will be involved in the analysis. Radiation from each of the N surfaces will, in general, be incident on all of the surfaces. If we consider the ith surface, the irradiation G_i of the surface will be the sum of the radiosities of N surfaces multiplied by the respective areas and view factors. That is,

$$A_i G_i = \sum_{j=1}^{N} A_j F_{ji} J_j \qquad (7.37)$$

Using the reciprocity relation $A_i F_{ij} = A_j F_{ji}$ [Eq. (7.26)], we get

$$A_i G_i = \sum_{j=1}^{N} A_i F_{ij} J_j \qquad (7.38)$$

The net rate of loss of radiant energy by the ith surface is

$$Q_{i,\, net} = A_i (J_i - G_i) = A_i J_i - \sum_{j=1}^{N} A_i F_{ij} J_j \qquad (7.39)$$

From Eq. (7.31), $1 = \sum_{j=1}^{N} F_{ij}$, i.e. $J_i = J_i \sum_{j=1}^{N} F_{ij}$

Therefore,

$$Q_{i,\,net} = A_i J_i \sum_{j=1}^{N} F_{ij} - \sum_{j=1}^{N} A_i F_{ij} J_j = \sum_{j=1}^{N} A_i F_{ij}(J_i - J_j) = \sum_{j=1}^{N} \left(Q_{ij}\right)_{net} \qquad (7.40)$$

Here $(Q_{ij})_{net}$ denotes the net rate of radiant energy loss from the ith surface because of radiation exchange with the jth surface. Writing Eq. (7.36) for the ith surface, we have

$$Q_{i,\,net} = \frac{E_{bi} - J_i}{(1 - \varepsilon_i)/\varepsilon_i A_i} \qquad (7.41)$$

From Eqs. (7.40) and (7.41), we get

$$Q_{i,\,net} = \frac{E_{bi} - J_i}{(1 - \varepsilon_i)/\varepsilon_i A_i} = \sum_{j=1}^{N} A_i F_{ij}(J_i - J_j) = \sum_{j=1}^{N} \frac{J_i - J_j}{1/A_i F_{ij}} \qquad (7.42a)$$

Also, the net rate of exchange of radiation between the surfaces i and j is

$$Q_{ij,\,net} = \frac{J_i - J_j}{1/A_i F_{ij}} \qquad (7.42b)$$

Equation (7.42a) is very important for the calculation of radiation exchange in a gray enclosure. It may be noted that J_i values should be known in order to calculate $Q_{i,\,net}$. But how to know J_i? For an N-surface enclosure, Eq. (7.42a) provides N simultaneous linear algebraic equations that may be solved (by the matrix method, or by the Gauss elimination method, etc.) to provide the N number of J_i values. Let us illustrate the application of Eq. (7.42a) by considering a two-surface enclosure.

7.7.2 Radiation Exchange in a Two-surface Gray Enclosure

Equation (7.42a), written for $i = 1$ and $i = 2$, gives

$$i = 1, \qquad \frac{E_{b1} - J_1}{(1 - \varepsilon_1)/\varepsilon_1 A_1} = A_1 F_{12}(J_1 - J_2) \qquad (7.43)$$

$$i = 2, \qquad \frac{E_{b2} - J_2}{(1 - \varepsilon_2)/\varepsilon_2 A_2} = A_2 F_{21}(J_2 - J_1) \qquad (7.44)$$

Subtracting Eq. (7.44) from Eq. (7.43) and putting $F_{12} = (A_2/A_1)F_{21}$, we get

$$E_{b1} - E_{b2} = (J_1 - J_2)\left[1 + F_{12}\frac{1 - \varepsilon_1}{\varepsilon_1} + F_{12}\frac{A_1}{A_2}\frac{1 - \varepsilon_2}{\varepsilon_2}\right]$$

The above equation can be solved for $(J_1 - J_2)$. Putting this value in place of $(J_1 - J_2)$ in Eq. (7.43) and simplifying, we get

$$Q_{1,\,net} = \frac{E_{b1} - J_1}{(1 - \varepsilon_1)/\varepsilon_1 A_1} = \frac{E_{b1} - E_{b2}}{(1 - \varepsilon_1)/\varepsilon_1 A_1 + 1/A_1 F_{12} + (1 - \varepsilon_2)/\varepsilon_2 A_2} \qquad (7.45)$$

Here $Q_{1,\,net}$ is the net rate of loss of radiant energy from surface-1. The right-hand side of Eq. (7.45) can be visualized as the ratio of radiation potential difference ($E_{b1} - E_{b2}$) and the radiation resistance. The resistance term (i.e. the denominator) is the sum of three individual resistances in series.

$$(1 - \varepsilon_1)/A_1\varepsilon_1 = \textit{radiation surface resistance} \text{ of surface-1.}$$

$$(1 - \varepsilon_2)/A_2\varepsilon_2 = \textit{radiation surface resistance} \text{ of surface-2.}$$

$$1/A_1F_{12} = \textit{space resistance} \text{ to radiation.}$$

It is to be noted that the above resistances are *notional* and should not be confused with the heat transfer resistances in conduction or convection. The network diagram for radiation exchange in a two-surface enclosure is shown in Fig. 7.26. If we put the expressions for E_{b1} and E_{b2} in Eq. (7.45), we get

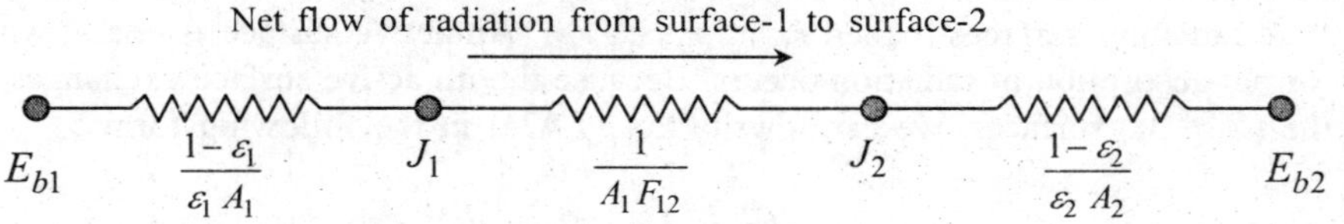

Net flow of radiation from surface-1 to surface-2

Fig. 7.26 Network diagram for radiant energy transport in a two-surface enclosure.

$$Q_{1,\,net} = \frac{\sigma(T_1^4 - T_2^4)}{(1 - \varepsilon_1)/\varepsilon_1 A_1 + 1/A_1F_{12} + (1 - \varepsilon_2)/\varepsilon_2 A_2} \qquad (7.46)$$

where T_1 and T_2 are the absolute temperatures of the surfaces 1 and 2, respectively.

The procedure discussed above can be extended to an enclosure consisting of a large number of surfaces and the corresponding network diagram can be prepared. But the algebraic exercise involved depends upon the geometry of the enclosure. For example, if some of the surfaces are flat or convex, their self-view factors F_{ii} are zero, and the solution will be a little simpler. Thus if we consider radiation exchange between two very large parallel flat surfaces, $F_{12} = 1$ (because each plate is in full view of the other), and $(A_1/A_2) = 1$. The net flux of radiant energy loss by surface-1 [from Eq. (7.46)] is

$$\frac{Q_{1,\,net}}{A_1} = \frac{\sigma(T_1^4 - T_2^4)}{1/\varepsilon_1 + 1/\varepsilon_2 - 1} \qquad (7.47)$$

7.7.3 Emissivity Factor

This is a term sometimes used in connection with the calculation of radiation exchange among diffuse gray surfaces. It is also called the *emissivity correction*. For an enclosure made up of two surfaces 1 and 2, it is defined as follows:

$$Q_{1,\,net} = F_{1\varepsilon}F_{12}A_1\sigma\left(T_1^4 - T_2^4\right) \qquad (7.48)$$

where

$Q_{1,\,net}$ is the net rate of radiation loss by surface-1

$F_{1\varepsilon}$ is the emissivity factor of surface-1

Comparing Eq. (7.48) with Eq. (7.46), we have

$$F_{1\varepsilon} = \frac{1}{1 + F_{12}(1 - \varepsilon_1/\varepsilon_1) + F_{12}(A_1/A_2)(1 - \varepsilon_2/\varepsilon_2)} \qquad (7.49)$$

The emissivity factor for any other surface can be defined similarly.

7.7.4 A Gray Enclosure with Re-radiating Surfaces

A *re-radiating surface* is one which radiates the entire amount of radiation that it receives from other surfaces. At steady state there is no *net* radiation absorption or *net* radiation emission at the re-radiating surface. In other words, the irradiation G_i and the radiosity J_i are equal for this type of surface, and $Q_{i,\,net} = 0$ [see Eq. (7.39)]. A re-radiating surface is functionally adiabatic. Well-insulated surfaces like those of furnace walls are sometimes assumed to be re-radiating.

Let us consider an enclosure consisting of N gray *active surfaces* (we call them 1, 2, 3, ..., N), and M re-radiating surfaces (called r_1, r_2, r_3, ..., r_M). An active surface is one at which a net emission or net absorption of radiation occurs. Because the ith active surface exchanges radiation with all the $(N + M)$ surfaces, we can rewrite Eq. (7.42a) in the following form:

$$Q_{i,\,net} = \frac{E_{bi} - J_i}{(1 - \varepsilon_i)/\varepsilon_i A_i} = \sum_j \frac{J_i - J_j}{1/A_i F_{ij}} \qquad (7.50)$$

where

$$j = 1, 2, ..., N, r_1, r_2, ..., r_M \quad \text{and} \quad i = 1, 2, ..., N.$$

Since the M re-radiating surfaces are adiabatic, it follows that

$$0 = \frac{E_{b,\,r_k} - J_{r_k}}{(1 - \varepsilon_{r_k})/\varepsilon_{r_k} A_{r_k}} = \sum_j \frac{J_{r_k} - J_j}{1/A_{r_k} F_{r_k,\,j}} \qquad (7.51)$$

where

$$k = 1, 2, ..., M \quad \text{and} \quad j = 1, 2, ..., N, r_1, r_2, ..., r_M.$$

The subscript r denotes a re-radiating surface.

Equations (7.50) and (7.51) together constitute a set of $(N + M)$ simultaneous linear algebraic equations which can be solved to determine the radiosities J_1, J_2, ..., J_N, J_{r1}, J_{r2}, ..., J_{rM} of the surfaces. Hence, the net rates of radiation exchange ($Q_{i,\,net}$) can be calculated from Eq. (7.50).

If we consider a three-surface enclosure with two active surfaces (1 and 2) and one re-radiating surface (3), Eq. (7.50) for the two active surfaces can be written as follows:

$$\frac{E_{b1} - J_1}{(1 - \varepsilon_1)/\varepsilon_1 A_1} = \frac{J_1 - J_1}{1/A_1 F_{11}} + \frac{J_1 - J_2}{1/A_1 F_{12}} + \frac{J_1 - J_3}{1/A_1 F_{13}} \qquad (7.52)$$

$$\frac{E_{b2} - J_2}{(1 - \varepsilon_2)/\varepsilon_2 A_2} = \frac{J_2 - J_1}{1/A_2 F_{21}} + \frac{J_2 - J_2}{1/A_2 F_{22}} + \frac{J_2 - J_3}{1/A_2 F_{23}} \qquad (7.53)$$

One term on the RHS of each of the above equations reduces to zero.

The radiation balance equation for the re-radiating surface-3 follows from Eq. (7.51). That is,

$$0 = \frac{J_3 - J_1}{1/A_3 F_{31}} + \frac{J_3 - J_2}{1/A_3 F_{32}} + \frac{J_3 - J_3}{1/A_3 F_{33}} \qquad (7.54)$$

Equations (7.52)–(7.54) can be solved for the unknown radiosities, J_1, J_2, and J_3. The net rate of energy exchange (loss or gain) at the active surfaces can be calculated using Eqs. (7.52) and (7.53). It follows from Eq. (7.51) that for a re-radiating surface

$$E_{b,\,rk} - J_{rk} = 0$$

or

$$E_{b,\,rk} = \sigma T_{rk}^4 = J_{rk} \qquad (7.55)$$

Once the radiosity of a re-radiating surface is known, its temperature can be easily calculated from Eq. (7.55). It is also seen that the radiosity of a re-radiating surface is equal to the emissive power of a blackbody at the same temperature.

7.7.5 Use of the Network Diagram to Calculate Radiation Exchange

The analogy between flow of current in an electrical network and radiation exchange in an enclosure has been mentioned in Section 7.7.2. It is based on the Kirchhoff's law for steady currents. The law states that *the algebraic sum of currents meeting at a point is zero*. Similarly, the algebraic sum of *heat currents* at each *node* of a radiation network diagram is zero at steady state. Every such node is connected to other nodes through radiation resistances (surface resistances or space resistances; see Section 7.7.2).

The radiation network diagram for a three-surface enclosure consisting of two active and one re-radiating surfaces is shown in Fig. 7.27. The nodes in the diagram (here a node is a point where two or more heat currents meet) are also denoted by J_1, J_2, and J_3. The algebraic sum of the 'heat currents' at each of the nodes is zero as represented by Eqs. (7.52), (7.53), and (7.54). For example, from Eq. (7.52), we have

$$\frac{E_{b1} - J_1}{(1 - \varepsilon_1)/\varepsilon_1 A_1} + \frac{J_2 - J_1}{1/A_1 F_{12}} + \frac{J_3 - J_1}{1/A_1 F_{13}} = 0 \qquad (7.56)$$

The LHS of Eq. (7.56) is the sum of all radiation currents reaching the node J_1. The total resistance between the nodes J_1 and J_2 can be obtained from Fig. 7.27, in which the resistances are in a series-parallel arrangement, and is given by

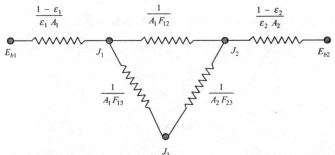

Fig. 7.27 Radiation network diagram for a three-surface gray enclosure (two active and one re-radiating surface).

$$\frac{1}{A_1 F_{12} + \dfrac{1}{(1/A_1 F_{13}) + (1/A_2 F_{23})}} = \frac{1}{A_1 F_{12} + \dfrac{A_1 A_2 F_{13} F_{23}}{A_2 F_{23} + A_1 F_{13}}}$$

The total resistance between the points E_{b1} and E_{b2} is

$$\frac{1 - \varepsilon_1}{\varepsilon_1 A_1} + \frac{1}{A_1 F_{12} + \dfrac{A_1 A_2 F_{13} F_{23}}{A_2 F_{23} + A_1 F_{13}}} + \frac{1 - \varepsilon_2}{\varepsilon_2 A_2} \qquad (7.57)$$

Hence the net rate of radiant heat flow from surface-1 to surface-2 (surface-3 is *adiabatic*) is given by

$$Q_{1, \, net} = \frac{E_{b1} - E_{b2}}{\text{Total resitance given by Eq. (7.57)}} \qquad (7.58)$$

Let the temperatures of the surfaces 1 and 2 be T_1 and T_2, respectively. If the surfaces are flat (or convex), $F_{11} = 0 = F_{22}$. That is,

$$F_{13} = 1 - F_{11} - F_{12} = 1 - F_{12} \quad \text{and} \quad F_{23} = 1 - F_{21} - F_{22} = 1 - F_{21}$$

Using these results along with the reciprocity relation, $A_1 F_{12} = A_2 F_{21}$, we can obtain from Eqs. (7.57) and (7.58) the following relation for the net rate of radiant energy flow.

$$Q_{1, \, net} = \frac{A_1 \sigma (T_1^4 - T_2^4)}{\dfrac{A_1 + A_2 - 2 A_1 F_{12}}{A_2 - A_1 (F_{12})^2} + \left(\dfrac{1}{\varepsilon_1} - 1 \right) + \dfrac{A_1}{A_2} \left(\dfrac{1}{\varepsilon_2} - 1 \right)} \qquad (7.59)$$

It appears from the above equation that the rate of radiation exchange *does not* depend upon the emissivity of the re-radiating surface.

The radiation network diagram for a five-surface enclosure consisting of three active surfaces (designated 1, 2, and 3) and two re-radiating surfaces (designated 4 and 5) is shown in Fig. 7.28.

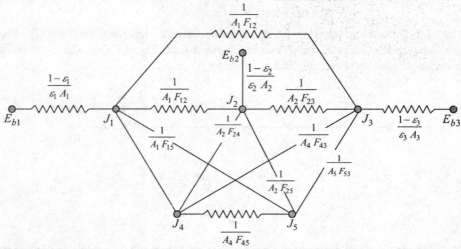

Fig. 7.28 Radiation network diagram for a five-surface gray enclosure (three active, two re-radiating surfaces).

There are five nodes. Each surface exchanges radiation with the remaining four surfaces. Radiosities of the surfaces can be calculated by solving five simultaneous linear equations derived from Eqs. (7.50) and (7.51).

Example 7.13 A schedule 40 pipe (o.d. = 114 mm) carrying saturated steam at 7.33 bar absolute runs through a room having a wall temperature 27°C. The insulation on a section of the pipe has been damaged exposing the pipe wall to the ambient. Calculate the net rate of heat loss from unit length of the bare pipe by radiation if (a) the pipe surface is considered black, (b) the pipe surface has an emissivity of 0.74 (which is the emissivity of oxidized carbon steel).

SOLUTION Since the room is big compared to the space occupied by the pipe, it can be considered to be a black enclosure. Also the view factor F_{12} (1: pipe surface; 2: room walls) is unity.

The net rate of radiative heat loss from the pipe is

$$Q_{12} = A_1 \varepsilon_1 F_{12} \sigma \left(T_1^4 - T_2^4 \right)$$

Given: $F_{12} = 1$; $\varepsilon_1 = 1$ (black), 0.74 (gray); $A_1 = \pi(114/1000)(1)$ m^2 = 0.358 m^2 per metre pipe length; T_1 = steam temp. (7.33 bar) = 440 K; T_2 = wall temperature = 300 K

(a) $Q_{12} = (0.358)(1)(1)(5.67 \times 10^{-8})(440^4 - 300^4) = \boxed{596.4 \text{ W}}$

(b) $Q_{12} = (0.358)(0.74)(1)(5.67 \times 10^{-8})(440^4 - 300^4) = \boxed{441.3 \text{ W}}$

Example 7.14 Consider an enclosure formed by two diffuse gray spherical surfaces. The outer radius of the inner sphere is r_1, and the inner radius of the outer sphere is r_2. The emissivities of the surfaces are ε_1 and ε_2 and the temperatures are T_1 and T_2.

(a) Determine (i) the view factors F_{12} and F_{21}, and (ii) the net rate of radiant energy gain by the inner surface. (b) Hence calculate the rate of loss of saturated liquid nitrogen at 1 atm pressure stored in a double-walled spherical Dewar flask.

The following data are given: r_1 = 15 cm r_2 = 15.5 cm, $\varepsilon_1 = \varepsilon_2 = 0.06$, temperature of the outer wall = 25°C, normal boiling point of nitrogen = –196°C, heat of vaporization of nitrogen at its normal boiling point = 200 kJ/kg. Convection in the annular space may be neglected.

SOLUTION (a) (i) Referring to Section 7.5 and also to Fig. 7.19, the view factors F_{12} and F_{21} are

$$F_{12} = 1 \quad \text{and} \quad F_{21} = \frac{A_1}{A_2} = \frac{4\pi r_1^2}{4\pi r_2^2} = \frac{r_1^2}{r_2^2} = \left(\frac{15}{15.5}\right)^2 = 0.936$$

where A_1 is the area of the inner sphere and A_2 is the area of the outer sphere.

(ii) The net rate of heat gain by the inner sphere can be easily determined by using Eq. (7.46).

The net rate of heat gain = $-Q_{1,\text{net}} = \dfrac{\sigma(T_2^4 - T_1^4)}{(1 - \varepsilon_1)/\varepsilon_1 A_1 + 1/A_1 F_{12} + (1 - \varepsilon_2)/\varepsilon_2 A_2}$

(b) *Given:* $T_1 = -196°C = 77$ K; $T_2 = 25°C = 298$ K; $r_1 = 0.15$ m; $r_2 = 0.155$ m; $\varepsilon_1 = \varepsilon_2 = 0.06$; $F_{12} = 1$; heat of vaporization of nitrogen, $h_v = 200$ kJ/kg.
Hence,

$$-Q_{1,\,net} = \frac{5.669 \times 10^{-8}(298^4 - 77^4)}{\dfrac{1 - 0.06}{(0.06)(4\pi)(0.15)^2} + \dfrac{1}{4\pi(0.15)^2} + \dfrac{1 - 0.06}{(0.06)(4\pi)(0.155)^2}} = \boxed{4.0 \text{ J/s}}$$

The rate of nitrogen loss $= \dfrac{-Q_{1,\,net}}{h_v} = \dfrac{4.0 \text{ J/s}}{200 \text{ kJ/kg}} = 20 \text{ mg/s} = \boxed{72 \text{ g/h}}$

Example 7.15 A brick wall of emissivity 0.8 is 6 m wide and 5 m high. It stands at a distance of 3 m from a 15 cm × 12 cm opening on a furnace in a boiler room. The centre line of the furnace opening lies 1 m left and 1 m below the centre point of the wall as shown in the Fig. 7.29. If the wall temperature is 35°C and the furnace temperature is 1400°C, calculate the net rate of radiant heat transfer to the wall. The furnace opening can be considered as a blackbody.

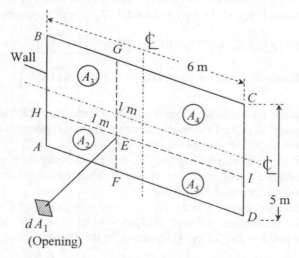

Fig. 7.29 Brick wall in a boiler room (Example 7.15).

SOLUTION The first step is to calculate the view factor between the opening and the wall. For this purpose, the wall is divided into four rectangles *AHEF*, *BGEH*, *CGEI* and *DFEI* (call them A_2, A_3, A_4 and A_5, respectively). Because the opening is small compared to each of these areas we denote it by dA_1; its area is also denoted by dA_1.

The opening is parallel to each of the rectangles, and the normal at the opening passes through *E*, the common corner of the rectangles. The view factor between the opening and any of the rectangles can be calculated using the following equation applicable to a small rectangular element parallel to a given rectangle such that the normal to the elementary area passes through a corner of the given rectangle (see Spiegel and Howell, 1981). For the dA_1–A_2 pair, the equation is

$$F_{dA_1-A_2} = \frac{1}{2\pi}\left[\frac{l_2}{\sqrt{l_2^2 + h^2}} \tan^{-1}\frac{l_1}{\sqrt{l_2^2 + h^2}} + \frac{l_1}{\sqrt{l_1^2 + h^2}} \tan^{-1}\frac{l_2}{\sqrt{l_1^2 + h^2}}\right]$$

For the dA_1-A_2 pair, we can obtain the following values of various quantities from Fig. 7.29.

$$l_1 = AF = 2 \text{ m}, \quad l_2 = AH = 1.5 \text{ m}, \quad h = E - dA_1 = 3 \text{ m}$$

where l_1 and l_2 are the two sides of the rectangle A_2, and h is the distance of the corner from dA_1. Thus,

$$F_{dA_1-A_2} = \frac{1}{2\pi}\left[0.447 \tan^{-1}(0.596) + 0.555 \tan^{-1}(0.416)\right] = 0.0731$$

Similarly,

$$F_{dA_1-A_3} = 0.1175; \qquad F_{dA_1-A_4} = 0.1641; \qquad F_{dA_1-A_5} = 0.0992$$

So the view factor between the opening and the wall (W) is

$$F_{dA_1-W} = F_{dA_1-A_2} + F_{dA_1-A_3} + F_{dA_1-A_4} + F_{dA_1-A_5} = 0.454$$

Calculation of the rate of radiant heat exchange

Let the area of the wall be denoted by A_W. The energy emitted by the black opening dA_1 is partly absorbed by the wall and the rest is reflected. The fraction of this *reflected* radiation intercepted by the opening is completely absorbed by it (because it is black). Also, the radiation *emitted* by the wall and *intercepted* by the opening is fully absorbed. The environment is assumed to be free from other sources of radiation.

Radiation emitted by dA_1 and absorbed by the wall $= dA_1 F_{dA_1-W} E_{b_1}\varepsilon_2$

The fraction of this radiation *reflected* by the wall and *reabsorbed* by the opening

$$= dA_1 F_{dA_1-W} E_{b_1}\varepsilon_2 (1 - \varepsilon_2) F_{W-dA_1}$$

Radiation emitted by W and absorbed by the opening $= A_W F_{W-dA_1}\varepsilon_2 E_{bW}$

All the above quantities are *rates*. Here E_{b1} is the emissive power of the opening, and E_{bW} is the emissive power of the blackbody at the wall temperature, and ε_2 is the emissivity (= absorptivity) of the wall.

Hence, the net rate of radiation exchange between the opening and the wall is

$$Q_{dA_1-W} = dA_1 F_{dA_1-W} E_{b_1}\varepsilon_2 - dA_1 F_{dA_1-W} E_{b1}\varepsilon_2(1 - \varepsilon_2)F_{W-dA_1} - A_W F_{W-dA_1}\varepsilon_2 E_{bW}$$

$$= dA_1 F_{dA_1-W} \varepsilon_2 E_{b1}[1 - (1 - \varepsilon_2)F_{W-dA_1}] - dA_1 F_{dA_1-W} \varepsilon_2 E_{bW}$$

$$= dA_1 F_{dA_1-W} \varepsilon_2 [E_{b1}\{1 - (1 - \varepsilon_2)F_{W-dA_1}\} - E_{bW}] \qquad \text{(i)}$$

Given: $dA_1 = 0.15 \text{ m} \times 0.12 \text{ m} = 0.018 \text{ m}^2$; $A_W = 6 \text{ m} \times 5 \text{ m} = 30 \text{ m}^2$; $\varepsilon_2 = 0.8$; $F_{dA_1-W} = 0.454$; $E_{b1} = \sigma T_1^4 = \sigma(1400 + 273)^4$; $E_{bW} = \sigma(35 + 273)^4$

$$F_{W-dA_1} = \frac{dA_1 F_{dA_1-W}}{A_W} = \frac{(0.018)(0.454)}{30} = 2.72 \times 10^{-4}$$

Therefore,

$$Q_{dA_1-W} = (0.018)(0.454)(0.8)(5.669 \times 10^{-8})[(1673)^4\{1 - (1 - 0.8)(2.72 \times 10^{-4})\} - (308)^4]$$

$$= \boxed{2900 \text{ W}}$$

Note: Since the area ratio dA_1/A_W is small, F_{W-dA_1} is also small; i.e.

$$(1 - \varepsilon_2)\, F_{W-dA_1} \ll 1.$$

Using this result in Eq. (i), we get the following simple equation for the rate of radiation exchange:

$$Q_{dA_1-W} = dA_1\, F_{dA_1-W}\, \varepsilon_2 (E_{b_1} - E_{bW})$$

Example 7.16 The base of a rectangular enclosure (3 m × 2 m; height = 3 m) is maintained at 400°C (Fig. 7.30). The top surface is held at 250°C. (a) If the side walls are perfectly insulated and the surfaces are diffuse gray with an emissivity 0.7, calculate the required net rate of heat supply to the base. (b) If the skin temperature of the outside of the top wall is 60°C and heat loss

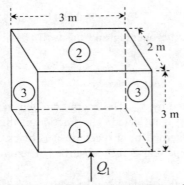

Fig. 7.30 Rectangular enclosure (Example 7.16).

from this surface occurs to a big factory shade at 30°C, what is the convective heat transfer coefficient at this surface?

SOLUTION Let the walls be named as: surface-1, the base (at 400°C); surface-2, the top (at 250°C); surface-3, all the insulated side walls (taken together).

(a) Surfaces 1 and 2 constitute a pair of parallel, directly opposed rectangles. The view factor F_{12} can be calculated from Fig. 7.12.

Here $W = 2$ m, $L = 3$ m, and $D = 3$ m; $R_2 = W/D = 2/3 = 0.666$, $R_1 = L/D = 3/3 = 1$; from Fig. 7.12, $F_{12} = 0.148$.

The net rate of radiant heat loss from the base can be calculated using Eq. (7.59) because surface-3 of the enclosure is insulated or *re-radiating*.

Given: $A_1 = 3$ m × 2 m = 6 m^2 = A_2; $\varepsilon_1 = 0.7 = \varepsilon_2$; $T_1 = 400°C = 673$ K, $T_2 = 523$ K

Thus,

$$Q_{1,\ net} = \cfrac{A_1 \sigma\, (T_1^4 - T_2^4)}{\dfrac{A_1 + A_2 - 2 A_1 F_{12}}{A_2 - A_1 (F_{12})^2} + \left(\dfrac{1}{\varepsilon_1} - 1\right) + \dfrac{A_1}{A_2}\left(\dfrac{1}{\varepsilon_2} - 1\right)}$$

$$= \frac{(6)(5.669 \times 10^{-8})(673^4 - 523^4)}{\dfrac{6 + 6 - (2)(6)(0.148)}{6 - (6)(0.148)^2} + \left(\dfrac{1}{0.7} - 1\right) + \dfrac{6}{6}\left(\dfrac{1}{0.7} - 1\right)} = \boxed{17.2 \text{ kW}}$$

Therefore, 17.2 kW of heat is required to be supplied to the base to maintain the base temperature at 400°C.

(b) At steady state, 17.2 kW heat must also leave the outer surface of the top wall by radiation and convection combined. The factory shed being big, it can be considered to be black. The rate of radiant heat loss from the top surface is

$$Q_{2, \text{ rad}} = A_2 \varepsilon_2 F_{24} \sigma (T_{2, o}^4 - T_4^4)$$

where the subscript '4' denotes the shed. Because all the radiation leaving surface-2 is intercepted by the shed, $F_{24} = 1$.

Given: $T_{2, o}$ = outer surface temperature of surface-2 = 60°C = 333 K; T_4 = ambient temperature = 30°C = 303 K

Thus,

$$Q_{2, \text{ rad}} = (6)(0.7)(1)(5.669 \times 10^{-8})(333^4 - 303^4) = 921 \text{ W}$$

Rate of convective heat loss (difference between the rate of heat input to the base and the rate of heat loss by radiation from the outside of the top wall)

$$= 17{,}200 \text{ W} - 921 \text{ W} = 16{,}279 \text{ W}$$

If the convective heat transfer coefficient is h, we have

$$A_2 h(T_{2, o} - T_4) = 16{,}279$$

or

$$(6)(h)(333 - 303) = 16{,}279$$

Therefore,

$$h = \boxed{90 \text{ W/m}^2 \text{ °C}}$$

Example 7.17 Two parallel directly opposed (i.e. coaxial) annular disks are maintained at temperatures 1000 K and 300 K, respectively. The disks have the following dimensions and emissivities.

Disk-1 (1000 K): inner radius = 10 cm, outer radius = 20 cm, emissivity = 0.8
Disk-2 (300 K): inner radius = 12 cm, outer radius = 25 cm, emissivity = 0.7

If the distance between the disks is 8 cm, what is the net rate of exchange of radiation between the disks? Assume that the disks are placed in an otherwise 'radiation-free' environment (that is, in an environment at absolute zero temperature).

SOLUTION First we determine the view factor, F_{12} (or F_{21}) which is required for the calculation of the radiation exchange rate between the disks shown in Fig. 7.31.

The areas of the disks are denoted by A_1 and A_2, while those of the central open circular regions are denoted by A_4 and A_3 as shown in the figure. Then, we have

$$F_{12} = F_{1(2, 3)} - F_{13} \qquad \text{(using the view factor algebra)}$$

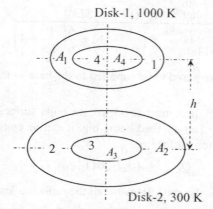

Disk-1, 1000 K

Disk-2, 300 K

Fig. 7.31   Two parallel directly opposed annular discs (Example 7.17).

$$= \frac{A_{2,3}}{A_1} F_{(2,3)1} - \frac{A_3}{A_1} F_{31} \qquad \text{(using the reciprocity relation)}$$

or

$$F_{12} = \frac{A_{2,3}}{A_1} \Big[F_{(2,3)(1,4)} - F_{(2,3)4} \Big] - \frac{A_3}{A_1} \Big[F_{3(1,4)} - F_{34} \Big] \qquad \text{(i)}$$

where $A_{2,3} = A_2 + A_3$, and $F_{(2,3)(1,4)}$ is the fraction of the radiation that is emitted by areas A_2 and A_3 and is intercepted by areas A_1 and A_4.

The view factors in Eq. (i) can be calculated by the procedure given in Example 7.9 for parallel coaxial disks or by using Fig. 7.14.

The notations used are: $r_{1,i}$ and $r_{1,o}$ are the inner and outer radii of disk-1;
$r_{2,i}$ and $r_{2,o}$ are the inner and outer radii of disk-2;
h is the distance between the disks.

Calculation of $F_{(2,3)(1,4)}$

Given: $r_{1,i} = 0.1$ m, $r_{1,o} = 0.2$ m; $r_{2,i} = 0.12$ m, $r_{2,o} = 0.25$ m; $h = 8$ cm $= 0.08$ m; we get

$$R_2 = \frac{r_{2,o}}{h} = \frac{0.25}{0.08} = 3.125; \qquad R_1 = \frac{r_{1,o}}{h} = \frac{0.2}{0.08} = 2.5;$$

$$X = 1 + \frac{1 + R_1^2}{R_2^2} = 1 + \frac{1 + (2.5)^2}{(3.125)^2} = 1.7424$$

Thus,

$$F_{(2,3)(1,4)} = \frac{1}{2}\Big[X - \sqrt{X^2 - 4(R_1/R_2)^2} \Big]$$

$$= \frac{1}{2}\Big[1.7424 - \sqrt{(1.7424)^2 - 4(2.5/3.125)^2} \Big] = 0.526$$

Calculation of $F_{(2,3)4}$

$$R_2 = \frac{r_{2,o}}{h} = \frac{0.25}{0.08} = 3.125; \quad R_1 = \frac{r_{1,i}}{h} = \frac{0.1}{0.08} = 1.25; \quad X = 1 + \frac{1 + R_1^2}{R_2^2} = 1.2624$$

Thus,

$$F_{(2,3)4} = \frac{1}{2}\left[X - \sqrt{X^2 - 4(R_1/R_2)^2}\right] = 0.143$$

Similarly,

$$F_{3(1,4)} = 0.815; \quad \text{and} \quad F_{34} = 0.4$$

The relevant areas are:

$$A_{2,3} = \pi r_{2,o}^2 = \pi (0.25)^2 = 0.1963 \text{ m}^2$$

$$A_3 = \pi r_{2,i}^2 = \pi(0.12)^2 = 0.04524 \text{ m}^2$$

$$A_1 = \pi(\pi r_{1,o}^2 - r_{1,i}^2) = \pi[(0.2)^2 - (0.1)^2] = 0.09425 \text{ m}^2$$

Using Eq. (i), we have

$$F_{12} = \frac{0.1963}{0.09425}(0.526 - 0.143) - \frac{0.04524}{0.09425}(0.815 - 0.4) = 0.5985$$

Calculation of the rate of radiative heat exchange

Under the given circumstances, net transfer of heat occurs from disk-1 to disk-2, but both the disks radiate to the environment too. It is a 'three-surface enclosure' problem. The third surface (which is the surroundings of the disks), we call it surface s, is at a temperature $T_s = 0$. Also, this 'surface' can be considered black, i.e. $J_s = E_{bs} = 0$.

Given: $T_1 = 1000$ K; $T_2 = 300$ K; $\sigma = 5.669 \times 10^{-8}$ W/m^2 K^4; $\varepsilon_1 = 0.8$; $\varepsilon_2 = 0.7$; $F_{12} = 0.5985$; $A_2 = \pi(r_{2,o}^2 - r_{2,i}^2) = 0.1511$ m^2; $F_{1s} = 1 - F_{11} - F_{12} = 1 - 0.5985 = 0.4015$; $F_{2s} = 1 - F_{21} - F_{22} = 1 - (A_1F_{12}/A_2) = 0.6267$

Now we calculate the following quantities in order.

$$\frac{1 - \varepsilon_1}{\varepsilon_1 A_1} = \frac{1 - 0.8}{(0.8)(0.09425)} = 2.652 \qquad \frac{1}{A_1 F_{12}} = \frac{1}{(0.09425)(0.5985)} = 17.73$$

$$\frac{1 - \varepsilon_2}{\varepsilon_2 A_2} = \frac{1 - 0.7}{(0.7)(0.1511)} = 2.836 \qquad \frac{1}{A_1 F_{1s}} = \frac{1}{(0.09425)(0.4015)} = 26.42$$

$$\frac{1}{A_2 F_{2s}} = \frac{1}{(0.1511)(0.6267)} = 10.56 \qquad E_{b1} = \sigma T_1^4 = 5.669 \times 10^{-8} (1000)^4 = 56,690$$

$$E_{b2} = \sigma T_2^4 = 5.669 \times 10^{-8}(300)^4 = 459.$$

We also have $A_1 F_{12} = A_2 F_{21}$.

Let us write Eq. (7.42a) for $i = 1$ and $2, j = 1, 2, 3$ and put the values of the known quantities (note that $J_3 = J_s = 0$). Thus,

$$\frac{E_{b1} - J_1}{(1 - \varepsilon_1)/\varepsilon_1 A_1} = \frac{J_1 - J_2}{1/A_1 F_{12}} + \frac{J_1}{1/A_1 F_{1s}}$$

or

$$\frac{56{,}690 - J_1}{2.652} = \frac{J_1 - J_2}{17.73} + \frac{J_1}{26.42}$$

and

$$\frac{E_{b2} - J_2}{(1 - \varepsilon_2)/\varepsilon_2 A_2} = \frac{J_2 - J_1}{1/A_2 F_2} + \frac{J_2}{1/A_2 F_{2s}}$$

or

$$\frac{459 - J_2}{2.836} = \frac{J_2 - J_1}{17.73} + \frac{J_2}{10.56}$$

Solving the above two equations, we get

$$J_1 = 46{,}023 \ \text{W/m}^2 \quad \text{and} \quad J_2 = 5484 \ \text{W/m}^2$$

The net rate of radiation exchange between disks 1 and 2 is [see Eq. (7.42b)]

$$Q_{12,\,\text{net}} = \frac{J_1 - J_2}{1/A_1 F_{12}} = \frac{46{,}023 - 5484}{17.73} = \boxed{2286 \ \text{W/m}^2}$$

7.8 RADIATION SHIELD

A radiation shield is a barrier wall of *low emissivity* placed between two walls in order to reduce the exchange of radiation between them. Radiation shields do not add or remove any energy from the system. They essentially put additional 'resistances' to transfer of radiant energy. Two adjacent walls, '1' and '2', without a radiation shield are shown in Fig. 7.32(a); a radiation shield placed between them is shown in Fig. 7.32(b).

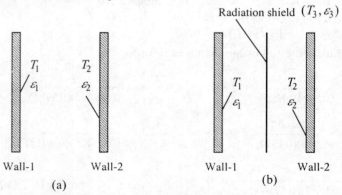

Fig. 7.32 Two parallel walls: (a) without any radiation shield and (b) with a radiation shield.

Let us determine the benefit in terms of reduction in the rate of radiation exchange, that may be accrued by using a radiation shield. We assume that the two walls are 'large' and have temperatures T_1 and T_2 and emissivities ε_1 and ε_2, respectively. The temperature and emissivity of the shield are T_3 and ε_3. If $T_1 > T_2$, it is expected that there will be a net loss of radiation by wall-1 and a net gain by wall-2. The radiation shield will have an intermediate temperature at *steady state*. As we assumed in the previous treatment, heat transfer in the system by conduction or convection is not taken into account. In the absence of the shield, the net rate of radiant energy loss per unit area of wall-1 is [see Eq. (7.47)]

$$\left[\frac{Q_{12}}{A}\right]_{net} = \frac{\sigma(T_1^4 - T_2^4)}{1/\varepsilon_1 + 1/\varepsilon_2 - 1} \tag{7.60}$$

In the presence of the radiation shield, at steady state, the net rate of transport of radiant energy from wall-1 to the shield (we call it wall-3), will be equal to that from the shield surface to wall-2. We may write these rates per unit area as follows:

$$\left[\frac{Q_{13}}{A}\right]_{net} = \left[\frac{Q_{32}}{A}\right]_{net} = \frac{\sigma(T_1^4 - T_3^4)}{1/\varepsilon_1 + 1/\varepsilon_3 - 1} = \frac{\sigma(T_3^4 - T_2^4)}{1/\varepsilon_3 + 1/\varepsilon_2 - 1} \tag{7.61}$$

Equation (7.61) may be solved for T_3 and the net rate of radiant energy loss from wall-1 may be determined. Then the percentage reduction in radiant exchange in the presence of the shield may be calculated.

As a special case if we assume that the emissivities of the walls and the shield are the same (i.e. if $\varepsilon_1 = \varepsilon_2 = \varepsilon_3 = \varepsilon$), the calculation becomes very simple. Putting this in Eq. (7.61), we get

$$\frac{\sigma(T_1^4 - T_3^4)}{1/\varepsilon + 1/\varepsilon - 1} = \frac{\sigma(T_3^4 - T_2^4)}{1/\varepsilon + 1/\varepsilon - 1}$$

or

$$T_3^4 = \frac{T_1^4 + T_2^4}{2} \tag{7.62}$$

Again, from Eq. (7.61), we have

$$\left[\frac{Q_{13}}{A}\right]_{net} = \frac{\sigma(T_1^4 - T_3^4)}{1/\varepsilon + 1/\varepsilon - 1} = \frac{\sigma[T_1^4 - (T_1^4 + T_2^4)/2]}{2/\varepsilon - 1} = \frac{\sigma(T_1^4 - T_2^4)/2}{2/\varepsilon - 1} \tag{7.63}$$

Dividing Eq. (7.63) by Eq. (7.60) [after putting $\varepsilon_1 = \varepsilon_2 = \varepsilon$ in the latter), we get

$$\frac{[Q_{13}]_{net}}{[Q_{12}]_{net}} = \frac{1}{2} \tag{7.64}$$

Therefore, there will be 50% reduction in the loss of radiant energy by putting a radiation shield between the walls if the emissivities of the walls and the shield are all equal. The result does not depend upon the exact position of the shield between the walls.

The emissivities of the two surfaces of a radiation shield may be equal or different depending upon the surface characteristics. Multiple radiation shields further reduce the radiant energy exchange. Radiation shields are extensively used in the cryogenic industry.

Example 7.18 A 2.54 cm o.d. tube is used to transport a cryogenic liquid at –196°C from a plant to an adjacent unit. The tube is enclosed in an evacuated concentric pipe of 52.5 mm i.d. having a wall temperature of –3°C. A thin-walled radiation shield is placed midway in the annular region between the tube and the pipe. Calculate the rate of heat gain by the liquid per metre length of the tube. The following emissivity data are available: tube wall = 0.05; pipe wall = 0.1; inner surface of the radiation shield = 0.02; outer surface of the radiation shield = 0.03.

SOLUTION We use the following subscripts for emissivity (ε), diameter (d), and temperature (T): i, inner tube; o, outer pipe; s, radiation shield; si, inner surface of the shield; so, outer surface of the shield. [For example, ε_{si} = emissivity of the inner surface of the shield.]

Using Eq. (7.45) for a two-surface gray enclosure, the net rate of radiant heat transfer from the outer pipe to the shield is

$$Q_{os} = \frac{A_o\sigma(T_o^4 - T_s^4)}{\dfrac{1 - \varepsilon_o}{\varepsilon_o} + \dfrac{1}{F_{os}} + \dfrac{A_o}{A_s}\left(\dfrac{1 - \varepsilon_{so}}{\varepsilon_{so}}\right)} \tag{i}$$

Similarly, the net rate of radiant heat transfer from the shield to the inner tube is

$$Q_{si} = \frac{A_s\sigma(T_s^4 - T_i^4)}{\dfrac{1 - \varepsilon_{si}}{\varepsilon_{si}} + \dfrac{1}{F_{si}} + \dfrac{A_s}{A_i}\left(\dfrac{1 - \varepsilon_i}{\varepsilon_i}\right)} \tag{ii}$$

Given: d_o = 0.0525 m; d_i = 0.0254 m; $d_s = (d_o + d_i)/2$ = 0.039 m; $l = 1$ m

Considering 1 m length (= l) of the pipe, we have

$$A_o = \pi d_o l = (\pi)(0.0525) = 0.165 \text{ m}^2$$

$$A_s = (\pi)(0.039)(1) = 0.1225 \text{ m}^2$$

$$A_i = \pi(0.0254)(1) = 0.0798 \text{ m}^2$$

$$\varepsilon_o = 0.1; \ \varepsilon_{so} = 0.03; \ \varepsilon_{si} = 0.02; \ \varepsilon_i = 0.05$$

$$T_o = -3 + 273 = 270 \text{ K}; \ T_i = -196 + 273 = 77 \text{ K}$$

View factors: For the long cylindrical enclosure made up of the outer pipe and the shield,

$$F_{so} = 1 \text{ (because the outer surface of the shield cannot 'see' itself.)}$$

$$A_o F_{os} = A_s F_{so} = A_s$$

$$F_{os} = A_s/A_o = d_s/d_o = 0.039/0.0525 = 0.743$$

Similarly,

$$F_{si} = d_i/d_s = 0.0254/0.039 = 0.651$$

Now,

$$\frac{1 - \varepsilon_o}{\varepsilon_o} + \frac{1}{F_{os}} + \frac{A_o}{A_s}\left(\frac{1 - \varepsilon_{so}}{\varepsilon_{so}}\right) = \frac{1 - 0.1}{0.1} + \frac{1}{0.743} + \frac{1 - 0.03}{(0.743)(0.03)} = 53.87$$

and

$$\frac{1 - \varepsilon_{si}}{\varepsilon_{si}} + \frac{1}{F_{si}} + \frac{A_i}{A_s}\left(\frac{1 - \varepsilon_i}{\varepsilon_i}\right) = \frac{1 - 0.02}{0.02} + \frac{1}{0.651} + \frac{1 - 0.05}{(0.651)(0.05)} = 79.72$$

Putting the values of the above groups in Eqs. (i) and (ii) and equating Q_{os} and Q_{si} at steady state, we have

$$\frac{(0.165)(5.669 \times 10^{-8})(270^4 - T_s^4)}{53.87} = \frac{(0.1225)(5.669 \times 10^{-8})(T_s^4 - 77^4)}{79.72}$$

Solving the above equation, we get

$$T_s = 244 \text{ K}$$

The net rate of heat gain per metre length of tube is

$$Q_{os} = \frac{(0.165)(5.669 \times 10^{-8})(270^4 - 244^4)}{53.87} = \boxed{0.307 \text{ W}}$$

7.9 RADIATION COMBINED WITH CONDUCTION AND CONVECTION

Industrial heat transfer by radiation is almost invariably accompanied by conduction and convection. There are numerous examples. Heat transfer occurs within a furnace mainly by radiation, but a substantial part of the heat transfer from the flowing combustion gases to the furnace wall or to the tubes or pipes in the furnace occurs by convection. Heat loss from a furnace wall occurs by combined conduction, convection, and radiation. Heat flows from inside the furnace to the outer surface by conduction. This heat is lost from the outer surface to the surroundings by convection and radiation. Similarly, heat loss from an insulated steam pipe also occurs by the above mechanism. This is a kind of series-parallel process schematically shown in Fig. 7.33. If l is the thickness of a plane wall of thermal conductivity k and surface temperatures

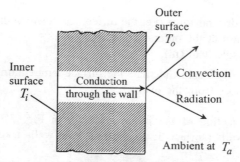

Fig. 7.33 Schematic of heat loss through a furnace wall.

T_i and T_o, we have

$$\frac{k}{l}(T_i - T_o) = h(T_o - T_a) + \varepsilon\sigma(T_o^4 - T_a^4) \qquad (7.65)$$

where h is the heat transfer coefficient at the outer surface due to natural and forced convection combined together, ε is the emissivity of the outer surface, and T_a is the ambient temperature. The use of the above equation is shown in Examples 7.19 and 7.21.

An important application of combined radiation and convection occurs in high temperature measurements. A thermometer bulb or a thermocouple tip placed in a flowing hot gas receives heat by convection from the gas and loses heat by radiation to the wall of the duct through which the gas is flowing. These two heat transfer rates are equal at steady state. Thus,

$$hA(T_g - T) = A\varepsilon\sigma(T^4 - T_w^4) \tag{7.66}$$

where A is the area of the bulb (or tip), ε is its emissivity, and T is the temperature. The temperatures of the bulk gas and of the wall of the duct are T_g and T_w, respectively. The temperature indicated by the instrument is different from the true gas temperature, T_G. The correction to the observed temperature can be obtained from the above equation.

A *radiation heat transfer coefficient* h_r is sometimes defined as the ratio of radiation heat flux and the temperature difference. Thus,

$$h_r = \frac{\left[Q_{12}/A\right]_{\text{net}}}{T_2 - T_1} \tag{7.67}$$

It may be noted that the radiation heat transfer coefficient depends strongly on the temperatures T_1 and T_2.

Example 7.19 A rectangular oven of 1.5 m × 1 m cross-section and 1 m height (all inside dimensions) is insulated with 7.5 cm thick fibre glass blanket (k = 0.08 W/m °C). The oven has outer walls made of carbon steel sheets which do not offer any appreciable heat transfer resistance. The bottom surface is maintained at 300°C. Heat loss occurs from the outer surfaces (the top and the side walls) by combined radiation and natural convection to an ambient at 30°C. Natural convection heat transfer coefficients can be estimated by using the following correlations:

$$h_t = 1.43(\Delta T)^{1/3} \text{ W/m}^2 \text{ °C} \qquad \text{(for the top surface)}$$

and

$$h_s = 0.601(\Delta T/l)^{1/5} \text{ W/m}^2 \text{ °C} \qquad \text{(for the side walls)}$$

where

ΔT is the temperature driving force (i.e. the difference between the skin temperature and the ambient temperature) at the outer surface (°C)

l is the height of a side wall (m).

All the surfaces are essentially black. Calculate the surface temperatures (both inside and outside) and the total rate of heat loss from the oven.

SOLUTION A vertical section of the oven is shown in Fig. 7.34. The inside top surface (surface-2) and the side walls (surface-3; all the four side walls taken together are collectively

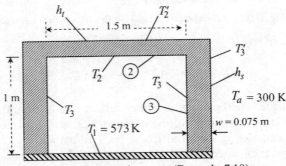

Fig. 7.34 Rectangular oven (Example 7.19).

called surface-3) receive radiant heat from the bottom (surface-1). Surfaces 2 and 3 also exchange heat with each other.

Notations used

Surface temperature — bottom = T_1, top (inside) = T_2, top (outside) = T_2', side walls (inside) = T_3, side walls(outside) = T_3', ambient temperature = T_a, insulation thickness = w.
Areas—surface-2 (inside) = A_2, surface-2 (outside) = A'_2, surface-2 (mean) = $A_{2,m} = (A_2 + A'_2)/2$.
Similar notations are used for surface-3 as well
The following heat balance equations can be written for the top surface.
The net rate of radiant heat inputs to surface-2:

$$\text{From the bottom } = A_2 F_{21}\, \sigma\left(T_1^4 - T_2^4\right) \tag{i}$$

$$\text{From the side walls } = A_2 F_{23}\, \sigma\left(T_3^4 - T_2^4\right) \tag{ii}$$

$$\text{Rate of heat flow through the top wall by conduction } = A_{2,m}(k/w)(T_2 - T_2') \tag{iii}$$

$$\text{Rate of heat loss from the top surface by convection } = A_2' h_t(T_2' - T_a) \tag{iv}$$

Rate of heat loss by radiation from the top surface
$$\text{to the ambient } = A_2'\sigma[(T_2')^4 - T_a^4] \tag{v}$$

In writing Eqs. (i) and (ii), the reciprocity relation $A_i F_{ij} = A_j F_{ji}$ has been used. Because the area of conduction changes from A_2 to A'_2 across the insulation thickness, the mean area $A_{2,m}$ has been used. The total rate of heat input to the top surface, i.e. [(i) + (ii)], is equal to the rate of heat loss, i.e. [(iv) + (v)], at steady state. Thus, we have

$$A_2 F_{21}\sigma(T_1^4 - T_2^4) + A_2 F_{23}\sigma(T_3^4 - T_2^4) = A_2' h_t(T_2' - T_a) + A_2'\sigma[(T_2')^4 - T_a^4] \tag{vi}$$

Also, the rate of heat conduction through the insulation as given by (iii) is equal to the rate of heat loss from the top surface by convection and radiation, i.e. [(iv) + (v)]. Hence, we have

$$(A_{2,m})(k/w)(T_2 - T_2') = A_2' h_t(T_2' - T_a) + A_2'\sigma[(T_2')^4 - T_a^4] \tag{vii}$$

Similar equations can be written for the side walls. Thus,

$$A_1 F_{13}\sigma(T_1^4 - T_3^4) - A_2 F_{23}\sigma(T_3^4 - T_2^4) = A_3' h_s(T_3' - T_a) + A_3'\sigma[(T_3')^4 - T_a^4] \tag{viii}$$

$$(A_{3,m})(k/w)(T_3 - T_3') = A_3' h_s(T_3' - T_a) + A_3'\sigma[(T_3')^4 - T_a^4] \tag{ix}$$

The values of the various quantities to be used in the above equations, (vi)–(ix), are either given in the problem statement or may be calculated as follows:

$T_1 = 300 + 273 = 573$ K, $T_a = 30 + 273 = 303$ K
$w = 0.075$ m, $k = 0.08$ W/m °C
$l = 1.075$ m, $\sigma = 5.669 \times 10^{-8}$ W/m^2 K^4
$A_1 = A_2 = 1$ m $\times$ 1.5 m = 1.5 m^2
$A_2' = 1.15$ m $\times$ 1.65 m = 1.9 m^2
$A_{2,m} = (1.5 + 1.9)/2 = 1.7$ m^2

A_3 = the total inside area of the four side walls = $2[(1)(1.5) + (1)(1)] = 5$ m^2
A_3' = total outside area of the four side walls = $2[(1.075)(1.15) + (1.075)(1.65)] = 6.02$ m^2
$A_{3, m}$ = mean area of the side walls = $(5 + 6.02)/2 = 5.51$ m^2
$h_t = 1.43(\Delta T)^{1/3} = 1.43(T_2' - T_a)^{1/3}$
$h_s = 0.601(\Delta T/l)^{1/5} = 0.601(T_3' - T_a)^{1/5}$

View factors: F_{21} = 0.252 [using Fig. 7.12 for two parallel directly opposed rectangles, area 1 m × 1.5 m placed at a distance of 1 m]

$$F_{23} = 1 - F_{21} - F_{22} = 1 - 0.252 - 0 = 0.748$$

$F_{13} = F_{23}$ (because of similar geometric configurations).
Substituting the above values in Eqs. (vi)–(ix), we get the following equations (in the same order)

$10^{-8}(573^4 - T_2^4) + (2.968 \times 10^{-8})(T_3^4 - T_2^4)$

$$= 1.2677 (T_2' - 303)^{4/3} + 5.0264 \times 10^{-8}\left[(T_2')^4 - (303)^4\right] \qquad \text{(x)}$$

$$T_2 - T_2' = 1.4984 (T_2' - 303)^{4/3} + 5.9411 \times 10^{-8}\left[(T_2')^4 - (303)^4\right] \qquad \text{(xi)}$$

$10^{-8}(573^4 - T_3^4) - 10^{-8}(T_3^4 - T_2^4)$

$$= 0.5687 (T_3' - 303)^{6/5} + 5.3654 \times 10^{-8}\left[(T_3')^4 - (303)^4\right] \qquad \text{(xii)}$$

$$T_3 - T_3' = 0.6156(T_3' - 303)^{6/5} + 5.8076 \times 10^{-8}\left[(T_3')^4 - (303)^4\right] \qquad \text{(xiii)}$$

The above set of four nonlinear equations (x)–(xiii) can be solved for the unknowns T_2, T_2', T_3 and T_3'. The solutions for the surface temperatures obtained by using *Mathcad Plus 6.0* are:

$$T_2 = 545.1 \text{ K}, \quad T_2' = 322.6 \text{ K}, \quad T_3 = 544.7 \text{ K}, \quad T_3' = 328.4 \text{ K}$$

The rate of heat loss from the top surface [Eq. (vii)] is

$$(A_{2, m})(k/w)(T_2 - T_2') = (1.7)(0.08/0.075)(545.1 - 322.6) = 403 \text{ W}$$

The rate of heat loss from the side walls [Eq. (ix)] is

$$(A_{3, m})(k/w)(T_3 - T_3') = (5.51)(0.08/0.075)(544.7 - 328.4) = 1271 \text{ W}$$

Total rate of heat loss = 403 + 1271 = $\boxed{1674 \text{ W}}$

Note: The calculation of heat transfer by combined conduction, convection, and radiation often involves trial-and-error or numerical computation. In the above solution, the result that T_2 and T_3 are nearly equal is not quite unexpected because in a small furnace the temperatures of the inner surface remain almost the same. However, the skin temperature of any of the side walls may not be uniform in a practical situation (what are the physical reasons?) and is likely to have an increasing temperature gradient upwards along the wall. But it is a common assumption to take a 'uniform' skin temperature (as it has been done in this solution). Natural convection heat transfer coefficient and heat flux from the top surface (facing upwards) happen to be larger than those for the vertical side walls. For this reason, the calculated outside temperature of the vertical walls is more than that of the top.

7.10 ABSORPTION AND EMISSION IN A GASEOUS MEDIUM

Our discussion has so far been confined to emission, absorption, and exchange of radiation at or between surfaces with non-absorbing intervening media. Several elementary gases like N_2 and O_2 and a few other gases of symmetrical non-polar molecular structure are practically 'transparent' to radiation at low temperatures. If two surfaces are separated by any such non-absorbing gas, the equations of radiation exchange given before are applicable. However, many gases (for example, CO_2, H_2O, NH_3, hydrocarbons, etc.) absorb and emit thermal radiation. This phenomenon considerably affects radiative heat transfer.

7.10.1 The Absorptivity and Emissivity of a Gas

The capability of a gas to absorb radiation strongly depends upon the wavelength. Figure 7.35 shows a layer of a gas of thickness L with an incident monochromatic radiation of wavelength λ

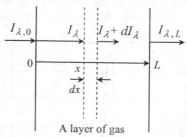

A layer of gas

Fig 7.35 Radiation transmission through an absorbing gas layer.

and intensity $I_{\lambda, 0}$ at one end. A part of the incident radiation is absorbed and the rest leaves from the other end of the layer of gas at an intensity, $I_{\lambda, L}$. Let us assume that the change in the radiation intensity is dI_λ as it passes through a thin layer of gas of thickness dx and located at a distance x from the impinging end. If the change in intensity is proportional to the local intensity I_λ and the thickness of the layer dx, then we can write

$$- dI_\lambda = a_\lambda I_\lambda dx \tag{7.68}$$

Integrating from $x = 0$ to $x = L$, we get

$$- \int_{I_{\lambda, 0}}^{I_{\lambda, L}} \frac{dI_\lambda}{I_\lambda} = a_\lambda \int_0^L dx$$

or

$$\frac{I_{\lambda, L}}{I_{\lambda, 0}} = \exp(-a_\lambda L) \tag{7.69}$$

where a_λ is a proportionality constant called the *monochromatic extinction coefficient*.

Equation (7.69) is known as *Beer's law*. The LHS of the equation represents the fraction of the incident radiation that is transmitted through a layer of the gas of thickness L. This fraction is the transmissivity τ_λ of the gas layer. Thus,

$$\tau_\lambda = \frac{I_{\lambda, L}}{I_{\lambda, 0}} = \exp\left(-a_\lambda L\right) \tag{7.70}$$

Let us recall Eq. (7.3). Because reflectivity ρ of a gas is negligible, the monochromatic absorptivity of the gas α_λ can be obtained as

$$\alpha_\lambda = 1 - \tau_\lambda = 1 - \exp\left(-a_\lambda L\right) \tag{7.71}$$

Further, if Kirchhoff's law holds, absorptivity is equal to the emissivity of the gas layer. Therefore,

$$\varepsilon_\lambda = \alpha_\lambda = 1 - \exp\left(-a_\lambda L\right) \tag{7.72}$$

where ε_λ is the spectral emissivity of the gas layer.

Experimental evidences on absorption of radiation by gases indicate that gases absorb radiation over narrow bands of the spectrum. Monochromatic absorptivity of carbon dioxide is shown in Fig. 7.36. The pressure (or the partial pressure) of the gas, temperature and the path length traversed by radiation determine the absorptivity (or emissivity). This last factor

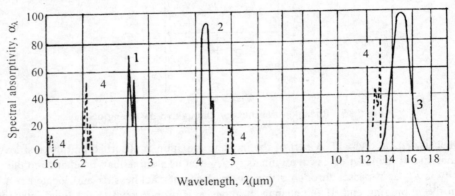

Fig. 7.36 Absorption bands of carbon dioxide [thickness of the gas layer: (1) 5 cm; (2) 3 cm; (3) 6.3 cm; (4) 100 cm].

considerably complicates the radiation calculation. If the surfaces which enclose the gas are diffuse emitters, the radiation incident on a surface moves at all possible angles through the gas. Therefore, the path length varies over a range. If the surfaces are parallel planes and the beam is normal to the surface, the maximum path length is the distance between them. An oblique radiant beam which traverses a longer distance gets stripped of more of its energy by absorption in the gas.

A practical strategy of solution of the problem has been developed by defining a *mean beam length*, L_e. The mean beam length depends upon the geometry of the enclosure and is given in Table 7.4 for a few cases. For an enclosure of arbitrary geometry, the approximate mean beam length can be obtained from the following equation

$$L_e = 3.6(V/A) \tag{7.73}$$

where V is the volume and A the area of the enclosure.

Table 7.4 Mean beam lengths for gas radiation

Shape	Characteristic dimension	Mean beam length, L_e
Sphere	Diameter, d	$0.63d$
Long cylinder	Diameter, d	$0.94d$
Cylinder with height = diameter		
(a) radiating to centre of base	Diameter, d	$0.71d$
(b) radiating to entire surface	Diameter, d	$0.65d$
Cube, radiation to all surfaces	Length of a side, L	$0.66L$
Infinite parallel planes	Separation distance, S	$1.80S$

Emissivity data for different gases have been correlated with pressure (p), temperature (T), and the path length (L), and are available as charts. The charts for CO_2 and H_2O vapours at 1 atm pressure are shown in Figs. 7.37 and 7.38 in which pL is the parameter. For any other pressure, the emissivity can be obtained by multiplying by a correction factor (see Spiegel and Howell, 1981). If the gaseous medium contains both CO_2 and H_2O, a further correction, $\Delta\varepsilon$, given in Fig. 7.39 is used to calculate the emissivity of the mixture (see Example 7.21).

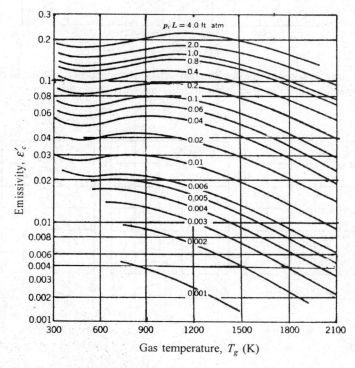

Fig. 7.37 Emissivity data for CO_2 for different values of the optical thickness, $p_c L$ (total pressure = 1 atm); the suffix c means CO_2.

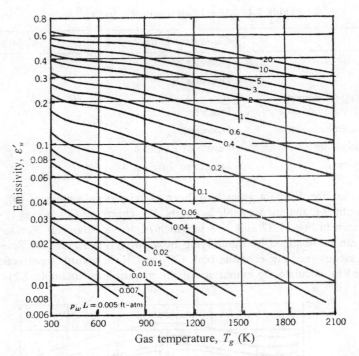

Fig. 7.38 Emissivity of water vapour for different values of the optical thickness, $p_w L$ (partial pressure of water vapour up to 1 atm); the suffix w means H_2O vapour.

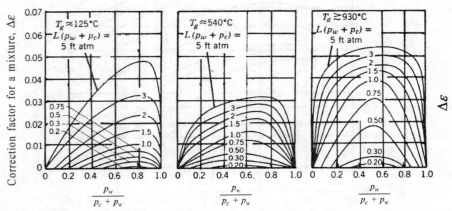

Fig. 7.39 Correction factor to be used when both CO_2 and water vapour are present.

Example 7.20 Carbon dioxide gas taken in a test cell at 300 K and 1 atm pressure is found to absorb 91% of incident monochromatic radiation at $\lambda = 4.2$ μm. The path length is 10 cm. Calculate the spectral extinction coefficient a_λ.

SOLUTION From Eq. (7.69), we have

$$I_{\lambda, L} = I_{\lambda, 0} \exp (-a_\lambda L)$$

Fraction of the incident radiation transmitted = $1 - 0.91 = 0.09$

The path length, $L = 10$ cm $= 0.1$ m.

Therefore,

$$\frac{I_{\lambda, L}}{I_{\lambda, 0}} = 0.09 = \exp [- a_\lambda (0.1)]$$

or

$$a_\lambda = \boxed{23.97 \ \text{m}^{-1}}$$

7.10.2 Radiation Exchange between a Non-luminous Gas and Black Surface of Enclosing Walls

Let us consider a gas at a temperature T_g enclosed by a surface at a temperature T_s. We assume that the surface is a blackbody. If A is the surface area of the enclosure, the rate at which the radiation emitted by the gas hits the surface is given by

$$Q_g = A \varepsilon_g(T_g) \sigma T_g^4 \tag{7.74}$$

where $\varepsilon_g(T_g)$ is the gas emissivity at temperature T_g and the existing pressure for the given enclosure geometry. The surface being 'black', the amount Q_g is totally absorbed.

The 'black' surface also emits radiation. The rate at which this radiation is absorbed by the gas depends upon T_s, T_g and the absorptivity of the gas. If the gas contains both CO_2 and H_2O vapours, the following equations can be used to calculate the radiation absorption rate, Q_a.

$$Q_a = A \ \alpha_g(T_g, \ T_s) \ \sigma T_s^4 \tag{7.75}$$

$$\alpha_g(T_g, \ T_s) = \alpha_c + \alpha_w - \Delta \alpha \tag{7.76}$$

$$\alpha_c = C_c \ \varepsilon_c' \ (T_g/T_s)^{0.65} \tag{7.77}$$

$$\alpha_w = C_w \ \varepsilon_w' \ (T_g/T_s)^{0.45} \tag{7.78}$$

$$\Delta \alpha = \Delta \varepsilon \ \text{(evaluated at the surface temperature, } T_s; \text{ see Example 7.21)} \tag{7.79}$$

The values of ε_c' (for CO_2) and ε_w' (for H_2O vapour) are obtained from Figs. 7.37 and 7.38 at the temperature T_s, taking $p_c L_e T_s/T_g$ and $p_w L_e T_s/T_g$ as the pressure–beam length parameters for the two gases, respectively. The net rate of radiation exchange between the surface and the gas is obtained by subtracting Eq. (7.75) from Eq. (7.74). Thus, we have

$$Q_{\text{net}} = Q_g - Q_a = A \sigma \left[\varepsilon_g(T_g) T_g^4 - \alpha_g(T_g, T_s) T_s^4 \right] \tag{7.80}$$

Example 7.21 Hot flue gas containing 8% carbon dioxide, 15.2% moisture, and the rest nitrogen and oxygen leaves a natural gas burner at 1100°C and flows through a 40 cm × 40 cm refractory-lined square duct having a wall temperature of 800°C. Calculate the rate of heat transfer from the gas to the walls of the duct per metre length by combined radiation and forced convection. The convective heat transfer coefficient is 15 W/m² °C. The walls can be assumed to be black and the total pressure is essentially atmospheric.

SOLUTION The net rate of radiant heat transfer can be calculated using Eq. (7.80).

$$Q_{rad} = A\sigma\left[\varepsilon_g(T_g)T_g^4 - \alpha_g(T_g,\,T_s)T_s^4\right] \tag{i}$$

Calculation of $\varepsilon_g(T_g)$

It is assumed that N_2 and O_2 do not absorb or emit radiation. Because the gas contains both CO_2 and moisture, the correction factor $\Delta\varepsilon$ is to be used.

   Partial pressure of CO_2, $p_c = (0.08)(1\text{ atm}) = 0.08$ atm
   Partial pressure of moisture, $p_w = (0.152)(1\text{ atm}) = 0.152$ atm
   Considering a length l of the duct,
      volume $= (0.4)(0.4)l = 0.16l$ m^3
      area $= 1.6l$ m^2

Hence,

$$\text{Mean beam length, } L_e = 3.6(\text{volume/area}) = 3.6(0.16l/1.6l)$$

$$= 0.36 \text{ m [see Eq. (7.73)]}$$

$$p_c L_e = (0.08)(0.36) \text{ atm-m} = 0.0945 \text{ atm-ft and similarly, } p_w L_e = 0.179 \text{ atm-ft}$$

$$T_g = 1100°C = 1373 \text{ K; and } T_s = 800°C = 1073 \text{ K}$$

   For $T_g = 1373$ K and $p_c L_e = 0.0945$ atm-ft, we use Fig. 7.37 to get, $\varepsilon'_c = 0.06$
   Similarly, at $T_g = 1373$ and $p_w L_e = 0.179$, $\varepsilon'_g = 0.048$ (from Fig. 7.38).
A correction $\Delta\varepsilon$ for the mixture can be obtained as follows from Fig. 7.39.

$$\frac{p_w}{p_c + p_w} = \frac{0.152}{0.08 + 0.152} = 0.655; \quad p_c L_e + p_w L_e = 0.0945 + 0.179 = 0.273; \quad T_g = 1100°C$$

From Fig. 7.39, we, therefore, get

$$\Delta\varepsilon = 0.003$$

Therefore,

$$\varepsilon_g(T_g) = \varepsilon'_c + \varepsilon'_w - 0.003 = (1)(0.06) + (1)(0.048) - 0.003 = 0.105$$

Calculation of $\alpha_g(T_g, T_s)$

Using Eqs. (7.76)–(7.79), ε'_c and ε'_w are obtained from Figs. 7.37 and 7.38 using $p_c L_e T_s/T_g$ and $p_w L_e T_s/T_g$ as the pressure-length parameters, and T_s as the abscissa.

$$p_c L_e T_s/T_g = (0.0945)(1073/1373) = 0.0738$$

and

$$p_w L_e T_s/T_g = (0.179)(1073/1373) = 0.14$$

   From Fig. 7.37, $\varepsilon'_c = 0.068$
   From Fig. 7.38, $\varepsilon'_w = 0.069$
   The correction factors $C_c = C_w = 1$ at 1 atm total pressure and $\Delta\alpha = \Delta\varepsilon = 0.003$.

Thus,

$$\alpha_g(T_g, T_s) = \alpha_c + \alpha_w - \Delta\alpha = C_c\varepsilon'_c(T_g/T_s)^{0.65} + C_w\varepsilon'_w(T_g/T_s)^{0.45} - \Delta\varepsilon$$

$$= (1)(0.068)(1373/1073)^{0.65} + (1)(0.069)(1373/1073)^{0.45} - 0.003 = 0.154$$

The radiant heat transfer rate can now be calculated from Eq. (i).

The area of the duct per metre length, $A = 4(0.4)(1) = 1.6$ m^2
Therefore,

$$Q_{rad} = (1.6)(5.669 \times 10^{-8})[(0.105)(1373)^4 - (0.154)(1073)^4] = 15.3 \text{ kW}$$

The convective heat transfer rate

$$Q_{conv} = hA(T_g - T_s) = (15)(1.6)(1373 - 1073) = 7.2 \text{ kW}$$

The total rate of heat transfer from the gas to the walls of the duct

$$Q = Q_{rad} + Q_{conv} = 15.3 + 7.2 = \boxed{22.5 \text{ kW}}$$

7.11 GREENHOUSE EFFECT

The sun has a surface temperature of about 5780 K, the flux of solar radiation on the earth's atmosphere has an annual average intensity of 1372 W/m^2 based on the projected area of the earth (i.e. πR^2, where R = radius of the earth). This is called the *solar constant*, S. If the earth is considered to be a blackbody receiving radiant energy from the sun and radiating back to the space (which is at absolute zero for all practical purposes), the theoretical steady state temperature of the earth T is given by

$$(\sigma)(4\pi R^2 T^4) = (S)(\pi R^2) = (1372)(\pi R^2)$$

Putting the value of σ, we get

$$T = 279 \text{ K}$$

The average temperature of the earth of about 288 K or 15°C is remarkably close to the above estimated figure. The match is rather fortuitous because it is known that about 30% of the solar radiation hitting the outer atmosphere is reflected back (this fraction is called *albedo*). Taking this factor into account, the effective temperature of the earth T_e may be calculated as

$$(\sigma)(4\pi R^2)T_e^4 = 1372(1 - 0.3)(\pi R^2)$$

or

$$T_e = 255 \text{ K}$$

This estimated effective temperature is about 33 K lower than the actual value. This difference is attributed to absorption and re-radiation of the long wavelength or thermal radiation associated with the temperature of the earth.

Absorption of *long wavelength radiation* by gases like carbon dioxide, methane, nitrous oxide, water vapour, and CFCs (this last category of compounds is much more infamous in causing the peril of ozone depletion in the stratosphere) gives rise to the most serious climatic threat to the mankind. For example, carbon dioxide shows a strong absorption band around 15 μm wavelength of radiation as also in narrow bands around 4.3 μm and 2.7 μm (see Fig. 7.36). These gases are practically transparent to *short wavelength radiation* that the earth receives from the sun, but are not so to the thermal radiation leaving the earth. This gives rise to radiant energy trapping or *greenhouse effect* and *global warming*.

The greenhouse effect has been a controversial scientific issue for quite some time. However, climatologists now generally agree that the average global temperature has risen by about 0.3 to 0.6°C over the past 100 years, and that the rate of rise is increasing alarmingly. Since the normal climatic conditions maintain such a balance, even a small increase in the average global temperature may lead to a series of catastrophic consequences. A number of sophisticated models have been proposed to explain and predict the facts and consequences of global warming (for a brief description, see Hileman, 1995). The issue has been a focus of deliberations in many conferences (both scientific and political). Reduced consumption of energy and enhanced efficiency of energy utilization have been identified as the feasible strategies for the reduction of emission of greenhouse gases, particularly of carbon dioxide, thereby mitigating the adverse effects of global warming. The first Earth summit in Rio de Janeiro in 1992 and the recent Kyoto conference (Cooney, 1997) dealt with the various aspects of this issue.

SHORT QUESTIONS

1. What are the ranges of wavelength of electromagnetic waves covering UV, visible, infrared, and thermal radiation?

2. What is the transmissivity of an opaque solid?

3. Does a piece of wood reflect radiation? If yes, what type of radiation is predominant in this case?

4. A piece of metal reflects 30% of the incident solar radiation. What are its absorptivity and transmissivity?

5. What do you mean by a blackbody? Why is it so called? Why does a cavity with a small hole behave as a blackbody?

6. What type of radiation is predominantly emitted by the earth?

7. What are emissive power and spectral emissive power of a blackbody? What are their units?

8. How do you distinguish between spectral emissive power and spectral radiation intensity?

9. How do you define radiation intensity?

10. What is Wein's displacement law? At what wavelength does a body at 2000 K emit maximum radiation?

11. What is a gray body? Define radiosity and irradiation.

12. A diffuse gray body of emissivity 0.5 and at a temperature 0°C radiates to an ambient at 0 K. What is its radiant heat flux?

13. If i and j denote the inner and the outer surfaces of a spherical shell, what are the values of F_{ii} and F_{jj}?

14. Consider a 'long' cylindrical cavity (diameter << length). If i and j denote the two flat ends, calculate F_{ij}.

15. What benefit can be derived by using a radiation shield and a re-radiating surface?

16. Why does a thermocouple used for measuring the temperature of a hot gas indicate a lower temperature than the true value?

17. What would be the form of Eq. (7.36) as $\varepsilon \to 1$?

PROBLEMS

7.1 A gray body at 2500 K has a total emissive power of 2.00×10^6 W/m^2. What is its maximum monochromatic emissive power and at what wavelength does it occur?

7.2 A furnace door has a small opening of diameter 3 cm. If the furnace is at 1200°C and the outside air temperature is 30°C, calculate the hourly radiant heat loss through the opening.

7.3 Radiant energy with an intensity of 700 W/m^2 strikes a flat plate normally. The absorptivity is twice the transmissivity and 2.9 times its reflectivity. Determine the rate of absorption, transmission, and reflection of energy in W/m^2.

7.4 A blackbody is at a temperature of 1000°C. Calculate:

(a) the wavelength at which the body has the maximum monochromatic emissive power, and the corresponding emissive power;

(b) the total emissive power of the blackbody;

(c) the fraction of the total radiant energy emission between the wavelengths 2.0 and 4.5 μm;

(d) the percentage reduction in the total emissive power when the temperature of the body falls down to 900°C; and

(e) the hemispherical emissive power.

7.5 A blackbody has a total hemispherical emissive power of 100 W/m^2. Calculate:

(a) its surface temperature;

(b) the wavelength above which (i) 50%, (ii) 75% of the radiant energy emission occurs; and

(c) the wavelength at which the body has the maximum monochromatic emissive power.

7.6 Saturated steam at 4 kg/cm^2 gauge pressure flows through a 2-inch schedule 40 (o.d. = 60 mm) bare horizontal pipe in a room at 28°C. Neglecting the resistance of the pipe wall, calculate the rate of heat loss by combined convection and radiation from the pipe and also the rate of condensation of steam. The emissivity of the pipe surface is 0.79 and that of the wall of the room is 0.75.

Calculate the percentage reduction in heat loss, and the corresponding radiation heat transfer coefficient, if the pipe is coated with an aluminium paint having $\varepsilon = 0.35$.

7.7 The distribution of spectral emissivity of a surface at 1000 K is given by

$$\varepsilon_\lambda(\lambda) = 0.3 \text{ for } 0 < \lambda < 1.5 \text{ μm}$$
$$\varepsilon_\lambda(\lambda) = 0.5 \text{ for } 1.5 < \lambda < 4 \text{ μm}$$
$$\varepsilon_\lambda(\lambda) = 0.4 \text{ for } \lambda > 4 \text{ μm}$$

Calculate:

(a) the total hemispherical emissive power;

(b) the total hemispherical emissivity; and

(c) the wavelength at which the body has the maximum emissive power.

7.8 The spectral emissivity of an alloy at 1450°F is approximately given by

$$\varepsilon_\lambda(\lambda) = 0.059\lambda - 0.27 \quad \text{for } 1 \le \lambda \le 10 \ \mu m$$

Calculate the emissive power of the surface at 1450°F within the range of wavelength $1 \le \lambda \le 10 \ \mu m$.

7.9 The emissivity of a gray surface is found to depend linearly on temperature as $\varepsilon(t) = 0.0003263t - 0.1284$, in the range $1000 \le t \le 1500$, where t is the temperature in °F. If the temperature of the body drops from 1500°F to 1000°F in 15 minutes, calculate the average flux of radiant energy loss from the surface.

7.10 (i) The dA_1 is a small element on a plane normal to the plane surface A_2 as shown in Fig. 7.40(a). Starting with the view factor integral [Eq. (7.25)] determine $F_{dA_1-A_2}$.

(ii) Using the view factor algebra, calculate the $F_{A_1-A_2}$ values for the surface pairs shown in Figs. 7.40(b), (c) and (d).

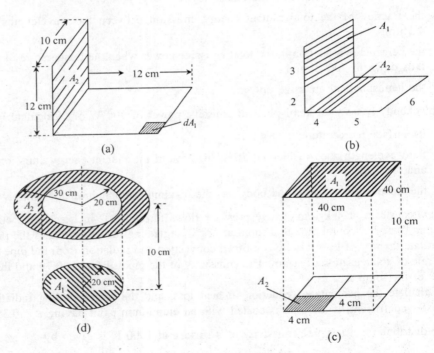

Fig. 7.40   Determination of view factors (Problem 7.10).

7.11 Heat radiates from the ceiling of a room to the walls and the floor at a rate of 250 kcal/h. The room is 3.2 m × 3.2 m square and 2.5 m high. Assume that the side walls and the floor are at the same mean temperature. If the emissivity of the ceiling (which has embedded heating coils in it) and that of the side walls is 0.6, and the emissivity of the floor is 0.55, determine the mean temperature of the walls and of the floor, if the ceiling temperature is 45°C.

7.12 A hot pipe (o.d. = 0.06 m and wall temperature = 150°C) runs centrally through a bigger pipe, 0.31 m i.d., held at 30°C. If the emissivities of the pipe surfaces are 0.6 and 0.75, respectively, calculate the net rate of heat loss per metre length of the hot pipe.

7.13 A hot plate (emissivity = 0.7, diameter = 20 cm) is used to keep the liquid in a cylindrical container warm. The diameter of the container is 25 cm and it is placed 3 cm directly above the hot plate. The system is kept in a large room at 25°C. If the hot plate temperature is 500°C, the temperature of the bottom of the container 80°C, and its emissivity 0.85, calculate the net rate of radiant heat gain by the container through its bottom surface. Draw the network diagram.

7.14 Two large parallel surfaces at temperatures 500°C and 200°C, and of emissivities 0.8 and 0.6 respectively, are separated by a polished metal plate of emissivity 0.1. Calculate the rate of heat exchange between the surfaces and also the temperature of the separating metal plate.

7.15 A gray body of emissivity 0.6 has a shape factor of 0.4 with respect to itself. The body has a mass of 2 kg, specific gravity 7.5, and specific heat 0.5 kJ/kg °C. If the body, initially at 200°C, is exposed to an environment at 30°C, calculate the time required for its temperature to drop down to 40°C. The convective heat transfer coefficient is 25 W/m² °C, and the heat transfer area is 300 cm².

7.16 Moulded plastic components in the form of cylinders, 6 cm diameter and 10 cm long, are cured by placing the material in an oven. The cylinders are at 30°C when these are fed to the oven. What should be the oven temperature so that the initial rate of temperature rise remains 1°C per minute? The following data are given : density of the cylinders = 1050 kg/m³, specific heat = 2 kJ/kg °C, emissivity = 0.8.

7.17 A 40 W incandescent electric bulb has a surface temperature of 100°C. The bulb is in a large room at 30°C. If the bulb is considered as a 6 cm diameter sphere having a surface emissivity of 0.8, calculate the rate of thermal energy loss by radiation and by natural convection from the bulb. If the filament temperature is 3000 K, calculate the average transmissivity of higher wavelength radiation through the glass. Glass may be assumed to be transparent to visible wavelength radiation.

7.18 A cylindrical furnace has a flat bottom and a hemispherical dome at the top. The bottom surface is maintained at 1000°C, and the top surface is at 700°C. If the cylindrical surface is perfectly insulated and the other surfaces are black, calculate (a) the required rate of heat input to the bottom surface for maintaining the temperature at the given value; (b) the required rate of heat input if the top and bottom surfaces were diffuse gray with an emissivity of 0.75. Also, calculate the corresponding radiosity and irradiation of the bottom surface. Draw the network diagram for the problem.

7.19 Consider the enclosure shown in Fig. 7.41. The top surface-1 is maintained at 500°C, and the bottom surface-2 at 200°C. Calculate the rates of heat supply or removal necessary to maintain the surfaces at the above temperatures if (i) the surfaces-1 and 2 are black and the curved surface-3 reradiating; (ii) the surfaces 1 and 2 have emissivities 0.8 and 0.7, respectively, and surface-3 is reradiating. If surface-3 is maintained at 400°C and has an emissivity of 0.65, what would be the rates of heat supply/removal at the three surfaces? What are the radiosities and irradiations at the surfaces? Draw the network diagram.

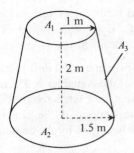

Fig. 7.41 An enclosure (Problem 7.19).

7.20 Two large parallel black surfaces are maintained at 600°C and 300°C, respectively. Two radiation shields are inserted between the surfaces having emissivities as shown in Fig. 7.42. Calculate the shield temperatures and the percentage reduction in radiant heat loss because of the shields. Also draw the network diagram.

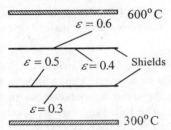

Fig. 7.42 Two large parallel black surfaces with radiation shields (Problem 7.20).

7.21 A furnace having a 6 m × 4 m rectangular base and a semi-cylindrical roof (Fig. 7.43) is used for heat treatment of steel sheets. Half of the base area is used for burning coal while the sheets are moved through the other half. If the coal bed temperature is 1200°C and the sheets are at 500°C, what is the maximum possible rate of radiant heat transfer to the sheets? Assume that the roof and the side walls of the furnace act as re-radiating surfaces.

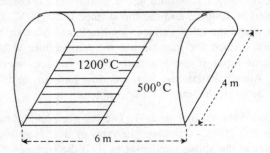

Fig. 7.43 A furnace (Problem 7.21).

7.22 A rectangular furnace of 3 m × 2 m × 1.5 m inside dimension has combustion gases (12% CO_2, 10% moisture, and the rest as N_2 and O_2) at 1500°C in it. If the walls are at 800°C, calculate the radiation heat transfer rate to the walls. The total pressure is 1.5 bar, and the walls are essentially black.

7.23 A cracking furnace has a number of tubes arranged in an equilateral triangular array. The tubes are 2-inch o.d. (51 mm), and the pitch is thrice the diameter. The combustion gases have 8% CO_2, 15.5% H_2O, and the rest as N_2 and O_2. If the gas temperature is 1300°C and the tube-wall temperature is 600°C, calculate the net rate of radiant heat transfer to the tubes from the gas. The pressure in the furnace is essentially atmospheric.

7.24 A thermocouple is inserted into a pipe carrying hot air at 250°C and flowing at a velocity of 5 m/s. The thermocouple tip is a 2 mm diameter sphere having an emissivity of 0.9. If the pipe-wall temperature is 150°C, estimate the error in the temperature reading of the thermometer.

7.25 Thermocouple tips are sometimes kept shielded in order to reduce the error in temperature measurement. A cylindrical thin-walled shield of 17 mm diameter and 10 cm length is placed in a large duct, oriented parallel to the direction of flow of hot air. The thermocouple

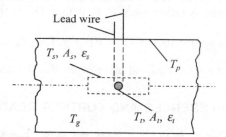

Fig. 7.44 Thermocouple in a large duct (Problem 7.25).

tip is positioned centarlly within the shield. If the thermocouple reads 300°C, what is the true temperature? What would have been the error in measurement in the absence of the shield? The following data are available:

Diameter of the thermocouple tip = 1.5 mm

Convective heat transfer coeffficient—at the shield = 180 W/m² °C, at the thermocouple tip = 150 W/m² °C

Temperature of the duct wall = 100°C

Emissivity—of the shield = 0.6, of the tip = 0.8.

Hints: Use the following heat balance equations:

$$h_t A_t (T_g - T_t) = \frac{E_{bt} - E_{bs}}{\dfrac{1 - \varepsilon_t}{A_t \varepsilon_t} + \dfrac{1}{A_t F_{t-s}} + \dfrac{1 - \varepsilon_s}{A_s \varepsilon_s}}$$

$$h_s (2 A_s)(T_g - T_s) + \frac{E_{bt} - E_{bs}}{\dfrac{1 - \varepsilon_t}{A_t \varepsilon_t} + \dfrac{1}{A_t F_{t-s}} + \dfrac{\varepsilon_s}{A_s (1 - \varepsilon_s)}} = A_s \varepsilon_s (E_{bs} - E_{bp})$$

7.26 A closed furnace door is 2 m high and 0.9 m wide. Its outside surface temperature is 70°C, and the emissivity is 0.8. What is the rate of heat loss by combined radiation and natural convection to an ambient at 32°C? What is the radiation heat transfer coefficient?

7.27 A hot ball, 5 cm in diameter, at 100°C is exposed to an environment at absolute zero. Given the following properties of the material of the ball, calculate its initial rate of cooling and the time for cooling down to 0°C. Consider radiant heat loss only. Properties: $\varepsilon = 0.7$; $c_p = 0.8$ kJ/kg K; $\rho = 8200$ kg/m³.

7.28 An annular upward-facing metal plate of the shape of annular disk is exposed to direct solar irradiation of 800 W/m². It looses heat by combined radiation and free convection to an ambient at 30°C. The emissivity of the plate is 0.9 for solar irradiation and 0.2 corresponding to the long wavelength region of the spectrum. What is the steady state temperature of the plate? The free convection heat transfer coefficient can be calculated using the equation $h = 1.35(\Delta T)^{0.25}$ W/m² °C, where ΔT is the temperature difference between the plate and the ambient in °C.

Hints: The plate absorbs solar radiation but emits only long wavelength radiation (because its temperature is rather low). If T_p is the plate temperature,

$$(0.9)(800) = (0.2)\sigma(T_p^4 - 303^4) + 1.35(T_p - 303)^{5/4}$$

Solve the above equation by trial-and-error or by the Newton-Raphson method.

REFERENCES AND FURTHER READING

Eckert, E.R.G. and R.M. Drake, *Analysis of Heat and Mass Transfer*, McGraw-Hill, New York, 1972.

Cooney, C.M., "Nations seek 'fair' greenhouse gas treaty in Kyoto", *Env. Sci. Technol.*, **31** (1997), 516A–518A.

Ellwood, P. and S. Dantos, "Process furnaces", *Chem. Eng.*, April 11, 1966.

Hileman, B., "Climate observations substantiate global warming models", *C.&E. News,* November 27, 1995, 18–23.

Incropera, F.P. and D.P. Dewitt, *Introduction to Heat Transfer*, 2nd ed., John Wiley, New York, 1996.

Jacob, M., *Heat Transfer*, Vol. 2, John Wiley, New York, 1957.

Osburn, J.O., "Simplified calculation of radiation shield", *Chem. Eng.*, September 22, 1969, 139.

Spiegel, R. and J.R. Howell, *Thermal Radiation Heat Transfer*, 2nd ed., McGraw-Hill, New York, 1981.

Sparrow, E.M. and R.D. Cess, *Radiation Heat Transfer*, Wadsworth Publ. Co., New York, 1966.

Wiebelt, J.A., *Engineering Radiation Heat Transfer*, Holt, Rinehart and Winston, Inc., New York, 1966.

Wimpress, R.N., "Rating fired heaters", *Chem. Eng.*, April 11, 1966, 151.

8

Heat Exchangers

In many process industries, it is a frequent necessity to heat, cool, vaporize or condense various fluid streams. Different types of heat transfer equipment are used for such purposes. In Chapter 1, we discussed the use of heat exchangers in a nitric acid plant. To give another example of a process industry in this context, let us consider a sulphuric acid plant. Hot gases leave the sulphur burner at about 1100°C and enter a waste heat boiler in which the gases are cooled to about 500°C and steam is generated. The gases then successively pass through multiple catalyst beds. In the process the hot gas mixture coming out of a catalyst bed is cooled to a required temperature exchanging heat with air (the air thus heated is used in the sulphur burner) or is used to superheat the steam obtained from the waste heat boiler or to preheat boiler-feed water in an 'economizer'. The product sulphuric acid leaves the absorption tower at about 90°C. It is cooled by heat exchange with water before it is pumped to the acid storage tank. This example cites a few of the uses of the heat exchange devices in a sulphuric acid plant. Some more varied types of heat transfer equipment are in use in fertilizer plants, petrochemical complexes and nuclear power plants.

A heat exchanger is a device in which two fluid streams, one hot and another cold, are brought into 'thermal contact' in order to effect transfer of heat from the hot fluid stream to the cold. It provides a relatively large area of heat transfer for a given volume of the equipment. The construction also caters for easy cleaning and maintenance of equipment.

8.1 CONSTRUCTION OF A SHELL-AND-TUBE HEAT EXCHANGER

A shell-and-tube heat exchanger is the most widely used heat exchange equipment. We shall first describe the constructional features of this type of heat transfer equipment. This will be followed by a discussion on a few other types of heat exchangers which are in frequent use in the chemical process industries as well as in the refrigeration, cryogenic, waste-heat recovery, metallurgical and manufacturing applications. Heat exchangers may be classified on the basis of the contacting technique, construction, flow arrangement, or surface compactness (Kakac, et al. 1981), as shown in Fig. 8.1. Most of the heat exchangers are of 'indirect contact' type in which the hot and the cold fluids are in thermal contact but physically separated by a barrier such as a tube wall, plate, etc. In a direct contact unit, the fluids are brought in physical contact for the purpose of heat exchange. A cooling tower in which utility water is cooled by contact with up-flowing air is a common example of direct contact heat exchange. Vapours leaving an evaporator are frequently condensed by direct contact with cooling water (see Chapter 9). Hot oils are sometimes cooled by direct contact with water*. Distilled fatty acid (in a low-grade vegetable oil-splitting and

---

*An example: Recovery of polycyclic aroamatics like naphthalene from coke oven gas is done by absorption in a mineral oil. The loaded oil leaving the absorption tower is steam-stripped to remove the absorbed compounds. The hot lean oil is cooled by direct contact with water and then recycled to the absorption tower.

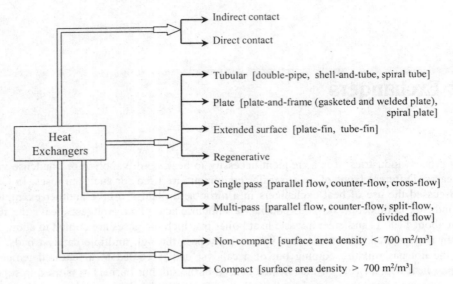

Fig. 8.1 Classification of heat exchangers.

refining plant) leaving the top of the refining column is condensed and cooled by direct contact with water. Solid-fluid heat transfer in a fluidized bed is another example of the direct contact technique.

A simplified diagram of a typical shell-and-tube, *fixed tube-sheet*, '1-2 pass' *equipment* is given in Fig. 8.2. It essentially consists of a shell to which the tube sheets are welded, one at each

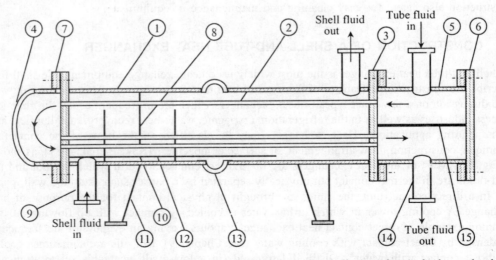

Fig. 8.2 Basic construction of a shell-and-tube heat exchanger. 1—shell; 2—tubes (only three tubes have been shown); 3—tube sheet; 4—stationary head bonnet; 5—stationary head channel; 6—channel cover; 7—flange; 8—baffles (only three baffles have been shown); 9—shell-fluid nozzle; 10—baffle spacer; 11—heat exchanger support; 12—tie rod for baffles; 13—shell expansion joint; 14—tube-fluid nozzle; 15—pass partition plate (for the tube-side fluid).

end. The heat exchanger tubes are fixed to the tube sheets. There are two 'heads', the bonnet and the channel, through which flows the tube side fluid. There are nozzles on both tube and shell sides for inlet and outlet of the two fluids. Figure 8.2 shows a '1-2 pass' heat exchanger, i.e. there is one shell pass, and there are two tube passes. The shell-side fluid enters at one end of the shell and leaves at the opposite end. The tube-side fluid, as shown in the figure, enters at the right end, flows through the tubes, reaches the bonnet on the left, takes a U-turn and returns to the other half of the partitioned channel and then leaves the exchanger. The channel here is divided into two compartments by a 'pass partition plate'. The number of tube- and shell-side passes can be increased by using more pass partition plates for both the sides (see Fig. 8.4). If no partition plate is used, we get a *unipass* system.

The basic construction and important components of a shell-and-tube exchanger (labelled in Fig. 8.2) are described below (the items in the parantheses refer to those in Fig. 8.2).

8.1.1 The Shell [item (1)]

The shell is the enclosure and passage of the shell-side fluid. It has a circular cross-section and is made by rolling a metal plate of suitable dimension into a cylinder and welding along the length. However, a section of a pipe of suitable thickness can be used as the shell if the required diameter is not large (up to about 60 cm). The selection of the material for the shell depends upon the corrosiveness of the fluid and the working temperature and pressure. Carbon steel is a common material for the shell under moderate working conditions.

8.1.2 The Tubes [item (2)]

The tubes provide the heat transfer area in a shell-and-tube heat exchanger. One of the fluids flows through the tubes and the other flows through the shell along the outside of the tubes. The fluids are brought into thermal contact through the tube wall. Both 'seamless' and welded tubes are used, and a wide variety of materials including low carbon steel, stainless steel, cupronickel, copper, brass, aluminium, etc. are used. The choice of the material depends upon the nature of the application. Finned tubes are sometimes used, and fin efficiency up to 90% may be attained.

Tubes of 19 mm (3/4 inch) and 25 mm (1 inch) diameter are more commonly used than larger tubes of 1-1/4 inch and occasionally 1-1/2 inch size. Narrower tubes may be used in smaller heat exchangers working with clean fluids. The tube wall thickness is designated in terms of BWG (Birmingham wire gauge). Tube specification in terms of SWG is also in practice. The selection of wall thickness (20 to 10 SWG are common) depends mainly upon three factors: (i) the maximum operating pressure, (ii) corrosion characteristics of both tube- and shell-side fluids, and (iii) thinning of the tube wall during bending when a U-tube exchanger is used. The length of the tubes is chosen during design depending upon factors such as heat load, room available for installation. Tubes of 8, 12, 16, 20 or 24 ft section were generally regarded as standard, but other tube lengths may also be used. Currently tube lengths of 3000, 4000, 6000 and 9000 mm (called nominal tube lengths) are generally used. Other sizes like 3500 mm, 4500 mm are also used.

Tubes are generally arranged in a triangular or a square pitch (see Fig. 8.12). Tube pitch, which is the centre-to-centre distance of two adjacent tubes, should not be less than 1.25 times the tube diameter. A square pitch arrangement allows easy cleaning of the outer surfaces of 'fouled' or 'scaled' tubes by using brushes (see Section 8.9).

8.1.3 The Tube Sheets [item (3)]

The tube sheets are circular, thick metal plates which hold the tubes at the ends. In the type of the exchanger shown in Fig. 8.2 (called *fixed head* or *fixed tube-sheet*), the tube sheets are welded to the shell at the ends. The tubes are inserted into holes of the two tube sheets. Perfect alignment of the holes is required; this is achieved by drilling tube holes through the sheets fastened together for the purpose, particularly for small-thickness tube sheets. The diameter of the tube hole is made slightly larger than the tube diameter. The arrangement of tubes on a tube sheet in a suitable pitch is called *tube-sheet layout*. Standard tables and charts of tube layout are available (Kern, 1950; Ludwig, 1983).

Two common techniques of fixing the ends of a tube to the tube sheets are: (i) expanded joints, and (ii) welded joints. Expanded joints may be made with or without grooves in the tube sheet holes. Groove joints are made by cutting small groove(s) in the tube sheet along the periphery of each hole [Fig. 8.3(a)] and expanding the tube into the groove(s). A mandrel with rollers mounted on a cage fitted to it, is inserted into the tube at one end. The mandrel is rotated at a high speed and the rollers press on the inside of the tube. This causes plastic flow of the tube metal into the grooves in the sheet. A very strong and durable joint can be made in this way.

Different types of non-groove joints are also shown in Figs. 8.3(b), (c), and (d). For ordinary services (small fluid pressure differential or low thermal stress), the plain joint [Fig. 8.3(b)] may be used. Here the tube is expanded to fill the small clearance between the tube and the hole in the tube sheet. Expansion may be caused by (i) a mandrel-roller as described above, (ii) hydraulic expander, or (iii) detonation. Other types of joints are the 'belled' or 'beaded' joint [Fig. 8.3(c)] or the welded joint. Groove-jointed tubes are quite often welded to the tube sheet as an additional protection against leakage particularly for high pressure services where at least one of the fluids is a gas or a vapour.

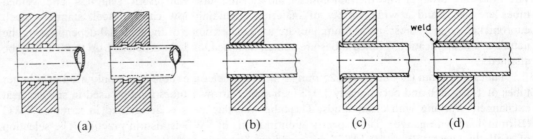

Fig. 8.3 A few common joints between the tube and the tube sheet: (a) grooved joint, (b) plain joint, (c) 'belled' or beaded joint, and (d) welded joint.

If a welded joint is used, the tubes and the tube sheets should be of the same or very similar materials so that they are 'corrosion compatible'. In fact, if the tube and the tube sheet are exposed to a corrosive environment, they must also be electrochemically compatible to each other even if a welded joint is avoided.

Low carbon steel tubes are used in non-corrosive services. Sometimes a thin layer of a corrosion-resistant alloy is metallurgically bonded to a low carbon steel tube sheet by using a 'detonation bonding' process. This reduces the cost while providing the necessary resistance to corrosion. When a grooved joint is chosen, the tube sheet should be at least as thick as the outside diameter of the tube. The cluster of tubes so fixed into the tube sheets is called a *tube bundle*.

8.1.4 The Bonnet and the Channel [items (4) and (5)]

Both the tube sheets of a shell-and-tube heat exchanger should have *closures* or *heads*. The space inside a closure is occupied by the tube-side fluid. A closure is called bonnet or channel depending upon its shape and construction. A bonnet has an integral cover, a channel closure has a removable cover.

The bonnet closure shown in Fig. 8.2 (item 4) consists of a short cylindrical section with a bonnet welded at one end and a flange welded at the other end. The flange is bolted to the tube sheet after inserting a suitable gasket to make the joint leak proof. A bonnet-type closure is used when it does not require to be fitted with any nozzle.

A channel-type closure is labelled as item 5 in Fig. 8.2. It is also made from a piece of cylindrical barrel with flanges welded to both ends. One of the flanges is bolted to a tube sheet using a gasket for sealing. The other flange of the channel is bolted to a flat channel cover plate (this is like a 'blind flange'). One may have access to the tubes after unbolting the channel cover plate.

Pieces of pipes of suitable diameter [either nominal bore (nb) or fabricated], called 'nozzles', are welded to the channel and serve as inlet and outlet for the tube-side fluid. The bonnet-type closure at the rear end of the exchanger in Fig. 8.2 should be replaced by a channel-type closure if a nozzle is required to be fitted at that end.

8.1.5 The Pass Partition Plate [item (15)]

A 'pass partition plate' or 'pass divider' is shown in the channel in Fig. 8.4. This plate causes two passes of the tube-side fluid. The number of passes in either the shell or the tube side indicates the number of times the shell- or the tube-side fluid traverses the length of the exchanger. For example, a '2-4 pass exchanger' has two shell passes and four tube passes as shown schematically in Fig. 8.4. There are two pass partition plates on the front end (in the channel) and one pass partition plate at the rear end (in the bonnet closure). The changes in the flow direction of the tube-side fluid are also shown in Fig. 8.4.

In order to have two passes of the shell-side fluid, a 'longitudinal shell pass baffle' is used. This is a flat plate, usually 6-13 mm (1/4 to 1/2 inch) thick, which runs axially along the shell to divide it into two semi-cylindrical halves. The plate is either welded to the shell or fitted into 'guide channels' that are welded longitudinally along the shell wall. A shell pass baffle is shown in Fig. 8.4.

One edge of the tube pass partition plate or pass divider is welded to the bonnet (or the channel cover as the case may be). The other edge presses against a gasket in a groove made on the tube sheet. This prevents leakage from one side of the plate to the other. Alternatively, both the edges may press against gasketed grooves—one groove on the tube sheet and the other on the bonnet or the channel cover plate.

For a given number of tubes, the area available for flow of the tube-side fluid is inversely proportional to the number of passes. By using two tube passes instead of one, the flow area is halved, and the fluid velocity and the Reynolds number are doubled. So by increasing the number of passes, the turbulence, and therefore the heat transfer coefficient can be increased but, of course, at the cost of a higher pressure drop. An even number of passes (in the case of multi-pass exchangers) on any side is generally used (for example, 1-2, 1-4, 2-4, 2-6 etc.; 1-3, 2-5, etc., are *not* used).

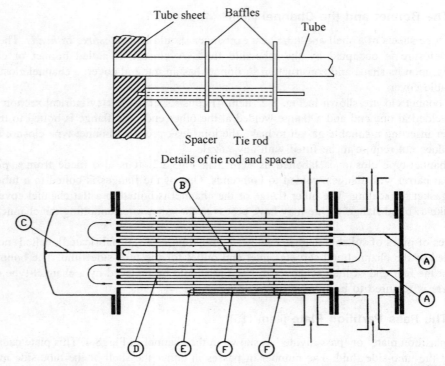

Fig. 8.4 A 2-4 pass shell-and-tube heat exchanger (fixed head). A: pass partition plate; B: longitudinal shell pass baffle; C: pass partition plate (at the rear end); D: baffle; E: tie rod; F: baffle spacer.

8.1.6 Nozzles

Small sections of pipes welded to the shell or to the channel which act as the inlet or outlet of the fluids are called *nozzles*. The end of a nozzle is flanged to connect it to the pipe carrying a fluid.

The shell-side inlet nozzle is often provided with an 'impingement plate' (Fig. 8.5). The

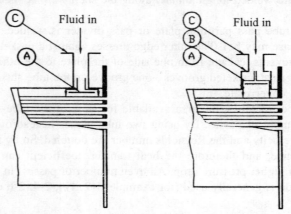

Fig. 8.5 Impingement plates—A: the plates; B: expanded nozzle; C: nozzle flange.

impingement plate prevents impact of the high velocity inlet fluid stream on the tube bundle. Such impact can cause erosion and cavitation of the tubes just in front of the nozzle, and can also cause vibration of the tubes. The erosion problem is aggravated (i) if the inlet liquid has suspended solid particles, or (ii) if the inlet gas stream has suspended dust or solid particles or even liquid droplets. (For example, when steam is used as the heating fluid, it may have suspended droplets of condensate. As steam enters the shell at a high velocity, which may be as high as 35 m/s or more, the water droplets strike the tubes at the front at a high speed. This causes erosion of the tubes which may develop leakage after a period of use.) An impingement plate is also called an *impingement baffle*.

Figure 8.5 shows two types of impingement plates. The impingement plate may be fitted in the shell just in front of the nozzle, or inside the expanded end of a nozzle. It is held in place by two or three narrow lugs or rods (not shown in the figure) welded to the shell or to the expanded nozzle. Other types of nozzles in a heat exchanger will be discussed later on.

8.1.7   Baffles [item (8)]

A baffle (or a shell-side baffle) is a metal plate usually in the form of the segment of a circle having holes to accommodate tubes. Shell-side baffles have two functions—(i) to cause changes in the flow pattern of the shell fluid creating parallel or cross flow to the tube bundle (thus increasing turbulence and, therefore, the heat transfer coefficient), and (ii) to support the tubes.

Segmental baffle

This is the most popular type of baffle. A segmental baffle may have horizontal or vertical cuts as shown in Fig. 8.6. The cut-out portion (called *baffle window*) provides the area for flow of the shell fluid. This area may vary from 15% to about 50% (but definitely *less than* 50%). A 20% cut segmental baffle means that the area of the cut-out portion is 20% of the area of the baffle. However, the net flow area of the shell fluid is calculated by subtracting the cross-sectional area of the tubes passing through the cut-out portion from the area of the cut-out portion itself. A small notch or segment is cut out at the lowest part of a baffle in a horizontal exchanger. This provides the passage to the liquid when the shell is required to be drained (may be during a shutdown or for cleaning or maintenance).

Other types of baffles

Among the other types of shell-side baffles, the double segmental and the disk-and-doughnut type baffles are shown in Fig. 8.6. The figure also explains their construction and the fluid flow patterns.

Baffles as tube supports

Segmental baffles also act as supports to the tubes. Vertically-cut segmental baffles provide better tube support than that provided by the horizontally-cut type. At least two baffles in between the tube sheets are required to support all the tubes. The baffles prevent the tubes from sagging and also from vibration caused by the flow of the fluid. A hole in a baffle is slightly larger (0.4 mm or 1/64 inch larger if the unsupported tube length is greater than 1 m; 0.8 mm larger when the unsupported length is equal to or smaller than 1 m) than the outer diameter of the tube. If the clearance is larger, the edge of a baffle hole may cut the tube because of the tube vibration. In

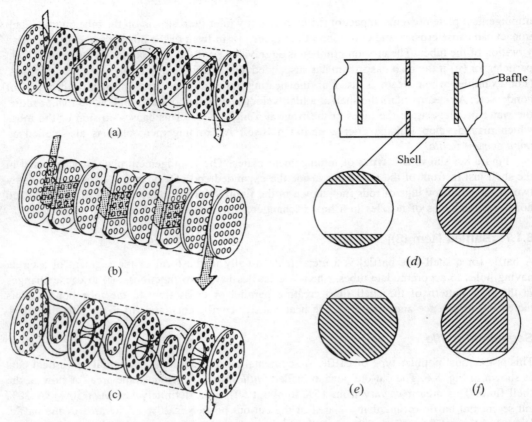

Fig. 8.6   A few types of baffles: (a) horizontal cut, (b) vertical cut, (c) disc and doughnut, (d) double segmental, (e) V-notch cut (liquid drains through the notch at the lowest level of a horizontal-cut baffle), and (f) cut-out area at the lowest level of a vertical-cut baffle (for liquid drainage). [In (d), (e) and (f) the shaded regions show the baffle area. The plain regions indicate the baffle windows.]

fact, 'flow-induced tube vibration' in a heat exchanger can turn out to be a serious problem and should be properly taken care of in the design.

A larger baffle spacing reduces the shell-side pressure drop, but simultaneously reduces both the turbulence and the shell-side heat transfer coefficient. A smaller baffle spacing increases both. A baffle spacing is selected in consideration of the allowable shell-side pressure drop and the heat transfer coefficient desired. The *minimum spacing* of segmental baffles is one-fifth of the shell diameter or 5 cm, whichever is larger.

8.1.8   Tie Rods and Baffle Spacers [items (12) and (10)]

A few tie rods having threaded ends are used to hold the baffles in position. One end of a tie rod is screwed to a tube sheet. A section of a tube or pipe (this does not mean the heat exchanger tube) of suitable diameter (the i.d. of the piece of tube should be slightly larger than the diameter of the tie rod) is slid over it. A baffle is then inserted followed by another section of the tube. These tube sections are called *spacers* because they maintain the spacing or distance between successive baffles. After the last baffle is inserted, locknuts are tightened at the threaded free end of the tie rod. This arrangement fixes all the baffles firmly and prevents their movement.

The tie rod and spacers are also shown in Fig. 8.4 (E and F). In this figure the tie rod is shown by dotted lines and the spacer tubes by firm lines.

8.1.9 Flanges and Gaskets [item (7)]

A shell-and-tube exchanger uses a number of flanges. Sometimes the tube sheets themselves act as shell flanges. The bonnet and the channel closures are fixed to the tube sheets by flanges.

Gaskets are placed between two flanges to make the joint leak-free. Compressed asbestos fibre (CAF), jacketed asbestos (JA), and PTFE gaskets are widely used at pressures up to 20 bar and temperatures up to 150°C (gaskets made of modified polymers are preferred to CAF now because asbestos is a carcinogen). The CAF gasket is made by bonding asbestos fibres with a polymeric binder. The CAF gasket may be impregnated with a fine mesh steel wire gauge for extra strength. Spirally-wound gaskets or metal gaskets are used in high pressure services.

Selection of gaskets has been briefly discussed by Kent (1978).

8.1.10 Expansion Joint [item (13)]

In many applications, there may be a substantial difference of expansion between the shell and the tubes because of the temperature difference between the two fluid streams. This causes thermal stress which may damage the exchanger. It is possible to avoid the problem by providing the shell with a suitable expansion device.

A common type of expansion joint shown in Fig. 8.2 acts as an expansion bellow. A piece of thin sheet expansion joint is made by cold forming (rolling or hydraulic forming) of a strip of metal or alloy. The piece is then welded, centrally placed, in-between two segments. Thus a shell with an expansion joint of single convolute is formed (Fig. 8.7). Multi-convolute and various other expansion joints are also used depending upon the operating conditions.

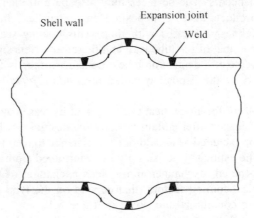

Fig. 8.7 A shell with an expansion joint (tubes not shown).

8.2 PROCESS DESIGN CONSIDERATIONS

Heat exchangers are extensively used in the chemical process industries and other engineering processes. Chemical engineers and mechanical engineers, in particular, ought to have a good understanding of the procedures of heat transfer design*. We shall now discuss the procedure of *thermal design* of shell-and-tube heat exchangers. This will be followed by a brief description of different types of shell-and-tube devices and also a few other types of heat exchangers.

As in the case of any other process equipment, the design of a heat exchanger may be divided into two parts—the *process design* part and the *mechanical design* part. Here, we shall discuss process design, more commonly called the thermal design, of heat exchangers only with the objective of sizing a heat exchanger. Within the constraints of allowable pressure drops, thermal design of a heat exchanger broadly includes the estimation of the heat transfer area, determination of the tube diameter, number and length of tubes, tube layout, shell diameter and the TEMA shell type, number of shell-side and tube-side passes of the fluids, number and size of baffles, shell and tube side pressure drops, etc. Mechanical design includes the selection of the materials of construction depending upon the process characteristics and conditions, determination of the thicknesses of the shell, the tube sheets, flanges, channels, baffles, etc., selection of gaskets, nozzle connections, design of the support, etc. Mechanical design calculations are done according to a 'standard' or 'stipulated' heat exchanger 'design code'. The most widely used code is that of Tubular Exchanger Manufacturers'Association (TEMA). This is a US code and is used together with ASME Section VIII (for the design of unfired pressure vessels). Thicknesses of the shell, bonnet, channel, etc. are calculated using the ASME code. The Indian code for heat exchanger design is IS 4503, and the British code is BS 3274.

Before going into the steps of design calculations, it is necessary to introduce two very important terms—the 'dirt factor' and the 'LMTD correction factor'.

8.2.1   Fouling of a Heat Exchanger—The Dirt Factor or Fouling Factor

Process fluid streams may contain suspended matters or dissolved solids. When such a fluid flows through a heat exchanger over a long period of time, deposition on the tube surfaces (and on the shell surface too) occurs. The surfaces may also be corroded by the fluids slowly and the resulting corrosion products also get deposited on the surface. In either case, a coating of deposit forms whose thickness increases with time. This coating (called *scale*) has a thermal conductivity much less than that of the tube wall and therefore offers significant resistance to heat transfer. Formation of a scale or a deposit on a heat transfer surface is called *fouling*, and the heat transfer resistance offered by the deposit is called the *fouling factor* or *dirt factor*, commonly denoted by R_d.

The fouling factor is zero for a new heat exchanger. It increases over the period of operation, and the tubes have to be cleaned after certain time. While designing a heat exchanger, a value of the fouling factor has to be included as an additional resistance to heat transfer. The fouling factor or the dirt factor *cannot* be estimated; it can only be determined from experimental data on heat transfer coefficient of a 'fouled' exchanger and a clean exchanger of similar design operated at identical conditions. Typical fouling factors (that depend on the type of fluid, the nature of the tube surface and operating conditions) are given in Table 8.1.

---

*'Engineering Design', as Sherwood (1963) describes it, 'is the process of applying the various techniques and scientific principles for the purpose of defining a device, a process, or a system in sufficient detail to permit its physical realization'.

Table 8.1 Fouling resistances for industrial fluids (as suggested by TEMA, 1988)

	Fouling factor or resistance	
	$\dfrac{\text{h ft}^2 \,^\circ\text{F}}{\text{Btu}}$	$\dfrac{\text{h m}^2 \,^\circ\text{C}}{\text{kcal}} \times 10^3$
Liquids		
Fuel oil	0.005	1.024
Quench oil	0.004	0.814
Transformer oil	0.001	0.205
Hydraulic fluid	0.001	0.205
Molten heat transfer salts	0.0005	0.102
Industrial organic heat transfer media	0.001	0.102
Refrigerant liquids	0.001	0.102
Mono- and di-ethanolamine solutions	0.002	0.409
Di- and tri-ethyleneglycol solutions	0.002	0.409
Caustic solutions	0.002	0.409
Vegetable oils	0.003	0.615
Gasoline, naptha, light distillates, kerosene	0.001	0.205
Light gas oil	0.002	0.409
Heavy gas oil	0.003	0.615
Absorption oil	0.002	0.409
Gases and Vapours		
Solvent vapours	0.001	0.205
Acid gases	0.001	0.205
Natural gas	0.001	0.205
Air	0.0005–0.001	0.102–0.205
Flue gases	0.001–0.003	0.205–0.615
Steam (saturated, oil free)	0.0005–0.0015	0.102–0.307
Water		
River water (treated and settled), velocity > 0.6 m/s	0.0015–0.002	0.205–0.409
Treated boiler feedwater	0.0005–0.001	0.102–0.205
Brackish water	0.002–0.003	0.407–0.615
Process water	0.001–0.002	0.205–0.409

Note: (a) Extended lists of fouling factors are available in TEMA and Ludwig (1983).

(b) Fouling resistances of both sides of a tube (or surface) should be considered in design calculations.

8.2.2 Log Mean Temperature Difference Correction Factor

We discussed in Chapter 4 how the temperature driving force varies with position in a cocurrent or countercurrent double-pipe heat exchanger. It has also been shown [see Section 4.7] that a log mean temperature difference (LMTD or ΔT_m) has to be used as the average driving force. A countercurrent device is more efficient for exchange of heat. But in many situations, a multi-pass exchanger gives a higher heat transfer coefficient than a single pass one because of larger fluid velocity. However, in a multi-pass exchanger like 1-2, 1-4, 2-4, etc. the fluids are not always in countercurrent flow. If one tube pass is countercurrent to the shell fluid, the other tube pass will be cocurrent. This deviation from truly countercurrent flow causes a change in the average driving force. A correction factor, denoted by F_T, is used to get the true mean temperature difference or the effective driving force. The heat transfer rate given by Eq. (4.37) then becomes

$$Q = U_d A F_T \Delta T_m \tag{8.1}$$

where U_d is the overall heat transfer coefficient that takes into account the fouling or the dirt factor R_d, and $F_T \Delta T_m$ is the true temperature difference. If U is the clean overall coefficient, then by addition of heat transfer resistances, we have

$$\frac{1}{U_d} = \frac{1}{U} + R_d \tag{8.2}$$

i.e. overall resistance of the fouled exchanger = overall resistance of the clean exchanger + heat transfer resistance due to dirt or scaling on both sides of the tubes.

The LMTD correction factor F_T can be directly obtained from available charts. These charts have been prepared from results obtained theoretically by solving for the temperature distribution in a multi-pass exchanger. We show below how the factor F_T can be calculated for a 1-2 exchanger (1 shell pass and 2 tube passes) shown in Fig. 8.8(a). The following notations are used:

T_h, T_c are the local (i.e. at position x) temperatures of the hot and the cold streams

T_{h1}, T_{c1} are the inlet temperatures of the hot and cold streams

T_{h2}, T_{c2} are the outlet temperatures of the hot and cold streams

a is the heat transfer area per unit length of each pass

U is the overall heat transfer coefficient

L is the length of the exchanger

W is the flow rate of a stream

c_p is the specific heat

Let us consider a differential length dx of the exchanger at a location x from the left end [see Fig. 8.8(a)]. One tube pass in the section is parallel to the shell fluid flow, and the other tube pass is in counterflow to it. The tube fluid temperature at x is taken as T_c' when it flows from left to right, and as T_c'' when it flows from right to left. The physical properties of the streams are assumed to remain constant.

The temperature of the hot fluid T_h increases with x. For the cold fluid, T_c' increases with x, but T_c'' decreases with x (because of a reversal of the direction of flow). A differential heat balance over the segment dx may be written as (note that dT_c'' is negative for the given configuration)

$$W_h c_{ph} (dT_h) = W_c c_{pc} (dT_c') + W_c c_{pc} (-dT_c'') \tag{8.3}$$

Temperature changes dT_c' and dT_c'' occur because of heat transfer from the hot fluid. Therefore,

$$W_c c_{pc} dT_c' = U(a\,dx)(T_h - T_c')$$

or

$$\frac{dT_c'}{dx} = \frac{U a}{W_c c_{pc}} (T_h - T_c') \tag{8.4}$$

and

$$W_c c_{pc}(-dT_c'') = U(a\,dx)(T_h - T_c'')$$

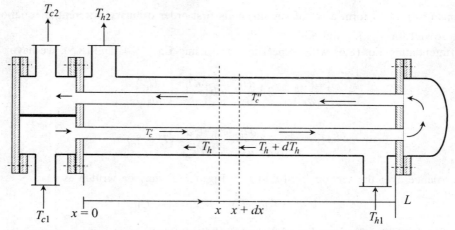

Fig. 8.8(a) Control volume for differential heat balance in a 1–2 pass heat exchanger.

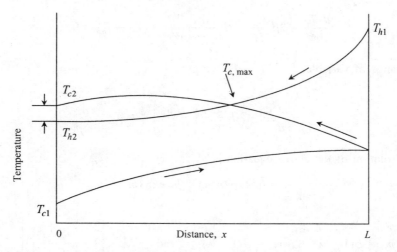

Fig. 8.8(b) Temperature profile and temperature cross $(T_{c2} - T_{h2})$ in a 1–2 pass counterflow exchanger.

or

$$-\frac{dT_c''}{dx} = \frac{U a}{W_c c_{pc}} (T_h - T_c'') \tag{8.5}$$

Substituting the expressions for dT_c' and dT_c'' from Eqs. (8.4) and (8.5) into Eq. (8.3) and rearranging, we get

$$\frac{dT_h}{dx} = \beta (2T_h - T_c' - T_c'') \tag{8.6}$$

where

$$\beta = \frac{U a}{W_h c_{ph}}$$

Equations (8.4)–(8.6) form a set of simultaneous first order ordinary differential equations that may be solved for T_h, T_c', and T_c''.

Differentiating Eq. (8.6) with respect to x and using Eqs. (8.4) and (8.5), we have

$$\frac{d^2 T_h}{dx^2} = 2\beta \frac{dT_h}{dx} - \beta \left(\frac{dT_c'}{dx} + \frac{dT_c''}{dx} \right) \tag{8.7}$$

$$= 2\beta \frac{dT_h}{dx} - \beta \frac{Ua}{W_c c_{pc}} (T_c'' - T_c') \tag{8.8}$$

A heat balance over the section from x to L in Fig. 8.8(a) may be written as

$$W_h c_{ph} (T_{h1} - T_h) = W_c c_{pc} (T_c'' - T_c') \tag{8.9}$$

Substituting for $(T_c'' - T_c')$ from Eq. (8.9) in Eq. (8.8), we get

$$\frac{d^2 T_h}{dx^2} - 2\beta \frac{dT_h}{dx} + \beta \frac{Ua}{W_c c_{ph}} \frac{W_h c_{ph}}{W_c c_{pc}} (T_{h1} - T_h) = 0 \tag{8.10}$$

Making a change of variable, $\xi = T_{h1} - T_h$, we get

$$\frac{d^2 \xi}{dx^2} - 2\beta \frac{d\xi}{dx} - \left(\frac{Ua}{W_c c_{pc}} \right)^2 \xi = 0 \tag{8.11}$$

The general solution of the above equation is

$$\xi = b_1 \exp (m_1 x) + b_2 \exp (m_2 x) \tag{8.12}$$

where

$$m_1 = \beta(1 + \lambda), \qquad m_2 = \beta(1 - \lambda) \qquad \text{and} \qquad \lambda = \left[1 + \left(\frac{W_h c_{ph}}{W_c c_{pc}} \right)^2 \right]^{1/2} \tag{8.13a}$$

The values of the integration constants, b_1 and b_2, can be obtained from the boundary conditions

$$x = 0, \; T_h = T_{h2}; \qquad x = L, \; T_h = T_{h1}$$

Thus,

$$b_1 = -(T_{h1} - T_{h2}) \frac{e^{m_2 L}}{e^{m_1 L} - e^{m_2 L}} \qquad \text{and} \qquad b_2 = (T_{h1} - T_{h2}) \frac{e^{m_1 L}}{e^{m_1 L} - e^{m_2 L}} \tag{8.13b}$$

The solution for $T_h \; (= T_{h1} - \xi)$ can be written from Eq. (8.12). Therefore,

$$T_h = T_{h1} - (T_{h1} - T_{h2}) \frac{e^{m_1 L + m_2 x} - e^{m_2 L + m_1 x}}{e^{m_1 L} - e^{m_2 L}} \tag{8.14}$$

Because $\xi = T_{h1} - T_h$, we may use Eq. (8.12) to write

$$\frac{d\xi}{dx} = -\frac{dT_h}{dx} = b_1 m_1 \exp(m_1 x) + b_2 m_2 \exp(m_2 x) \tag{8.15}$$

Substituting for $\dfrac{dT_h}{dx}$ from Eq. (8.6), we get

$$-\beta(2T_h - T_c' - T_c'') = b_1 m_1 e^{m_1 x} + b_2 m_2 e^{m_2 x}$$

But, at $x = 0$, $T_h = T_{h2}$, $T_c' = T_{c1}$, $T_c'' = T_{c2}$. Therefore,

$$-\beta(2T_h - T_c' - T_c'') = b_1 m_1 + b_2 m_2 \tag{8.16}$$

Substituting for b_1 and b_2 from Eq. (8.13b), we get

$$-\beta(2T_{h2} - T_{c1} - T_{c2}) = \frac{\beta(T_{h1} - T_{h2})}{e^{m_1 L} - e^{m_2 L}} \left[(1 - \lambda)e^{m_1 L} - (1 + \lambda)e^{m_2 L} \right]$$

or

$$T_{h1} + T_{h2} - T_{c1} - T_{c2} = \lambda(T_{h1} - T_{h2}) \frac{e^{m_1 L} + e^{m_2 L}}{e^{m_1 L} - e^{m_2 L}}$$

$$= \lambda(T_{h1} - T_{h2}) \frac{e^{(m_1 - m_2)L} + 1}{e^{(m_1 - m_2)L} - 1}$$

or

$$(m_1 - m_2)L = \ln\left[\frac{T_{h1} + T_{h2} - T_{c1} - T_{c2} + \lambda(T_{h1} - T_{h2})}{T_{h1} + T_{h2} - T_{c1} - T_{c2} - \lambda(T_{h1} - T_{h2})} \right] \tag{8.17}$$

The overall heat balance equation is

$$W_h c_{ph}(T_{h1} - T_{h2}) = W_c c_{pc}(T_{c2} - T_{c1}) \tag{8.18}$$

Using the above relation in the expression for λ given by Eq. (8.13a), we get

$$\lambda = \left[1 + \left(\frac{W_h c_{ph}}{W_c c_{pc}} \right)^2 \right]^{1/2} = \frac{\left[(T_{h1} - T_{h2})^2 + (T_{c2} - T_{c1})^2 \right]^{1/2}}{T_{h1} - T_{h2}}$$

or

$$\lambda(T_{h1} - T_{h2}) = [(T_{h1} - T_{h2})^2 + (T_{c2} - T_{c1})^2]^{1/2} \tag{8.19}$$

Also from Eq. (8.13a), we may write

$$(m_1 - m_2) = 2\beta\lambda = \frac{2Ua}{W_h c_{ph}} \lambda \tag{8.20}$$

The total heat transfer area in the exchanger is $2aL$. If ΔT_{cm} is the 'corrected mean driving force, the total rate of heat transfer is

$$Q = U(2aL)\Delta T_{cm} = W_h c_{ph}(T_{h1} - T_{h2}) \tag{8.21}$$

or

$$\Delta T_{cm} = \frac{W_h c_{ph}}{2UaL}(T_{h1} - T_{h2}) = \frac{\lambda(T_{h1} - T_{h2})}{(m_1 - m_2)L} \text{ [from Eq. (8.20)].}$$

Substituting for $\lambda(T_{h1} - T_{h2})$ from Eq. (8.19), and $(m_1 - m_2)L$ from Eq. (8.17), we get

$$\Delta T_{cm} = \frac{[(T_{h1} - T_{h2})^2 + (T_{c2} - T_{c1})^2]^{1/2}}{\ln\left[\dfrac{T_{h1} + T_{h2} - T_{c1} - T_{c2} + \lambda(T_{h1} - T_{h2})}{T_{h1} + T_{h2} - T_{c1} - T_{c2} - \lambda(T_{h1} - T_{h2})}\right]} \tag{8.22}$$

The LMTD correction factor F_T is defined as [see Eq. (8.1)]

$$\Delta T_{cm} = F_T \, \Delta T_m$$

or

$$F_T = \frac{\Delta T_{cm}}{\Delta T_m} \tag{8.23}$$

But how to define ΔT_m for a multi-pass exchanger? We define it on the basis of the inlet and outlet temperatures of the streams to and from the exchanger. The terminal driving forces on this basis are

$$(\Delta T)_1 = T_{h1} - T_{c2} \quad \text{and} \quad (\Delta T)_2 = T_{h2} - T_{c1}$$

The corresponding log mean driving force is

$$\Delta T_m = \frac{(\Delta T)_1 - (\Delta T)_2}{\ln[(\Delta T)_1/(\Delta T)_2]} \tag{8.24}$$

Substituting for ΔT_m from Eq. (8.24) and for ΔT_{cm} from Eq. (8.22) in Eq. (8.23), we get

$$F_T = \frac{\dfrac{\sqrt{R^2 + 1}}{R - 1} \ln\left[\dfrac{1 - \tau_c}{1 - R\tau_c}\right]}{\ln\left[\dfrac{(2/\tau_c) - 1 - R + \sqrt{R^2 + 1}}{(2/\tau_c) - 1 - R - \sqrt{R^2 + 1}}\right]} \tag{8.25}$$

where
the temperature ratio, $\tau_c = (T_{c2} - T_{c1})/(T_{h1} - T_{c1})$ (8.26)
the capacity ratio, $R = (T_{h1} - T_{h2})/(T_{c2} - T_{c1})$ (8.27)

Once ΔT_m and F_T are known, the corrected mean temperature driving force, ΔT_{cm}, is calculated from Eq. (8.23).

The lengthy algebraic exercise was necessary in order to express the correction factor F_T in terms of the terminal temperature differences only. Such expressions for this factor for the cases of exchangers having a large number of passes can be obtained by following a similar procedure. However, the exercise is very tedious because more number of temperature variables will have to be handled (in the case of a 1-2 exchanger analysed above, the number of temperature variables is three). However, charts for values of F_T for different modes of flow are readily available and can be used directly.

8.2.3 Temperature Distribution in Multi-pass Exchangers and Temperature Cross

Temperature profiles of the hot and the cold streams in a parallel or counterflow exchanger have been shown in Fig. 8.8(b). For complex flow modes, the temperature distributions will predictably be different. Considering a 1-2 exchanger in which the hot and the cold fluid streams enter the unit at the same end [Fig. 8.9(a)], the temperature profile is given in Fig. 8.9(b). If the fluid streams enter at the opposite ends [Fig. 8.8(a)], then a temperature cross may sometimes occur. Temperature cross means that the difference between the cold and the hot fluid temperatures is positive when these streams leave the exchanger [$T_{c2} - T_{h2}$ in Fig. 8.8(b)]. If it occurs, the cold fluid temperature reaches the maximum at a point inside the exchanger and not at its exit. This point also coincides with the point of intersection of the temperature profiles of the hot fluid and the cocurrent zone of the cold fluid [Fig. 8.8(b)]. The quantity, $T_{c2} - T_{h2}$, is called the

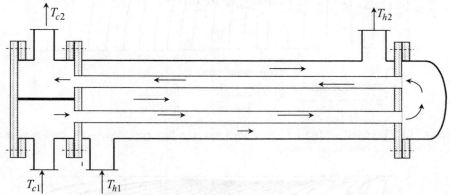

Fig. 8.9(a)   Flow in a 1–2 pass cocurrent heat exchanger.

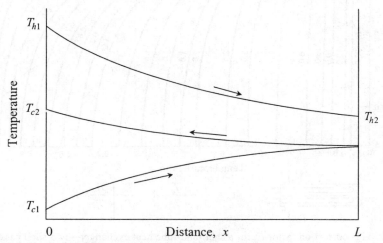

Fig. 8.9(b)   Temperature profile in a 1–2 cocurrent exchanger (no temperature cross).

temperature cross of the exchanger. However, if the temperature cross does not appear (i.e. if $T_{c2} < T_{h2}$), then $T_{c2} - T_{h2}$ is called the *approach*. (In a 1-2 counterflow exchanger, the difference between the hot fluid inlet and the cold fluid outlet temperatures is called the *approach*.)

Typical charts for the LMTD correction factor F_T are given in Fig. 8.10. One of the factors governing the choice of numbers of shell and tube passes is the value of the correction factor F_T. A value of F_T less than 0.8 is generally not acceptable.

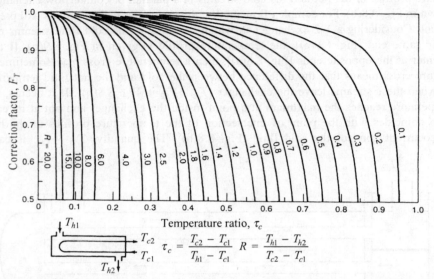

Fig. 8.10(a)   LMTD correction factor F_T for a shell-and-tube heat exchanger—one shell pass and two or multiple of two tube passes.

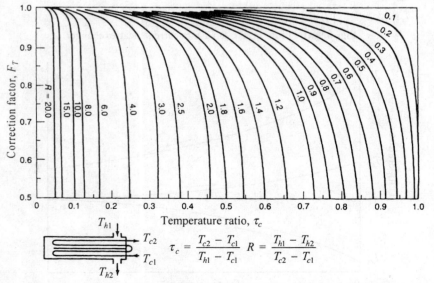

Fig. 8.10(b)   LMTD correction factor F_T for a shell-and-tube heat exchanger—two shell passes and four or multiple of four tube passes.

8.2.4   The Caloric Temperature

The heat transfer coefficient in a heat exchanger varies from one point to another because of the continuous variation in the relevant thermo-physical properties of the fluid streams. Among these properties, viscosity is usually a strong function of temperature. For cases where a large change in the viscosity occurs, it is suggested (see Kern, 1950) that the physical properties of a stream are evaluated at the 'caloric temperature' rather than at the 'arithmetic mean temperature'. Procedures and charts for estimation of the caloric temperature are available (Kern, 1950; Ludwig, 1983).

8.2.5   Individual and Overall Heat Transfer Coefficients

Equation (8.1) is the basic relation for the calculation of the heat transfer area in an exchanger. We have already discussed how to calculate ΔT_m (i.e. LMTD) and the correction factor F_T. The other quantity that remains to be determined is U_d (i.e. the overall heat transfer coefficient that includes the fouling factor). A relation for the 'clean' overall heat transfer coefficient can be written on the basis of the outside or the inside heat transfer area as given in Chapter 3 [Eqs. (3.9) and (3.10)]. The additional resistances can be easily incorporated in these equations. In general, there may be two fouling resistances (on the inner and the outer surface of the tubes), and the total number of resistances in series becomes five. The relation for the overall heat transfer coefficient based on the outside tube area then becomes

$$\frac{1}{U_d} = \frac{1}{h_o} + R_{do} + \left(\frac{A_o}{A_m}\right)\left(\frac{r_o - r_i}{k_w}\right) + \left(\frac{A_o}{A_i}\right)R_{di} + \left(\frac{A_o}{A_i}\right)\left(\frac{1}{h_i}\right) \tag{8.28}$$

The dirt factors, R_{do} and R_{di}, are known from experience or can be obtained from the available literature (typical values are given in Table 8.1). The inside and outside heat transfer coefficients can be determined by using standard correlations or charts.

Inside-film coefficient

A number of correlations for estimation of heat transfer coefficients have been cited in Chapter 4 (Section 4.5). For a tubular exchanger, the Dittus-Boelter equation, Eq. (4.9), or the Sieder-Tate equation, Eq. (4.10), may be used depending upon whether the viscosity ratio (μ/μ_w) is significantly different from unity or not. A chart for the Colburn j_H factor given in Fig. 8.11(a) may be used over a wide range of Reynolds number covering both laminar and turbulent flow. A number of correlations are available in the literature for specific types of fluids.

Outside-film coefficient

For a double-pipe heat exchanger, the outside-film coefficient can be calculated using the Dittus-Boelter or the Sieder-Tate equation on the basis of *equivalent diameter* of the annulus. A similar approach may be adopted for a shell-and-tube exchanger *if there are no baffles on the shell side*. However, baffles are invariably used to enhance turbulence and thereby to increase the shell-side heat transfer coefficient, although the presence of baffles increases the pressure drop on the shell side. Given below are the methods of determination of shell-side heat transfer coefficient in a baffled heat exchanger. Kern (1950) suggested the following equation for the purpose.

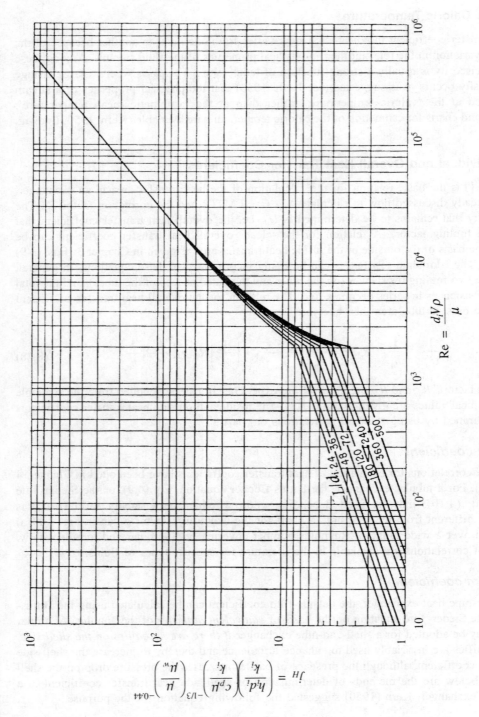

Fig. 8.11(a)   Colburn factor (j_H) for tube-side heat transfer.

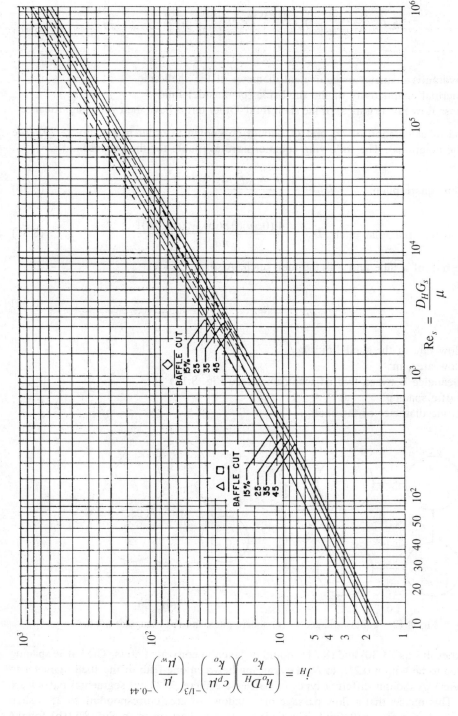

Fig. 8.11(b) Colburn factor (j_H) for shell-side heat transfer with segmental baffles.

$$\frac{h_o D_H}{k_o} = 0.36 \left(\frac{D_H G_s}{\mu}\right)^{0.55} \left(\frac{c_p \mu}{k_o}\right)^{0.33} \left(\frac{\mu}{\mu_w}\right)^{0.14}$$ (8.29)

where

D_H is the hydraulic diameter or equivalent diameter of the shell side

k_o is the thermal conductivity of the shell-side (or outside) fluid

G_s is the mass flow rate (kg/m^2 s) of the shell fluid. Other terms have their usual significances.

The method of calculation of D_H and G_s should be known in order to use Eq. (8.29) (see Fig. 8.12 for the notations). The following equations are available for the purpose.

For tubes on square pitch: $$D_H = \frac{4(p^2 - \pi d_o^2/4)}{\pi d_o}$$ (8.30)

For tubes on 60° triangular pitch: $$D_H = \frac{4[(0.5p)(0.86p) - (\pi d_o^2/8)]}{\pi d_o/2}$$ (8.31)

where d_o is the o.d. of a tube and p is the tube pitch. Also,

$$G_s = \frac{W}{a_s}; \qquad a_s = \frac{c'B D_s}{p}$$ (8.32)

where

W is the flow rate of the shell fluid, kg/s

a_s is the flow area, m^2

c' is the clearance between two adjacent tubes (see Fig. 8.12)

B is the baffle spacing

D_s is the inside diameter of the shell.

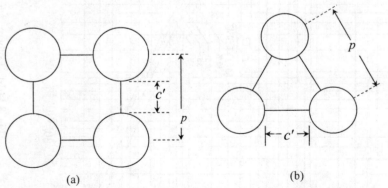

(a)   (b)

Fig. 8.12   Tubes arranged in (a) square pitch and (b) triangular pitch.

The quantities in Eqs. (8.30) and (8.31) should be taken in consistent units. The baffle spacing is usually chosen to be within $0.2D_s$ to D_s (i.e. varying between one-fifth of the shell diameter to the shell diameter). Although different types of baffles are there, 25% cut segmental baffles are very common. This means that a flow passage or 'window' of area corresponding to 25% shell diameter is available for the shell fluid. The Colburn factor chart given in Fig. 8.11(b) for the shell-side coefficient may also be conveniently used for estimation of the shell-side heat transfer coefficient. Several other correlations and charts for shell-side coefficient are available in the literature (Ludwig, 1983).

Once the individual coefficients are known, U_d can be calculated. The typical ranges of individual and overall coefficients for several common systems are given in Table 8.2. The overall coefficients are generally expressed on *outside tube area basis*.

Table 8.2 Typical ranges of overall heat transfer coefficients (Ludwig, 1983)

Hot fluid	Cold fluid	Btu/h ft^2 °F	kcal/h m^2 °C*
Condensation			
Steam (under pressure)	Water	350–750	1750–3800
Steam (vacuum)	Water	300–600	1500–3000
Saturated organic solvents	Water	100–200	500–1000
Organic solvents (atmospheric, some non-condensibles)	Water, brine	50–120	250–600
Organic solvents (vacuum, high non-condensibles)	Water, brine	10–50	50–250
Organic solvents (low boiling, atmospheric)	Water	80–200	400–1000
Hydrocarbons (low boiling, vacuum)	Water	10–30	50–150
Heating			
Steam	Water	250–750	1250–3800
Steam	Light oils	50–150	250–750
Steam	Organic solvents	100–200	500–1000
Steam	Heavy oils	10–80	50–400
Steam	Gases	5–50	25–250
Vaporization			
Steam	Water	350–750	1750–3800
Steam	Organic solvents	100–200	500–1000
Steam	Light oils	80–180	400–900
Steam	Heavy oils	25–75	125–370
Water	Refrigerants	75–150	370–750
Organic solvents	Refrigerants	30–100	150–500
Heating/cooling (no change of phase)			
Water	Water	150–300	750–1500
Organic solvents	Water	50–150	250–750
Gases	Water	3–50	15–250
Light oils	Water	60–160	250–800
Heavy oils	Water	10–50	50–250
Organic solvents	Light oil	20–70	100–350
Water	Brine	100–200	500–1000
Organic solvents	Brine	30–90	150–450
Organic solvents	Organic solvents	20–60	100–300
Heavy oils	Heavy oils	8–50	40–250

* Multiply by 1.163 to get the value in W/m^2 °C

8.2.6 Pressure Drop Calculation

There are quite a few correlations and charts available for the calculation of pressure drops over the tube and the shell sides of a heat exchanger. One method is described below.

Tube-side pressure drop

Use the following equation (Kern, 1950)

$$\Delta P_t = \frac{f\, G_t^2\, Ln}{2g\rho_t d_i \varphi_t} \tag{8.33}$$

where

    f is the friction factor (as obtained from Fig. 8.13)
    G_t is the mass velocity of the tube fluid, kg/m² s
    L is the tube length, m
    n is the number of tube passes
    g is the gravitational acceleration, 9.8 m/s²
    ρ_t is the density of the tube fluid
    d_i is the inside diameter of a tube
    φ_t is the dimensionless viscosity ratio
    ΔP_t is the pressure drop, kg/m².

The factor φ_t is related to the viscosity ratio as follows:

$$\varphi_t = (\mu/\mu_w)^m; \quad m = 0.14 \text{ for Re} > 2100; \quad m = 0.25 \text{ for Re} < 2100. \tag{8.34}$$

In addition to frictional loss for flow through the tubes, the head loss owing to change of direction of the fluid in a multi-pass exchanger has to be taken into account. This is sometimes called the *return loss*. Kern (1950) suggests that the return loss per pass be taken four times the velocity head. The pressure drop owing to the return loss ΔP_r is given by

$$\Delta P_r = 4n\left(\frac{V^2}{2g}\right)\rho_t \tag{8.35}*$$

where n is the number of the tube passes and V is the linear velocity of the tube fluid. Equation (8.35) is based on the assumption that the end losses amount to four velocity heads per tube pass. The total tube-side pressure drop is

$$\Delta P_T = \Delta P_t + \Delta P_r \tag{8.36}$$

Shell-side pressure drop

For an unbaffled shell the following equation may be used.

$$\Delta P_s = \frac{f_s G_s^2 LN}{2g\rho_s D_H \varphi_s} \tag{8.37}$$

where

    L is the shell length, m
    N is the number of the shell passes
    ρ_s is the shell fluid velocity, m/s
    G_s is the shell-side mass velocity, kg/m² s [see Eq. (8.32)]
    D_H is the hydraulic diameter of the shell, m
    φ_s is the viscosity correction factor for the shell-side fluid.

---

* The coefficient 4 in Eq. (8.35) is quite a bit on the higher side; see Sinnott (1996), p. 610.

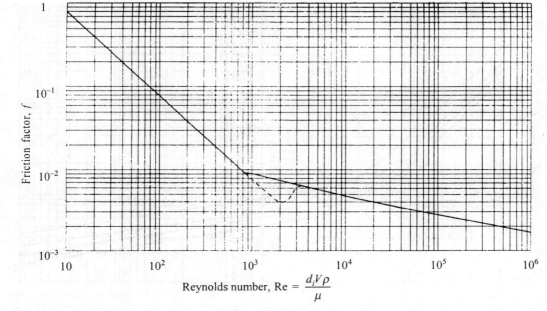

Fig. 8.13 Tube-side friction factor chart.

$$D_H = \frac{4[(\pi D_s^2/4) - (\pi d_o^2 N_t)]}{\pi d_o N_t + \pi d_s} \tag{8.38}$$

where

d_o is the o.d. of the tube, m

D_s is the inside diameter of the shell, m

N_t is the number of tubes in the shell

and

$$\varphi_s = \left(\frac{\mu}{\mu_w}\right)^{0.14} \tag{8.39}$$

For a shell with segmental baffles, a modified form of Eq. (8.37) can be used.

$$\Delta P_s = \frac{f_s G_s^2 D_s (N_b + 1)}{2g\rho_s D_H \varphi_s} \tag{8.40}$$

where

N_b is the number of baffles

D_H is the hydraulic diameter of the shell, m [see Eq. (8.30) or (8.31)].

The Reynolds number of shell-side flow is given by

$$Re_s = \frac{D_H G_s}{\mu} \tag{8.41}$$

The friction factor for the shell-side flow f_s may be obtained from Fig. 8.14 corresponding to the Re_s calculated from Eq. (8.41).

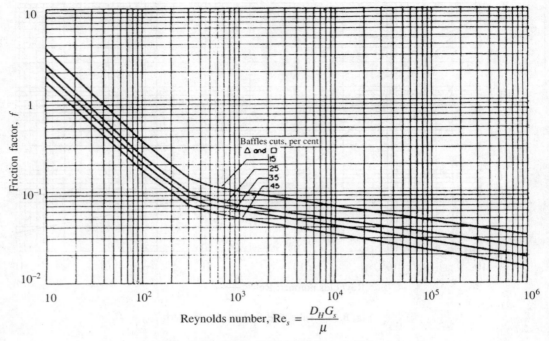

Fig. 8.14   Shell-side friction factor chart.

8.3   DOUBLE-PIPE HEAT EXCHANGER DESIGN PROCEDURE

Basic construction and operation of double-pipe heat exchangers have been discussed in Chapter 4. These exchangers are used when the flow rates of the fluids and the heat duty are small (less than 500 kW). These are also suitable for high pressure services. A typical double-pipe heat exchanger basically consists of a tube or pipe fixed concentrically inside a larger pipe or tube. The outer pipe acts as a jacket. A straight construction with single sections of inner and outer pipes as shown schematically in Chapter 4 is possible. But a *hairpin* construction with two sections each of the inner and the outer pipes is more convenient because it requires less space.

The hairpin construction is shown in Fig. 8.15(a). A packing and gland arrangement near the inlet and exit ends of the assembly provides sealing to the annulus and supports the inner pipe. The opposite ends are joined by a U-bend through union or welded joints. Support lugs may be fitted at these ends to hold the inner pipe in position. The outer pipes are joined by flanges at the return ends in order that the assembly may be opened or dismantled for cleaning and maintenance. Alternatively, the return ends may be sealed by packing and the gland with the inner pipe sections fitted with a U-bend [as shown in Fig. 8.15(b)]. In the latter arrangement, the U-bend of the inner pipe remains outside the jacket.

If the required heat transfer area is large, several hairpins may be connected in series. Such a connection is shown in Fig 8.15(b). All the return bends of the inner pipe are kept outside the jacket and do not contribute to the heat transfer area.

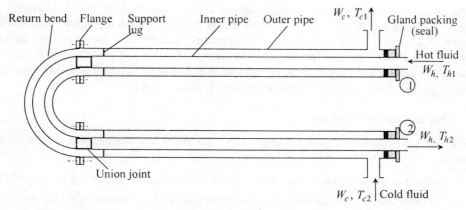

Fig. 8.15(a)   Double-pipe heat exchanger—the hairpin construction.

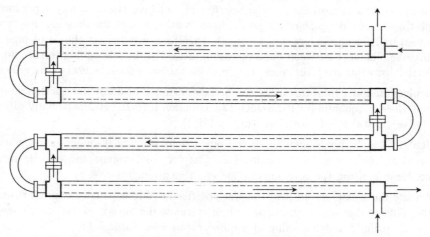

Fig. 8.15(b)   Double-pipe heat exchanger—two hairpins in series.

8.3.1 Energy Balance and Heat Duty Calculation

A heat exchanger design problem is recommended to be defined well as far as possible. The characteristics of the hot and the cold fluids, their flow rates, the terminal temperatures, the inlet pressure of the gaseous stream, if any, and the maximum allowable pressure drop for each stream should be specified. In addition, the thermophysical properties of the fluids (density, viscosity, specific heat, thermal conductivity, etc.) should be given or found out from the literature. Feedback data on the performance of an existing equipment for a similar application are greatly useful in improving the design not only of heat exchangers but also of a variety of other process equipment.

An overall heat balance for the countercurrent double-pipe exchanger shown in Fig. 8.15(a) may be written as follows:

$$Q = W_c c_{pc}(T_{c1} - T_{c2}) = W_h c_{ph}(T_{h1} - T_{h2}) \tag{8.42}$$

The meanings of the subscripts and notations are: c, cold fluid; h, hot fluid; 1 and 2, the terminals shown in Fig. 8.15(a) (in this figure, end 1 is at a higher temperature than end 2); W, flow rate of a stream; c_p, specific heat; T, temperature; and Q, *heat load* or *heat duty* of the exchanger. The above equation has six variables in it— W_c, W_h, T_{c1}, T_{c2}, T_{h1}, T_{h2}. If any five of them are known, the remaining one can be calculated from Eq. (8.42). In this calculation, the heat exchange (gain or loss) with the ambient medium, if any, is neglected.

8.3.2 The Design Procedure

The following procedure may be followed step-by-step for the design of a double-pipe heat exchanger.

  (i) From the known terminal temperatures, calculate the log mean driving force, LMTD.

 (ii) Select the diameters of the inner and the outer pipes. There are no rules for this purpose. It primarily depends upon the flow rates of the streams. If a small diameter pipe is selected, the fluid velocity will be large, i.e. the Reynolds number will also be large. This will no doubt provide a larger heat transfer coefficient, but give rise to a larger pressure drop as well. Conversely, if a pipe of bigger diameter is used, though the pressure drop will be less, the heat transfer coefficient will be small. Therefore, a judicious choice of pipe diameters will strike a balance between these opposing factors. If the allowable pressure drops for the individual streams are given, these may provide the basis for selection of the pipe diameters.

(iii) Calculate the inner fluid Reynolds number; estimate the heat transfer coefficient h_i from the Dittus-Boelter equation [or the Sieder-Tate equation if (μ/μ_w) is substantially different from unity] or from the j_H-factor chart [Fig. 8.11(a)].

 (iv) Calculate the Reynolds number of the outer fluid flowing through the annulus. Use the equivalent diameter of the annulus (see Chapter 4). Estimate the outside heat transfer coefficient h_o using the equations or the chart mentioned above.

  (v) Calculate the clean overall heat transfer coefficient (this is usually expressed on the basis of the outside diameter of the inner tube); calculate the design overall coefficient U_d from Eq. (8.2) using a suitable value of the *dirt factor* (see Table 8.1).

 (vi) Calculate the heat transfer area A from Eq. (8.1) (note that for a counterflow double-pipe exchanger, the question of LMTD correction does not arise; take $F_T = 1$). Determine the length of the pipe that will provide the required heat transfer area. Use a number of hairpins in series, if the length is large.

(vii) Calculate the pressure drops of the fluids from Eq. (8.33). Use the Reynolds number calculated above to determine the friction factor from Fig. 8.13.

The procedure given above is illustrated in Example 8.1. If the change in viscosity of any stream is large, it is desirable that the liquid properties be calculated at the 'caloric temperature' [the procedure described in Kern (1950) or in Ludwig (1983) may be used].

It has been mentioned before that if some of the information required for heat balance is not given in the problem, the designer has to make reasonable assumptions for these. We illustrate this by an example. Let us consider that the heat exchanger in Fig. 8.15(a) uses water from a cooling tower in the jacket. But the problem does not specify the inlet or the exit temperature of water. During the summer season in our country, a cooling tower gives water at about 30°C (the temperature will be less in the winter, but may be as high as 35°C in some regions in summer).

So the design temperature of the inlet water to the exchanger can be assumed to be 30°C. A lower inlet water temperature must not be chosen because in that case the heat exchanger will not perform according to the design in the summer season. The exit temperature of water should normally be around 40°C (if a higher exit temperature is allowed, there will be rapid deposition of scale on the tube surface). So, in this example, the inlet and outlet temperatures may be assumed to be 30°C and 40°C (or less than this) depending upon the acceptable driving force at the hot end.

Example 8.1 Benzene from the condenser at the top of a distillation column is cooled at a rate of 1000 kg/h from 75°C to 50°C in a countercurrent double-pipe heat exchanger (Fig. 8.16). The construction of the heat exchanger is of hairpin type with an effective length of 15 m. The inner

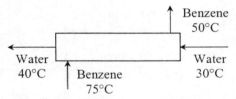

Fig. 8.16 Countercurrent double-pipe heat exchanger (Example 8.1).

tube is of carbon steel, 25 mm o.d., 14 BWG. The outer pipe is schedule 40, 1-1/2 inch nb (nominal bore). Benzene flows through the annulus. Water which flows through the inner tube, entering at 30°C and leaving at 40°C, is the coolant.

(a) Calculate the heat duty of the exchanger and the water flow rate.

(b) Calculate the individual film coefficients and the overall coefficient based on both inside and outside areas.

(c) Do you think that the tube walls have gathered scale and have been fouled? If so, esimate the fouling factor.

The following data are available:

Inner tube: i.d. = 21 mm; o.d. = 25.4 mm; wall thickness = 2.2 mm; thermal conductivity of the tube wall = 74.5 W/m K.

Outer pipe: i.d. = 41 mm; o.d. = 48 mm.

Thermophysical properties of benzene at the average temperature (62.5°C): specific heat = 1.88 kJ/kg °C; viscosity = 0.37 cP; density = 860 kg/m³; thermal conductivity = 0.154 W/m K.

Properties of water at the average temperature (35°C): viscosity = 0.8 cP; thermal conductivity = 0.623 W/m K.

SOLUTION (a) 1000 kg of benzene is cooled from 75°C to 50°C per hour. Therefore,

$$\text{Heat duty} = (1000 \text{ kg/h})(1.88 \text{ kJ/kg °C})(75 - 50)°C = \boxed{47{,}000 \text{ kJ/h}}$$

Water is heated from 30°C to 40°C.

Therefore,

$$\text{Water rate} = \frac{47,000}{(4.187)(10)} = \boxed{1122 \text{ kg/h}}$$

(b) Tube-side (water) calculations

Specific heat of water = 4.187 kJ/kg °C; d_i = 21 mm = 21×10^{-3} m; μ for water = 0.8 cP = 8×10^{-4} kg/m s

Flow area = $(\pi/4)(21 \times 10^{-3})^2 = 3.46 \times 10^{-4}$ m^2

Flow rate = 1122 kg/h = 1.122 m^3/h

$$\text{Velocity} = \frac{1.122}{(3.46 \times 10^{-4})(3600)} = 0.9 \text{ m/s}$$

$$\text{Reynolds number, Re} = \frac{(21 \times 10^{-3})(0.9)(1000)}{8 \times 10^{-4}} = 23,625$$

$$\text{Prandtl number, Pr} = \frac{c_p \mu}{k} = \frac{(4.187)(1000)(8 \times 10^{-4})}{0.623} = 5.37$$

Use of Dittus-Boelter equation to calculate h_i

$$\text{Nu} = \frac{h_i d_i}{k} = 0.023(\text{Re})^{0.8}(\text{Pr})^{0.3} = (0.023)(23,625)^{0.8}(5.37)^{0.3} = 120$$

Thus,

$$h_i = (120)\left(\frac{k}{d_i}\right) = \frac{(120)(0.623)}{(21)(10^{-3})} = 3560 \text{ W/m}^2 \text{ °C}$$

Outer side (benzene) calculation

Flow area = inner cross-section of the pipe – outer cross-section of the tube

$$= (\pi/4)(41 \times 10^{-3})^2 - (\pi/4)(25.4 \times 10^{-3})^2 = 8.13 \times 10^{-4} \text{ m}^2$$

Wetted perimeter = $\pi(d_i + d_o) = \pi(0.041 + 0.0254) = 0.2086$ m

Hydraulic diameter of the annulus,

$$d_H = \frac{(4)(\text{area})}{\text{wetted perimeter}} = \frac{(4)(8.13 \times 10^{-4})}{0.2086} = 0.0156 \text{ m}$$

Benzene flow rate = 1000 kg/h = (1000)/(860) m^3/h = 1.163 m^3/h

$$\text{Velocity} = \frac{1.163}{(8.13 \times 10^{-4})(3600)} = 0.397 \text{ m/s}$$

For benzene: μ = 0.37 cP = 3.7×10^{-4} kg/m s; c_p = 1.88 kJ/kg °C; k = 0.154 W/m K

Reynolds number, $Re = \dfrac{(0.0156)(0.397)(860)}{3.7 \times 10^{-4}} = 14{,}395$

Prandtl number, $Pr = \dfrac{(1880)(3.7 \times 10^{-4})}{0.154} = 4.51$

Calculation of h_o from the Dittus-Boelter equation (taking $(Pr)^{0.4}$ for cooling)

$$Nu = (0.023)(14395)^{0.8}(4.51)^{0.4} = 89.12$$

$$h_o = (89.12)\left(\frac{k}{d_H}\right) = \frac{(89.12)(0.154)}{0.0156} = \boxed{879.8 \text{ W}/\text{m}^2 \, ^\circ\text{C}}$$

Calculation of clean overall heat transfer coefficient, outside area basis

Using Eq. (8.28), $\dfrac{1}{U_o} = \dfrac{1}{h_o} + \left(\dfrac{A_o}{A_m}\right)\left(\dfrac{r_o - r_i}{k_w}\right) + \dfrac{A_o}{A_i}\left(\dfrac{1}{h_i}\right)$

Values of various quantities:

$$A_o = (\pi)(0.0254)(l)$$

$$A_i = (\pi)(0.021)(l)$$

$$A_m = \frac{(0.0254 - 0.021)(\pi l)}{\ln(0.0254/0.021)} = (0.0231)(\pi l)$$

$$A_o/A_m = \frac{0.0254}{0.0231} = 1.098$$

$$A_o/A_i = \frac{0.0254}{0.021} = 1.21$$

Therefore,

$$\frac{1}{U_o} = \frac{1}{879.8} + (1.098)\left(\frac{0.0254 - 0.021}{(2)(74.5)}\right) + \frac{1.21}{3560} = 0.00151$$

or

$$U_o = \boxed{662.3 \text{ W}/\text{m}^2 \text{ K}}$$

Note that the wall resistance is negligible and the tube-side resistance is small.

Calculation of U_i

$$U_o A_o = U_i A_i$$

or

$$U_i = U_o \frac{A_o}{A_i} = (662.3)(1.21) = \boxed{801.4 \text{ W}/\text{m}^2 \text{ K}}$$

Calculation of LMTD

$$\Delta T_1 = 75 - 40 = 35; \quad \Delta T_2 = 50 - 30 = 20^\circ\text{C}$$

Therefore,

$$\text{LMTD} = \frac{35 - 20}{\ln{(35/20)}} = 26.8°C$$

Now calculate the required area from, $Q = U_o A_o \Delta T_m$

$$Q = \frac{(47,000)(1000)}{3600} = 13,055 \text{ W}$$

Therefore,

$$A_o = \frac{Q}{U_o \Delta T_m} = \frac{13,055}{(662.3)(26.8)} = 0.74 \text{ m}^2 = \text{required area}$$

$$\text{Tube length necessary, } l = \frac{0.74}{(\pi)(0.0254)} = 9.3 \text{ m}$$

(c) The actual length of the tube in the hairpin = 15 m, i.e. the cooling process needs an area considerably larger than the theoretical value. In other words, the actual heat transfer coefficient in the existing hairpin is less than that computed. The exchanger surface must have been fouled.

Calculation of the fouling factor, R_d

Heat transfer area of the hairpin = $(\pi)(0.0254)(15) = 1.197 \text{ m}^2$

Overall heat transfer coefficient with dirt factor (outside area basis),

$$U_{do} = \frac{13,055}{(1.197)(26.8)} = 407 \text{ W/m}^2 \text{ °C}$$

Using Eq. (8.2), we get

$$R_{do} = \frac{1}{U_{do}} - \frac{1}{U_o} = \frac{1}{407} - \frac{1}{662.3} = \boxed{0.000947 \text{ m}^2 \text{ °C/W}}$$

8.4 SHELL-AND-TUBE HEAT EXCHANGER—DESIGN PROCEDURE

It is clear from the foregoing discussion that the design of a double-pipe heat exchanger involves a bit of trial in the selection of the diameters of the inner and the outer pipes. If the heat transfer coefficients are too low or the pressure drop is too high for a set of assumed pipe diameters, the calculations have to be repeated for another set of values of the pipe diameters. The design procedure of a shell-and-tube heat exchanger also involves trials, but is not as simple as that of the double-pipe configuration. The more important tasks in the design process of a shell-and-tube heat exchanger are:

  (i) to determine the heat transfer area required for the given heat duty;
 (ii) to specify the tube diameter, length, and number;
(iii) to find the shell diameter;
(iv) to find the number of shell-and-tube passes;

(v) to arrange the tubes on the tube sheet, or to determine the tube layout; and

(vi) to find the type, size, number, and spacing of the baffles.

The energy balance and heat duty calculations are similar to those of a double-pipe unit. From the known terminal temperatures (some of the quantities may be specified in the problem, some may be selected), the LMTD can be calculated. But the quantities U_d (the overall heat transfer coefficient of the 'dirty' exchanger) and A (the heat transfer area) have considerable mutual dependence. This is because U_d depends, besides other quantities, upon the Reynolds number Re and, for a given liquid flow rate, Re depends upon the size and the number of tubes. Therefore, the heat transfer coefficient, in turn, depends upon the diameter and the number of the tubes. The latter quantities give the heat transfer area. The coefficient U_d is also determined partly by the shell-side coefficient which depends upon the shell-side Reynolds number and which in turn is a function of the tube number, diameter and pitch. Therefore, the interdependence of U_d and A is not fully explicit, and a trial-and-error method of calculation is adopted for the design. One has to (i) start with an *assumed value* of the overall coefficient heat transfer U_d, (ii) calculate the heat transfer area and select the diameter, length, and the number of tubes, and (iii) select the tube layout, shell diameter, baffle type, number, and spacing of baffles. The inside and outside heat transfer coefficients for the tube bundle can now be calculated by using the above data and information. So, the overall coefficient including the fouling factor can now be calculated. The calculated value is checked against the assumed value of U_d. A step-wise description of the design procedure is given below. A variety of softwares (e.g. ASPEN/SP, Design II, HYSIM) are available for the simulation and design of process equipment including heat exchangers.

(i) Perform the energy balance and calculate the exchanger heat duty.

(ii) Obtain the necessary thermophysical properties of the hot and the cold fluid streams at the mean temperature. (Take these properties at the caloric temperature of the hot and the cold fluids, if the variation of the viscosity is large.) The thermophysical properties of many fluids are available from sources like Perry (1984), International Critical Tables (1927), etc. If the necessary data are not available, these may be estimated by using suitable correlations (Reid et al., 1988).

(iii) Select the tentative number of shell and tube passes; calculate the LMTD and the correction factor F_T.

(iv) Assume a reasonable value of the overall coefficient U_d *on the outside tube area basis*. The value of the overall heat transfer coefficient for the given service may be obtained from the literature (see Table 8.2). Calculate the heat transfer area A from Eq. (8.1).

(v) Select the tube diameter, its wall thickness (in terms of BWG or SWG), and the tube length. Calculate the number of tubes required to provide the area A calculated above.

(vi) Select the tube pitch. Select the shell diameter that can accommodate the required number of tubes. A tube-sheet layout table (tube counts) is used for this purpose (see Table 8.3).

(vii) Select the type, size (e.g. percentage cut), number, and spacing of baffles.

(viii) Estimate the tube-side and the shell-side heat transfer coefficients using the methods given in Section 8.2.5. If the estimated shell-side coefficient appears to be small, a closer baffle spacing should be tried and the outside-film coefficient (and also the pressure drop) should be recalculated. If the tube-side coefficient is low, adjust the number of tube passes to increase the Reynolds number, and thereby the heat transfer coefficient. However, this is subject to allowable pressure drop across the exchanger.

Table 8.3   Tube-sheet layout and tube count of a shell-and-tube heat exchanger (Ludwig, 1983)

37	35	33	31	29	27	25	23¼	21¼	19¼	17¼	15¼	13¼	12	10	8	I.D. of Shell (In.)	Tube	Pass
1269	1143	1019	881	763	663	553	481	391	307	247	193	135	105	69	33	3/4″ on 15/16″ Δ	Fixed Tubes	One-Pass
1127	1007	889	765	667	577	493	423	343	277	217	157	117	91	57	33	3/4″ on 1″ Δ		
965	865	765	665	587	495	419	355	287	235	183	139	101	85	53	33	3/4″ on 1″ □		
699	633	551	481	427	361	307	247	205	163	133	103	73	57	33	15	1″ on 1¼″ Δ		
595	545	477	413	359	303	255	215	179	139	111	83	65	45	33	17	1″ on 1¼″ □		
1242	1088	964	846	734	626	528	452	370	300	228	166	124	94	58	32	3/4″ on 15/16″ Δ	Fixed Tubes	Two-Pass
1088	972	858	746	646	556	468	398	326	264	208	154	110	90	56	28	3/4″ on 1″ Δ		
946	840	746	644	560	486	408	346	280	222	172	126	94	78	48	26	3/4″ on 1″ □		
688	608	530	462	410	346	292	244	204	162	126	92	62	52	32	16	1″ on 1¼″ Δ		
584	522	460	402	348	298	248	218	172	136	106	76	56	40	26	12	1″ on 1¼″ □		
1126	1008	882	768	648	558	460	398	304	234	180	134	94	64	34	8	3/4″ on 15/16″ Δ	U Tubes²	
1000	882	772	674	566	484	406	336	270	212	158	108	72	60	26	8	3/4″ on 1″ Δ		
884	778	688	586	506	436	362	304	242	188	142	100	72	52	30	12	3/4″ on 1″ □		
610	532	466	396	340	284	234	192	154	120	84	58	42	26	8	XX	1″ on 1¼″ Δ		
526	464	406	356	304	256	214	180	134	100	76	58	38	22	12	XX	1″ on 1¼″ □		
1172	1024	904	788	680	576	484	412	332	266	196	154	108	84	48	XX	3/4″ on 15/16″ Δ	Fixed Tubes	Four-Pass
1024	912	802	692	596	508	424	360	292	232	180	134	96	72	44	XX	3/4″ on 1″ Δ		
880	778	688	590	510	440	366	308	242	192	142	126	88	72	48	XX	3/4″ on 1″ □		
638	560	486	422	368	308	258	212	176	138	104	78	60	44	24	XX	1″ on 1¼″ Δ		
534	476	414	360	310	260	214	188	142	110	84	74	48	40	24	XX	1″ on 1¼″ □		
1092	976	852	740	622	534	438	378	286	218	166	122	84	56	28	XX	3/4″ on 15/16″ Δ	U Tubes²	
968	852	744	648	542	462	386	318	254	198	146	98	64	52	20	XX	3/4″ on 1″ Δ		
852	748	660	560	482	414	342	286	226	174	130	90	64	44	24	XX	3/4″ on 1″ □		
584	508	444	376	322	266	218	178	142	110	74	50	36	20	XX	XX	1″ on 1¼″ Δ		
500	440	384	336	286	238	198	166	122	90	66	50	32	16	XX	XX	1″ on 1¼″ □		
1106	964	844	732	632	532	440	372	294	230	174	116	80	XX	XX	XX	3/4″ on 15/16″ Δ	Fixed Tubes	Six-Pass
964	852	744	640	548	464	388	322	258	202	156	104	66	XX	XX	XX	3/4″ on 1″ Δ		
818	224	634	536	460	394	324	266	212	158	116	78	54	XX	XX	XX	3/4″ on 1″ □		
586	514	442	382	338	274	226	182	150	112	82	56	34	XX	XX	XX	1″ on 1¼″ Δ		
484	430	368	318	268	226	184	154	116	88	66	44	XX	XX	XX	XX	1″ on 1¼″ □		
1058	944	826	716	596	510	416	358	272	206	156	110	74	XX	XX	XX	3/4″ on 15/16″ Δ	U Tubes²	
940	826	720	626	518	440	366	300	238	184	134	88	56	XX	XX	XX	3/4″ on 1″ Δ		
820	718	632	534	458	392	322	268	210	160	118	80	56	XX	XX	XX	3/4″ on 1″ □		
562	488	426	356	304	252	206	168	130	100	68	42	30	XX	XX	XX	1″ on 1¼″ Δ		
478	420	362	316	268	224	182	152	110	80	60	42	XX	XX	XX	XX	1″ on 1¼″ □		
1040	902	790	682	576	484	398	332	258	198	140	94	XX	XX	XX	XX	3/4″ on 15/16″ Δ	Fixed Tubes	Eight-Pass
902	798	694	588	496	422	344	286	224	170	124	82	XX	XX	XX	XX	3/4″ on 1″ Δ		
760	662	576	490	414	352	286	228	174	132	94	XX	XX	XX	XX	XX	3/4″ on 1″ □		
542	466	400	342	298	240	190	154	120	90	66	XX	XX	XX	XX	XX	1″ on 1¼″ Δ		
438	388	334	280	230	192	150	128	94	74	XX	XX	XX	XX	XX	XX	1″ on 1¼″ □		
1032	916	796	688	578	490	398	342	254	190	142	102	68	XX	XX	XX	3/4″ on 15/16″ Δ	U Tubes²	
908	796	692	600	498	422	350	286	226	170	122	82	52	XX	XX	XX	3/4″ on 1″ Δ		
792	692	608	512	438	374	306	254	194	146	106	70	48	XX	XX	XX	3/4″ on 1″ □		
540	464	404	340	290	238	190	154	118	90	58	38	24	XX	XX	XX	1″ on 1¼″ Δ		
456	396	344	300	254	206	170	142	98	70	50	34	XX	XX	XX	XX	1″ on 1¼″ □		
37	35	33	31	29	27	25	23¼	21¼	19¼	17¼	15¼	13¼	12	10	8	I.D. of Shell (In.)		

[1]Allowance made for tie rods.

[2]R.O.B. = 2½ × tube dia. Actual number of "U" tubes is one-half the above figures.

[a]3/4 in. tube has 0.1963 sqft/ft, a 1 in. OD has 0.2618 sqft/ft. Allowance made for tie rods.

[b]R.O.B. = 2½ × tube dia. Actual number of "U" tubes is one-half the above figures.

(ix) Calculate the 'clean' overall coefficient U on the 'outside tube area basis'. Select the 'dirt factor' R_d applicable to the given system and service. Calculate the overall coefficient U_d from Eq. (8.2). Calculate the area A based on this U_d value from Eq. (8.1).

(x) Compare the calculated U_d and A values with the values assumed in step (iv) above. If the assumed heat exchanger configuration gives about 10% excess area than that required, it may be acceptable. This provides a reasonable *overdesign* which may be desirable in many cases (however, overdesign is not desirable in some cases). Otherwise, a new configuration in terms of the number and size of the tubes and tube passes, shell diameter, etc. is assumed and the calculations are repeated starting from step (iii).

(xi) Calculate the tube-side and the shell-side pressure drops using the methods given in Section 8.2.6. If a pressure drop value is more than the corresponding allowable value, further adjustments in the heat exchanger configuration will be necessary. The procedure is illustrated in the following example.

Example 8.2 Ethylbenzene* is manufactured by alkylation of benzene with ethylene. The reaction products are separated in a series of distillation columns. Ethylbenzene is obtained as the distillate from the last column (called the ethylbenzene column).

In a 50 tpd ethylbenzene plant, the final product leaves the overhead condenser at 135°C. This is required to be cooled to 40°C before pumping it to the storage tank.

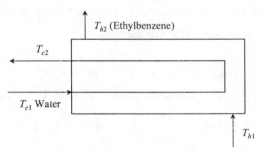

Fig. 8.17 Ethylbenzene cooler (Example 8.2).

Cooling water is available at 30°C for cooling. Design a shell-and-tube heat exchanger (Fig. 8.17) for this purpose. The allowable pressure drop is 0.15 kg/cm² on both tube and shell sides. The necessary thermophysical property data are given below:

Ethylbenzene at the mean liquid temperature (87.5°C)
   density = 840 kg/m³; specific heat = 2.093 kJ/kg K
   viscosity, $\ln \mu = -6.106 + 1353T^{-1} + 5.112 \times 10^{-3}T - 4.552 \times 10^{-6}T^2$ (μ in cP and T in K)
   thermal conductivity, $k = 0.2142 - 3.44 \times 10^{-4}T + 1.947 \times 10^{-7}T^2$ (k in W/m °C, T in K).
Water (at 35°C)
   density = 993 kg/m³
   viscosity = 8×10^{-4} kg/m s
   specific heat = 4.175 kJ/kg K = 1.0 kcal/kg K
   thermal conductivity = 0.623 W/m K = 0.536 kcal/h m² K

---

*Ethylbenzene is partially dehydrogenated to produce styrene which is the starting material for making polystyrene and a number of synthetic rubbers.

SOLUTION   We will follow the steps given in Section 8.4.

(i) Energy balance:

Plant capacity = 50 tons per day = $\dfrac{50,000}{24}$ = 2083 kg/h

Heat duty, $Q = \dot{m}c_p\Delta T$ = (2083)(2.093)(135 − 40) = 414,173 kJ/h = 98,920 kcal/h

Water flow rate = $\dfrac{414,173}{(4.175)(40 - 30)}$ = 9920 kg/h = 9.99 m³/h

(ii) Ethylbenzene is a light organic. Its properties do not change much with temperature. So its thermophysical properties may be taken at the arithmetic average temperature (87.5°C) of the liquid.

Viscosity at 87.5°C (360.5 K):

$\ln \mu = -6.106 + (1353/360.5) + (5.112 \times 10^{-3})(360.5) - (4.552 \times 10^{-6})(360.5)^2$

μ = 0.33 cP = 3.3 × 10⁻⁴ kg/m s

Thermal conductivity at 87.5°C:

$k = 0.2142 - (3.44 \times 10^{-4})(360.5) + (1.947 \times 10^{-7})(360.5)^2$

= 0.1156 W/m K = 0.0994 kcal/h m K

The properties of water at its mean temperature (35°C) are given in the problem.

(iii) Let us try a 1-1 pass countercurrent exchanger.

LMTD calculation

$$\Delta T_1 \text{ (hot end)} = 135 - 40 = 95°C$$

$$\Delta T_2 \text{ (cold end)} = 40 - 30 = 10°C$$

$$\text{LMTD} = \frac{95 - 10}{\ln (95/10)} = 38°C \; (= 68°F)$$

Assume an overall coefficient (on outside area basis), U_{do} = 350 kcal/h m² °C

$$\text{Area required} = \frac{Q}{U_{do}\text{LMTD}} = \frac{98,920}{350(38)} = 7.44 \text{ m}^2$$

(iv) Select 3/4 inch, 16 BWG tubes (i.d. = 15.7 mm, o.d. = 19 mm), 3000 mm long. Put ethylbenzene on the shell-side and cooling water on the tube-side (this is because a surface in contact with cooling water fouls more quickly; cleaning of the inner surface of a tube can be done easily).

Outer surface area of a tube = $(\pi)(0.019)(3.0)$ = 0.179 m²

Number of tubes required = 7.44 m²/0.179 m² = 42

$$\text{Linear velocity of water} = \frac{\text{Water flow rate}}{\text{Flow area}} = \frac{9.99}{(3600)(0.00813)} = 0.341 \text{ m/s}$$

Flow area = $(42)(\pi/4)(0.0157)^2$ = 0.00813 m²

This linear velocity is too low for water (a velocity above 1 m/s should be maintained). It should be increased at least by about four times. So, let us go for a 1–4 pass exchanger.

(v) Trial exchanger: 1–4 pass: tubes: 3/4 inch, 16 BWG; 3000 mm

$$\text{Area per tube} = 0.179 \text{ m}^2$$

From the tube count table (Table 8.3) select a 1–4 pass 10 inch shell (254 mm) having 3/4 inch, 16 BWG tubes on 1 inch (25 mm) 60° triangular pitch. The total number of such tubes that can be accommodated in a 10 inch shell is 44. Take 44 tubes, 11 in each pass.

(vi) Select 25% cut segmental baffles with 0.15 m (6 inch) baffle spacing.

(vii) Estimation of heat transfer coefficients:

Tube side (water)

$$\text{Flow area} = (\pi/4)(0.0157)^2(11) = 0.00213 \text{ m}^2$$

$$\text{Velocity} = \frac{9.99}{(3600)(0.00213)} = 1.303 \text{ m/s}$$

$$\text{Re} = \frac{d_i u \rho}{\mu} = \frac{(0.0157)(1.303)(993)}{8 \times 10^{-4}} = 2.54 \times 10^4$$

From Fig. 8.11(a),

$$j_H = \frac{h_i d_i}{k} \left(\frac{c_p \mu}{k}\right)^{-1/3} = 85$$

Now

$$\frac{c_p \mu}{k} = \frac{(4.175)(1000)(8 \times 10^{-4})}{0.623} = 5.36$$

Therefore,

$$h_i = (85)(0.536/0.0157)(5.36)^{1/3} = 5075 \text{ kcal/h m}^2 \text{ °C}$$

Shell side (organic)

$c' = (25.4 - 19) \text{ mm} = 6.4 \times 10^{-3} \text{ m}; \quad B = 0.5 \text{ m}; \quad p = 2.54 \text{ cm}; \quad D_s = 10 \text{ inch} = 0.254 \text{ m}.$

From Eq. (8.32), we have

$$\text{Flow area} = a_s = \frac{c' B D_s}{p} = \frac{(6.4 \times 10^{-3})(0.15)(0.254)}{0.0254} = 0.0096 \text{ m}^2$$

$$G_s = \frac{W}{a_s} = \frac{2083}{0.0096} = 2.17 \times 10^5 \text{ kg/m}^2 \text{ h}$$

$$d_o = 19 \text{ mm} = 1.9 \text{ cm}$$

From Eq. (8.31), equivalent diameter for shell-side cross flow,

$$D_H = \frac{4[0.5p)(0.86p) - (\pi d_o^2/8)]}{\pi d_o/2}$$

$$= \frac{4[0.5)(2.54)(0.86)(2.54) - \pi(1.9)^2/8]}{\pi(1.9/2)}$$

$$= 1.82 \text{ cm} = 1.82 \times 10^{-2} \text{ m}$$

$$\text{Re} = \frac{D_H G_s}{\mu} = \frac{(1.82 \times 10^{-2})(2.17 \times 10^5)}{(3600)(3.3 \times 10^{-4})} = 3324$$

From Fig. 8.11(b), $j_H = 32$

$$h_o = (j_H)\left(\frac{k}{D_H}\right)\left(\frac{c_p \mu}{k}\right)^{1/3} = \frac{(32)(0.0994)}{1.82 \times 10^{-2}}(6)^{1/3} = 317 \text{ kcal/h m}^2 \text{ °C}$$

(viii) and (ix) Calculation of U_{do}:

From Eq. (8.28), we have (neglecting tube-wall resistance)

$$\frac{1}{U_{do}} = \frac{1}{h_o} + R_{do} + \frac{A_o}{A_i}R_{di} + \frac{A_o}{A_i h_i}$$

Here $\dfrac{A_o}{A_i} = \dfrac{d_o}{d_i} = \dfrac{19}{15.7} = 1.21$

Take $R_{do} = 0.21 \times 10^{-3}$ h m^2 °C/kcal and $R_{di} = 0.35 \times 10^{-3}$ h m^2 °C/kcal

Thus,

$$\frac{1}{U_{do}} = \frac{1}{317} + 0.21 \times 10^{-3} + (1.21)(0.35 \times 10^{-3}) + \frac{1.21}{5075}$$

or

$$U_{do} = 248 \text{ kcal/h m}^2 \text{ °C}$$

This value of the 'dirty' overall coefficient is considerably lower than the trial overall coefficient 350 kcal/h m^2 °C. A second trial is therefore necessary.

Second trial

Based on the above results we modify the trial exchanger considering the following:

　(i) Use a smaller baffle spacing to increase the shell-side Reynolds number and hence h_o.

　(ii) Use 4500 mm tube length to increase the area per tube.

　(iii) From the tube count table (Table 8.3) it is seen that 48 tubes can be accommodated if a square-pitch tube layout is selected.

Trial exchanger: 1–4 pass, 19 mm tubes on square pitch in a 254 mm (10 inch) shell, 4500 mm tube length. Number of tubes = 48 (Table 8.3), 12 tubes per pass. Also select 25% cut segmental baffles with 0.1 m (4 inch) spacing.

Estimation of heat transfer coefficients:

Tube side

Flow area per pass = $(\pi/4)(0.0157)^2(12) = 0.00232$ m^2

$$\text{Velocity} = \frac{(9.99)}{(0.00232)(3600)} = 1.196 \text{ m/s}$$

$$\text{Re} = \frac{d_i u \rho}{\mu} = \frac{(0.0157)(1.196)(993)}{8 \times 10^{-4}} = 2.33 \times 10^4$$

From Fig. 8.11(a), we have $j_H = \dfrac{h_i d_i}{k}\left(\dfrac{c_p \mu}{k}\right)^{-1/3} = 83$

Therefore,

$$h_i = (83)(0.536/0.0157)(5.36)^{1/3} = 4960 \text{ kcal/h m}^2 \text{ °C}$$

Shell side

$c' = (25.4 - 19)$ mm $= 6.4 \times 10^{-3}$ m; $B = 0.10$ m; $p = 2.54$ cm; $D_s = 0.254$ m

From Eq. (8.32), we have

$$\text{Flow area} = a_s = \frac{c'BD_s}{p} = \frac{(6.4 \times 10^{-3})(0.10)(0.254)}{0.0254} = 0.0064 \text{ m}^2$$

$$G_s = \frac{W}{a_s} = \frac{2083}{0.0064} = 3.25 \times 10^5 \text{ kg/m}^2 \text{ h}$$

$$d_o = 19 \text{ mm} = 1.9 \text{ cm}$$

With tubes on square pitch, the equivalent shell diameter is [see Eq. (8.30)]

$$D_H = \frac{4[(2.54)^2 - \pi(1.9)^2/4]}{\pi(1.9)} = 2.42 \text{ cm} = 2.42 \times 10^{-2} \text{ m}$$

From Eq. (8.41), the Reynold number for shell-side cross flow is

$$\text{Re} = \frac{D_H G_s}{\mu} = \frac{(2.42 \times 10^{-2})(3.25 \times 10^5)}{(3600)(3.3 \times 10^{-4})} = 6620$$

From Fig. 8.11(b), $j_H = 48$. Thus,

$$h_o = (j_H)\left(\frac{k}{D_H}\right)\left(\frac{c_p \mu}{k}\right)^{1/3} = \frac{(48)(0.0994)}{2.42 \times 10^{-2}}(6)^{1/3} = 358 \text{ kcal/h m}^2 \text{ °C}$$

[It may be found that the correlation (8.29) gives $h_o = 340$ kcal/h m^2 °C.]

Overall coefficient (take dirt factors as before):

$$\frac{1}{U_{do}} = \frac{1}{358} + 0.21 \times 10^{-3} + (1.21)(0.35 \times 10^{-3}) + \frac{1.21}{4960}$$

or

$$U_{do} = 272 \text{ kcal/h m}^2 \text{ °C}$$

This is less than the assumed overall coefficient but further calculation is necessary to determine whether the design is acceptable. We have to take into account the LMTD correction factor for the 1–4 pass exchanger. So we shall now calculate the area required and the area available.

Calculation of the area required (on the basis of $U_{do} = 272 \text{ kcal/h m}^2 \text{ °C}$):

We need the LMTD correction factor, F_T. This can be done by using Fig. 8.10(a).

$$\tau_c = \frac{T_{c2} - T_{c1}}{T_{h1} - T_{c1}} = \frac{40 - 30}{135 - 30} = 0.095$$

$$R = \frac{T_{h1} - T_{h2}}{T_{c2} - T_{c1}} = \frac{135 - 40}{40 - 30} = 9.5$$

$$F_T = 0.8$$

$$\text{Area required} = \frac{Q}{U_{do} F_T \text{LMTD}} = \frac{98920}{(272)(0.8)(38)} = 12 \text{ m}^2$$

$$\text{Area available (48 tubes, 19 mm o.d., 4.5 m long)} = \pi(0.019)(4.5)(48) = 12.9 \text{ m}^2$$

$$\text{Per cent excess area} = \frac{12.9 - 12}{12} = 7.5\%$$

Therefore, the design is acceptable. (Note that the small parts of each tube that lie within the tube sheets do not contribute to the heat transfer area.)

Pressure drop calculations:

Tube side
Use Eq. (8.33) to calculate the tube pressure drop.
Flow area (per pass) = 0.00232 m²

$$G_t = \frac{9920 \text{ kg/h}}{(3600 \text{ s/h})(0.00213 \text{ m}^2)} = 1187.7 \text{ kg/m}^2 \text{ s}$$

Tube length, $L = 4.9$ m, $n = 4$ (four tube passes), $\rho_t = 993$ kg/m³, $d_i = 0.0157$ m

$\varphi_t = 1.0$, $g = 9.8$ m/s², Re $= 2.33 \times 10^4$, and friction factor, $f = 0.0037$ [Fig. 8.13]

Tube side pressure drop (Eq. 8.33), $\Delta P_t = \dfrac{f G_t^2 Ln}{2g\rho_t d_i \varphi_t}$

$$= \frac{(0.0037)(1187.7)^2(4.5)(4)}{(2)(9.8)(993)(0.0157)}$$

$$= 307.4 \text{ kg/m}^2 = (307.4)(9.8)$$

$$= 3013 \text{ N/m}^2 = 0.444 \text{ psi}$$

$$\text{Return loss (Eq. 8.35),} \quad \Delta P_r = 4n\frac{V^2\rho_t}{2g}$$

where

$$V = 1.196 \text{ m/s}, \ n = 4, \ g = 9.8 \text{ m/s}^2, \ \rho_t = 993 \text{ kg/m}^3$$

Therefore,

$$\Delta P_r = \frac{(4)(4)(1.196)^2(993)}{(2)(9.8)} = 1159 \text{ kg/m}^2 = 1.136 \times 10^4 \text{ N/m}^2$$

Total tube side pressure drop

$$\Delta P_T = \Delta P_t + \Delta P_r = 307.4 + 1159 = 1466 \text{ kg/m}^2 \ (= 1.44 \times 10^4 \text{ N/m}^2)$$

Shell side
Using Eq. (8.40), we have

$$\Delta P_s = \frac{f_s G_s^2 D_s (N_b + 1)}{2g\rho_s D_H \varphi_s}$$

Shell side, Re = 6620, f_s = 0.052 [from Fig. 8.14], D_s = 0.254 m

$$\rho_s = 840 \text{ kg/m}^3, \ \varphi_s = 1, \text{ and } G_s = 3.25 \times 10^5/3600 \text{ kg/m}^2 \text{ s}$$

Baffle spacing = 0.10 m. Therefore, the number of baffles in the 4.9 m long shell

$$N_b = \frac{4.5}{0.10} - 1 = 45 - 1 = 44$$

Therefore,

$$\Delta P_s = \frac{(0.052)(3.25 \times 10^5/3600)^2(0.254)(44 + 1)}{(2)(9.8)(840)(0.0242)(1)} = 12.2 \text{ kg/m}^2 = 120 \text{ N/m}^2$$

The heat exchanger at a glance

Shell, 0.254 m; tubes, 19 mm dia, 16 BWG; 48 tubes on 25.4 mm square pitch; tube length, 4500 mm; 1 shell pass, 4 tube passes.

Pressure drop: tube side, 1.136×10^4 N/m^2; shell side, 120 N/m^2

Both tube- and shell-side pressure drop values are within the allowable limit.

The more important results of the design of a heat exchanger are presented in the form of a 'Heat Exchanger Data Sheet'. The design of the exchanger of the foregoing example is given in the data sheet (in the format prescribed by TEMA) on page 316. (The data sheet can be completely filled after both thermal and mechanical designs are done.)

The Bell method of shell-side calculations

The 'Bell method' (Schlunder, 1983; Kakac et al. 1981; Sinnott, 1996) is an alternative technique of calculation of the shell-side heat transfer coefficient and pressure drop in a shell-and-tube heat exchanger. The 'Kern method' of calculation described before is based on cross-flow of the shell-side fluid in each section between the baffles. However, true cross-flow does not occur in the shell. The Bell method takes into account the deviations from cross-flow owing to leakage of shell fluid through the gaps between the tubes and the baffle holes, and between the baffles and the

Heat Exchanger Data Sheet (Example 8.2)

1				Job No. **PS-015**		
2	Customer	*ABC Petrochemicals*		Reference No.		
3	Address	*123 Dreamchem, Calcutta*		Proposal No.		
4	Plant Location	*HABRA*		Date	Rev.	
5	Service of Unit	*ETHYLBENZENE COOLER*		Item No.		
6	Size *250–4500* mm	Type	(Hor/Vert) *HORIZONTAL*	Connected in	Parallel	Series
7	Surf/Unit (Gross/Eff.)	*12* m² Shells/Unit	Surf/Shell(Gross/Eff.)			m²
8		PERFORMANCE OF ONE UNIT				
9	Fluid Allocation		Shell Side		Tube Side	
10	Fluid Name		*ETHYLBENZENE*		*WATER*	
11	Fluid Quantity, Total	kg/h	*2083*		*9920*	
12	Vapour (in/out)					
13	Liquid		*2083*	*2083*		
14	Steam					
15	Water				*9920*	*9920*
16	Noncondensable					
17	Temperature (in/out)	°C	*135*	*40*	*30*	*40*
18	Specific Gravity *mean*		*0.84*	*0.84*	*0.99*	*0.99*
19	Viscosity, Liquid *mean*	cP	*0.33*	*0.33*	*0.80*	*0.80*
20	Molecular Weight, Vapour					
21	Molecular Weight, Noncondensable					
22	Specific Heat *mean*	kcal/kg °C	*0.50*	*0.50*	*1.0*	*1.0*
23	Thermal Conductivity *mean*	kcal/h m °C	*0.0994*	*0.0994*	*0.536*	*0.536*
24	Latent Heat	kcal/kg @ °C				
25	Inlet Pressure	kg/cm²				
26	Velocity	m/s				
27	Pressure Drop, Allow./Calc.	kg/cm²	*0.15 / 0.15*		*0.15 / 0.0012*	
28	Fouling Resistance (Min.)		*0.00021*		*0.00035*	
29	Heat Exchanged	*98920*	kcal/h MTD (Corrected)		*30.5*	°C
30	Transfer Rate, Service	*272*	Clean	*298*		kcal/h m² °C
31	CONSTRUCTION OF ONE SHELL				Sketch (Bundle/Nozzle Orientation)	
32			Shell Side	Tube Side		
33	Design/Test Pressure kg/cm²		/	/		
34	Design Temperature °C					
35	No. Passes per Shell		*1*	*4*		
36	Corrosion Allowance mm					
37	Connections	In				
38	Size &	Out				
39	Rating	Intermediate				
40	Tube No. *48* OD *19* mm	Thk (min/avg) *16 BWG* mm	Length *4500* mm;	Pitch *25.4* mm. ◁ 30 △ 60 ⊠ 90 ◇ 45		
41	Tube Type		Material			
42	Shell	ID *254* OD	mm	Shell Cover		(Integ.)(Remov.)
43	Channel or Bonnet		Channel Cover			
44	Tubesheet-Stationary		Tubesheet-Floating			
45	Floating Head Cover		Impingement Protection			
46	Baffles-Cross	Type *SEGMENTAL*	% Cut (Diam/Area) *25*	Spacing: c/c *100* Inlet		mm
47	Baffles-Long		Seal Type			
48	Supports-Tube	U-Bend		Type		
49	Bypass Seal Arrangement		Tube-Tubesheet joint			
50	Expansion Joint		Type			
51	Inlet Nozzle	Bundle Entrance		Bundle Exit		
52	Gaskets-Shell Side		Tube Side			
53	Gaskets-Floating Head					
54	Code Requirements		TEMA Class			
55	Weight/Shell	Filled with Water	Bundle			kg
56	Remarks					
57						
58						
59						
60						
61						

shell, and the bypass flow in the baffle window zone. Suitable correction factors for these phenomena are taken into account. The Bell method, although more involved than the much simpler Kern method, is reported to give more accurate estimates of the shell-side heat transfer coefficient and the pressure drop.

8.5 THE EFFECTIVENESS-NTU METHOD OF HEAT EXCHANGER ANALYSIS

We have discussed the method of design of a heat exchanger for a specified heat duty and known terminal temperatures of the hot and the cold fluids. But it may sometimes be necessary to operate a heat exchanger under conditions different from the design conditions. This will definitely affect the performance of the equipment. Let us consider an example. A heat exchanger has been designed to cool a stream of hot oil from 110°C to 45°C using water that gets heated from 30°C to 38°C. But in the month of December, the cooling water temperature may come down to 20°C. So, the cooling water flow rate has to be reduced in order that the oil does not get cooled below the specified temperature of 45°C. How do we determine the flow rate of cooling water required under the changed operating condition?

This and many other questions about heat exchanger performance under varied conditions may be answered by *the effectiveness-NTU method of heat exchanger analysis*. The method is also useful for comparison between various types of heat exchangers. The effectiveness of a heat exchanger is defined as

$$\text{Effectiveness, } \eta = \frac{\text{Actual heat transfer}}{\text{Maximum possible heat transfer}} \tag{8.43}$$

In the following discussion we shall consider the counterflow situation [see Fig. 8.15(a)] for which the actual heat transfer rate can be calculated using Eq. (8.42).

$$Q = W_c c_{pc}(T_{c1} - T_{c2}) = W_h c_{ph}(T_{h1} - T_{h2}) \tag{8.44}$$

Referring to the counterflow heat exchanger shown in Fig. 8.15(a), it is seen that the maximum value of the difference in temperature is $(T_{h1} - T_{c2})$, i.e. the difference between the highest temperature of the hot fluid and the lowest temperature of the cold fluid. If any of the streams (hot or cold) undergoes this change in temperature, then maximum transfer of heat is possible. But, in any case, the energy received by one stream must be equal to the energy lost by the other. So the stream which undergoes the maximum temperature change must have the lower value of Wc_p, otherwise the energy balance will not be satisfied. Therefore, the maximum possible heat transfer rate may be written as

$$Q_{\max} = (Wc_p)_{\min} (T_{h1} - T_{c2}) \tag{8.45}$$

The actual rate of heat transfer is given by Eq. (8.42) for the flow rates W_c and W_h of the hot and the cold streams, respectively. Now, if the maximum change of temperature occurs to the hot fluid, the heat exchanger effectiveness is given by the ratio

$$\eta_h = \frac{Q}{Q_{\max}} = \frac{W_h c_{ph}(T_{h1} - T_{h2})}{W_h c_{ph}(T_{h1} - T_{c2})} = \frac{T_{h1} - T_{h2}}{T_{h1} - T_{c2}} \tag{8.46}$$

Similarly, if the maximum change of temperature occurs to the cold fluid, the effectiveness is given by

$$\eta_c = \frac{Q}{Q_{max}} = \frac{W_c c_{pc}(T_{c1} - T_{c2})}{W_c c_{pc}(T_{h1} - T_{c2})} = \frac{T_{c1} - T_{c2}}{T_{h1} - T_{c2}} \tag{8.47}$$

The effectiveness η lies between 0 and 1.

We will now develop a relation for the effectiveness for the case of counterflow of the fluids. Let us assume that the cold fluid undergoes the maximum temperature change (i.e. $W_c c_{pc}$ is minimum). The effectiveness is then given by Eq. (8.47).

For the countercurrent flow situation [Fig. 8.15(a)], the equation corresponding to Eq. (4.34) becomes

$$\ln \frac{T_{h1} - T_{c1}}{T_{h2} - T_{c2}} = -UA \left(\frac{1}{W_c c_{pc}} - \frac{1}{W_h c_{ph}} \right) = -\frac{UA}{W_c c_{pc}} \left(1 - \frac{W_c c_{pc}}{W_h c_{ph}} \right)$$

or

$$\frac{T_{h1} - T_{c1}}{T_{h2} - T_{c2}} = \exp\left[-\frac{UA}{W_c c_{pc}} \left(1 - \frac{W_c c_{pc}}{W_h c_{ph}} \right) \right] \tag{8.48}$$

or

$$\frac{T_{h2} - T_{c2}}{T_{h1} - T_{c1}} = \exp\left[\frac{UA}{W_c c_{pc}} \left(1 - \frac{W_c c_{pc}}{W_h c_{ph}} \right) \right] \tag{8.49}$$

From the energy balance equation (8.42), we have

$$T_{h2} = T_{h1} - \frac{W_c c_{pc}}{W_h c_{ph}} (T_{c1} - T_{c2}) \tag{8.50}$$

Substituting for T_{h2} [Eq. (8.50)] on the LHS of Eq. (8.49), we get

$$\frac{T_{h2} - T_{c2}}{T_{h1} - T_{c1}} = \frac{T_{h1} - (W_c c_{pc}/W_h c_{ph})(T_{c1} - T_{c2}) - T_{c2}}{T_{h1} - T_{c1}}$$

$$= \frac{(T_{h1} - T_{c1}) - \frac{W_c c_{pc}}{W_h c_{ph}} (T_{c1} - T_{c2}) + (T_{c1} - T_{c2})}{T_{h1} - T_{c1}} \tag{8.51}$$

Equating the RHSs of Eqs. (8.49) and (8.51), we get

$$1 + \left(1 - \frac{W_c c_{pc}}{W_h c_{ph}} \right)\left(\frac{T_{c1} - T_{c2}}{T_{h1} - T_{c1}} \right) = \exp\left[\frac{UA}{W_c c_{pc}} \left(1 - \frac{W_c c_{pc}}{W_h c_{ph}} \right) \right]$$

Let us call $W_h c_{ph} = C_{max}$ and $W_c c_{pc} = C_{min}$. Then, we have

$$\frac{T_{c1} - T_{c2}}{T_{h1} - T_{c1}} = \frac{\exp\left[(UA/C_{min})(1 - C_{min}/C_{max}) \right] - 1}{(1 - C_{min}/C_{max})}$$

As assumed in this analysis and stated before, the maximum temperature change occurs to the cold fluid. So, $W_c c_{pc}$ shall be minimum and $W_h c_{ph}$ shall be maximum. Therefore,

$$\frac{T_{c1} - T_{c2}}{(T_{h1} - T_{c1}) + (T_{c1} - T_{c2})} = \frac{\exp\left[\dfrac{UA}{C_{min}}\left(1 - \dfrac{C_{min}}{C_{max}}\right)\right] - 1}{\left(1 - \dfrac{C_{min}}{C_{max}}\right) + \exp\left[\dfrac{UA}{C_{min}}\left(1 - \dfrac{C_{min}}{C_{max}}\right)\right] - 1}$$

or

$$\eta_c = \frac{1 - \exp\left[-\dfrac{UA}{C_{min}}\left(1 - \dfrac{C_{min}}{C_{max}}\right)\right]}{1 - \left(\dfrac{C_{min}}{C_{max}}\right)\exp\left[-\dfrac{UA}{C_{min}}\left(1 - \dfrac{C_{min}}{C_{max}}\right)\right]} \tag{8.52}$$

Similarly, it can be shown that the effectiveness based on the maximum temperature change of the hot fluid is the same as above except that now $C_{min} = W_h c_{ph}$ and $C_{max} = W_c c_{pc}$. Therefore, we can use a single expression for both types of effectiveness, keeping in mind that C_{min} means Wc_p of the stream undergoing maximum temperature change.

Also, we call

$$UA/C_{min} = \text{NTU}; \qquad C_{min}/C_{max} = C_r \tag{8.53}$$

where NTU is a dimensionless parameter that stands for the 'number of transfer units' for heat transfer. Then, the common relation for effectiveness becomes

$$\eta = \frac{1 - \exp\left[-\text{NTU}(1 - C_r)\right]}{1 - C_r \exp\left[-\text{NTU}(1 - C_r)\right]} \tag{8.54}$$

For a cocurrent or parallel flow heat exchanger, the effectiveness is given by

$$\eta = \frac{1 - \exp\left[-\text{NTU}(1 + C_r)\right]}{1 + C_r} \tag{8.55}$$

The relations and charts for various other configurations (like multi-pass, crossflow, etc.) are available in the literature. The ratio $C_r = C_{min}/C_{max}$ is called the *capacity ratio*. The above relations, Eqs. (8.54) and (8.55), are applicable to 1-1 pass shell-and-tube exchangers too.

For a countercurrent heat exchanger having one-shell pass and 2, 4, 6, ..., tube passes (with a flow configuration shown in Fig. 8.10(a) for a 1-2 pass unit, for example), the effectiveness and NTU are given by

$$\eta = 2\left[(1 + C_r) + \{(1 + C_r^2)^{1/2}\}\,\frac{1 + \exp\{-(1 + C_r^2)^{1/2}\,\text{NTU}\}}{1 - \exp\{-(1 + C_r^2)^{1/2}\,\text{NTU}\}}\right]^{-1} \tag{8.56}$$

$$\text{NTU} = -(1 + C_r^2)^{-1/2}\,\ln\left[\frac{(2/\eta) - (1 + C_r) - (1 + C_r^2)^{1/2}}{(2/\eta) - (1 + C_r) + (1 + C_r^2)^{1/2}}\right] \tag{8.57}$$

Algebraic relations as well as charts relating NTU and effectiveness of exchangers of other configurations are available (Kays and London, 1984).

The NTU method is simple and useful for calculation of the rate of heat exchange attainable with a particular heat exchanger configuration, if the overall heat transfer coefficient and the area of heat exchange are known. No trial-and-error or iterative method is necessary for the calculation, as it is the case when the 'log mean temperature difference' is used.

Example 8.3 A heat transfer oil (specific heat = 0.454 kcal/kg °C) leaving a hydraulic system at a rate of 10,000 kg/h at 85°C has to be cooled to 50°C before it can be circulated back to the system. It is desired to determine whether an existing 1-2 pass exchanger having an area of 15 m² would be suitable. Water is available at 30°C and must not be heated to above 38°C. An overall heat transfer coefficient (including the dirt factor) of 400 kcal/h m² °C can be attained in the exchanger. Water flows through the shell and the oil through the tubes.

How will the heat transfer rate and the exit oil temperature be affected if the water flow rate is increased by 20%? The overall heat transfer coefficient U_d is assumed to remain unchanged.

SOLUTION   *Given:* For the hot stream (oil), W_h = 10,000 kg/h, c_{ph} = 0.454 kcal/kg °C, T_{h1} = 85°C, T_{h2} = 50°C.

For the cold stream (water), W_c = ?, c_{pc} = 1 kcal/kg °C, T_{c2} = 30°C, T_{c1} = 38°C.

Heat balance: $(W_c)(1)(38 - 30) = (10,000)(0.454)(85 - 50)$

or

$$W_c = 19,862 \text{ kg/h}$$

For the hot stream: $W_h\, c_{ph}$ = (10,000)(0.454) = 4540 kcal/h °C

For the cold stream, $W_c\, c_{pc}$ = (19,862)(1) = 19,862 kcal/h °C

Take the hot stream as the 'minimum stream'. Then, we have

$$C_{min} = 4540 \text{ kcal/h °C}; \qquad C_{max} = 19,862 \text{ kcal/h °C}$$

$$C_r = \frac{C_{min}}{C_{max}} = \frac{4540}{19,862} = 0.2286$$

Effectiveness, $\eta = \dfrac{T_{h1} - T_{h2}}{T_{h1} - T_{c2}} = \dfrac{85 - 50}{85 - 30} = \dfrac{35}{55} = 0.636$

Using Eq. (8.57), we get

$$\text{NTU} = -\{1 + (0.2286)^2\}^{-1/2} \ln \left[\frac{(2/0.636) - (1 + 0.2286) - \{1 + (0.2286)^2\}^{1/2}}{(2/0.636) - (1 + 0.2286) + \{1 + (0.2286)^2\}^{1/2}} \right]$$

$$= 1.1652$$

Given: the overall 'dirty' heat transfer coefficient, U_d = 400 kcal/h m² °C
From Eq. (8.53), we have

$$A = \frac{(\text{NTU})(C_{min})}{U_d} = \frac{(1.1652)(4540)}{400} = 13.2 \text{ m}^2$$

Therefore, the area required is 13.2 m^2. As the area required is about 13.5% more than the area available (15 m^2), therefore, the given heat exchanger will perform the required heat duty.

If the water rate is increased by 20%, the new water rate would be

$$W_c' = 19,862 + (19,862)(0.2) = 23,834 \text{ kg/h}$$

Even under the changed operating condition, oil would be the minimum stream, because C_{min}, U_d, and A all remain unchanged. So, the NTU will also remain unchanged [see Eq. (8.53)].

$$C_{max} = W_c' \, c_{p, \text{water}} = (23,834)(1) = 23,834 \text{ kcal/h } °C$$

$$C_r = \frac{C_{min}}{C_{max}} = \frac{4540}{23,834} = 0.1905$$

Using Eq. (8.56), we get

$$\eta = 2 \left[(1 + 0.1905) + \{1 + (0.1905)^2\}^{1/2} \frac{1 + \exp\left[-\{1 + (0.1905)^2\}^{1/2} (1.1652)\right]}{1 - \exp\left[-\{1 + (0.1905)^2\}^{1/2} (1.1652)\right]} \right]^{-1}$$

$$= 0.645$$

Now,

$$\eta = \frac{T_{h1} - T_{h2}}{T_{h1} - T_{c2}} = \frac{85 - T_{h2}}{85 - 30} = 0.645$$

or

$$T_{h2} = \boxed{49.5°C}$$

So, the temperature of the exit oil will reduce by 0.5°C. This calculation also indicates that the exit oil temperature is not very sensitive to the change in water flow rate.

Previous rate of heat transfer = (10,000)(0.454)(85 − 50) = 158,900 kcal/h

The new rate of heat transfer = (10,000)(0.454)(85 − 49.5) = $\boxed{161,170 \text{ kcal/h}}$

Thus, there will be a 1.5% increase in the rate of heat transfer.

8.6 OTHER TYPES OF SHELL-AND-TUBE EXCHANGERS

Different types of shell-and-tube exchangers have been described in the standard of the Tubular Exchanger Manufacturers' Association (TEMA). A few important types are shown in Fig. 8.18. More details are available in TEMA (1988), Ludwig (1983) and Saunders (1988).

The various items labelled in Fig. 8.18 show the constructional features and components of the exchangers. Several items of the list have already been discussed. Other important items are listed in the 'Key to Fig. 8.18'. Besides shell-side and tube-side nozzles for inlet and outlet of the two streams, a heat exchanger is provided with instrument connection taps or nozzles (item 34) for the measurement of temperature, pressure, level, etc. in the equipment. A drain and a vent are provided in the shell (items 32 and 33). A heat exchanger empties through the drain during shutdown. When the exchanger initially fills, the gases present in it leave through the vent. In a condenser, noncondensible gases tend to accumulate with time and have to be vented out from time to time; otherwise, the rate of condensation may decrease drastically. Lifting lugs (item 36) are used to hold and lift the equipment during shipping, installation or shifting from one place to another. Saddles (item 35) support a heat exchanger. Other items relate to the floating head design.

8.6.1 Floating-head Exchangers

The 'fixed-head' or 'fixed tube-sheet' heat exchanger has been shown in Figs. 8.2 and 8.4. In this type of exchanger, the tube sheets are welded to the shell and no relative motion between the shell and the tube bundle is possible. Considerable thermal stress is generated if the temperature differential between the shell and the tube sides is large. As stated before, an expansion joint or bellow is used on the shell to absorb the thermal stress. An alternative arrangement is the *floating-head design* in which one tube sheet is bolted to a shell flange [Fig. 8.18(a)], while the other tube

1. Stationary Head—Channel	20. Slip-on Backing Flange
2. Stationary Head—Bonnet	21. Floating Head Cover—External
3. Stationary Head Flange—Channel or Bonnet	22. Floating Tube Sheet Skirt
4. Channel Cover	23. Packing Box
5. Stationary Head Nozzle	24. Packing
6. Stationary Tube Sheet	25. Packing Gland
7. Tubes	26. Lantern Ring
8. Shell	27. Tie Rods and Spacers
9. Shell Cover	28. Transverse Baffles or Support Plates
10. Shell Flange—Stationary Head End	29. Impingement Plate
11. Shell Flange—Rear Head End	30. Longitudinal Baffle
12. Shell Nozzle	31. Pass Partition
13. Shell Cover Flange	32. Vent Connection
14. Expansion Joint	33. Drain Connection
15. Floating Tube Sheet	34. Instrument Connection
16. Floating Head Cover	35. Support Saddle
17. Floating Head Flange	36. Lifting Lug
18. Floating Head Backing Device	37. Support Bracket
19. Split Shear Ring	38. Weir
	39. Liquid Level Connection

Key to Fig. 8.18

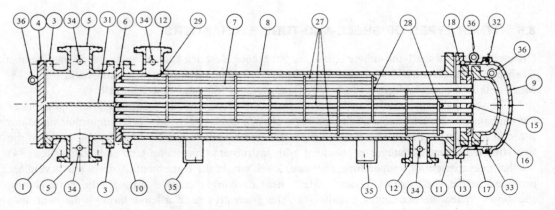

Fig. 8.18(a)   Floating-head heat exchanger.

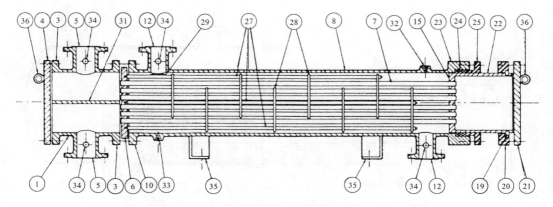

Fig. 8.18(b)   Outside-packed floating-head exchanger.

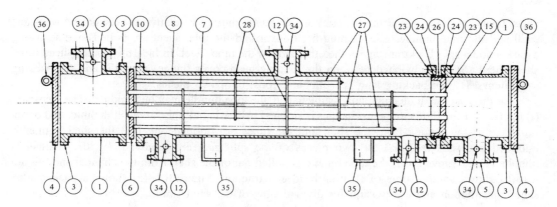

Fig. 8.18(c)   Divided-flow packed tube sheet exchanger.

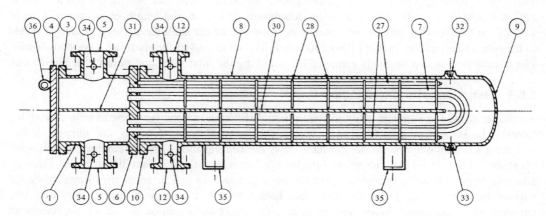

Fig. 8.18(d)   Heat exchanger with a removable U-bundle.

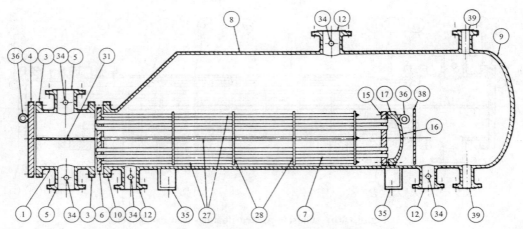

Fig. 8.18(e)   Kettle type reboiler.

sheet floats or slides inside the channel or bonnet permitting a relative movement between the shell and the tube bundle. The floating-head cover and the tube sheet are bolted together using a gasket and a split backing ring on the other side of the tube sheet. In fact, the floating head tube-sheet (item 15) is sandwiched between the floating-head cover flange (item 17) and the backing ring (item 18). The backing ring is 'split' and consists of two pieces.

For the purpose of cleaning and maintenance, the tube bundle may be removed. The steps are: (i) the flow of the fluids to the exchanger is stopped by closing all the valves in the inlet and outlet nozzles; (ii) the liquids from the shell and the tubes are drained out; (iii) the shell bonnet is unbolted and removed, and the two pieces of the split ring are removed; (v) the channel is unbolted and removed; (vi) the tube bundle is pulled out of the shell from the channel end. Figure 8.18(b) shows another type of floating-head construction. Figure 8.18(c) shows a packed tube sheet construction and a provision for divided flow of the shell fluid.

8.6.2   Heat Exchanger with a U-bundle

This type of exchanger [Fig. 8.18(d)] has a single tube sheet. Each tube is bent to form a 'U', and both the ends of the tube are fixed to the tube sheet. However, the length of an 'inner' tube in a bundle is smaller than that of an 'outer' tube. One end of the shell has a flange which is bolted to the tube sheet and the channel flange together. The other end of the shell has a welded closure. The U-tube bundle can be easily removed by unbolting the tube sheet and pulling the bundle out.

8.6.3   Reboilers and Condensers

A reboiler is used to vaporize a liquid stream in a distillation column (or sometimes in a stripping tower). The kettle-type reboiler [Fig. 8.18(e)] is a common device used for this purpose. It has a floating-head tube bundle through which flows the heating fluid. The U-bundles are also used. Most part of the shell has a diameter considerably larger than the tube bundle diameter. Thus, an adequate vapour space is provided above the boiling liquid. Liquid droplets are formed when the vapour bubbles burst. These droplets fall back through the vapour space, thus avoiding entrainment of the liquid. Whenever necessary, the tube bundle can be pulled out together with the floating head. After vaporization, the residual liquid flows over a weir (item 38) and leaves through the liquid outlet nozzle. An impingement plate is usually provided at the vapour outlet [not shown in Fig. 8.18(e); the arrangement is similar to that shown in Fig. 8.5] in order to arrest any fine liquid droplet that may not settle in the vapour space.

Another common type of reboiler, called the *thermosyphon reboiler*, is frequently used in the process industry. A thermosyphon reboiler is a natural circulation type vaporizer generally used to vaporize the bottom liquid of a distillation column. In a horizontal thermosyphon reboiler [Fig. 8.19(a)], a part of the liquid vaporizes in the shell. The vapour–liquid mixture, as the name suggests, flows to the bottom of the column, where the phases get separated. Partial vaporization of the liquid occurs in the tubes of the vertically mounted shell-and-tube exchanger of a vertical thermosyphon reboiler.

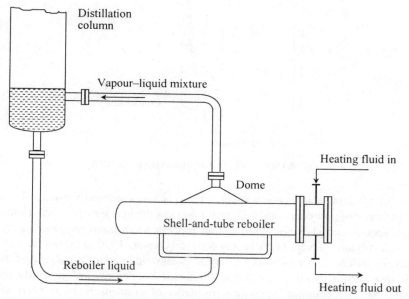

Fig. 8.19(a)   Horizontal thermosyphon reboiler.

Horizontal thermosyphon reboilers are mostly used in the petroleum refining industry, whereas the vertical type [Fig. 8.19(b)] is most common in the chemical industry for liquids of viscosity less than 0.5 centipoise.

About 10–20% of the liquid entering a thermosyphon reboiler gets vaporized (the maximum is about 30% so as to avoid larger pressure drop). The allowable pressure drop across the exchanger is 0.015 to 0.03 kg/cm^2, and the mean temperature driving force is maintained within 50°C. The boiling heat transfer coefficient for non-aqueous systems normally ranges between 1000 to 2500 kcal/h m^2 °C, and the heat flux is maintained within about 50,000 kcal/h m^2. Details of design, operation, and selection of thermosyphon reboilers have been addressed by Kern (1950), Lord, et al. (1970), Frank and Prickett (1973), Collins (1976), Smith (1986), Love (1992), Martin and Sloley (1995), and Sloley (1997).

Surface condensers may be horizontal or vertical, and the vapour may condense in the tube or in the shell (Kern, 1950; Kakac, 1991). The presence of non-condensables and sub-cooling of the condensate are two important factors in the design of condensers. Lord et al. (1970) pointed

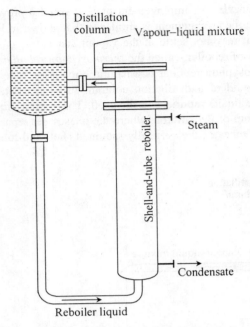

Fig. 8.19(b)   Vertical thermosyphon reboiler.

out a number of advantages of condensation in the tubes in a vertically mounted shell-and-tube device. The presence of non-condensables greatly reduces the heat transfer coefficient. The design of condensers requires an elaborate computation process when non-condensables are present or when there is a mixture of vapour to be condensed (Gloyer, 1970; Mazzarotta and Sebastiani, 1995). A partial condenser, in which only a part of the entering vapour is condensed, is sometimes preferred for use with a distillation column. Shell-side condensers are generally of vertically-mounted type. Several common operating problems of condensers have been discussed by Steinmeyer and Mueller, 1974.

8.6.4   The RODbaffle Exchanger

The RODbaffle heat exchanger (Fig. 8.20), developed in the late seventies, uses arrays of rods inserted between alternate tube rows of the bundle. The rods, having diameter equal to the tube-to-tube spacing, are arranged in the same layout as the tube pitch. These rods are welded to support rings. The first ring has a set of rods, spaced two tube rows apart, welded to it. The next ring has rods filling the alternate spaces to the first ring but oriented at a 90° angle to the first set of rods. The pattern repeats itself along the length of the exchanger. The normal ring spacing is 150 to 200 mm. The rings are fastened to skid bars with suitable spacers.

The RODbaffle technology was originally aimed at eliminating flow-induced tube vibrations (and the resultant possible tube failure). But the device has shown enhanced thermal performance and much lower shell-side pressure drop. This feature of low shell-side pressure drop is an added advantage when the device is used as a condenser. RODbaffles are more commonly used in TEMA *E*- and *X*-shell configurations (see below).

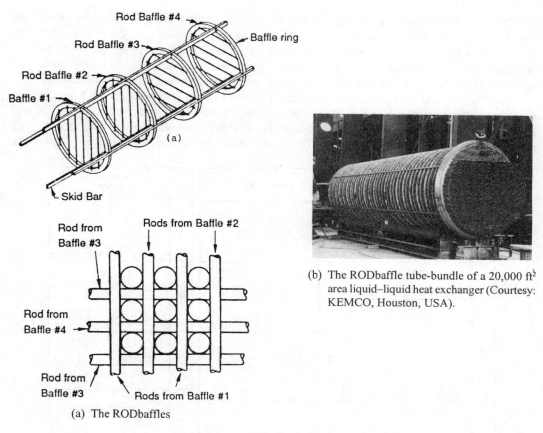

(a) The RODbaffles

(b) The RODbaffle tube-bundle of a 20,000 ft^2 area liquid–liquid heat exchanger (Courtesy: KEMCO, Houston, USA).

Fig. 8.20 The RODbaffle heat exchanger.

8.6.5 Air-cooled Exchangers

An air-cooled heat exchanger consists of a bank of finned tubes carrying a hot fluid with air blowing across the tube bundle. The flow of air may be forced draught or induced draught. An air-cooled exchanger needs a larger area than that needed by the water-cooled shell-and-tube variety and is more expensive. But it has a lower operating cost and is less prone to fouling. Details of air-cooled exchangers are available in the literature (see, for example, Baker, 1980; Saunders, 1988; Mukherjee, 1997).

8.7 CLASSIFICATION OF SHELL-AND-TUBE EXCHANGERS

It has been stated before that the most widely used standard or code for the design of the above type of heat exchangers is that of the Tubular Exchanger Manufacturers' Association (TEMA). The TEMA code specifies the mechanical design procedure, tolerances allowed and the dimensions of the various parts of an exchanger. Three basic classes of standards are given: **C**, **B** and **R**.

The class **C** standards are meant for general purpose exchangers (like water-to-water, light duty hydrocarbon services, air-heaters or coolers, less corrosive light duty services, etc.). The Class **B** standards meet the general requirements of heat exchangers of chemical process industries. The Class **R** standards are meant for rugged construction, heavy duty applications and services. These satisfy the requirements of petroleum processing and related applications.

The TEMA code also classifies the common types of heads or closures (both front and rear heads) and shells as shown in Fig. 8.21. The front heads may be of five types: *A*, *B*, *C*, *N*, and *D*. The type *A* head is virtually a removable channel with a removable cover. This is used with a fixed-tube-sheet or U-tube exchanger. This is the most common type of stationary head. The

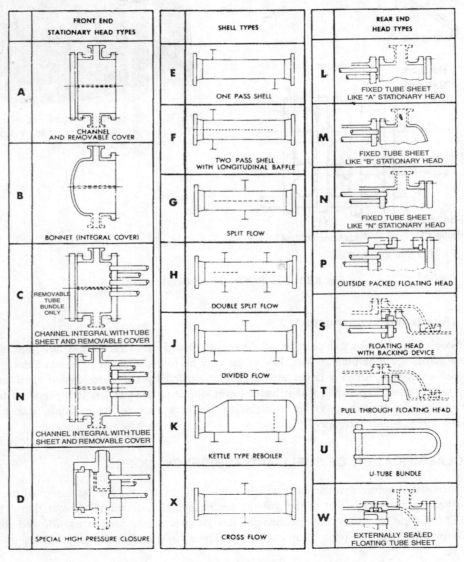

Fig. 8.21 Types of shells and ends (TEMA).

type *B* is a removable channel with an integral cover used with fixed-tube-sheet, U-tube and removable-bundle type exchangers. The type *C* channel is integral to the tube sheet and has a removable cover. There are two variations of this type—one is fitted to the shell by a flanged joint (this is used for the fixed-tube-sheet design). The removable bundle design under type *C* is not convenient. In type *D*, the channel and the tube sheet are of integral forged construction. The channel cover is removable. This is a special type of channel suitable for very high pressure service.

The TEMA standard specifies seven types of shells—the *E, F, G, H, J, K,* and *X* types. The *E*-type shell (*single shell pass*) is the most common design because of its low cost and simplicity. But the *E*-shell is restricted to operating conditions where a 'temperature cross' (Section 8.2.3) does not occur. But if a temperature cross cannot be avoided, the *F*-type shell with two shell passes and two tube passes is used. The *F*-type shell is provided with a longitudinal shell baffle to allow two shell passes. A purely countercurrent flow is achieved in this 2-2 pass configuration. *Split-* and *divided-flow* (types *G, H, J,* and *K*) are recommended for services in which the shell-side heat transfer resistance is small and also when the shell-side pressure drop has to be kept low. The *G*-shell is frequently used in thermosyphon reboilers where the longitudinal baffle reduces the possibility of flashing of the lighter components of the shell liquid. The divided-flow *J*-shell has a pressure drop about one-eighth of that of an *E*-shell of similar size. The *J*-type shell is preferred for condensation services. The *K*-shell is used as a reboiler for pool boiling. The *X*-shell has cross-flow arrangement of the two fluids. It offers a very low pressure drop for the shell fluid. It is often used with finned tubes for gas cooling and vapour condensation.

The design of the *rear head* is more varied. Eight types of rear heads (*L, M, N, P, S, T, U,* and *W*) are described. Types *L, M,* and *N* are similar to front head types *A, B, C* (fixed-tube-sheet type). The type *P* [also see Fig. 8.18(c)] has an outside packed floating head with stuffing box (or lantern ring) sealing. The type *S* [also see Fig. 8.18(a)] has a floating tube sheet held between a split ring and a tube sheet cover. The tube sheet assembly can move freely within the shell cover. The bundle can be taken out after removing the shell cover, the tube sheet cover and the split ring. In type *T*, the floating tube sheet is bolted to the tube-sheet cover. The tube bundle along with the tube-sheet cover can be pulled through while it is removed. The type *U* is used where the tube bundle is made of U-tubes [see also Fig. 8.18(d)]. The type *W* is another floating-head design which uses a packing to separate the tube-side and shell-side fluids. The packing, compressed between the shell flange and the rear cover flange, presses against the tube sheet, thereby providing the sealing.

8.8 MATERIALS OF CONSTRUCTION

A variety of materials—carbon steel, alloy steels, copper alloys, and a few exotic materials (for example, tantalum)—are used for the construction of heat exchangers. A single material need not necessarily be used to build an exchanger. In many heat exchangers, the shell, the tubes, the tube sheets, and the channels are made of different materials. The selection of materials depends upon the types of fluids handled and the operating temperature and pressure.

Water is the most common coolant in heat exchangers and condensers. Water quality, particularly the concentrations of dissolved salts, chlorides, sulphides and oxygen plays an important part in the selection of the tube material. Different types of alloy steels and copper alloys form common tube materials. Type AISI 304 stainless steel resists crevice corrosion up to 200 ppm chlorides and type AISI 316 up to 1000 ppm. The 4-1/2% molybdenum alloys resist such attacks up to 2000 ppm chlorides. An alloy forms a protective coating on the wall in contact with a corrosive liquid (for stainless steel, it is chromic oxide; for a copper alloy in contact with water, it is cuprous hydroxychloride). Both stainless steel and copper alloys show the tendency to corrode under deposited scales or sediments.

Tube sheets may be made of carbon steel, copper alloys, or alloy steels. The problem of galvanic corrosion has to be taken care of while selecting compatible materials for the tubes and the tube sheets. It is not desirable to use carbon steel or stainless steel tube sheets with tubes of copper alloys, because the former is cathodic to the latter. An impressed current cathodic protection unit may be used to overcome the galvanic corrosion problem. Similar considerations should be given to selection of materials for the channels and pass partition plates. Some aspects of material selection for heat exchangers have been discussed by Tuthill (1990). A few practical guidelines regarding materials selection, fabrication, inspection and testing of heat exchanges have been suggested by Greene (1999). Materials specification and testing of heat exchanger tubes have been discussed by Whitcraft (1999).

The selection of materials for a few typical heat exchange services is given in Table 8.4. Nowadays heat exchangers made of special materials are available for highly corrosive liquids (e.g. Wharry, 1996). An example is the all-teflon heat exchanger made by AMETEK (Fig. 8.22).

Table 8.4 Materials of construction for a few typical industrial heating, cooling and condensing services

Application	Fluid pairs handled	Operating temperature (°C)	Recommended material of construction
Heating	Tannin liquor, 10% total solid/ Steam, low pressure	30 – 80 100 – 100	SS to AISI 316
	Urea-ammonia solution/ Ammonia solution	120 – 180 200 – 140	SS to AISI 316L
	Spent (caustic) effluent/ Treated spent caustic	60 – 110 120 – 70	SS to AISI 316
	Methanol/ D&S* steam (low pressure)	30 – 80 100 – 100	SS to AISI 304
Cooling	Ammonia carbamate/ Cooling water	110 – 40 30 – 35	SS to AISI 316
	Treated effluent/ Cooling water	75 – 50 30 – 35	SS to AISI 304
	Sodium dichromate/ Cooling water	80 – 50 30 – 35	SS to AISI 321
	Diethyl oxalate Cooling water	80 – 50 35 – 35	SS to AISI 316
Condensing	Vapour from synthetic resin reactor Cooling water	160 – 160 30 – 35	SS to AISI 316L
	Process vapour (Ammonium sulphate) Cooling water	80 – 80 30 – 35	SS to AISI 316L

*D&S steam means dry and saturated steam.

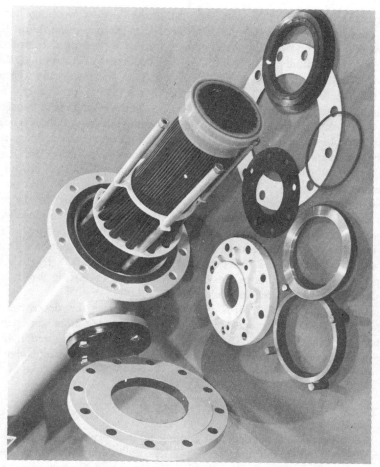

Fig. 8.22 An all-teflon shell-and-tube heat exchanger (Courtesy: AMETEK, Wilmington, Delaware).

8.9 CLEANING OF HEAT EXCHANGERS

Deposition of scales on both inner and outer surfaces of tubes continues to occur when a heat exchanger is in service. The scale thickness grows gradually with time and the fouling resistance increases. After some period of time (depending upon the fluids handled) the heat exchanger cannot be operated at the rated capacity. It has to be cleaned.

The inner tube surface of a heat exchanger is cleaned by poking with a wire brush. The outside tube surface of a fixed-head exchanger has to be cleaned chemically (i.e. by using a solution which dissolves the scale) because the tube bundle cannot be removed. In the case of a floating-head or U-bundle exchanger, the tube bundle is pulled out. Cleaning is done by driving a wire brush across the array of tubes. Better cleaning is achieved if the tubes are on a square pitch rather than on a triangular pitch, because easy movement of the cleaning brush is then possible.

The mechanism of fouling of heat transfer surfaces is not well understood. Six categories of fouling have been identified (Mukherjee, 1998): (i) precipitation fouling—by deposition of dissolved solids or salts (Dissolved calcium and magnesium salts deposit on a surface if the surface is warmer than the water; this occurs because of the "inverse solubility" of these salts.); (ii) particulate fouling—by deposition of suspended particles present in the fluids; (iii) corrosion fouling; (iv) chemical reaction fouling—by deposits formed by chemical reactions among the components present in the fluids; (v) bio-fouling—by attachment of microbiological species on the surface; (vi) solidification fouling—by solidification of some substance because of sub-cooling of the fluid(s). A minimum liquid velocity of 1 m/s (3 ft/s), preferably 1.5 m/s, should be maintained in the tubes to create turbulence not only to achieve a high heat transfer coefficient, but also to reduce deposition of sediments and scales. Fouling occurs rapidly if a heat exchanger, full of liquid, is kept out of service for a considerable period. Fouling in refinery shell-and-tube heat exchangers and its mitigation have been discussed recently by Joshi (1999).

One direct way of reducing fouling is to use very clean water. This is not always a practicable option. Anti-fouling treatment of cooling water has been used and improved over the years. A few novel techniques in the recent times have been used for reducing fouling. In one method, called the SPIRELF system, fine wire springs are put inside the tube and held in place by wires at the ends of the tube. Both axial and radial vibrations are induced in the springs by the flowing fluid. It is claimed that this technique reduces fouling and augments the tube-side heat transfer coefficient by a factor as high as 1.8. Another kind of tube insert, called HiTRAN, consisting of fine wires shaped into a series of loops, has been used. Besides reducing fouling, the device is claimed to enhance the heat transfer coefficient by 25 folds.

Fouling, corrosion (including corrosion that occurs under the deposited scale, called underdeposit corrosion), tube vibration, thermal stress, etc. are the more important factors that may lead to heat exchanger failure. These phenomena and problems and maintenance and refurbishing of heat exchangers have been addressed in an excellent recent review by Yokell (1999).

8.10 HEAT TRANSFER IN AN AGITATED VESSEL

Agitated vessels and tanks with heating and cooling arrangements are frequently used in the process industries as reactors, liquid mixers, crystallisers, etc. Various polymeric resins (phenolic resins, alkyd resins) are prepared in agitated kettles. Emulsion and suspension polymerization (of styrene, vinyl chloride, etc.) are carried out in agitated reactors. Hydrogenation of vegetable oils is carried out in stirred vessels in which the agitator augments the rate of heat transfer and also keeps the catalyst in suspension. The mixing phenomenon in an agitated vessel has been discussed at length in several standard books (e.g. Nagata, 1975; Oldshue, 1983).

8.10.1 Heating and Cooling Arrangements

Heating and cooling arrangements in an agitated vessel may be made in a number of ways: (i) a jacket on the vessel wall; (ii) a helical coil inside the vessel; (iii) a half-pipe or 'limpet coil' on the outer wall; (iv) a dimple jacket on the wall. These are shown in Figs. 8.23 and 8.24. Very often a batch agitated reactor (for example, the alkyd resin reactor) is charged with the feed followed by heating to a desired temperature. As soon as the reaction starts (it may be initiated by adding a catalyst or an initiator) a coolant is pumped through the coil (or jacket) at a prescribed rate in order to remove the heat of reaction and to maintain the desired reaction temperature. In such a case, the heating medium and the cooling fluid (generally the same fluid) are passed through the coil or the jacket in succession.

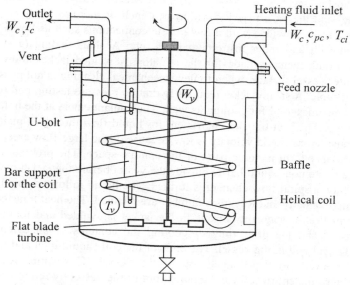

Fig. 8.23 A stirred vessel with a helical heating/cooling coil.

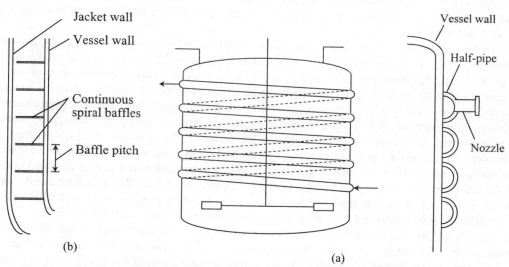

Fig. 8.24 Arrangements for heat transfer in an agitated vessel: (a) limpet (half-pipe) coil on the outer wall and (b) continuous spiral baffles in the jacket of a vessel.

If steam is used as the heating medium, a jacket welded to the vessel wall is convenient. Steam is admitted into the annular space between the vessel wall and the jacket, and the condensate is removed through a steam trap. A vent is provided at the jacket top to remove the non-condensables. A steam jacket is rarely used if the required heating temperature demands a steam pressure above 5 kg/cm^2 gauge. If high temperature (and, therefore, high pressure) steam is used, both the vessel and the jacket are made thick-walled. (Heat transfer fluids are alternative media of heat transfer; see Section 8.14.) This increases the cost substantially. The clearance between the vessel and the jacket (jacket width) is 1 inch or more.

An agitated vessel is provided with a few nozzle connections such as liquid inlet and outlet, vent, drain, thermowell and other instrument ports, etc. A shaft sealing (a stuffing box or a mechanical seal) arrangement is made for the agitator. A welded jacket allows only chemical cleaning of the outer vessel surface, when fouled; mechanical cleaning is not possible because the surface is not accessible. A steam jacket requires hydraulic pressure testing before it is put to use.

A jacket is sometimes used for cooling as well. The coolant enters at the bottom of the jacket and leaves at the top. The heat transfer coefficient for liquid flow through a plain jacket is rather low for two reasons: (i) the liquid velocity is low because of the large flow area, and (ii) stagnant fluid pockets invariably form in some parts of the annular space. The problem can be overcome by spot-welding a continuous spiral baffle outside the vessel before the jacket is welded to it [Fig. 8.24(b)]. In this way, a spiral flow channel is formed. The liquid follows a spiral path (except for leakage), and a high velocity and Reynolds number are attained. The heat transfer coefficient also increases, the formation of stagnant pockets in the jacket is avoided and fouling is reduced.

An alternative technique of heating or cooling the contents of an agitated vessel is the installation of a helical coil in the vessel. The ends of the coil (liquid inlet and outlet) are welded (or flanged or screwed) to the flanged top of the vessel. This simplifies fabrication and maintenance, because the entire coil can be pulled out of the vessel by removing the flanged top. In order to guard against possible damage of a coil by vibration (vibration is mainly created by the agitator), the coil is fastened to a support bar by U-bolts. Small brackets welded to the vessel wall can also be used to hold a coil. Helical coils are generally made of 25 to 50 mm diameter (1 to 2 inch) tubes.

Instead of an external jacket that embraces almost the whole vessel, 'limpet coils' are used more frequently. A limpet coil consists of a spiral half-pipe welded to the outside of the vessel [Fig. 8.24(a)]. This design is quite popular particularly when a liquid (a thermic fluid or water) is used for heating or cooling. A limpet coil can be designed to have a few flow channels running parallel, so as to allow a larger coolant flow and a higher cooling load. The half-pipe channels are generally made of 25 to 75 mm diameter pipes (or tubes). A 'half-pipe' is made by cutting a pipe through its centre-line. For an expensive material like stainless steel, a half-pipe is, however, made from a metal strip of appropriate size. A spacing of three-fourth of an inch is generally maintained between adjacent half-pipes. The jacket fluid wets a part of the outside area of the vessel. But the vessel wall being a very good conductor of heat, the outer heat transfer area would effectively be larger than the wetted area. A common practice is to add 60% of the 'unwetted' area to the wetted area to calculate the effective outside heat transfer area.

It will be relevant to compare the relative advantages and limitations of the above three arrangements of heating and cooling of an agitated vessel. A jacketed vessel is a good choice if low pressure steam is used for heating. A jacket with a spiral baffle is preferably used in a small vessel (less than 2.5 m^3). For large vessels, the limpet design is claimed to be substantially cheaper, particularly when stainless steel or a similar material of construction is used. A helical coil is cheaper than a limpet, but is unsuitable in a corrosive environment (for example, if a rubber or glass-lined vessel is used), or if the liquid in the vessel is highly viscous.

The content of a liquid storage tank (for example, tar) is sometimes required to be kept warm in order to prevent solidification or to keep it pumpable. A simple arrangement to do this is to insert a U-tube bundle and bolt it to a flange fitted to the tank at the bottom (Fig. 8.25). This is also known as *bayonet exchanger*. Steam or any other heating fluid may be used. This heating technique is also used in kettle-type reboilers.

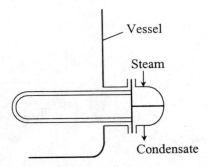

Fig. 8.25   The bayonet heater for a storage tank.

8.10.2   Thermal Design of an Agitated Vessel

Thermal design of an agitated vessel is based on the heating or cooling load calculated by consideration of the process requirements. The flow rate of the jacket or coil fluid is first determined. The heat transfer area is calculated from the heat load, mean temperature driving force and the overall heat transfer coefficient. For an agitated vessel in continuous operation, the calculation of the heat transfer area can be done by using the following well-known equation

$$Q = U_d A \ \Delta T_m \tag{8.58}$$

where ΔT_m is the log mean driving force between the liquid in the tank and the heating (or cooling) medium flowing through the coil or the jacket.

The above simple equation cannot be used for batch heating or cooling where the content of the tank has to be heated or cooled to a desired temperature in a prescribed time, and the driving force is a function of time.

Let us refer to Fig. 8.23 and use the following notations. *Coil liquid:* W_c = flow rate; c_{pc} = specific heat; T_{ci} = inlet temperature (held constant); T_c = outlet temperature at any time t (this is a function of time). *Vessel fluid:* W_v = mass of the fluid; T_{vi} = initial temperature; T_{vf} = final temperature; T_v = temperature at any time t; c_{pv} = specific heat; U_d = overall 'dirty' heat transfer coefficient; A = area of heat transfer; ΔT_m = log mean temperature difference at time t.

We arbitrarily assume that the vessel liquid is being heated. If the temperature of the vessel liquid changes by dT_v within a small time dt, then we may write the following differential heat balance equation:

$$W_v c_{pv} \frac{dT_v}{dt} = W_c c_{pc} (T_{ci} - T_c) = U_d A \ \Delta T_m \tag{8.59}$$

where

$$\Delta T_m = \frac{(T_{ci} - T_v) - (T_c - T_v)}{\ln \dfrac{T_{ci} - T_v}{T_c - T_v}} = \frac{T_{ci} - T_c}{\ln \dfrac{T_{ci} - T_v}{T_c - T_v}} \tag{8.60}$$

From Eqs. (8.59) and (8.60), we get

$$W_c \, c_{pc} \, (T_{ci} - T_c) = U_d \, A \ \frac{T_{ci} - T_c}{\ln \dfrac{T_{ci} - T_v}{T_c - T_v}}$$

On simplification, we get

$$T_{ci} - T_c = (T_{ci} - T_v) \frac{\exp\left(\dfrac{U_d A}{W_c c_{pc}}\right) - 1}{\exp\left(\dfrac{U_d A}{W_c c_{pc}}\right)} \tag{8.61}$$

Putting Eq. (8.61) in (8.59), we have

$$\frac{W_v c_{pv}}{W_c c_{pc}} \frac{dT_v}{dt} = (T_{ci} - T_v)\frac{\beta - 1}{\beta} \qquad \text{where } \beta = \exp\left(\frac{U_d A}{W_c c_{pc}}\right)$$

Integrating we get the time required for heating the vessel liquid from T_{vi} to T_{vf}.

$$\int_{T_{vi}}^{T_{vf}} \frac{dT_v}{T_{ci} - T_v} = \left(\frac{W_c c_{pc}}{W_v c_{pv}}\right)\left(\frac{\beta - 1}{\beta}\right)\int_0^t dt \tag{8.62}$$

Therefore,

$$\ln \frac{T_{ci} - T_{vi}}{T_{ci} - T_{vf}} = \left(\frac{W_c c_{pc}}{W_v c_{pv}}\right)\left(\frac{\beta - 1}{\beta}\right) t \tag{8.63}$$

If the heating fluid temperature remains constant at T_{ci} (for example, if steam condenses in the jacket), the above equation takes the following form.

$$\frac{T_{ci} - T_{vi}}{T_{ci} - T_{vf}} = \exp\left(\frac{U_d A t}{W_v c_{pv}}\right) \tag{8.64}$$

Equation (8.63) or (8.64) can be used to calculate the heat transfer area, if the charge in the vessel is to be heated (or cooled) in a given time t.

8.10.3 Correlations for Individual Coefficients

A number of correlations are available for the estimation of the heat transfer coefficients for heat transfer in agitated vessels depending upon the geometry and the type of the agitator such as the flat-blade turbine, the anchor agitator, etc. (Bondy and Lippa, 1983; Oldshue, 1983; Holland and Chapman, 1966).

For a vessel provided with a six flat-blade turbine agitator and internal helical coil, the following correlations may be used to estimate the 'outside' heat transfer coefficient (h_v).

Baffled or unbaffled vessel

$$\text{Nu} = \frac{h_v D_T}{k} = 0.54(\text{Re}_d)^{0.67} (\text{Pr})^{0.33} \left(\frac{\mu}{\mu_w}\right)^{0.14} \qquad \text{for } \text{Re}_d < 400 \tag{8.65}$$

Normal baffled vessel

$$\text{Nu} = \frac{h_v D_T}{k} = 0.74(\text{Re}_d)^{0.67} (\text{Pr})^{0.33} \left(\frac{\mu}{\mu_w}\right)^{0.14} \qquad \text{for } \text{Re}_d > 400 \tag{8.66}$$

where

$$\text{Re}_d = \frac{d^2 N \rho}{\mu} = \text{the agitator Reynolds number}$$

d is the impeller diameter
D_T is the inside diameter of the vessel
N is the speed of the agitator in rps

For the estimation of the inside heat transfer coefficient in a jacket having spiral baffles, the Sieder-Tate equation with a 'coil correction factor' may be used.

$$\frac{h_i d_e}{k} = 0.027 \, (\text{Re})^{0.8}(\text{Pr})^{0.33}\left(\frac{\mu}{\mu_w}\right)^{0.14}\left(1 + 3.5\frac{d_i}{d_c}\right) \tag{8.67}$$

where

h_i is the heat transfer coefficient for flow through the coil (or a spiral channel)
d_i is the inside diameter of the tube (or the hydraulic diameter in the case of a spiral channel)
d_c is the diameter of the spiral (in the case of a jacket with a spiral channel, take $0.6W_c$ as the 'effective flow rate' of the liquid considering leakage; W_c is the flow rate of the jacket liquid).

The tube-side heat transfer coefficient for a spiral coil can be estimated by using the following correlation.

$$\text{Nu} = \frac{h_c d_i}{k} = 0.021(\text{Re})^{0.85} \, (\text{Pr})^{0.4}\left(\frac{d_i}{D_c}\right)^{0.1}\left(\frac{\mu}{\mu_w}\right)^{0.14} \tag{8.68}$$

where d_i is the inside diameter of the tube and D_c the diameter of the coil.

Thermal design of agitated jacketed vessels and the available correlations have been reviewed recently by Dream (1999).

Example 8.4 A batch polymerization reactor, 1500 mm in diameter and 1800 mm high, has a limpet coil of 18 turns. The inner diameter of the half-pipe is 52.5 mm and the pitch of the coil is 79.5 mm. In each batch, 2200 kg of the monomer (density = 850 kg/m³; specific heat = 0.45 kcal/kg °C; thermal conductivity = 0.15 kcal/h m °C) at 25°C is charged to the reactor that has to be heated to 80°C before the initiator is added to start the polymerization. Heating is done by a thermic fluid available at 120°C (density = 900 kg/m³; specific heat = 0.5 kcal/kg °C; thermal conductivity = 0.28 kcal/h m °C). The average viscosity of the heat transfer fluid may be taken as 4 cP, and that of the monomer as 0.7 cP. The vessel is provided with a flat-blade turbine agitator (six blade, 0.5 m diameter) which rotates at 150 rpm. The volume of the charge is such that, the liquid surface remains nearly at the level of the top of the limpetted region. The height of the limpetted section = 1464 mm. A fouling factor of 0.0002 h m² °C/kcal may be taken for both the vessel and the coil side. Calculate the time required to heat the charge.

SOLUTION Pitch of the coil = 0.0795 m; coil diameter = 0.0525 m; outside area of the vessel covered by the limpetted region = $(\pi)(1.5)(1.464)$ = 6.9 m².

Adding 60% of the 'unwetted' area to the wetted area, the effective outside heat transfer area of the vessel is

$$A_o = \frac{0.0525 + (0.6)(0.0795 - 0.0525)}{0.0794}(6.9) = 5.96 \text{ m}^2$$

Inside heat transfer area of the vessel (same as the outside area, if the wall thickness is small),

$$A_i = 6.9 \text{ m}^2$$

Vessel-side heat transfer coefficient

$$\text{Agitator diameter} = 0.5 \text{ m}; \quad \text{rpm} = 150$$

$$\text{Re}_d = \frac{d^2 N \rho}{\mu} = \frac{(0.5)^2 (150/60)(850)}{(0.7)(10^{-3})} = 7.59 \times 10^5$$

$$\text{Pr} = \frac{(0.45)(0.7 \times 10^{-3})(3600)}{0.15} = 7.56$$

Use Eq. (8.66) to calculate the heat transfer coefficient (take $\mu/\mu_w = 1$)

$$\text{Nu} = (0.74)(7.59 \times 10^5)^{0.67} (7.56)^{0.33} = 12,560$$

$$h_i = (\text{Nu})(k/D_T) = (12,560)(0.15/1.5) = 1256 \text{ kcal/h m}^2 \text{ °C}$$

Coil-side heat transfer coefficient

Take the linear velocity of the heat transfer fluid = 1.5 m/s

$$\text{Flow area of the coil} = (\pi/4)(0.0525)^2 = 2.165 \times 10^{-3} \text{ m}^2$$

$$\text{Flow rate of the fluid} = (1.5)(2.165 \times 10^{-3})(3600) = 11.69 \text{ m}^3/\text{h}$$

$$= (11.69)(850) = 9936 \text{ kg/h} = W_c$$

Hydraulic diameter of the limpet coil, $d_H = \dfrac{(4)(\pi/8)(d_i)^2}{d_i + (\pi/2)d_i} = \dfrac{\pi}{2 + \pi}(0.0525) = 0.0321 \text{ m}$

$$\text{Coil Reynolds number, Re} = \frac{V d_H \rho}{\mu} = \frac{(1.5)(0.0321)(900)}{4 \times 10^{-3}} = 10,820$$

$$\text{Prandtl number of the coil fluid, Pr} = \frac{(0.5)(4 \times 10^{-3})(3600)}{0.28} = 25.7$$

Use Eq. (8.68) to calculate the coil-side coefficient (take $\mu/\mu_w = 1$)

$$\text{Nu} = (0.021)(10,820)^{0.85}(25.7)^{0.4}(0.0321/1.5)^{0.1} = 141$$

$$h_c = (141)(k/d_H) = (141)(0.28/0.0321) = 1230 \text{ kcal/h m}^2 \text{ °C}$$

Overall heat transfer coefficient (based on inside or vessel-side area, A_i)

$$\frac{1}{U_i} = \frac{1}{h_i} + \frac{A_i}{h_c A_o} + R_{di} + R_{dc} = \frac{1}{1256} + \frac{6.9}{(1230)(5.96)} + 0.0002 + 0.0002$$

or

$$U_i = 467 \text{ kcal/h m}^2 \text{ °C}$$

Use Eq. (8.63) to calculate the time of batch heating.

Given: $W_c = 9936$ kg/h; $c_{pc} = 0.5$ kcal/kg °C; $W_v = 2200$ kg; $c_{pv} = 0.45$ kcal/kg °C; inlet temperature of the coil fluid, $T_{ci} = 120$°C; initial temperature of the vessel liquid, $T_{vi} = 25$°C; final temperature, $T_{vf} = 80$°C

$$\beta = \exp\left(\frac{U_i A_i}{W_c c_{pc}}\right) = \exp\left(\frac{(467)(6.9)}{(9936)(0.5)}\right) = 1.913$$

Putting the values of various quantities in Eq. (8.63), we get

$$\ln\frac{120-25}{120-80} = \frac{(9936)(0.5)}{(2200)(0.45)}\left(\frac{1.913-1}{1.913}\right)t$$

or

$$t = \boxed{22 \text{ min}}$$

8.11 COMPACT HEAT EXCHANGERS

Compact heat exchangers constitute a new class of exchangers which emerged as a viable alternative to the conventional shell-and-tube design. As the name implies, a compact heat exchanger accommodates a large heat transfer area in a small volume. Although several types of such heat exchangers have been developed, the more important types are the *plate* and the *spiral* heat exchangers. The latter type includes the *spiral-plate* and the *spiral-tube* exchangers. Some of the modern compact heat exchangers offer heat transfer areas up to 1800 m² and can accommodate liquid flow rates over 40 m³/min while being at least three times lighter than conventional shell-and-tube devices performing the same heat duty. The size-comparison is evident from Fig. 8.26, that shows a conventional tubular exchanger (7 ft 4-1/2 inch in length, 7-7/8 inch in diameter) and a compact brazed plate heat exchanger (20-1/2 inch by 4-1/2 inch) for the same heat duty. For these reasons and for some other advantages the plate and spiral units are now the favoured alternatives to shell-and-tube varieties.

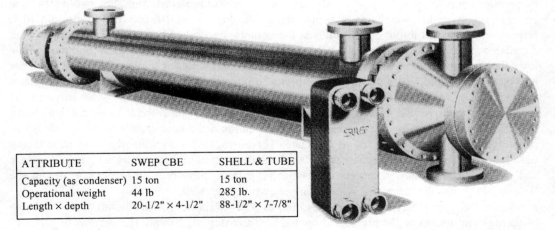

ATTRIBUTE	SWEP CBE	SHELL & TUBE
Capacity (as condenser)	15 ton	15 ton
Operational weight	44 lb	285 lb.
Length × depth	20-1/2" × 4-1/2"	88-1/2" × 7-7/8"

Fig. 8.26 A shell-and-tube heat exchanger and a compact brazed heat exchanger (SWEP CBE) for identical service. (Courtesy: SWEP Inc., New York.)

8.11.1 Plate Heat Exchangers

The plate heat exchanger (PHE) was first introduced in Germany in the 1930s to meet the hygienic demand of the dairy industry. Of all the available alternatives to conventional shell-and-tube exchangers, the most widely used variety is the plate heat exchanger (PHE). A PHE (Fig. 8.27 gives an exploded view) is made of thin, cold-pressed, corrugated metal plates or sheets fitted into a frame. The plates are supported and aligned by an upper and a lower carrying bar.

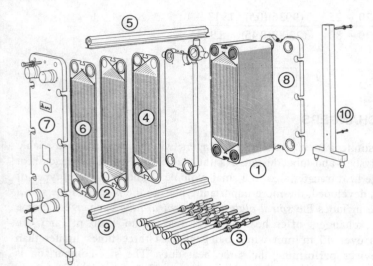

1. Plate pack
2. Corrugated plates with inlet and outlet ports
3. Tie bar
4. Gasket and gasket groove
5. Top bar
6. Flow plate
7. Fixed head with inlet and outlet ports
8. Followerplate with four ports
9. Bottom bar (guide bar)
10. Column

Fig. 8.27 An exploded view of a plate heat exchanger. (Courtesy: APV Baker, Richmond, Virginia.)

In addition to the corrugation, each plate has a depression in the form of a channel (also made by pressing) near the periphery of a plate into which a gasket is placed. Thus, any two plates have an intervening gasket which prevents mixing of the liquids and directs the fluids into their respective flow paths through the portals at the corners. Each plate has four corner holes. When the plates are packed (with gaskets in between), these corner holes form four continuous flow lines for the two fluids; two of these serve as the inlet and outlet of the hot fluid, and the other two of the cold fluid. After putting all the plates in position, two cover plates or pressure plates are put at the two ends and the entire assembly is fastened by tie rods in order to press the corrugated plates over the gaskets to provide a leak-free unit. Typical fluid flow paths are shown in Fig. 8.28. A plate acts as a barrier between the hot and the cold fluid streams which are in countercurrent flow on two sides of the plate. The corrugated design greatly helps in inducing turbulence in the fluids, increasing the effective surface area, and imparting additional strength and rigidity to the plates. A variety of corrugated patterns have been introduced. Plates of the 'herringbone' and the 'chevron' patterns are shown in Fig. 8.29.

Individual plates may have an area as high as 3 m² or more. The depth, angle, and the type of corrugation determine the pressure drop and heat transfer coefficient. The number of plates to be used depend upon the heat duty. The nominal gap between the plates ranges from 2 mm to 5 mm giving hydraulic mean diameter in the range between 4 mm to 10 mm. Big PHE units, 15 ft or more in height, are built nowadays having an enormous heat exchange capacity (Fig. 8.30).

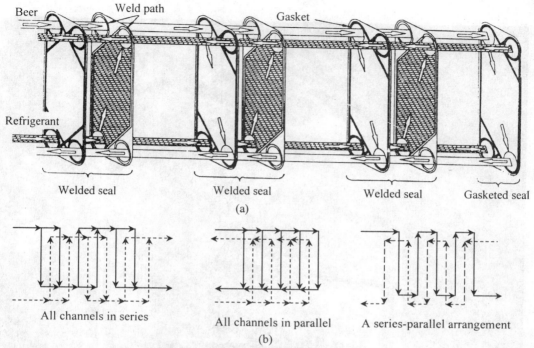

Fig. 8.28 Flow configuration in a plate heat exchanger: (a) countercurrent flow in a PHE with welded plate pairs and (b) typical series-parallel flow arrangements.

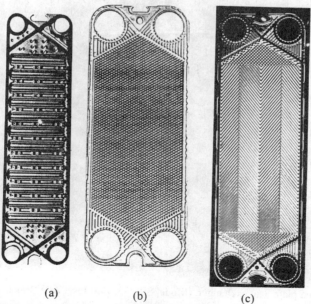

Fig. 8.29 Typical plate designs: (a) troughed plate, (b) chevron plate, and (c) herringbone-type plates. (Courtesy: APV Baker, Virginia.)

Fig. 8.30   An assembled plate heat exchanger. (Model SR 235, Courtesy: APV Baker, Richmond, Virginia.)

Materials of construction

The PHEs are now considered suitable for virtually all kinds of applications, the limitations being the operating pressure and the acceptable material of construction. The plates are made from thin-gauge sheets varying from 0.5 mm to 1 mm in thickness. Because of a large number of contact points between two adjacent plates, a large difference of pressure of the fluids may be allowed even with very thin plates. However, the materials must be corrosion resistant in the given environment, as the maximum allowable corrosion rate is not more than 2 mils per year. Stainless steel is the lowest grade material suitable for PHEs, and even then such an exchanger may cost less than a shell-and-tube unit made of carbon steel for the same service and heat duty. Besides, special alloys like Hestalloy, Incoloy, Inconel, aluminium brass, titanium, tantalum and similar exotic materials are used depending upon the corrosiveness of the fluids. But the plates are *not* made from carbon steel.

Gaskets

Elastomeric materials are used as gaskets. The common gaskets materials are given below:

- Nitrile rubber—which can be used up to a temperature of 110°C for handling liquids such as mineral oils, aqueous solutions, dilute mineral acids, and aliphatic hydrocarbons.
- EPDM (obtained from ethylene-propylene-diene monomer)—which can be used up to a temperature of 160°C for fluids such as mineral and organic acids or bases, aqueous solutions or steam.
- Viton—which is a duPont product based on the copolymer of vinylidene fluoride and hexafluoro-propylene. It can be used up to a temperature of 100°C for mineral oils, hydrocarbons, chlorinated hydrocarbons, etc.
- Resin-cured butyl rubber, neoprene, hypalon, silicone rubber, etc.

Gaskets are traditionally glued to the plates during assembly in order to make a PHE free from leakage. In modern construction, clips and studs are used to secure gaskets to the plates. This new glueless system results in better sealing. It cuts the cost of service and maintenance, as the plates can be easily cleaned or gaskets replaced after removing the tie rods (that act as bolts) and making a space between two adjacent plates. The unit need not be dismantled. A typical gasket seal is shown in Fig. 8.31.

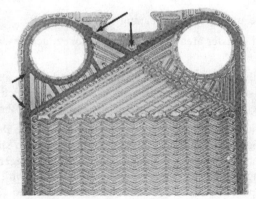

Fig. 8.31 APV double seal of the plate heat exchanger port.

Applications

PHEs are now suitable for a wide variety of applications. Besides liquid-to-liquid heat exchange, these are also used for evaporation, reboiling or condensing services and even for slurries. However, the maximum acceptable solids loading is 40%. Plugging and erosion are the major problems with slurries. Corrosive liquids can be handled by selecting suitable materials of construction.

The maximum temperature, which is limited by the gasket material used, is about 190°C, and the maximum pressure is about 25 kg/cm^2.

Fouling

As in conventional exchangers, fouling in a PHE may occur because of scaling, deposition of solids by crystallization, corrosion, and even by biological materials. But the effect of fouling is much more significant in a PHE. A PHE offers a clean heat transfer coefficient three to five times more than a shell-and-tube type heat exchanger. Consequently, fouling resistance plays a more important role in a PHE. For example, for a given heat duty, a fouling resistance of 1.76×10^{-4} m^2 °C/W (0.001 ft^2 h °F/Btu) will increase the required surface area of a typical water-to-air shell-and-tube unit by about 35%, but will increase the required area of a PHE by about 100%. The allowable fouling resistance in a PHE is about *one-tenth* of that in a shell-and-tube exchanger for similar applications.

Advantages of a PHE

Because of a large number of advantages of the PHEs over their shell-and-tube counterparts, the use of these units has grown over the recent years. The major advantages are listed below:

- A PHE offers a very high heat transfer coefficient on both sides of the plate. For clean water-to-water service, an overall coefficient of about 6000 kcal/h m^2 °C (more than 1000 Btu/h ft^2 °F) is achievable compared to about 2000 kcal/h m^2 °C for a shell-and-tube exchanger at the same pressure drop. Increase in heat transfer coefficient is about three to five times.

- A PHE is suitable even for a close approach temperature as low as 2°C, and for a large temperature cross. (In a shell-and-tube device the approach temperature is rarely allowed to go below 5°C.)

- A PHE offers ease of inspection, cleaning and maintenance.

- In a PHE, the heat transfer area can be increased or reduced by adding or removing some plates.

- A PHE conveniently performs multiple heat exchange duties in a single exchanger. In other words, a PHE construction and arrangement is possible in which two or more streams can be heated or cooled in different sections of the same unit at the same time.

- A PHE requires much less floor space—about one-fifth of the floor space of that required for a shell-and-tube unit for the same heat duty. In addition, a shell-and-tube unit requires additional free space (of about the same length as the exchanger) for removal of the tube bundle for cleaning and maintenance. A PHE needs only a little additional space for this purpose. A PHE gives a heat transfer area between 120 m^2 to 225 m^2 per cubic metre of exchanger volume.

- A PHE offers low hold-up volumes of the fluids.

- A PHE costs less than a shell-and-tube exchanger especially when an expensive material of construction is used.

Table 8.5 shows a cost comparison of a shell-and-tube exchanger with the PHE.

Table 8.5 Cost comparison of a shell-and-tube (tubular) exchanger with a plate heat exchanger

(*Basis:* Identical duty of preheating 0.4 m³ per hour of a process liquor from 30 to 80°C using 0.3 m³ of a hot liquid stream at 104°C.)

Parameter	PHE	Tubular
Flow configuration	Cold and hot streams through alternate passages of the plate pack	Cold fluid in the tube-side, hot fluid in the shell
Number of passes	3–3	1–1
Heat transfer area	1.1 m²	5.0 m²
Pressure drop	0.2 bar on either side	tube-side : 0.2 bar shell-side: 0.15 bar
Fouling	10% area overdesign	tube-side : 0.005 shell-side: 0.005
Material of construction	SS-316 pressed heat transfer plates fitted with nitrile rubber gaskets; SS-316 end connections; other parts, including frame, of carbon steel	Both tube and shell sides made of SS-316
Installation area	0.25 m²	2.35 m²
Installed cost (relative)	1	3 to 4

The major limitations of a PHE concern temperature and pressure as discussed in a previous section. There are a few other limitations too. If steam is used as the heating fluid, rapid fluctuations in steam temperature and pressure reduce the gasket life. Because the pressure drop is relatively high, PHEs are not generally attractive for low pressure and high volume gas heating or cooling applications.

Design considerations

There are only a few correlations and a limited amount of data available in the open literature on the design of PHEs. However, the manufacturers have extensive data, collected by them by laboratory tests of their own *proprietary* plates. They use design correlations and methodologies developed on the basis of their data. Data or correlations for one kind of plate design are not applicable to any other design. The corrugated design is pretty common in the PHEs as it helps to create a high degree of turbulence even at a low Reynolds number. The *critical Reynolds number* (for transition from laminar to turbulent flow pattern) of a PHE ranges between 10 to 400 depending upon the type of plate (i.e. the type of corrugation). The Reynolds number is defined as

$$Re = GD_H/\mu \tag{8.69}$$

where

G is the mass flow rate = W/bs (W is the flow rate per channel, b is the breadth of a plate, s is the nominal gap between two adjacent plates)

D_H is the hydraulic diameter = $4bs/2(b + s)$.

Given below is a correlation for one of the most widely used plates (Marriott, 1971).

$$\text{Nu} = 0.374 \, (\text{Re})^{0.67} \, (\text{Pr})^{1/3} \, (\mu/\mu_w)^{0.15} \qquad (8.70)$$

Charts or correlations for the friction factor are also proprietary information. The steps followed for the design of a PHE are similar to those for the design of tubular exchangers. However, the design is principally based on the 'plates' and their characteristics, developed and patented by giant manufacturers in the field. The general design procedure is as follows:

(i) Calculate the heat load, volumetric flow rates of the streams, terminal temperatures across the exchanger, and LMTD.

(ii) Select the specific construction of the plates suitable for the required service (for example, a wider-gasketed plate for cooling of a viscous liquid such as 50% caustic soda solution).

(iii) Select suitable pass and flow arrangement on the basis of allowable pressure drop for either stream.

(iv) From an estimated flow rate per passage (i.e. the plate velocity), the clean overall heat transfer coefficient and pressure drop per passage can be directly obtained from the plate characteristic curves.

(v) Select fouling resistance and obtain the 'dirty' overall heat transfer coefficient. Calculate the total area requirement from the known heat load, calculated LMTD, and the 'dirty' overall heat transfer coefficient.

(vi) Determine the effective number of plates from the known heat transfer area per plate. Then, calculate the number of plates required which is one more than the effective number of plates.

APV Co. Ltd., Alfa Laval, Trenton (USA) and several others are among the leading manufacturers of plate heat exchangers. Newer and modified versions of the PHE are being developed with claimed advantages. One interesting modification is the APV 'plate-and-shell' heat exchanger (Fig. 8.32). It is built with a stack of plates enclosed in an outer shell. The plates

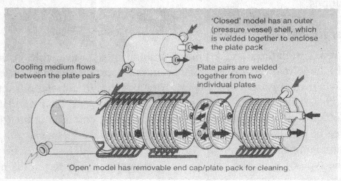

Fig. 8.32   Flow configuration of the APV plate-and-shell heat exchanger.

are welded in pairs and one of the fluids flows through the passages between the welded pairs. The other fluid circulates through the region between the adjacent plate pairs. The advantage of the plate-and-shell construction is claimed to be its higher working pressure (up to 70 bar) and temperature (up to 350°C) limits, while having the compactness of a PHE.

Some recent developments in plate heat exchangers

A number of novel modifications of the PHEs have been made in the recent times. One of these modifications is the *welded-plate exchanger* in which the field gasket (the peripheral part of the gasket) is replaced by a welded joint. The plates are welded at the periphery, but elastomeric or teflon ring gaskets are used to divert the flow of hot and cold fluids to appropriate channels. The use of welded joints reduces the total gasket area by about 90% on the aggressive fluid side and eliminates the possibility of leakage. In a *brazed-plate exchanger*, which is a very recent development, the elastomeric gaskets are replaced by brazed joints, thus eliminating the possibility of leakage. A copper or nickel brazing material can be used for plates made of AISI 316 stainless steel.

A novel, modified form of PHE is the printed circuit heat exchanger (PCHE) developed by Heatric Ltd. (UK) in the late 1980s. A PCHE is a supercompact heat exchanger (comparable with the one shown in Fig. 8.26). It is built from flat metal plates in which fluid passages are made by the chemical milling technique. The technique is similar to that used to prepare an electronic printed circuit board, and hence the name PCHE. The plates are joined by 'diffusion bonding' to make blocks. Diffusion bonding is a welding technique in which the metal surfaces are held together at a very high temperature allowing crystal growth that ultimately causes a bond of strength equal to that of the parent metal. The blocks thus made are welded together to form a heat exchanger core. The depth of the flow passages in a PCHE typically varies from 0.3 to 1.5 mm which is considerably less than that of an ordinary PHE. Such a unit offers a heat transfer area ranging from 500 to 2500 m^2 per cubic metre of core volume, which is several times greater than that of a gasketed PHE, and over an order of magnitude greater than that of a shell-and-tube unit. It is claimed that a PCHE is suitable over a remarkably wide temperature range of about –200°C to 900°C, and pressures up to 150 bar or even higher. A PCHE is typically made from SS 316L sheets, but higher alloys may be used depending upon the service. It can be used for a variety of heavy duty liquid–liquid, gas–liquid, gas–gas heat exchange services, and also for condensation and vaporization duties.

The recent trends in heat transfer devices are very much in favour of PHEs. In Western Europe, the PHEs account for about 65% of the heat exchanger market. However, in the US the figure is about 15–20%.

8.11.2 Spiral-plate and Spiral-tube Heat Exchangers

Besides PHEs, spiral-plate and spiral-tube units are also two important types of compact heat exchangers. The spiral design, as described below, offers the advantage of larger heat transfer coefficient because of the *secondary flow* created in a fluid in spiral flow. They are particularly suitable for handling viscous fluids and slurries, offer reduced fouling rate and can be cleaned relatively easily.

A spiral-plate exchanger is fabricated from two long metal sheets or strips rolled around a mandrel. The sheets are maintained at a short distance apart by spacer studs welded to the sheets. A uniform gap or distance between the plates is thus maintained, forming a pair of concentric spiral passages of uniform cross-section, one for the hot and the other for the cold fluid. The channels are alternately welded on their opposite ends. One of the fluids, normally the hot fluid, enters almost axially and flows spirally outwards and leaves through a nozzle at the periphery. The cold fluid, on the other hand, enters the unit through a peripheral nozzle, flows spirally inwards (that is, countercurrent to the hot fluid), and leaves through the central passage. The configuration is shown in Fig. 8.33. The concentric channel assembly has removable covers with full face gaskets that seal the open ends of the alternate channels so that the liquids are kept separated. The heat transfer surfaces become accessible (for cleaning or maintenance) when the covers are removed.

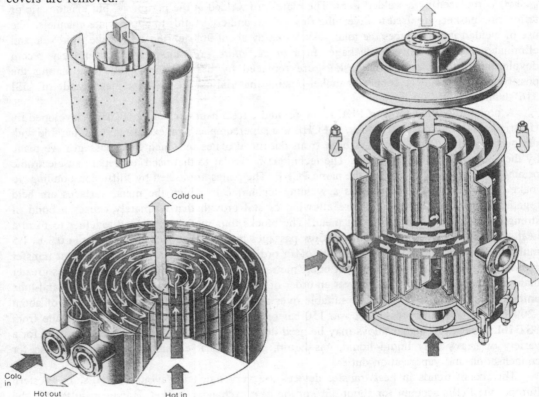

Fig. 8.33   Spiral-plate heat exchanger. (Courtesy: Alfa Laval, Sweden.)

The spiral-plate exchanger has a high thermal efficiency. It operates in truly countercurrent mode. No LMTD correction is required for calculation of the effective temperature driving force. No temperature cross occurs. For this reason, a spiral unit can be operated even at a close temperature approach as low as 2°C. A high degree of turbulence is created by 'secondary flows' and eddies that result from spiral flow path of the fluids. At a velocity that keeps the flow in a tubular exchanger in the laminar regime, the flow in a spiral unit becomes turbulent. This greatly reduces fouling. In fact, the spiral flow pattern has a scouring effect on the wall. The major advantages of a spiral-plate exchanger are: (i) higher heat transfer coefficient; (ii) less fouling, the fouling factor is about one-third or less than that for a tubular exchanger; (iii) well-suited to handle slurries; (iv) well-suited for viscous liquids; (v) can be designed for low flow rates and smaller heat duties compared to the tubular exchangers.

Carbon steel, stainless steel, alloys of nickel, copper or aluminium, Hastelloy B and C are a few materials commonly used for the construction of the spiral-plate heat exchangers. Thermal stresses do not cause any problem for the spiral exchanger because of its spiral construction that can absorb thermal expansion. The normal working pressure limit is 350 psig. Sheets of thickness ranging from 12 gauge (0.105 inch) to 3 gauge (0.25 inch) are used. Heat transfer rates up to 5×10^5 kcal per hour can be achieved. A surface area as high as 400 m^2 can be obtained in a single shell.

A comparison between the characteristics and performance of the spiral-plate units and the shell-and-tube exchangers is given in Table 8.6 (Bailey, 1994).

Table 8.6 Comparison between spiral-plate heat exchangers and shell-and-tube exchangers (Bailey, 1994)

Parameter	Spiral	Shell-and-tube
Operating requirements		
Design temperature, °C	−30 to 800	−70 to 1100
Pressure limit, kg. cm^2	up to 25	up to 200
Maximum flow per shell, m^3/min	10	130
Maximum heat transfer area, m^2	400	2000
Fouling factor, h m^2 °C/kcal	6×10^{-5} to 2×10^{-4}	2×10^{-4} to 2×10^{-3}
Heat Transfer requirements		
Liquid to liquid	Good	Good
Gas to liquid	Good	Good
Gas to gas	Fair	Excellent
Condensation	Excellent	Excellent
Vaporization	Good	Excellent
Temperature cross	Excellent	Fair
Low approach temperature (LMTD < 10°C)	Excellent	Fair
Slurries, fibrous, viscous liquids	Excellent	Fair
Highly fouling services	Excellent	Fair
Mechanical requirements		
Ease of cleaning, one side	Excellent	Excellent (tube-side)
Ease of cleaning, both sides	Excellent	Fair
Chemical cleaning	Excellent	Fair
High temperature differential	Excellent	Fair
Ease of repairing	Fair	Good

A spiral-tube exchanger consists of a pair of concentric tubes spiralled into coils clamped between a cover plate and a casing. A few such coils are arranged one over another, and the assembly is held together by the casing and the cover plate. Both ends of each coil are connected by welding to manifolds for inlet and outlet liquids. Carbon steel, alloy steel, and alloys of nickel and copper are the preferred materials of construction. The working pressure limit is more than that for the spiral-plate units.

8.12 OTHER COMMON HEAT EXCHANGE DEVICES

There are quite a few other types of heat exchanger devices which are preferred for specific heating and cooling services. Finned-tube and coil-type exchangers and jacketed vessels are typical examples. A finned-tube exchanger generally consists of a bundle of finned tubes enclosed in a casing or a box. These are widely used for air heating or cooling, waste heat recovery and as fired heaters. The design procedure for finned-tube bundles is available in the literature (Ganapathy, 1990, 1996). A coil-type exchanger consists of a helical tube or pipe that may be (i) immersed in a pool of hot or cold liquid, (ii) installed in a reactor or a tank for heating or cooling purpose, or (iii) cooled by a spray of liquid over it. These are preferred for cooling services involving small heat load and requiring a heat transfer area less than 2 m². In contrast, a double-pipe exchanger may be used if the required heat transfer area is less than about 5 m². For similar or somewhat larger area requirement, a spiral-plate or spiral-tube exchanger may also be a good choice. Jacketing is a common practice for keeping a reactor vessel hot or cold. For higher heat duties (and consequently larger heat transfer area requirement) one has to go for a shell-and-tube or a PHE.

It may be useful to cite a few typical applications, guided by process requirements, of the different types of exchangers described before. Preheating of a solution of explosive grade ammonium nitrate before evaporation is done in a shell-and-tube exchanger by compressed hot air. A steam-heated finned tube device is used for heating the air. Because controlled heating is required, steam is *not* directly used for this purpose. A coil heater (or cooler) has many typical uses. It is widely used in stirred reactors (for example, in resin kettles), in cooling of a thermic fluid (by spraying water on a coil carrying the hot fluid) and similar applications. Table 8.7 briefly summarizes the application strategy of common types of heat exchangers.

Table 8.7 Some practical applications of exchangers other than the shell-and-tube type

Exchanger type	Applications
Finned tube	Air preheating for evaporation of the feed liquor in an ammonium nitrate concentration plant.
	Air heating for hot air drying operations.
Double pipe	Water cooling of distillate in an alcohol plant.
	Cooling of condensate in a water-based resin plant (the condensate is reused in the system).
Cooling coil	Water (or chilled water) cooling of reaction mass (for removal of heat of reaction) in a stirred reactor.
	Water cooling of heat transfer oil (as an efficient alternative to tubular aftercooler) in a synthetic resin plant.
	Water cooling of spent wash and biodegradable mass in anaerobic reactors of liquid effluent treatment plants.
Plate heat exchanger	Water (or chilled water) cooling of products (hydrogen, chlorine, caustic soda) in a caustic-chlorine plant.
	Low pressure steam-preheating of process liquor in evaporation units.
	Steam-heating of edible oils in refining plants.
	Process-service water heat exchange in power plants.
	Milk chilling and pasteurization in a dairy plant.

8.13 PIPE TRACING

When a highly viscous or molten material is pumped through a pipe, it may be necessary and useful to *trace* the pipe in order to reduce the viscosity (to reduce the pumping cost or to prevent freezing of the material). Here, *tracing* means a suitable arrangement to keep the pipe hot. Thus, a pipe carrying molten sulphur in a sulphuric acid plant is kept hot by tracing. Similarly, a pipe carrying a heavy residue in a refinery, or a syrup in a food industry, has to be traced.

There are three techniques of supplying heat to a pipeline. These are:

- Steam jacketing
- Electrical tracing
- Steam tracing

8.13.1 Steam Jacketing

A *jacketed* pipe consists of an inner pipe carrying the liquid and an outer concentric pipe welded to the inner one, thus forming an annulus. Steam is fed to the jacket to keep the inner pipe hot. The condensate is removed through a steam trap. Jacketed pipes are made in flanged sections (Fig. 8.34), and the diameter may be as high as 500 mm.

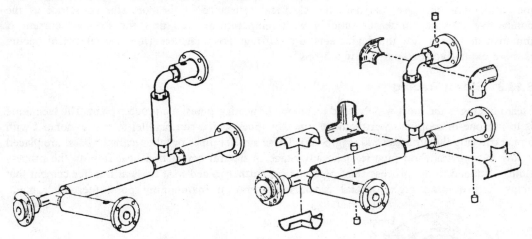

Fig. 8.34 Sections of steam-jacketed pipes (House, 1968).

The steam-jacketed pipe has good heat transfer characteristics. But the installation and maintenance costs are high. A common problem is the formation of cracks that may develop from differential thermal stresses. The problem aggravates if the inner and the outer pipe materials have substantially different coefficients of expansion. If a crack develops, the high pressure steam contaminates the process fluid, and when steam flow is stopped for some reason, the process fluid comes out in the jacket. Serious problems may arise if a crack occurs in a steam traced pipe carrying a highly corrosive liquid.

Among the advantages of the jacketed pipes, fast heat-up rate and even temperature distribution are the major ones. A jacketed system is usually preferred if the liquid viscosity is 500 cP or more. The pipe lines carrying molten sulphur in a sulphuric acid plant are all steam-traced.

In an *internally jacketed* (or *traced*) *system,* a narrow pipe carrying steam runs through the centre of the pipe that carries the process fluid. Though an internal tracing offers as good heat transfer characteristics as the jacketed system, it is associated with several problems such as: (i) the increased pressure drop in the annulus, (ii) the use of low pressure steam which is usually not preferred, (iii) overheating of the liquid in the annulus, (iv) possible corrosion of the inner pipe by the process fluid, and (v) the possibility of failure or crack.

8.13.2 Electrical Heating

Electrical resistance heating was once an important method of keeping a pipeline hot. However, heating wires are now preferred if electrical heating is adopted. For operating temperatures up to about 100°C, one or two resistance wires are moulded into a flat ribbon or strip. For heating to higher temperatures, a metal-clad cable containing electrical resistance wires surrounded by a mineral-filled insulating material is used. Such a cable is attached to the pipe surface by heat transfer cement. The most important advantage of electrical heating is its accurate temperature control. The self-regulating electrical heat tracing is a recent development (Thompson, 1997). A self-regulating heat tracing cable consists of power supply to a conductive polymer core through bus wires. The conductive core is surrounded by an extruded polymer insulator, a metal braid and an outer jacket. When current flows through the conductive polymer core, heat is generated. As the temperature increases, so does the electrical resistance of the core (the resistance of the conductive polymer increases quickly with temperature), and thus the flow of current is diminished. This is how the device acts in a self-regulatory fashion. However, electrical tracing is more expensive for quite a few reasons.

8.13.3 Steam Tracing

Steam tracing is the most widely used technique of heating pipes in a process plant. The technique is to use one or more smaller diameter pipes or tubes which run parallel to and in contact with a pipe carrying the process fluid (Fig. 8.35). These smaller pipes or tubes, called *tracer,* are placed below the process fluid pipe (to take advantage of natural convection heat flow to the process fluid), strapped to the process fluid pipe and are further joined to it by heat transfer cement that helps in providing high thermal conductivity and in maintaining good thermal contact.

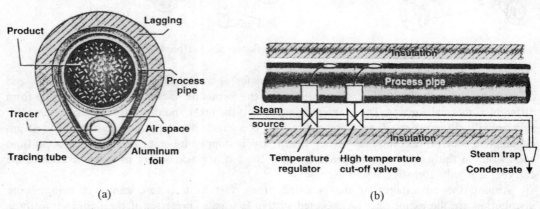

Fig. 8.35   (a) Typical configuration of a steam tracing system and (b) temperature regulation of a steam tracer (Radle, 1997).

Alternatively, the strapped pipe assembly is wrapped by aluminium foil to achieve good thermal contact. The assembly is then insulated as a whole (Baen and Barth, 1994). Steam enters at one end of a tracer and the condensate leaves through a steam trap at the other end.

Copper tubes are preferred for steam tracing applications because they can be not only bent easily but also connected easily by brass fittings. However, at steam pressures above 250 psi, stainless steel tubes are used. Carbon steel tubes are not used for the purpose. Tubes of diameter less than 10 mm (3/8 inch) are not used. Larger diameter tubes are used for tracing runs more than 15 m. The tracer tube should be adequately slopped for flow of the condensate to the steam trap (steam traps are discussed in Chapter 9). Special arrangements need to be done for vertical process pipes. The maximum length of a steam tracer should be limited to 60 m.

Arrangement for temperature control of a traced pipe should be done to avoid overheating. This is especially important for process pipes carrying heat sensitive materials. A temperature control system (temperature-actuated steam valve) and a high temperature cut-off valve may be used for better control [see Fig. 8.35(b)].

More details of steam jacketing and steam tracing are available in House (1968), Bertram et al. (1972), and Radle (1997).

8.14 HEAT TRANSFER FLUIDS

Steam is the most common and low cost heat transfer medium. But with the rise in heating temperature, the pressure of steam increases rapidly. As a result, thick-walled vessels, pipes, and fittings are required for high temperature steam heating. This increases the equipment cost substantially. Heat transfer fluids (also called 'thermic fluids'), in combination with fired heaters, offer an attractive alternative to steam heating. In fact, small oil-fired units that operate with heat transfer fluids in closed circuit are now readily available and used for small and medium heating requirements (Magee and Adams, 1998). Most polymerization reaction kettles use heat transfer fluids for heating.

Heat transfer fluids or media may be petroleum-derived, synthetic organic compounds (Cuthbert, 1994; Guffey, 1997) or molten salts. The first type, or the heat transfer oil, has an operating temperature limit of about 315°C (600°F). Synthetic heat transfer fluids may be used up to 400°C (750°F) without appreciable degradation. The more common heat transfer fluids and their maximum operating temperatures are given in Table 8.8.

Table 8.8 Common heat transfer fluids

Material	Maximum operating temp, °C	Phase	Particulars
Dowtherm A* (a eutectic mixture of diphenyl and diphenyl oxide, 26.5% DP, 73.5% DPO)	400	Liquid/vapour	Maximum temperature = 425°C; auto-ignition temperature = 600°C; freezing point = 12°C; can be used as a vapour under a maximum pressure of 10 atm (abs.).
Partially hydrogenated terphenyl mixtures	340	Liquid	Maximum film temperature = 370°C; considerably viscous at room temperature.

Table 8.8 Common heat transfer fluids (Cont.)

Material	Maximum operating temp, °C	Phase	Particulars
Diphenylethane mixtures	315–340	Liquid	Maximum film temperature = 34–370°C; low viscosity even at room temperature
Dimethyl silicones (Dow 'Syltherm 800')	400	Liquid	Low toxicity
Chlorinated biphenyls (Monsanto 'Therminol')	315	Liquid	Toxic degradation products
Servotherm (Indian Oil Corporation)		Liquid	
Molten salt (eutectic mixture) (NaNO$_3$–27%, NaNO$_2$–20%, KNO$_3$–53%)		Liquid	Melting point = 142°C; alloy steel has to be used at higher temperatures
Molten salt (eutectic mixture) (NaNO$_3$–46%, KNO$_3$–54%)		Liquid	
Mercury	540	Liquid/vapour	Toxic and expensive
Sodium-potassium eutectic	750	Liquid	

*Other varieties of Dowtherm are: Dowtherm G (max. temp. 370°C), Dowtherm Q (max. temp. 340°C).

SHORT QUESTIONS

1. What factors are responsible for the deterioration of performance of a heat exchanger with time?

2. What causes tube vibration? Why is it detrimental? How can it be checked?

3. What type of problem may occur if the shell liquid (i) enters at a high velocity, (ii) has suspended solid particles, or if the entering gas-vapour mixture has entrained droplets? What is the remedy?

4. What routine tests are performed on a heat exchanger before it is installed?

5. What problems are likely to occur with a substantial overdesign say, 40% excess area of (i) a reboiler, (ii) a condenser?

6. Mention a few reasons which can cause leakage between the tube-side and the shell-side.

7. What are the causes of leakage at a pass partition plate? What is the result?

8. What are the functions of the vent and the drain in a heat exchanger?

9. What problem may arise in a condenser if it is not provided with a vent?

10. What safety measure should be taken if the shell-side pressure is high?

11. Name the type of service for which condensation is preferred inside tubes rather than in the shell.

12. A heat exchanger was designed on the basis of a fouling factor corresponding to the cooling water of a particular quality. It was later found that the cooling water contained more dissolved solids than expected. Can you suggest any kind of adjustment in the operating conditions that may help in reducing scaling?

13. What type of problems may arise if the holes in a baffle plate (or tube support plate) are somewhat larger than the tube's outer diameter?

14. A heat exchanger has been designed on the basis of calculated U_c (clean heat transfer coefficient) equal to 50 Btu/h ft^2 °F and a dirt factor of 0.005. No overdesign or excess area is provided. Do you think that the exchanger will have an 'apparent overdesign' for some period of time after installation? What is the apparent per cent excess area at the start?

15. You have performed design calculations of a heat exchanger so as to provide 10% excess area. Now, your boss asks you to provide 20% excess area rather than 10%. How will you do it? By increasing the tube length or by increasing the tube count. Which method is reasonable?

16. A heat exchanger has a leaking tube. What is the simplest practical solution to the problem?

17. What may be the effect of overdesign (or excess area) of an overhead condenser on the performance of a distillation column?

18. To avoid the problem resulting from overdesign of a condenser attached to a distillation column, the operator decides to reduce the cooling water flow rate. As a result, the exit temperature of the cooling water rises to 55°C in place of 46°C (which was the exit temperature for the original cooling water flow rate). Can this adjustment lead to any other problem?

19. Can you deliberately reduce the vapour-side coefficient in a condenser? How?

20. Mention a few symptoms that indicate underdesign of a condenser.

21. Do you think that maldistribution of the tube-side fluid may occur under some circumstances in a heat exchanger?

22. Describe with reasons the types of leakage that can occur in shell-and-tube heat exchangers?

23. How is a heat exchanger cleaned? Give your answer with respect to the tube-side/shell-side of a fixed-head/floating-head exchanger.

24. What attachment is provided for lifting a heat exchanger for the purpose of transportation or installation?

25. A condenser of a hydrocarbon vapour is designed for horizontal installation. Is its capacity likely to be affected if it is installed vertically? Why and how?

26. In a sulphuric acid plant, hot gases leave the sulphur burner at around 1100°C and enter the waste heat boiler. This is simply a shell-and-tube heat exchanger with hot gases flowing through the tubes and water boiling in the shell to generate steam at about 8-10 kg/cm^2. Why is the gas passed through the tube-side and not through the shell-side?

27. Qualitatively, show the temperature profiles of both the hot and cold fluids for a heat exchanger that has an effectiveness of 100% for both parallel and counterflow operations.

28. The TEMA standard stipulates that a horizontal heat exchanger should be installed level (i.e. horizontally, subject to allowable tolerance). Do you think that this may create practical problems during shutdown?

29. A vertical condenser condenses a vapour at above 200°C using cooling water. The nozzle for cooling water exit is located a little below the upper tube sheet. In reality, the narrow region on the shell-side between the water exit nozzle and the upper tube sheet will remain as an air pocket filled with air and water vapour. Do you think that the presence of this empty space below the upper tube sheet may create any problem?

30. What are the relative advantages of using a thermic fluid over steam for heating reactors?

31. Derive Eqs. (8.30) to 8.32.

PROBLEMS

8.1 The demineralized feed water to a reactor for continuous suspension polymerization of acrylonitrile is cooled from 31°C to 25°C in a heat exchanger at a rate of 3.35 m³ per hour. Calcium chloride brine is used as the coolant that enters the exchanger at 7.5°C and leaves at 11.5°C. In consideration of the not-too-high heat load and high heat transfer coefficient attainable for water-to-water heat exchange, it is felt that a double-pipe heat exchanger may be suitable. The working pressure of both the liquids is about 3.5 kg/cm² and the allowable pressure drop is 0.3 kg/cm². Design the exchanger. The following data are available for the cold stream: μ (at the mean temperature) = 1.33 cP; k = 0.51 kcal/h m °C. Fouling factors: 0.0002 kcal/h m² °C (for the cold stream), 0.0001 kcal/h m² °C (for the hot stream).

[*Hint:* Put the chilled water in the jacket and demineralized water in the tube; inner and outer pipes of 1 inch and 2 inch dia may work. In this case the outer pipe or jacket has to be insulated. Length of the exchanger = about 3 metre.]

8.2 The 'green liquor' from a wood pulping unit is required to be preheated from 78°C to 95°C by steam in a tubular heat exchanger before it is fed to the evaporator. It is required to design a shell-and-tube heat exchanger for the purpose. The following data and information are available.

Green liquor: flow rate = 9900 kg/h; properties at the mean fluid temperature: density = 1100 kg/m³, specific heat = 0.94 kcal/kg °C, viscosity = 4 cP, thermal conductivity = 0.3 kcal/h m °C.

Steam: dry and saturated at 140°C.

(*Suggested design:* 1–6 pass, 38 mm tube, 3000 mm long, heating area = 30 m². Also try other configurations like 1–2 and 1–4 pass.)

8.3 Carbon dioxide is compressed in a multi-stage centrifugal compressor for a 2400 ton per day double stream urea plant. Because the gas temperature increases by about 150°C during compression in one stage, an interstage cooler is used to cool the gas before it goes for the next stage of compression. Design a shell-and-tube interstage cooler. The following data are supplied.

Carbon dioxide: flow rate = 44,000 kg/h; inlet temperature = 205°C; outlet temperature = 45°C; inlet gas pressure = 24 kg/cm².

Cooling water (shell side): inlet temperature = 32°C; allowable maximum outlet temperature = 42°C; inlet pressure = 3 kg/cm². Allowable pressure drop: shell side, 0.7 kg/cm²; tube side, 0.25 kg/cm².

Note: Carbon dioxide at the inlet contains about 1% moisture and 75% of this moisture condenses in the cooler. The heat load for moisture condensation should be taken into account, but its contribution to the total heat transfer area may not be large. Thermophysical properties of carbon dioxide at the given operating condition may be obtained from the literature or may be estimated by using the available correlation (see Reid et al. 1988).

A probable design: 3/4 inch tube, 12 ft, 440 nos. in a 27 inch shell; put CO_2 in the tubes; suggested tube material: AISI 316L; 1–2 pass floating head. The shell must be provided with a rupture disk. Why?)

8.4 Feed liquid ammonia to a urea reactor is to be preheated from 25°C to 70°C by using steam condensate water available at 105°C. Design a U-tube shell-and-tube exchanger for the purpose.

Given: flow rate of ammonia (tube side) = 30 tons per hour; working pressure = 210 kg/cm²; allowable pressure drop is 0.5 kg/cm² on both sides. The properties of the hot and the cold streams are available in the literature. [*Hint:* use 19 mm tubes, 6 m long].

8.5. In a petrochemical plant, acrylonitrile is stored in a tank, cooled and at above atmospheric pressure. To remove the heat absorbed by the liquid from the ambient, a stream of acrylonitrile is continuously passed through an external heat exchanger. Heat transfer calculations show that 9.6 tons of the liquid has to be cooled from 25 to 13°C per hour. Chilled water ($CaCl_2$ brine, see Problem 8.1) is available at 7.5°C that may get heated to 11.5°C in the exchanger. The allowable pressure drop is about 0.5 kg/cm² on both sides. Size a shell-and-tube exchanger based on the above data.

[*Hint:* Use a 1–6 pass exchanger, acrylonitrile in the tubes, 19 mm tubes, about 3 metre long; no. of tubes: about 80. Also try a 1–4 pass exchanger.]

8.6 A spent caustic liquor is heated at the rate of 6 m³ per hour from 60°C to 100°C before feeding it to an evaporator, using a hot liquid stream available at 105°C. The properties of the streams are like those of water. The allowable pressure drop for each stream must be limited to 0.7 kg/cm². Design a shell-and-tube heat exchanger for this service.

[*Hint:* a suggested design—19 mm tubes, 6000 mm long, 60 tubes, about 23 m².]

8.7 Calculate the exit temperature of the coil fluid in Example 8.4 at the beginning and at the end of the heating cycle of the monomer in the agitated vessel.

REFERENCES AND FURTHER READING

ASME Boiler and Pressure Design Code, Section VIII, Div. 1, American Society of Mechanical Engineers, New York, 1980.

Baen, P.R. and R.E. Barth, "Insulate heat tracing systems correctly", *Chem. Eng. Progr.*, September 1994, 41–46.

Bailey, K.M., "Understand spiral heat exchangers", *Chem. Eng. Progr.*, May 1994, 59–63.

Baker, W.J., "Selecting and specifying air-cooled heat exchangers", *Hydrocarbon Proc.*, May 1980, 173–178.

Bayer, J. and G. Trumpfheller, "Specification tips to maximise heat transfer", *Chem. Eng.*, May 1993, 90–97.

Bertram, C.G., V.J. Desai, and E. Interess, "Designing steam tracing", *Chem. Eng.*, April 3, 1972, 74–80.

Bondy, F. and S. Lippa, "Heat transfer in agitated vessels", *Chem. Eng.*, April 4, 1983, 62–71.

Burley, J.R., "Don't overlook compact heat exchangers", *Chem. Eng.*, August 1991, 90–96.

Chapman, F.S. and F. A. Holland, "Heat transfer correlations in jacketed vessels", *Chem. Eng.*, February 15, 1965, 175–182.

Collins, G.K., "Horizontal thermosyphon reboiler design", *Chem. Eng.*, July 19, 1976, 149–152.

Cowan, C.T., "Choosing materials of construction for plate heat exchangers", *Chem. Eng.*, June 9, 1975, 100–103.

Cross, P.H., "Preventing fouling in plate heat exchangers", *Chem. Eng.*, January 1, 1979, 87–90.

Cuthbert, J., "Choose the right heat transfer fluid", *Chem. Eng. Progr.*, July 1994, 29–37.

Devore, A., G.J. Vago, G.J. Picozzi, and G.E. Lummus, "Specifying and selecting heat exchangers", *Chem Eng.*, October 6, 1980, 133–148.

Dream, R.F., "Heat transfer in agitated jacketed vessels", *Chem. Eng.*, January 1999, 90–96.

Escoe, A.K., *Mechanical Design of Process Systems*, Vols. 1 and 2, Gulf Publ. Co., Houston, 1986.

Frank, O. and R.D. Prickett, "Designing vertical thermosyphon reboilers", *Chem. Eng.*, September 3, 1973, 107–110.

Ganapathy, V., "Design and evaluate finned tube bundle", *Hydrocarbon Proc.*, September, 1996, 103–111.

Gentilcore, M.J., "Estimate heating and cooling times for batch reactors", *Chem. Eng. Prog.*, March 2000, 41–45.

Gloyer, W., "Thermal design of mixed vapour condensers", *Hydrocarbon Proc.*, June 1970, 103–108, July 1970, 107–110.

Greene, B., "A practical guide to shell-and-tube heat exchangers", *Hydrocarbon Proc.*, January 1999, 79-86.

Guffey, G.E., "Sizing heat transfer fluids and heaters", *Chem. Eng.*, October 1997, 126–131.

Gulley, D., "Troubleshooting shell-and-tube heat exchangers", *Hydrocarbon Proc.*, September 1996, 91–102.

Holland, F.A. and F.S. Chapman, *Liquid Mixing and Processing, in Stirred Tanks*, Reinhold Publ. Corp., New York, 1966.

House, F.F., "Pipe tracing and insulation", *Chem. Eng.*, June 17, 1968, 243–246. *International Critical Tables*, 1927.

Joshi, H.M., "Mitigate fouling to improve heat exchanger reliability", *Hydrocarbon Proc.*, January 1999, 93–95.

Kakac, S., R.K. Shah, and W. Aung, *Handbook of Single Phase Convective Heat Transfer*, John Wiley, New York, 1987.

Kakac, S., A.E. Bergles, and F. Mayinger, *Heat Exchangers, Thermal-Hydraulic Fundamentals and Design*, Hemisphere Publ. Corp., New York, 1981.

Kakac, S. (Ed.), *Boilers, Evaporators and Condensers*, John Wiley, New York, 1991.

Kays, W.M. and A.L. London, *Compact Heat Exchangers*, 3rd ed., McGraw-Hill, New York, 1984.

Kent, G.R., "Selecting gaskets for flanged joints", *Chem. Eng.,* March 1978, 125-128.

Kern, D.Q., *Process Heat Transfer*, McGraw-Hill, New York, 1950.

Kerner, J., "Sizing plate heat exchangers", *Chem. Eng.*, November 1993, 177–180.

Lawry, F.J., "Plate-type heat exchangers", *Chem. Eng.*, June 29, 1959, 89–94.

Lord, R.C., P.E. Minton, and R.P. Slusser, (i) "Design of heat exchangers", *Chem. Eng.*, January 26, 1970, 96–118; (ii) "Design parameters for condensers and reboilers", *Chem. Eng.*, March 23, 1970, 127–134; (iii) "Guide to trouble-free heat exchangers", *Chem. Eng.*, June 1, 1970, 153–160.

Love, D.L., "No hassle reboiler selection", *Hydrocarbon Proc.*, October 1992, 41–47.

Ludwig, E.E., *Applied Process Design*, Vol. 3, 2nd ed., Gulf Publ. Co., Houston, 1983.

Magee, B. and J.L. Adams, "Choosing a fired heater", *Chem. Eng.*, February 1998, 84–91.

Marcovitz, R.E., "Picking the best vessel jacket", *Chem. Eng.*, November 15, 1971, 156–162.

Marriott, J., "Where and how to use plate heat exchangers", *Chem. Eng.*, April 5, 1971, 127–134.

Martin, G.R. and A.W. Sloley, "Effectively design and simulate thermosyphon reboiler systems", *Hydrocarbon Proc.*, June 1995, 101–110.

Mazzarotta, B. and E. Sebastiani; "Process design of condensers for vapour mixtures in the presence of non-condensible gases", *Can. J. Chem. Eng.*, *73,* August 1995, 456–461.

Mehra, D.K., "Shell-and-tube heat exchangers", *Chem. Eng.*, July 25, 1983, 47–56.

Minton, P.E., (i) "Designing spiral-plate heat exchangers", *Chem Eng.*, May 4, 1970, 103–112; (ii) "Designing spiral-tube heat exchangers", *Chem. Eng.*, May 18, 1970, 145–152.

Mukherjee, R., "Effectively design air-cooled heat exchangers", *Chem. Eng. Progr.*, February 1997, 26–46.

Mukherjee, R., "Effectively design shell-and-tube heat exchangers", *Chem. Eng. Progr.*, February 1998, 21–37.

Mukherjee, R., "Broaden your heat exchanger design skill", *Chem. Eng. Progr.*, March 1998, 35–43.

Nagata, S., *Mixing: Principles and Applications,* Kodansha, Tokyo, 1975.

Oldshue, J.Y., *Fluid Mixing Technology,* McGraw Hill, New York, 1983.

Perry, R.H. and D.W. Green (Eds.), *Perry's Chemical Engineers' Handbook*, 6th ed., McGraw-Hill, New York, 1984.

Radle, J., "Steam tracing keeps fluids flowing", *Chem. Eng.*, February 1997, 94–97.

Reid, R.C., J.M. Prausnitz, and B.E. Poling, *The Properties of Gases and Liquids*, McGraw-Hill, New York, 1988.

Rubin, F.L., "Heat transfer topics often overlooked", *Chem. Eng.*, August 1992, 74–85.

Samdani, G., "Heat exchange—the next wave", *Chem. Eng.*, June 1993, 30–37.

Saunders, E.A.D., *Heat Exchangers*, Longman Scientific & Technical, UK, 1988.

Schlunder, E.V. (Ed.), *Heat Exchange Design Handbook*, Hemisphere, New York, 1983.

Sherwood, T.K., *A Course in Process Design*, MIT Press, Cambridge, MA, 1963.

Sinnott, R.K., *Coulson & Richardson's Chemical Engineering*, Vol. 6, Butterworth-Hienemann, Oxford, 1996.

Sloley, A.W., "Properly design thermosyphon reboilers", *Chem. Eng. Progr.*, March 1997, 52–64.

Smith, R.A., *Vaporisers*, Longman Scientific & Technical, U.K, 1986.

Steinmeyer, B.E. and A.C. Mueller, "Why condensers don't operate as they are supposed to?" *Chem. Eng. Progr.*, July 1974, 78–82.

Steinmeyer, D., "Understand ΔP and ΔT in turbulent flow heat exchangers", *Chem. Eng. Progr.*, June 1996, 49–55.

Taborek, J., G.F. Hewitt, and N. Afgan (Eds.), *Heat Exchangers Theory and Practice*, Hemisphere Publ. Co., New York, 1983.

The Tubular Exchanger Manufacturers' Association Standards, 7th ed., New York, 1988.

Thompson, J.C., "Evaluating heat tracing", *Hydrocarbon Proc.*, September 1997, 75–78.

Trom, L., "Heat exchangers: is it time for a change? *Chem. Eng.*, February 1996, 70–77.

Tuthill, A.H., "The right metal for heat exchanger tubes", *Chem. Eng.*, January 1990, 120–124.

Usher, J.D., "Evaluating plate heat exchangers", *Chem. Eng.*, February 23, 1970, 90–94.

Wharry, S.R., "Transfer heat in a resin sheath", *Chem. Eng.*, February 1996, 74–75.

Whitcraft, P.K., "Utilizing heat exchanger tubing specifications", *Hydrocarbon Proc.*, January 1999, 87–92.

Yokell, S., "Refurbishing worn-out heat exchangers", *Chem. Eng.*, June 1999, 78–88.

Yohell, S., "Troubleshooting shell-and-tube heat exchangers", *Chem. Eng.*, June 25, 1983, 57–75.

9

Evaporation and Evaporators

Evaporation ordinarily means vaporization of a liquid or that of a solvent from a solution. Thus, water 'evaporates' from a river, or cane-sugar juice is 'evaporated' to concentrate it. In chemical engineering terminology, *evaporation* means *removal of a part of the solvent from a solution of a non-volatile solute by vaporization.* The objective of vaporization is to concentrate the solution. It is one of the more important operations in the process industries. Typical examples are: concentration of cane-sugar juice in a sugar plant, concentration of an aqueous solution of ammonium sulphate in a fertilizer plant, concentration of dilute recycled sodium hydroxide in an alumina plant and many others. However, if the entire solvent is vaporized out from a solution (or dispersion) leaving a solid residue as the product, the operation is called *drying*. Thus, milk is 'spray-dried' to produce milk powder, and a detergent formulation is 'spray-dried' to get the detergent powder. Evaporation differs from distillation in the way that in the case of distillation, a solution containing more than one 'volatile' compound is vaporized (in a reboiler) and the components are separated thereafter in a distillation column.

It should be noted that a solution undergoing 'evaporation' may not contain a solute that is strictly non-volatile. Here 'non-volatile' means that the solute has a negligible volatility. Take the case of evaporation of a dilute solution of glycerine (called 'sweet water') obtained from a fat saponification unit for the manufacture of soap. It is called 'evaporation' because glycerine has a much lower volatility compared to that of water.

Evaporation of a solution is an essential step in the operation of a 'crystallization' unit. In crystallization, the solution is evaporated to make it supersaturated. Crystals grow in the supersaturated solution.

9.1 TYPES OF EVAPORATORS—THEIR CONSTRUCTION AND OPERATION

Evaporators used in the process industry mostly have tubular heating surface. An adequate number of tubes are provided through which the solution circulates. The tubes are heated by steam that condenses on their outer surface. The velocity of circulation of the solution through the tubes should be reasonably high so that—(i) a high inside heat transfer coefficient is attained, and (ii) formation of deposits or scales on the inner surface is reduced. Circulation may be caused by density gradient of the solution in the vertical tubes, or by an external mechanical means like a pump. Accordingly, most evaporators are broadly classified as: (1) *natural circulation*, and (2) *forced circulation*. There are a few other types of widely used evaporators which will be discussed later.

9.1.1 Natural-circulation Evaporators

The important types of natural-circulation evaporators are described below. All these evaporators use vertical tubes. The horizontal tube types are obsolete nowadays.

Short-tube vertical evaporators or calandria type evaporators

A short-tube vertical (STV) evaporator has a short tube bundle enclosed in a shell. This is called a *calandria* (Fig. 9.1). The calandria is of annular construction, i.e. there is an open cylindrical region at the centre. The lower tube sheet of the calandria is bolted to the flange of the bottom

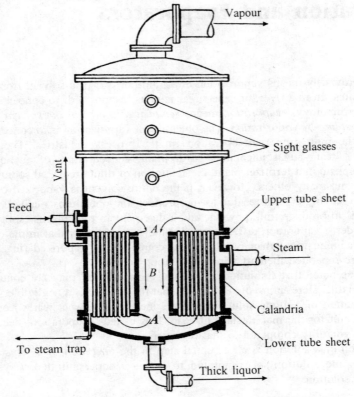

Fig. 9.1(a) Short-tube vertical evaporator: A, Tube sheet; B, downtake. (Courtesy: Swenson.)

end of the evaporator (the bottom is usually a dished or a conical end). The upper tube sheet is similarly bolted to the cylindrical chamber above. The feed is supplied through a nozzle above the upper tube sheet and steam is supplied to the shell or steam chest of the calandria. The solution is heated and partly vaporized in the tubes. The liquid flows up because of the prevailing density gradient. Vapour–liquid disengagement occurs above the upper tube sheet. The liquid flows down through the central open space of the calandria called the *downtake* or the *downcomer*. Thus, a continuous natural recirculation of the solution occurs. Thick product liquor is withdrawn from the bottom.

The diameter of a STV evaporator ranges from one metre to a few metres. Tubes of 50 mm to 76 mm (2 to 3 inches) diameter and 1.2 to 2 m length are commonly used. The cross-sectional area of the downtake is liberally sized and generally varies from 50% to 100% of the flow area of the tubes. This ensures low frictional resistance to flow through the downtake.

Other nozzle connections and accessories of the evaporator are described below. Besides the

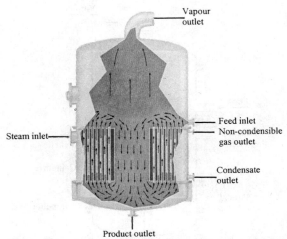

Fig. 9.1(b) Illustrated cut-out of vapour and liquid flow in a calandria. (Courtesy: Swenson Process Equipment, Harvey, Illinois.)

thick liquor outlet pipe, a drain connection is provided at the bottom. The steam condensate leaves through a drain nozzle connected to a steam trap. A bleed or vent line is provided in the shell for the release of the non-condensables[1] in the steam. There is a vapour outlet at the top of the evaporator body. An entrainment separator in this line arrests the liquid droplets in the vapour. A gauge glass is used to see the liquid level in the evaporator. One or more sight glasses on the body gives a view of the interior of the evaporator. A manhole is provided at the top (not shown in the figure) to allow a workman to get into the evaporator for cleaning or maintenance purposes. Tube cleaning is usually done by poking the tubes with a wire brush.

The short-tube vertical evaporators are used to concentrate a variety of solutions. A common example is concentration of the sugar solution. However, these evaporators are not suitable for solutions in which precipitation or salting out of a solid may occur. The natural circulation velocity in the evaporator is not sufficient to keep the solid particles in suspension. The problem may be overcome by installing a propeller in the downtake pipe to increase the circulation rate. This construction is called a *propeller calandria* (Fig. 9.2).

The STV evaporators are also known as *standard evaporators*. Their main advantages are low cost, easy cleaning, and low 'headroom' requirement. However, they are unsuitable for viscous liquids; and they offer a low heat transfer coefficient and need a larger floor space.

Basket-type vertical evaporator

This type of evaporator (Fig. 9.3) has operational features similar to those of the STV or calandria type. However, the construction is considerably different. The heating element is a tube bundle with fixed tube sheets welded to the shell piece, thus forming the 'basket'. This tube-bundle assembly or basket is supported by thick internal brackets. The brackets are welded to the evaporator body and the tube bundle, in turn, is bolted to the brackets, thus allowing removal of the bundle by opening the flanged top of the evaporator for the purpose of maintenance and

---

[1]This is very important. There may be some non-condensable gases in steam. If these are not vented out, the steam condensation heat transfer coefficient reduces drastically.

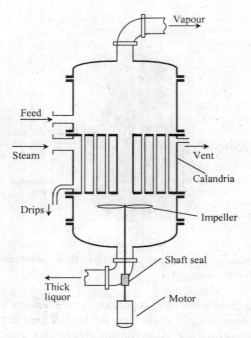

Fig. 9.2   Propeller calandria evaporator.

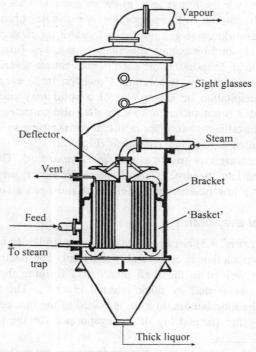

Fig. 9.3   Basket-type vertical evaporator (Swenson).

cleaning. The diameter of the tube bundle is smaller than that of the evaporator vessel, thus providing an annular space. The liquid gets heated and boils in the vertical tubes. Recirculation occurs through the annular space. Steam enters the chest through a pipe located centrally to the upper tube-sheet. A vent line is also connected to the upper tube-sheet for purging the non-condensibles. The condensate drips from the bottom of the basket through a steam trap. The vapour generated by boiling of the liquid strikes a deflector plate fixed to the steam pipe. The entrained liquid droplets are thereby removed from the vapour. The concentrated liquor leaves through an outlet pipe at the conical bottom of the evaporator.

Like the calandria-type device, the basket evaporator is also unsuitable for viscous liquids because the natural circulation forces are not strong enough to circulate viscous liquids. However, it can handle better scale-forming solutions because of the convenience with which it can be cleaned. A basket evaporator uses tubes similar to those of the calandria type. The main advantage of the basket-type evaporator is its ease of cleaning and maintenance because the basket can easily be taken out for this purpose. Other features are similar to those of the STV evaporator.

Long-tube vertical (LTV) evaporator

This is another widely used natural circulation evaporator. It has a long vertical tube bundle fitted with a shell. The shell is projected into a larger diameter chamber or vapour head at the top (Fig. 9.4). The feed enters the tube bundle at the bottom, flows through the tubes *once*, while undergoing vigorous boiling, and discharges into the vapour head and impinges on a deflector

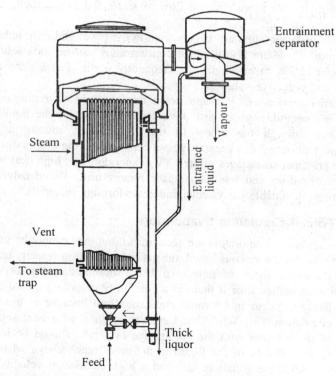

Fig. 9.4 Long-tube vertical evaporator (Swenson).

plate above the free top end of the tube bundle. This removes most of the entrained liquid droplets from the vapour. The concentrated liquor leaves the vapour chamber through a pipe and is withdrawn.

Before going into further details of the equipment, let us examine the heat transfer advantage of the LTV design. In any evaporator, the liquid side heat transfer resistance virtually controls the rate of heat transfer, because the resistance of the condensing steam on the shell-side is very small. The heat transfer rate can be enhanced by increasing the liquid-side coefficient. This depends on the liquid Reynolds number (see Chapter 6, for boiling in a vertical tube). After the feed enters a long vertical heated tube, it gains sensible heat to reach the boiling point. This occurs over a small portion of the length at the bottom of the tube. When the liquid starts boiling, the two-phase mixture flows at a higher velocity because of its larger specific volume. As boiling continues in the tube, more and more vapour is generated and the velocity of the mixture increases greatly. The net result is a large increase in the heat transfer coefficient and a high heat transfer rate. This is the theoretical basis of high capacity of an LTV evaporator. It should be noted that the driving force for heat transfer to the boiling liquid in the lower part of a tube is less than that in the upper region. The reason is that the liquid boils at a higher temperature at the lower part of a long tube because of the higher pressure caused by hydrostatic and frictional heads. So, the driving force is less there.

An LTV evaporator uses tubes of 25 mm to 50 mm in diameter and 5 m to 10 m in length. The liquid level in the tubes, as indicated by a gauge glass, should not be above 1 m. The diameter of the tube bundle is less than that in the short-tube evaporator. Baffles are sometimes provided in the shell-side for more uniform flow of steam fed to the shell. The steam condensate drips through a steam trap.

The LTV evaporators are largely used in the paper and pulp industries for concentrating the black liquor. These are also used for concentrating caustic soda solutions (nickel tubes, 25 mm in diameter and 6 m in length, are frequently used). These evaporators are also suitable for foaming and corrosive solutions.

A variation of the LTV evaporator is the *long-tube recirculation evaporator* (Fig. 9.5). The flow is not once-through or one pass. Instead, a part of the liquid leaving the vapour space recirculates through the tubes. In the construction shown in the figure, vapour–liquid disengagement occurs in a separate vessel connected to the evaporator by a vapour pipe at the top.

The principal advantages of the LTV evaporators are: high heat transfer coefficient, low cost, low liquid hold-up and less floor space requirement. The disadvantages are: high headroom requirement, unsuitable for viscous and scale-forming materials.

9.1.2 Forced-circulation Evaporators

Natural-circulation evaporators are economical, but they cannot be used in quite a few situations: (i) If the solution is a viscous liquid, the natural circulation velocity remains very low. This greatly reduces the heat transfer coefficient. (ii) If the solution contains suspended solid particles (it may occur in a crystallizer), or if there is a possibility of salting out of a solute, settling of particles is very likely to occur in a natural-circulation unit because of too low a velocity attainable in natural circulation. (iii) While handling the solution of a heat sensitive material, the time of contact of the solution with the hot surface of a tube should be low. This is possible only by increasing the velocity of the liquid in the tubes much above what can be achieved in natural circulation. All these situations demand a high circulation velocity in the tubes which can be achieved by using a pump. An evaporator that employs a pump to enhance circulation is called a *forced-circulation evaporator*.

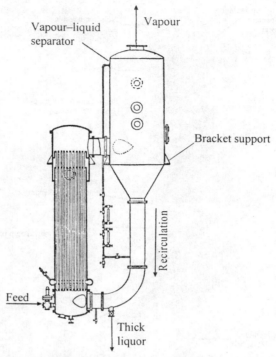

Vapour–liquid separator

Vapour

Bracket support

Recirculation

Feed

Thick liquor

Fig. 9.5 Long-tube recirculation evaporator.

A forced-circulation evaporator has a tubular exchanger for heating the solution without boiling. The superheated solution flashes in a vapour chamber where it gets concentrated. Two types of constructions are common—one with a vertical and the other with a horizontal heat exchanger. The former type (Fig. 9.6) is somewhat similar to the long-tube recirculation evaporator (Fig. 9.5). In the construction shown in Fig. 9.6, a pump draws the liquor from the flash chamber through a heat exchanger and forces it at a high velocity back to the flash chamber which has a top-mounted 'direct-contact' condenser (details of contact condensers are given in Section 9.2.1). However, the heat exchanger may be placed after the pump discharge as well. A part of the concentrated liquor is continuously taken out as product. The feed is introduced at a point before the pump suction. Heating steam is introduced to the shell side of the heat exchanger. A forced-circulation unit with a horizontal 1-2 pass heat exchanger is shown in Fig. 9.7.* This latter construction allows easy removal of the tube bundle for maintenance or cleaning.

As mentioned before, forced-circulation evaporators are used for handling viscous or heat-sensitive materials and also in the crystallization operation. Liquid velocities in tubes usually range from 1.2 m/s to 3 m/s or even more. The combined total head (static head *plus* frictional head) must remain sufficiently high so that the liquid does not boil in the heat exchanger. The

*This unit of Swenson design has a special feature. It has a top-mounted stripping column provided with 'ballast trays' for removing volatile compounds from the water vapour. A stream of concentrated liquor is fed to the top tray as 'reflux'. In the conventional design, this top part is absent and the vapour flows directly to the condenser.

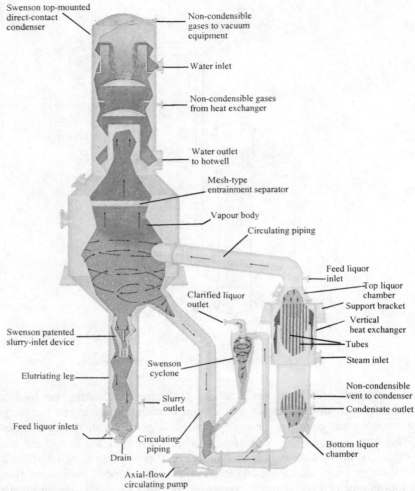

Swenson top-mounted direct-contact condenser

Non-condensible gases to vacuum equipment

Water inlet

Non-condensible gases from heat exchanger

Water outlet to hotwell

Mesh-type entrainment separator

Vapour body

Circulating piping

Feed liquor inlet

Top liquor chamber

Support bracket

Vertical heat exchanger

Tubes

Steam inlet

Clarified liquor outlet

Swenson patented slurry-inlet device

Non-condensible vent to condenser

Condensate outlet

Elutriating leg

Swenson cyclone

Slurry outlet

Feed liquor inlets

Circulating piping

Bottom liquor chamber

Drain

Axial-flow circulating pump

Fig. 9.6   Forced-circulation, submerged-inlet vertical-tube evaporator. (Courtesy: Swenson Process Equipment, Harvey, Illinois.)

static head in the exchanger may be increased as desired by placing it sufficiently below the flash vessel. Tangential entry of the superheated liquid into the flash vessel reduces the entrainment of liquid droplets in the vapour.

Typical applications of these evaporators are in the evaporation of brine (copper tubes may be used in the heat exchanger) and in the concentration of caustic soda solutions. Because of a high heat transfer coefficient attainable, the area of the heating surface can be small. As a result, these evaporators are preferred for corrosive solutions that demand special materials of construction.

The major advantages of this type of evaporator have been indicated before. The disadvantages are: high capital cost, high headroom requirement, and high floor area requirement if a horizontal heat exchanger is used.

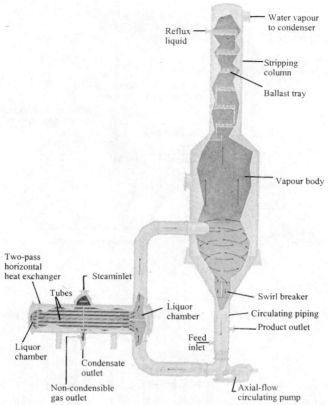

Fig. 9.7 Forced-circulation, submerged-inlet, horizontal-tube (1–2 pass) evaporator with stripping column. (Courtesy: Swenson Process Equipment, Harvey, Illinois.)

9.1.3 Falling-film Evaporators

In a falling-film evaporator [Fig. 9.8(a)], the liquid flows down the inner walls of the tubes in a vertical tube bundle. The tubes are heated by condensing steam or any other hot liquid on the shell side. As the feed liquor flows down the tube wall, the water vaporizes and the liquid concentration gradually increases. The thick liquor is withdrawn from the bottom and the vapour goes to a separator where the entrained liquid droplets are separated.

The most important factor for satisfactory operation of a falling-film unit is liquid distribution. In one arrangement, each of the tubes has a number of notches, usually rectangular, cut at the periphery at the top end [Fig. 9.8(b)]. The notched part of a tube projects above the tube sheet. The feed enters at a point above the upper tube sheet and forms a liquid pool on it. An 'impingement plate' may be provided at the feed inlet. The feed liquor enters the tubes through the notches or weirs and forms a falling film in each of the tubes. Other types of distributors are also used.

The number of tubes in the bundle is determined by the liquid rate. A minimum liquid rate should be maintained in order that a continuous film is formed instead of rivulets. The film Reynolds number [$Re_f = 4\Gamma/\mu$, where Γ is the liquid rate per unit circumference of a tube and

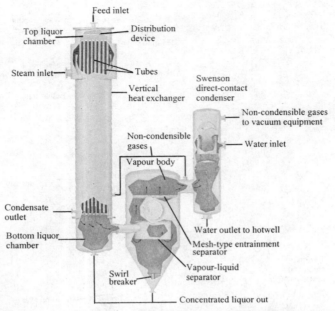

Fig 9.8(a)   Falling-film evaporator. (Courtesy: Swenson Process Equipment, Harvey, Illinois.)

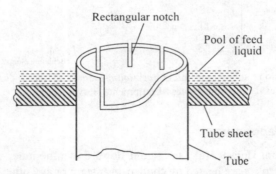

Fig. 9.8(b)   Notches at the top end of a tube for liquid distribution.

μ is the liquid viscosity] should be above 2000. At higher Reynolds numbers, ripples appear on the free surface of the film. The liquid rate must be high enough to ensure complete wetting of the tube wall so that no dry spot is formed. The minimum liquid rate for complete wetting may be calculated using the equation

$$\Gamma = 0.128(\mu\, s\, \sigma)^{0.2}$$

where Γ is in kg/(metre circumference)(s), μ is the liquid viscosity in kg/m s, s is the specific gravity of the liquid, and σ is the surface tension in N/m.

The heat transfer rate in a falling-film device is pretty high. The contact time of the liquid with the hot wall remains low. For this reason, the device is suitable for heat-sensitive materials. However, the solution viscosity must not be high (otherwise, the film of the solution falling down the wall will be too thick). The upper limit of viscosity may vary between 100 to 500 cP. The device is thus widely used for concentration of fruit juices and many other solutions (for example, this type of evaporator is used for concentration of a solution of ammonium nitrate). It is not suitable if salting out of the solute occurs. Evaporation can be accomplished by application of vacuum, or by using an inert gas fed below the lower tube sheet and flowing through the tubes countercurrent to the film. The vapour passes through a wire mesh entrainment separator [called a 'demister' pad, see Fig. 9.16(b)] and leaves through the top of the evaporator.

9.1.4 Climbing- or Rising-film Evaporator

A climbing-film evaporator (Fig. 9.9) looks similar to the long-tube vertical evaporator. It has a vertical tube bundle heater and a vapour chamber. The operation can be easily understood by

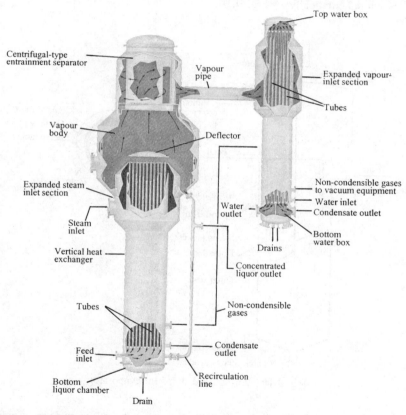

Fig. 9.9   Rising-film evaporator with vertical-tube surface condenser. (Courtesy: Swenson Process Equipment, Harvey, Illinois.)

referring to the discussion on boiling of a liquid in upflow through a vertical heated tube (Section 6.6). After gaining sensible heat at the lower part of the tube, the liquid starts boiling. If the heat transfer rate is sufficiently high and the tube is long, the volume fraction of the vapour in the two-phase mixture increases and the bubbles form slugs rising through the tubes at a high velocity. Thin liquid slugs remain sandwiched between vapour slugs. A thin liquid film is dragged along the tube wall by rising slugs, thus generating the climbing or rising film. The heat transfer coefficient in such a device is very high. The Swenson design shown in Fig 9.9 is provided with a shell-and-tube surface condenser. It has also a 'deflector' plate placed above the upper tube sheet of the evaporator which helps in reducing entrainment of liquid droplets.

9.1.5 Agitated Thin-film Evaporators

If the viscosity of the liquid is very high or the viscosity increases very sharply with the concentration of the feed, even a forced-circulation evaporator may not offer an adequate rate of heat transfer. A falling-film evaporator is not suitable either, because the film thickness of a highly viscous liquid becomes high. A solution to the problem is offered by an *agitated thin-film evaporator* shown schematically in Fig. 9.10.

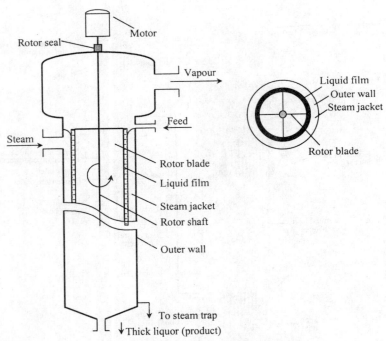

Fig. 9.10 An agitated thin-film evaporator.

It consists of a vertical steam-jacketed cylinder along the inner surface of which the feed liquid flows as a film. Four vertical blades mounted on a central shaft continuously agitate the film. The blades maintain a small clearance (typically 1.5 mm or less) from the wall. The action of the agitator greatly increases the heat transfer coefficient, reduces fouling of the surface, and keeps the film thickness small. Any liquid scrapped or thrown away from the film returns to the film surface again. Some designs of the agitator (or the rotor) cause the liquid to follow a spiral path down the heated wall. The vapour generated flows up through the core and leaves through a vapour pipe in the unjacketed upper section. This section has a larger diameter so that the vapour velocity is reduced and the entrained liquid droplets return to the evaporation zone.

The agitated thin-film evaporator can concentrate liquids having a viscosity as high as 100,000 cP. It offers a low residence time and liquid hold-up, and produces a uniform concentrate. Highly viscous, heat-sensitive, and fouling liquids can be evaporated in this equipment. Typical applications include concentration of tomato paste, candies, beer malt, meat extract, tannin extract, vitamins, enzymes, amino acids, glue, gelatin, water-soluble polymers, etc. The main disadvantages are high capital cost, moving internals, and high maintenance cost.

9.1.6 The Plate Evaporator

Plate evaporators are being increasingly used in the recent time, because of their compact construction and flexibility. The basic construction is similar to that of the plate heat exchangers. The liquid flows as a thin film over one surface of a plate, while the other side remains in contact with the heating medium or condensing steam. The common configurations include rising film, falling film, and the combination of both. The film flow generally remains turbulent because of the high vapour velocity. As a result, a very high overall heat transfer coefficient is achieved. The plate evaporators are especially suitable for high viscosity (20,000 cP or even higher), heat sensitive and foaming solutions. Typical applications include concentration of malt extract, glue, gelatin, emulsions, detergents, milk and other food products, pharmaceuticals, etc. The flow arrangement in a plate evaporator is shown schematically in Fig. 9.11.

The important characteristics of different types of evaporators are summarized in Table 9.1.

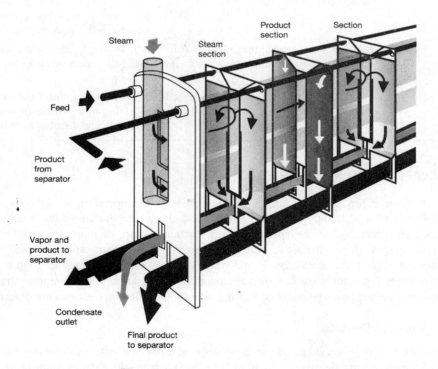

Fig. 9.11 Falling-film plate evaporator flow arrangement. (Courtesy: APV Heat Exchanger, Goldsboro, NC.)

Table 9.1   Characteristics of different types of evaporators

Evaporator	Typical products handled	Comments
Calandria	Salt; glycerine from spent soap lye.	Suitable for batch or continuous operation in single or multiple effects.
Forced circulation	Salting or scale-forming materials depending on steam-chest configuration; caustic soda solution, sodium sulphate, tomato juice to 30% concentration, etc.	Available with: (1) horizontal steam-chest with external vapour-separator (less used now); (2) vertical steam-chest with external separator; (3) vertical steam-chest with integral vapour head. Operates with either submerged or partially filled tubes in single or multiple effects.
Falling film	Low to medium viscosity materials, heat sensitive products, fruit juices, and pharmaceuticals.	Single or multiple effects; can be operated on single pass or with partial recycle of concentrated products.
Natural circulation (Thermosyphon)	Foaming liquids; less viscous materials; black liquor from the pulp industry; spent soap lye; electroplating solutions; spin bath liquid.	External separator provides some holding time adjustment; integral vapour-head type with downcomer gives minimum hold-up.
Agitated film	Handles the full range of feed viscosities; gelatin; fruit puree; glue.	Available: (1) vertical with integral vapour separator; (2) vertical with external separator, cocurrent flow; (3) horizontal with tapered shell, countercurrent flow.
Rising film	Caprolactum; ammonium nitrate; fruit juices; for crystal producing solutions with suspended solids.	Allows single-pass operation with high liquid and vapour velocities; minimum liquid hold-up
Plate type	Fruit juices, extracts; gelatin; condensed and whole milk.	Liquid and vapour flow essentially as in rising-and falling-film evaporators without liquid distribution problems.

9.2   EVAPORATOR AUXILIARIES

There are quite a few common accessories required for the operation of an evaporator. If an evaporator is operated under vacuum to reduce the boiling temperature of the liquid, there must be a vacuum producing device, for example, a vacuum pump, or more commonly a steam-jet ejector coupled with a barometric condenser. Surface condensers are also used for vapour condensation. The steam condensate in the steam chest or shell is drained through a steam trap. An entrainment separator is used in the vapour outlet line in order to remove the entrained liquid droplets from the vapour. The construction and operation of these accessories are described below.

9.2.1   Vacuum Devices

Evaporators, particularly the last one or two effects of a multiple-effect evaporator, are frequently operated under varying degrees of vacuum. The vapour from the final evaporator of the train is drawn by pulling vacuum. The vapour, which is generally water vapour, is condensed (by direct or indirect contact with cooling water) and the non-condensables, which include dissolved gases in the feed liquor as well as the air that leaks into the vacuum evaporators, are removed. Vacuum devices are, therefore, the common auxiliaries of evaporators. They may be broadly classified into two categories—the vacuum pumps and the steam-jet ejectors.

Vacuum pumps

There are a few types of common vacuum pumps. A 'reciprocating vacuum pump' works on the principle of one or more pistons in to-and-fro motion in a cylinder(s). Both 'dry' and 'wet' machines are used. In a dry pump, the vapour enters the pump directly. But a wet pump is provided with a surface condenser (or with a direct contact condenser). A reciprocating vacuum pump can produce reasonably high vacuum and has a good efficiency. However, the presence of a large number of moving parts increases the maintenance cost. The 'sliding vane' unit may be used for small capacity but high vacuum services. The centrifugal devices are suitable for higher capacities. The liquid ring pumps are suitable for withdrawing corrosive vapours at a moderate degree of vacuum.

Steam-jet ejector and barometric condenser

Installation of a vacuum pump to maintain the desired low pressure by withdrawing the vapour often proves prohibitive in cost. A steam-jet ejector combined with a condenser of the vapour is a simple and cheap method of maintaining vacuum. Condensation may be done by using a *surface condenser* or a *barometric condenser*. The latter is the more common and cheap device used for maintaining vacuum in air evaporator.

A barometric condenser (which is a 'direct-contact condenser' of the vapour) coupled with a steam-jet ejector is shown in Fig. 9.12(a). Let us first discuss the construction and the operating principles of the ejector. A typical ejector has three basic parts—the nozzle, the mixing chamber, and the diffuser. High pressure steam (the motive fluid) enters the steam chest at the end and expands in passing through the steam nozzle. Steam issues through the nozzle into the suction chamber at a high velocity. The pressure in this chamber remains very low (because the pressure head of steam is converted to velocity head according to the Bernaulli's equation). Steam and non-condensibles are drawn into the suction chamber. The gas–vapour mixture moves through a venturi-shaped converging-diverging section. The *diverging section* is called the *diffuser* in which the velocity head is converted to pressure head so that the mixture can be discharged into the atmosphere. A multi-nozzle steam-jet ejector is shown in Fig. 9.12(b).

Water vapour with non-condensibles, if any, from the evaporator first enters the barometric condenser as shown in Fig. 9.12(a). It is a contact condenser, i.e. the vapour condenses in direct contact with water fed to the condenser. A number of splash plates are provided in the condenser to enhance condensation. Water leaves the condenser through a *tail pipe* or *barometric leg* of a height of about 34 ft (atmospheric pressure is nearly equal to that of a water column 34 ft high). It discharges into a sump called *hot well*. The hot well is so called because the temperature of water may be as high as 50 to 55°C. (The rise in water temperature is due to the latent heat released by the condensed water vapour. It depends upon the relative rates of water supply and vapour input.) In an ideal situation, the pressure inside the barometric condenser should be equal to the vapour pressure of water at the condenser temperature. However, there may be some non-condensables in the inlet vapour which accumulate in the condenser chamber and tend to raise the pressure gradually. So, the barometric condenser is connected to the suction of the steam-jet ejector to withdraw the non-condensables and water vapour continuousiy. In this way, the vacuum in the condenser and in the evaporator is stabilized. The vapour and non-condensables sucked

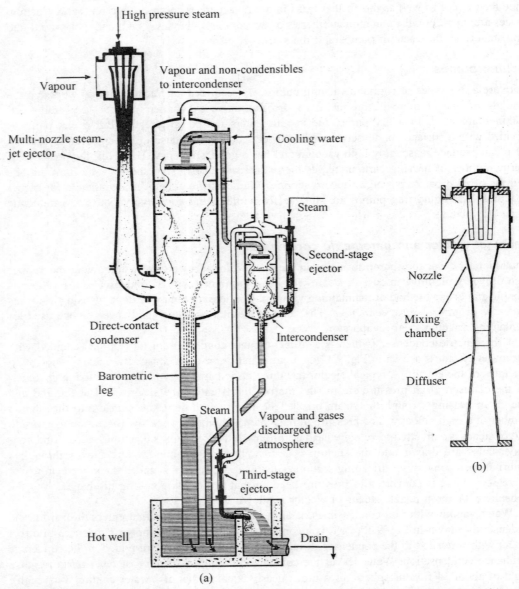

Fig. 9.12   (a) Three-stage ejector with barometric condenser (after Ludwig, 1977) and (b) multi-nozzle steam ejector. (Courtesy: Croll-Reynolds, Westfield, NJ.)

from the condenser are ultimately discharged into the atmosphere. The condenser in Fig. 9.12(a) is of counter-flow type. A parallel-flow condenser is shown in Fig. 9.13.

The water used for direct-contact condensation of water vapour may get contaminated with volatiles or entrained solutes leaving the evaporator. In many cases, it becomes undesirable to pump this water through the cooling water circuit of the plant. It is better discarded. This means a huge consumption of water. When water is in short supply or is expensive, a surface condenser

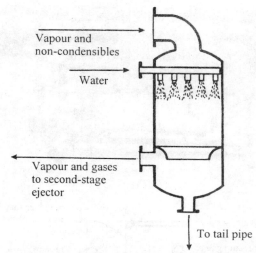

Vapour and
non-condensibles

Water

Vapour and gases
to second-stage
ejector

To tail pipe

Fig. 9.13 Parallel-flow barometric condenser.

(shell-and-tube type) may be used and the cooling water may be sent back to the water cooling tower of the plant.

Steam ejectors can be classified into single-stage and multi-stage types. Multi-stage ejectors may be further classified into condensing and non-condensing types. The single-stage ejector is generally recommended for suction pressures from atmospheric to about 80 mm (about 3 inch) of mercury pressure (i.e. up to 680 mm Hg vacuum). Multi-stage condensing ejectors are available in two to six stages. Intercondensers (surface condenser or direct contact condenser) between stages condense steam from the preceding stage, reducing the 'load' to be compressed in succeeding stages. Four- five- or six-stage ejectors are used to achieve pressures as low as 5 microns. In high vacuum applications, only the final two stages are fitted with condensers. The steam and non-condensables from the last stage are discharged to the atmosphere. The unit shown in Fig. 9.12(a) is a two-stage one. Two tail pipes are used, one each for the barometric condenser and the intercondenser. A five-stage steam-jet ejector unit is shown in Fig. 9.14(a). Typical piping layout for a three-stage steam ejector system is shown in Fig. 9.14(b). Details of ejectors and vacuum equipment are available in Ludwig (1977), Mangnall (1989), Power (1994), and Croll (1998).

A combination of a steam ejector and a vacuum pump is also used in some cases. Croll-Reynolds (New Jersey, USA) makes such a device, a combination of a steam-jet ejector and a liquid ring vacuum pump, which they call a 'Rotajector'.

9.2.2 Steam Traps

In all types of steam-heated equipment like evaporators, reboilers, steam-jacketed vessels, etc. steam condenses after giving away its latent heat. The condensate must be removed continuously without allowing the steam to blow out. A *steam trap* is an automatic valve that allows the condensate to discharge but *traps* (or prevents the flow of) steam. It distinguishes between the condensate and the steam on the basis of (i) difference in density, (ii) difference in temperature, or (iii) difference in flow characteristics. A few common types of steam traps are described below.

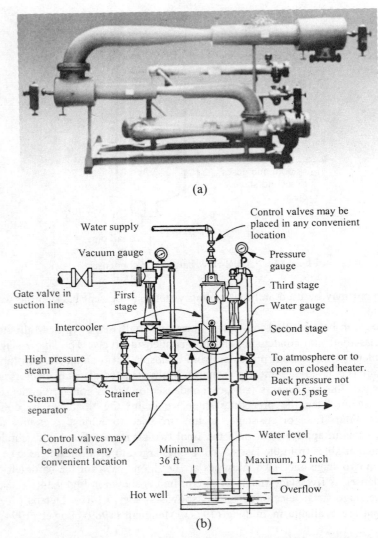

Fig. 9.14   (a) A five-stage steam ejector and (b) typical piping layout for a three-stage steam ejector system. (Courtesy: Croll-Reynolds, Westfield, NJ.)

Mechanical steam traps

This type of trap functions on the basis of the density difference. These include the 'float-and-thermostatic trap' and the 'inverted bucket trap'.

The 'float-and-thermostatic trap' or F&T trap [Fig. 9.15(a)] consists of a condensate chamber, a float, and a valve for the discharge of the condensate. When the condensate flows into the chamber, the float rises and opens up the valve connected to it. If the depth of the condensate increases, the float rises further and the valve opening increases simultaneously. This ensures a continuous and modulated discharge of the condensate which is desirable in some applications.

Because the discharge valve always remains submerged in the condensate, it cannot let the non-condensables out of the trap. However, non-condensables are invariably present in steam and must be removed along with the condensate. For this purpose, the F&T trap is fitted with a thermostatically regulated vent above the condensate level. This thermostatic vent remains open below the saturation temperature of the steam used and allows the air to bleed out during start-up. Under normal operating conditions, as the non-condensables go on accumulating above the condensate in the trap, the temperature drops locally and the vent opens. When the non-condensables leave, steam occupies the space and the temperature rises. The vent closes again and remains so till non-condensables accumulate again significantly.

The 'inverted bucket trap' [Fig. 9.15(b)] has a float of the shape of an inverted cylinder or bucket open at the bottom. The orifice in the condensate discharge passage has a valve seat. The valve operates by the movement of a lever connected to the bucket. When there is no or little condensate in the trap, the bucket is kept floating by the entering steam. As the condensate enters and accumulates in the chamber, the steam flow stops and the bucket sinks in the condensate. The valve opens and the condensate (with some steam and non-condensables) is discharged. There is a small 'weep hole' at the top of the inverted bucket through which the steam trapped in the bucket bleeds. The weep hole also prevents locking of the valve and the resulting sluggish operation. An inverted bucket trap discharges the condensate intermittently.

Thermostatic traps

There are two important types of thermostatic traps — the 'liquid expansion thermostatic trap' and the 'balanced pressure thermostatic trap'.

A 'liquid expansion thermostatic trap', shown in Fig. 9.15(c) has a closed metal cartridge filled with an oil or a suitable liquid. A major part of the cartridge is corrugated (and, therefore, contractible), and the other end has a valve fixed to it. The condensate collects in the trap and also in the line leading to it. As the liquid in the trap cools down, the cartridge contracts, the valve opens and the condensate is discharged. When the trap empties, the cartridge comes in contact with saturated steam (which is actually at a higher temperature than that of the 'cooled' condensate), it then expands, closes the valve and stops the flow of steam out of the trap.

The 'balanced pressure thermostatic trap' [Fig. 9.15(d)] consists of a flexible bellows, filled with a suitable liquid, with a valve fitted to its end. The valve is actuated by boiling of the liquid and the resultant rise in pressure and expansion of the bellows. As the condensate collects in the trap, it cools down and the temperature of the bellows also drops. This lowers the vapour pressure of the liquid in the bellows, and the bellows contracts. The valve opens and the condensate (with non-condensables) is discharged. After the trap is emptied, live steam enters, and this raises the temperature and pressure in the bellows. The bellows expands and the valve closes again.

Another type of this category is the 'bimetallic trap' in which the valve fitted with a bimetallic strip, opens or closes depending upon the temperature and the resultant bending action of the strip.

Thermodynamic traps

The 'thermodynamic' (TD) trap shown in Fig. 9.15(e) has a coin-sized flat disc on a flat, smoothly machined seat acting as the valve. The disc, when seated, covers the orifices for condensate flow as shown in the figure. The collected condensate, cooled to a temperature below the saturation temperature of the steam, leaves the disc and goes out through the orifices. When most of the condensate is discharged, the remaining liquid is nearly at the steam temperature. As this residual condensate flows out, it flashes and the high velocity 'flash steam' creates a lower pressure below the disc. The disc settles down on the seat and closes the orifices.

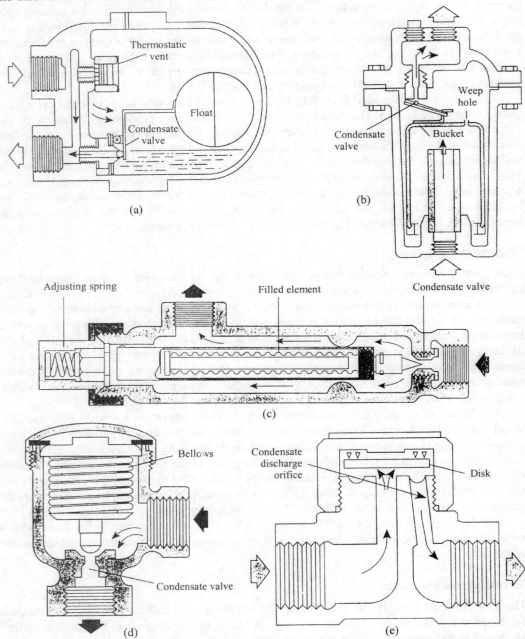

Fig. 9.15 Steam traps of different types: (a) float-and-thermostatic trap, (b) inverted bucket trap, (c) liquid expansion thermostatic trap, (d) balanced pressure thermostatic trap, and (e) thermodynamic trap.

The TD trap has a single moving part (the disc), is rugged in construction and operates relatively smoothly. A·snapping noise of lifting and settling of the disc is indicative of the trap function. Silence of the trap indicates its malfunction. (In fact, the 'sound method' is the simple common way of checking the operation of many steam traps.)

More details of construction, operation, selection, and maintenance of steam traps are given by Radle (1992), Haas (1990), Mackay (1992), and O'Dell (1992).

9.2.3 Entrainment Separators

In an equipment like an evaporator, vaporizer or reboiler, vapour bubbles burst vigorously on reaching the liquid surface. In that process, a lot of liquid droplets are thrown up in the vapour space above the boiling liquid. The droplets vary widely in size. The bigger ones immediately fall back into the liquid, and the medium-sized droplets remain suspended in the liquid a little longer before coming down. Most of the small droplets, however, are carried away or entrained by the outgoing vapour. This phenomenon is called *entrainment*. The entrained liquid particles must be removed before the vapour finally leaves the system such that any loss of liquid or solution is prevented. An entrainment separator is a device that retains the liquid droplets when the vapour passes through it. The capture of the tiny droplets may occur through three mechanisms—(i) diffusion, (ii) interception, and (iii) inertial compaction.

Different types of entrainment separators are in use; two of them are shown in Fig. 9.16. The *helical baffle* or *catchall* type [Fig. 9.16(a)] has two turns of helical baffles fitted to the central vapour inlet pipe. While the vapour flows through the baffled annular space, it attains a swirling motion and strikes the wall. The liquid particles are mostly retained on the wall. The liquid flows down the wall, accumulates at the bottom and from where it leaves through the drain.

A simpler type of entrainment separator is made of several layers of wire-mesh to form a pad [Fig. 9.16(b)]. This is supported on brackets below the vapour exit pipe. As the vapour flows through the pad, the liquid droplets impinge on the wires of the pad and get separated from the vapour by impaction mechanism. The retained liquid drips back to the boiling liquid. The device is also known as *demister* (because it removes 'mist'* or fine droplets). A typical wire-mesh is

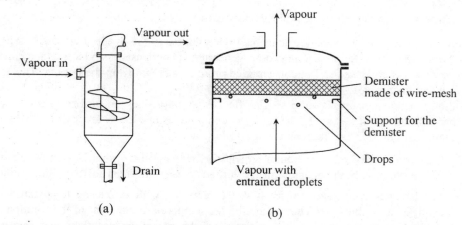

Fig. 9.16 Entrainment separators: (a) helical baffle (catchall) type and (b) wire-pad type (demister).

*Droplets smaller than 10 micron diameter are called *mist*; *sprays* refer to droplets greater than this size.

made from 0.011 inch wire at a density of about 9 lb/ft^3. Such a mist pad can remove droplets larger than 6 microns in size at 99% efficiency. The centrifugal entrainment separator is also used in many applications for retaining liquid droplets.

Other types of entrainment separators are 'impingement separators' in which the suspended droplets in the gas or vapour strike a set of baffle plates kept fixed in a vessel while changing the flow direction. As the droplets hit the plates, they are retained by them. In a 'knockout drum', the velocity of the gas or vapour is greatly reduced as it enters an empty drum or vessel and the droplets settle by gravity. Details of entrainment separation have been discussed by Fabian et al. (1993), Capps (1994) and Ziebold (2000).

The problem of entrainment becomes critical for those liquids which are highly foaming. Foaming is caused by the presence of surface active agents or colloidal particles in the solution that accumulate at the liquid surface. A stable blanket of bubbles forms on the liquid surface as a result of foaming. The foam must not be allowed to enter the vapour pipe because it becomes an extremely difficult task to separate the liquid in it at a later stage. A sufficiently high vapour space is provided above the boiling liquid to hold the foam which breaks continuously as it is generated. Another practical method of reducing foaming is to maintain the boiling liquid surface well inside the heated tubes where the bubbles expand and burst. A very small amount of an anti-foaming agent (e.g. silicone oil, vegetable oils, fatty acids, etc.) may be added, if permissible, to reduce foaming.

9.3 PRINCIPLES OF EVAPORATION AND EVAPORATORS

A solution boils in an evaporator to give off the vapour, and thereby becomes concentrated. However, the physical phenomenon and principles involved in the evaporation operation are more varied than those discussed under 'Boiling' in Chapter 6. For example, chemical evaporators are often built and operated as multiple-effect units for achieving steam economy. Solutions often exhibit boiling point elevation, thus reducing the temperature driving force for heat transfer. It is necessary to consider these and a few more features before we go for evaporator calculations.

9.3.1 Single- and Multiple-effect Evaporators

The simplest method of evaporation is to feed the solution to the evaporator which is provided with sufficient heat transfer area. The vapour generated is condensed using a 'surface condenser' or a 'direct contact condenser'. The concentrated product is drawn from the bottom. This is called a *single-effect evaporator*.

Although simple in operation, a single-effect evaporator does not utilize steam efficiently. More than a kilogram of steam (1.1 to 1.3 kg is common) is needed in order to vaporize 1 kg of water from the solution. The reasons are the following:

(i) Very often, the feed temperature remains below its boiling point. A part of the steam is utilized to supply the sensible heat required to raise the feed to its boiling point.

(ii) The latent heat of vaporization of water decreases with increasing temperature. Steam condensing in the steam chest is at a higher temperature than that of the solution. So the latent heat released by condensation of 1 kg steam is less than that required for vaporization of 1 kg of water from the boiling solution.

(iii) Some amount of heat loss from the evaporator to the ambient always occurs.

Now let us consider an arrangement in which two evaporators are put in series such that the vapour generated in one is fed to the steam chest of the second for heating. Partly concentrated solution flows from the first to the second where it attains the desired final concentration. The vapour generated in the second evaporator is sent to a condenser. The arrangement shown in Fig. 9.17 is called the *double-effect evaporator*.

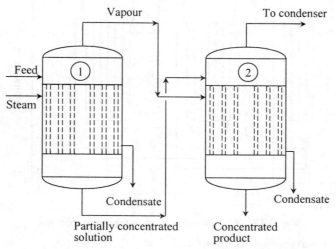

Fig. 9.17   A double-effect evaporator (forward feed).

One important point is to be noted in this connection. The vapour leaving evaporator-1 in Fig. 9.17 is at the boiling temperature of the liquid leaving the first effect. In order that transfer of heat occurs from the condensing vapour (from evaporator-1) to the boiling liquid in evaporator-2 (i.e. the second effect), the liquid in evaporator-2 must boil at a temperature considerably less than the condensation temperature of the vapour in order to ensure reasonable driving force for the transfer of heat. A method of achieving this is to maintain a suitable lower pressure in the second effect (evaporator-2) so that the liquid boils in it at a lower temperature. If the first evaporator operates at atmospheric pressure, the second one must do so under vacuum.

For 1 kg of steam supplied, about 1 kg of water is evaporated in each of the two evaporators in series. Thus, the steam economy in such an arrangement is nearly double that of a single-effect evaporator. The arrangement can be extended by putting more evaporators in series and we have a triple-effect or a quadruple-effect evaporator. In the multiple-effect operation, several evaporators are connected by suitable piping in order that the vapour from one effect passes to the next in series. The vapour produced in the first effect is used to vaporize water from the solution in the second effect. This goes on sequentially until the vapour from the last effect is discarded in the condenser. In most situations, the solution also flows from one effect to the next and gets partially concentrated in each effect. The net result of this arrangement is multiple reuse of heat supplied to the first effect, thereby improving the 'steam economy' (defined later). Pressure augmentation of the vapour flowing from one effect to the next by thermocompression (see Section 9.7) becomes very effective in enhancing the temperature driving force. A schematic of a triple-effect unit is shown in Fig. 9.18. The flow diagram of a triple-effect falling-film evaporator unit for the concentration of tannin extract is shown in Fig. 9.19. The unit is provided

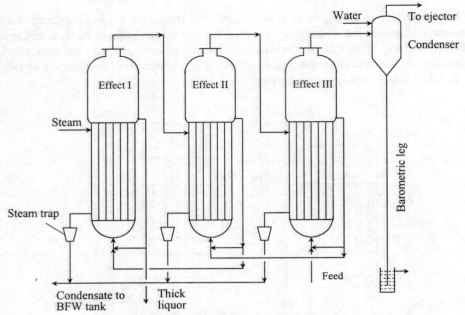

Fig. 9.18 A triple-effect evaporator (backward feed).

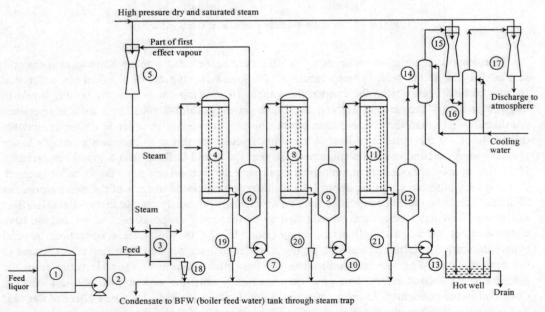

Fig. 9.19 Flow diagram of a triple-effect forward feed falling-film evaporator (with thermocompression of a part of the first effect vapour). 1: Feed balance tank; 2: Feed pump; 3: Feed preheater (plate type); 4: First-effect evaporator; 5: Thermocompressor; 6: First-effect vapour separator; 7: First-effect (product) pump; 8: Second-effect evaporator: 9: Second-effect separator; 10: Second-effect (product) pump; 11: Third-effect evaporator; 12: Third-effect separator; 13: Product pump; 14: Barometric contact condenser; 15: First-stage ejector; 16: Intermediate contact condenser; 17: Second-stage ejector; 18–21: Steam traps.

with a thermocompressor for the first-effect vapour and a two-stage steam-jet ejector and intercondenser. Photograph of a triple-effect falling-film evaporator for the concentration of membrane-cell caustic soda liquor is shown in Fig. 9.20.

Fig. 9.20 Triple-effect, falling-film membrane-cell caustic soda evaporators. (Courtesy: Swenson Process Equipment, Harvey, Illinois.)

The 'effect' of an evaporator unit to which external steam is supplied, is called the *first effect*. The meanings of *forward feed* and *backward feed* will be explained later.

9.3.2 Capacity and Economy

The performance of a steam-heated evaporator is measured in terms of its *capacity* and *economy*. The capacity of an evaporator means its vaporization capacity, that is, the number of kilograms of water it can vaporize per hour. The economy (or *steam economy*) is the number of kilograms of water vaporized from all the effects taken together for each kilogram of steam fed to the first effect. For a single-effect unit, it is less than unity (approximately 0.8). For an n-effect unit, the capacity is roughly n-times that of a single-effect unit, and the steam economy is about $0.8n$. However, pumps and interconnecting pipelines are required to transfer the liquid and vapour from one effect to another and this adds to the capital cost. The ratio of 'capacity' to 'economy' gives the steam consumption of the evaporator per hour.

9.3.3 Boiling Point Elevation (BPE)

Most evaporators produce concentrated solutions having a boiling point substantially higher than that of water (or the solvent in question) at the pressure prevailing in the vapour space. This difference of boiling points is called the *boiling point elevation* or the *boiling point rise*. Consider a strong solution of caustic soda that boils at 115°C at atmospheric pressure. If this solution is obtained from an evaporator operating under atmospheric pressure, the boiling point elevation of the solution is 115° − 100° = 15°C. The boiling point elevation (BPE) of the thick liquor must be known before the calculations for evaporator design are done, because the BPE reduces the effective temperature driving force compared to the case of boiling the pure solvent.

The boiling point of a solution can be calculated if the activity coefficient of the solvent in solution and the fugacity coefficient of the vapour are known. Although a lot of research work has been done on the thermodynamics of concentrated solutions, it is still difficult to theoretically calculate the BPE of concentrated solutions, particularly for solutions of non-electrolytes. Experimental boiling points data for solutions of different concentrations and pressure are, therefore, necessary.

However, in many practical situations, only a limited volume of boiling point data are available. The best use of these data can be made by taking help of an empirical rule, called the 'Duhring rule'. The rule states that the boiling point of a solution of given concentration is a linear function of the boiling point of water. Thus, if we plot the boiling points of the solution at several pressures against the corresponding boiling points of pure water, we get a straight line. A set of straight lines will be obtained if such plots are made for solutions of different concentrations. However, the lines are not parallel, the slope being larger at a higher concentration. The rule works good over moderate ranges of pressures, but does not give satisfactory correlation at a high pressure. Nevertheless it is a practically useful rule. The main advantage is that if the boiling points of a solution at two different pressures are known and the corresponding boiling points of water are read from the steam tables, a 'Duhring line' through the points can be drawn. This line can be used to predict the boiling point of a solution at any other pressure. If a few such lines are drawn, the boiling point of any solution at any pressure can be estimated by interpolation. Duhring plots of caustic soda solutions (aq. solution of sodium hydroxide) are shown in Fig. 9.21.

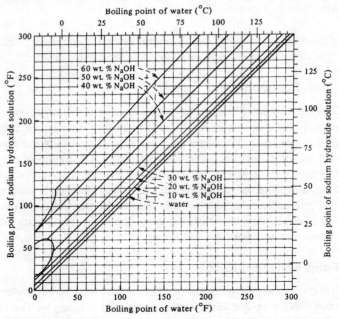

Fig. 9.21 Duhring plots for aqueous solutions of sodium hydroxide.

9.3.4 Temperature Driving Force

The boiling phenomenon in a vertical tube has been described in Chapter 6. After flashing of the superheated liquid and after vapour–liquid disengagement in the vapour space of an evaporator, the recycled liquid flows down the external pipe (see Fig. 9.4). A part of this liquor is withdrawn as the product and the remaining part gets mixed with the feed and enters the evaporator tubes. Let T_b be the boiling point of the liquid in the evaporator at the pressure prevailing in the vapour chamber, i.e. the temperature of the surface liquid in the vapour chamber is T_b. The temperature of the recycled liquid entering the tubes is equal to T_b (or a little less if the feed is cold). As the liquid flows up the tubes, its temperature rises. The total local pressure at any position in the tube is the sum of (i) vapour chamber pressure, (ii) hydrostatic head, and (iii) frictional head loss. The liquid starts boiling at a level where its temperature rises to its saturation temperature under the local pressure of the liquid.* Further up the tubes, the local temperature drops because of a reduction in the local pressure. The liquid temperature profile in the tubes is shown in Fig. 9.22. At a low liquid velocity, the liquid temperature rises in the tube up to a certain height before it

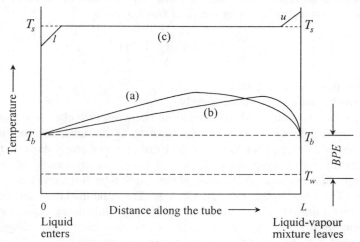

Fig. 9.22 Liquid temperature profile in the tubes: (a) liquid temperature at a low velocity, (b) liquid temperature at a high velocity, and (c) the shell-side temperature.

starts boiling. The temperature gradually drops thereafter because of loss of superheat [curve (a)]. If the liquid flows at a higher velocity the temperature rise is less and the liquid does not boil until it reaches near the top of the tube [curve (b)]. The shell-side temperature is represented by the curve (c). If the steam enters the shell at the top with a little superheat, it looses superheat along the line segment u. Similarly, if there is a sub-cooling of the condensate at the lower end of the shell, the change in the condensate temperature is represented by the line segment l. Along most part of the tube length, the shell-side temperature remains at T_s —the saturation temperature of steam at the shell pressure. The boiling temperature of water at the pressure of the vapour chamber is T_w.

---

*The temperature of a boiling liquid remains a little above the saturation temperature because of the superheat in the liquid (see Chapter 6).

The 'boiling point rise' of the liquid in this case is $T_b - T_w$. The *true temperature driving force* is the distance between the curve (c) and the liquid temperature curve (a) or (b). It is obvious from Fig. 9.22 that the driving force varies along the tube length. However, it is difficult to estimate this temperature variation. So, in design calculations $T_s - T_b$ is taken as the temperature difference or driving force. If the overall heat transfer coefficient in an evaporator is determined from experimental data on heat transfer rate and using $T_s - T_b$ as the driving force, this overall coefficient is called the *corrected coefficient*.

9.3.5 Heat Transfer Coefficient

The heat transfer coefficient for condensation of steam in the shell is very high. As a result, the liquid-side heat transfer coefficient virtually controls the rate of heat transfer. However, prediction of the liquid-side heat transfer coefficient is difficult. The entering liquid velocity is generally low, and, therefore, the liquid-side coefficient remains low. The coefficient increases greatly as the liquid starts boiling after reaching a certain height in the tubes. If the evaporator operates at a very high liquid velocity in the tubes and no boiling of the liquid occurs in the tubes, the heat transfer coefficient h_i can be estimated by using the following equation.

$$\frac{h_i d}{k} = 0.0278 \, (\text{Re})^{0.8} \, (\text{Pr})^{0.4} \tag{9.1}$$

where
  d is the inner diameter of the tube
  k is the thermal conductivity of the liquid
  Re and Pr are the Reynolds and the Prandtl number, respectively.
  There is no practically useful method of estimation of the tube-side coefficient if there is boiling in the tubes of an evaporator. Design calculations are frequently done on the basis of data obtained from an operating evaporator. Table 9.2 gives typical values of the heat transfer coefficients in industrial evaporators. Fouling of the inside tube surface is a major problem with most evaporators. The tubes require periodic cleaning. The scale thickness, and hence the fouling resistance, grows with time. This factor must be considered while designing an evaporator and an adequate cushion for the fouling resistance has to be provided. The change in the overall coefficient with time may be correlated by the following equation.

Table 9.2 Typical values of overall heat transfer coefficients in evaporators

Evaporator type	Overall coefficient	
	kcal/h m² °C	Btu/h ft² °F
Short-tube vertical or calandria evaporators	750–2500	150–500
Long-tube vertical evaporators		
Natural circulation	1000–3000	200–600
Forced circulation	2000–10,000	400–2000
Agitated thin-film evaporators		
Low to medium viscosity	1800–2700	350–500
High viscosity (10^4 cP and above)	1500	300
Falling-film evaporators (low viscosity)	500–2500	100–500
Rising-film evaporators	2000–5200	400–1100
Plate evaporators	1800–2700	350–550

Note: 1 Btu/h ft² °F = 4.88 kcal/h m² °C   1 W/m² °C = 0.86 kcal/h m² °C

$$\frac{1}{U^2} = \frac{1}{U_0^2} + \beta\theta \tag{9.2}$$

where

U_0 is the overall coefficient for a clean tube

U is the overall coefficient at any time θ

θ is the time for which the evaporator is in operation

β is a constant for a particular liquid.

If such a plot is prepared, the magnitude of the fouling resistance at any time can be determined.

9.3.6 Enthalpy of a Solution

Enthalpy-concentration data of solutions are required for making the energy balance calculation of an evaporator. These data may be determined experimentally by 'calorimetry', or may be calculated if the values of 'integral heat of solution' at different solution concentrations are known. The integral heat of solution is defined as the change in enthalpy of a solution when one mole of solute is dissolved in n_1 moles of solvent at 25°C and atmospheric pressure. The enthalpy of a solution at temperature T relative to the pure solvent and solute at temperature T_o is expressed as

$$H_s = n_1 H_1 + n_2 H_2 + n_2 \Delta H_{s2} \tag{9.3}[1]$$

That is,

$$H_s = n_2 \Delta H_{s2}^o + (n_1 + n_2)(S)(T - T_o) \tag{9.4}$$

where

H_s is the enthalpy of $(n_1 + n_2)$ moles of solution at temperature T

n_1, n_2 are the moles of solvent and solute, respectively

H_1, H_2 are the molar enthalpies of pure solvent and solute at temperature T

ΔH_{s2} is the integral heat of solution of the solute (component 2) at temperature T

ΔH_{s2}^o is the integral heat of solution of the solute (component 2) at temperature 25°C

S is the molar specific heat of the solution

Enthalpy-concentration data for many solutions are available in the literature. The integral heat of solution data are also available for various systems.

9.4 SINGLE-EFFECT EVAPORATOR CALCULATION

Material and energy balance equations are written for an evaporator in order to calculate the rate of solvent vaporization and the rate of required heat input. Referring to Fig. 9:23, the balance equations are

Total material balance, $\qquad W_f = W_p + W_v \tag{9.5}$

and

Energy (or enthalpy) balance,

$$W_f i_f + W_s i_s = W_p i_p + W_v i_v + W_s i_l + \text{heat losses} \tag{9.6}$$

---

[1]Equation (9.3) is written on the basis that the solvent and the solute are individually heated from T_o to T and then mixed at that temperature. Equation (9.4) assumes that the solvent and the solute are mixed at the temperature T_o and the resulting solution is brought to the temperature T.

where
  W is the flow rate of a stream, kg/h
  i is the enthalpy of a stream (with respect to a suitable reference temperature, usually 25°C), kJ/kg
  the subscripts f, p, v, s and l refer to the feed, product, vapour, steam, and condensate, respectively.

From the enthalpy data of the solutions, steam and condensate, the rate of heat input or the rate of steam flow can be calculated.

If sufficient enthalpy-concentration data are not available, approximate calculations are done. The following approximations, each of which neglects the heat of solution, are suggested:

(i) The specific heat of a dilute solution can be taken as that of the solvent. If both feed and product are assumed to be dilute, their enthalpies (with respect to the reference temperature, T_o) can be written as

$$i_f = c_{po}(T_f - T_o) \quad \text{and} \quad i_p = c_{po}(T_b - T_o) \tag{9.7}$$

where
  c_{po} is the specific heat of the solvent
  T_f is the feed temperature
  T_s is the temperature of the thick liquor (= boiling temperature in the evaporator).

(ii) If the specific heat of the solution is known at any concentration, its value at another concentration can be calculated by using the following equation.

$$c_{p2} = c_{po} - (c_{po} - c_{p1})\left(\frac{w_2}{w_1}\right) \tag{9.8}$$

where c_{p_1} is the known specific heat of a solution having w_1 weight fraction of the solute, and c_{p_2} is the specific heat at a weight fraction w_2 of the solute. Equation (9.8) assumes that the specific heat of a solution is a linear function of its concentration in weight fraction.

(iii) If the required data for an aqueous solution are not available, the following equation may be used.

$$c_p = c_{po}(1 - w) \quad \text{for } w < 0.2 \tag{9.9}$$

where c_{po} is the specific heat of the solvent.

(iv) If no specific heat data of the solution are available but the specific heat of the anhydrous solute is known, the following equation may be used.

$$c_p = c_{po}(1 - w) + \bar{c}_p w \tag{9.10}$$

where $\bar{c}_p$ is the specific heat of the anhydrous solute.

If the specific heats of the feed (c_{pf}) and of the product (c_{pp}) are estimated by the above equation(s), their enthalpies can be calculated from the following equation.

$$i_f = c_{pf}(T_f - T_o) \quad \text{and} \quad i_p = c_{pp}(T_p - T_o) \tag{9.11}$$

Further, if the entering steam is saturated and there is no sub-cooling of the condensate, only the latent heat of the steam becomes available for evaporation. Then we can put $(i_s - i_l) = \lambda_s =$ latent heat of the steam at the given steam pressure.

Neglecting heat losses, the required steam rate and the rate of heat transfer $Q \, (= W_s \lambda_s)$ can be calculated. The area of heat transfer A can now be obtained from the equation

$$Q = U_D A \, \Delta T \tag{9.12}$$

where $\Delta T = T_s - T_b$ [here T_s is the saturation temperature of steam in the shell; T_b is the boiling point of the solution at the pressure of the vapour chamber (T_b includes the boiling point elevation)]. The overall coefficient U_D should either be known from the performance data of an operating evaporator of the same type and processing the same solution, or a reasonable value can be selected from Table 9.2. With this information, the area A can be calculated. The tube length and the tube layout are specified later.

Example 9.1 An aqueous solution of a high molecular weight solute is concentrated from 5% to 40% at a rate of 100 m³/day. The feed temperature is 25°C, and the concentrated product leaves at its boiling point. Calculate the rate at which heat must be supplied if evaporation occurs at (i) 1 atmosphere pressure, (ii) a vacuum of 650 mm Hg. What advantage of this operation under vacuum is apparent from the answers?

Given: density of the feed = 1020 kg/m³; specific heat of the feed = 4.1 kJ/kg °C, and the specific heat of the product = 3.9 kJ/kg °C.

SOLUTION The basis of the following calculations is 1 day of operation.

Material balance

$$\text{Feed entering} = (100 \text{ m}^3)(1020 \text{ kg/m}^3) = 1.02 \times 10^5 \text{ kg} \ (= W_f)$$

$$\text{Mass of the solute} = (1.02 \times 10^5)(0.05) = 5.1 \times 10^3 \text{ kg}$$

$$\text{Mass of the water} = (1.02 \times 10^5)(0.95) = 9.69 \times 10^4 \text{ kg}$$

$$\text{Feed concentration} = 5\% \text{ solute, i.e. } 95/5 = 19 \text{ kg water per kg solute}$$

$$\text{Product concentration} = 40\% \text{ solute, i.e. } 60/40 = 1.5 \text{ kg water per kg solute}$$

$$\text{Water leaving with the product} = (5.1 \times 10^3 \text{ kg solute})(1.5 \text{ kg water/kg solute})$$

$$= 7.65 \times 10^3 \text{ kg}$$

$$\text{Water evaporated} = 9.69 \times 10^4 - 7.65 \times 10^3 = 8.925 \times 10^4 \text{ kg} \ (= W_s)$$

$$\text{Product} = 7.65 \times 10^3 + 5.1 \times 10^3 = 1.275 \times 10^4 \text{ kg} \ (= W_p)$$

Energy balance

$$\text{Take the reference temperature for enthalpy calculation} = 0°C$$

$$\text{Enthalpy of the feed (at 25°C)} = (4.1)(25 - 0) = 102.5 \text{ kJ/kg solution} \ (= i_f)$$

Case (i): *Evaporation at 1 atmosphere pressure.*

$$\text{Temperature of the product} = 100°C \text{ (because the solute has a 'high molcular}$$
$$\text{weight', the boiling point elevation is neglected)}$$

$$\text{Enthalpy of the product} = (3.9)(100 - 0) = 390 \text{ kJ/kg} \ (= i_p)$$

$$\text{Enthalpy of the vapour generated at 100°C and 1 atm pressure (from the steam table)}$$

$$= 2680 \text{ kJ/kg} \ (= i_v)$$

Let us refer to Fig. 9.23.

Energy balance equation:

$$W_f \, i_f + q_s = W_v \, i_v + W_p \, i_p \qquad [\text{compare with Eq. (9.6)}, \ q_s = W_s(i_s - i_l)]$$

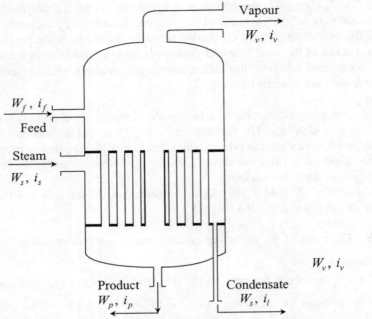

Fig. 9.23   Material and energy balance in a single-effect evaporator (Example 9.1).

or

$$(1.02 \times 10^5)(102.5) + q_s = (8.925 \times 10^4)(2680) + (1.275 \times 10^4)(390)$$

Therefore,

$$q_s = \text{rate of supply of heat} = \boxed{2.337 \times 10^8 \text{ kJ per day}}$$

Case (ii): *Evaporation at* 650 mm Hg *vacuum*, or 110 mm Hg *pressure* (reference barometer = 760 mm Hg)

Boiling point of water at 110 mm Hg pressure = 53.5°C

$$= \text{temperature of the concentrated product}$$

Enthalpy of the product at 53.5°C = (3.9)(53.5 − 0) = 208.6 kJ/kg (= i_p)

Enthalpy of the saturated steam at 53.5°C = 2604 kJ/kg (from steam table).

Using the previous energy balance equation, we have

$$(1.05 \times 10^5)(102.5) + q_s = (1.275 \times 10^4)(208.6) + (8.925 \times 10^4)(2604)$$

Therefore,

$$q_s = \text{rate of heat supply} = \boxed{2.246 \times 10^8 \text{ kJ per day}}.$$

It is seen that 4% less heat is required in the second case, where the evaporator operates under vacuum. This is the apparent advantage. Also, a smaller heat transfer area will be required because of the greater driving force available (this is because the liquid now boils at a lower temperature). However, the capital and operating cost of the vacuum producing device may undo part or whole of this advantage.

Example 9.2 A single-effect, vertical short-tube evaporator is used to concentrate a syrup from 10% to 40% solids at the rate of 2000 kg of feed per hour. The feed enters at 30°C and a reduced pressure of 0.33 kg/cm² is maintained in the vapour space. At this pressure, the liquor boils at 75°C. Saturated steam at 115°C is supplied to the steam chest. No sub-cooling of the condensate occurs. Calculate the steam requirement and the number of tubes (0.0254 m, 16 BWG) if the height of the calandria is 1.5 m. The following data are given: specific heat of liquor = 0.946 kcal/kg °C; latent heat of steam at 0.33 kg/cm² = 556.5 kcal/kg; boiling point of water at this pressure = 345 K. The overall heat transfer coefficient = 2150 kcal/h m² °C.

SOLUTION Feed rate, W_f = 2000 kg/h with 10% solids

Solids in = (2000)(0.1) = 200 kg/h;   water in = 1800 kg/h

Products out (at 40% concentration), W_p = 200/0.4 = 500 kg/h;   water out = 300 kg/h

Evaporation rate, $W_v = W_f - W_p$ = 2000 − 500 = 1500 kg/h

Neglecting heat losses and noting that $i_s - i_l = \lambda_s$, if no sub-cooling of the condensate occurs, the energy balance over the evaporator becomes [see Eq. (9.6)]

$$W_f i_f + W_s \lambda_s = W_p i_p + W_v i_v$$

Taking the boiling temperature of the product liquor as the reference temperature, i.e. T_o = 75°C, we have

$$i_f = c_p(T_f - T_o) = 0.946(30 - 75) = -42.5 \text{ kcal/kg}, \qquad i_p = 0$$

$$\lambda_s \text{ (at 115°C)} = 529.5 \text{ kcal/kg}, \qquad i_v = 556.5 \text{ kcal/kg}$$

Boiling point elevation = (273 + 75) − 345 = 3 K. Because the BPE is low, we neglect the superheat of the vapour produced.

Substituting all the quantities in the energy balance equation, we get

$$(2000)(-42.5) + W_s (529.5) = (500)(0) + (1500)(556.5)$$

or

$$W_s = \boxed{1737 \text{ kg/h}} = \text{steam rate}$$

Rate of heat transfer, $Q = W_s \lambda_s$ = (1737)(529.5) = 9.20 × 10⁵ kcal/h

Given: U = 2150 kcal/h m² °C;   ΔT = 115 − 75 = 40°C

$$\text{Evaporator area, } A = \frac{Q}{U\Delta T} = \frac{9.20 \times 10^5}{(2150)(40)} = 10.7 \text{ m}^2$$

Calculation of the number of tubes

Heat transfer coefficient of an evaporator is usually taken on *inside area basis*.

Inside diameter of 0.0254 m o.d., 16 BWG tube = 0.0221 m

Area of a single tube (1.5 m long) = π(0.0221)(1.5) = 0.104 m²

Number of tubes = (10.7 m²)/(0.104 m²) = $\boxed{102}$

Example 9.3 An 8% solution of a water soluble polymer (molecular weight = 2000) is concentrated to 40% in a single-effect calandria evaporator at the rate of 2000 kg of feed per hour.

The feed enters at 30°C and a vacuum of 660 mm Hg is maintained in the evaporator. Saturated steam is available at a pressure of 8.0 bar absolute, but pressure may be reduced to a lower value, if necessary, by throttling. The following test data on the overall heat transfer coefficient are available.

ΔT (°C) :	30	35	40	45	50	55	60	70	80	100
U_D (W/m² °C):	820	1145	1270	1325	620	780	860	1000	640	350

Calculate (i) the steam pressure to be used in the calandria, (ii) heat transfer rate required, and (iii) the steam requirement.

SOLUTION   Feed rate = 2000 kg/h (= W_f), concentration = 8% by weight
Solid rate = (2000)(0.08) = 160 kg/h
Concentrated product rate (at 40% concentration) = 160/0.4 = 400 kg/h (= W_p)
Absolute pressure in the evaporator = 760 − 660 = 100 mm Hg
Boiling temperature of water at this pressure = 325 K, λ_s = 2380 kJ/kg

Calculation of the boiling point elevation

The boiling point elevation,

$$\Delta T_b = \frac{RT_b^2}{\lambda_s} \cdot \frac{wM}{Wm} = \frac{(8.303)(325)^2(40)(18)}{(2380)(60)(2000)} = 2°C$$

where
    ΔT_b is the boiling point elevation
    T_b is the boiling point of water
    λ_s is the latent heat of steam
    w is the mass of the solute
    W is the mass of the solvent
    m is the molecular weight of the solute
    M is the molecular weight of the solvent
    R is the gas constant.
The boiling point of 40% solution at 660 mm Hg vacuum = 325 + 2 = 327 K.
From the given data on heat transfer coefficient, it appears that U_D first increases with ΔT and then declines. For maximum heat flux (and minimum area), we should select the ΔT that gives the maximum $U_D \Delta T$.

ΔT, °C :	30	35	40	45	50	55	60	70	80
$U_D\Delta T$, W/m² :	24,600	40,075	50,800	59,625	31,000	42,900	51,600	70,000	51,200

It is seen that $U_D\Delta T$ or the flux becomes maximum at a temperature drop, ΔT = 70°C, and we therefore select this value.
Saturation temperature of steam in the steam chest = liquor temperature + ΔT

$$= 327 + 70 = 397 \text{ K}$$

Saturation pressure of steam at this temperature (from the steam table) = $\boxed{2.15 \text{ bar (abs)}}$ So, the steam pressure has to be reduced from 8 bar to 2.15 bar (the steam has also to be *desuperheated after throttling*).

Calculation of the steam rate

Take the specific heat of the liquor equal to that of water.

Enthalpy of the feed (reference temperature = 0°C) at 30°C = (4.2)(30 − 0) = 126 kJ/kg = i_f

Enthalpy of the product at 327 K or 54°C = (4.2)(54) = 227 kJ/kg

Enthalpy of vapour produced at 54°C (from the steam table) = 2607 kJ/kg = i_v

From Eq. (9.6), we have

$$Q = W_s(i_s - i_l) = W_p i_p + W_v i_v - W_f i_f$$

$$= (400)(227) + (1600)(2607) - (2000)(126) = \boxed{4.01 \times 10^6 \text{ kJ/h}}$$

Heat of vaporization of saturated steam at 397 K = 2188 kJ/kg

$$\text{Rate of steam supply} = \frac{4.01 \times 10^6}{2188} = \boxed{1833 \text{ kg/h}}$$

9.5 MULTIPLE-EFFECT EVAPORATORS

A multiple-effect evaporator has been shown in Fig. 9.18, and its operating principles have been briefly described in Section 9.3.1. The type of the multiple-effect evaporator shown in Fig. 9.18 is called the *backward feed* because the steam and the liquor flow in opposite directions. There are other types of feeding arrangements too. Evaporators up to fifteen effects are known to be in use. Multiple-effect evaporators allow high steam economy.

9.5.1 Classification Based on the Mode of Feed Supply

Depending upon the directions of flow of the heating medium and of the feed or the liquor, multiple-effect evaporators can be classified into four categories: (i) forward feed, (ii) backward feed, (iii) mixed feed, and (iv) parallel feed.

Forward feed

A forward feed unit has already been shown in Fig. 9.19. The feed is introduced to the first effect. Partly concentrated liquor flows to the second effect, and then from the second to the third effect. Thick liquor is withdrawn from the third or from the last effect. Steam is fed to the shell of the first effect, and the vapour generated therein flows to the shell of the second effect and acts as the heating medium there. Vapour generated in the second effect supplies heat for boiling the liquor in the third effect. The vapour from the third effect is condensed in a barometric condenser connected to a steam-jet ejector. The final condensate is discarded with the water supplied to the barometric condenser. Non-condensables (which arise out of any dissolved gases in the liquid and from leakage of air into the system that usually operates under vacuum) are withdrawn by the ejector. The flow of liquor from one effect to the other occurs spontaneously because of the difference of pressure maintained in the successive effects. Valves in the connecting pipelines regulate the flow rates.

Backward feed

The backward feed arrangement has been shown in Fig. 9.18. Here the feed is introduced to the last effect. Partly concentrated liquor flows to the second and then to the first effect from which the thick liquor is withdrawn. Pumps are used to maintain the flow of liquor against a positive pressure.

Mixed feed

In the mixed feed arrangement, the dilute feed enters an intermediate effect and flows to the next higher effect till it reaches the last effect. On this section, the liquid flow occurs in the forward feed mode. Partly concentrated liquor is then pumped to the effect before the one to which the feed is introduced. It then flows towards the first effect in the backward feed mode. Thick liquor is withdrawn from the first effect. This feeding arrangement for the case of a triple-effect evaporator is shown in Fig. 9.24(a).

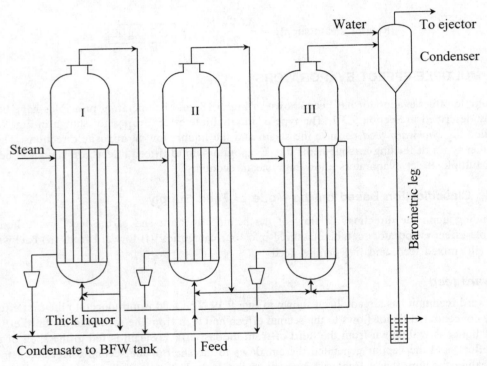

Fig. 9.24(a)   A triple-effect evaporator (mixed feed).

Parallel feed

In a crystallization operation, the evaporator is fed with recycled mother liquor mixed with fresh feed. The liquid from the evaporator flows to the crystallizer. There is no question of withdrawal of thick liquor as in the case of other evaporators when concentration of the feed is the only objective. In such an application [Fig. 9.24(b)], the feed is divided into a number of streams and each is fed to an effect in the unit. Concentrated liquor is withdrawn from each effect separately. The successive effects, however, operate under gradually decreasing pressure so that the vapour generated in one can act as the heating medium in the next.

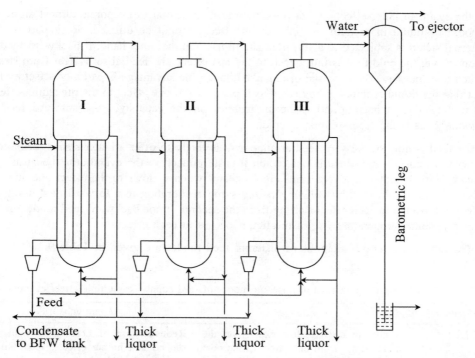

Fig. 9.24(b)   A triple-effect evaporator (parallel feed).

9.5.2 Comparison between the Forward and Backward Feed Modes

Of the above two modes of liquid flow in an evaporator, the former one is operationally simple. It does not need intermediate pumps for transfer of the liquid from one effect to the next; the liquid flows automatically because the pressure in the next effect is lower. The backward feed technique offers some advantages over the forward feed arrangement, although at higher capital and operating costs because of intermediate pumping. Three other important situations in connection with these two feeding modes are discussed below.

(i) Consider the case when the thick liquor or the final product is highly viscous. In a forward-feed evaporator, the last effect that performs the final concentration job operates at the lowest temperature. So, the viscosity of the liquor in the last effect will be very high and the heat transfer coefficient low. Conversely, if backward feeding is used, the liquor of the highest concentration will remain in the first effect in which the heating steam temperature is the highest. Therefore, the viscosity of the thick liquor will be substantially reduced and the heat transfer coefficient, and hence the capacity, will be much higher than in the former case. This is a definite advantage of using backward feed if the final thick liquor is highly viscous but not very heat sensitive.

(ii) Now let us consider the situation when the feed is cold. If this feed is admitted to the first effect, a considerable amount of external steam supplied to this effect is consumed to raise the temperature of the feed to the boiling point. Because this amount of steam does not contribute to generation of vapours (it only supplies sensible heat to the feed), multiple reuse of the steam in latter effects is not possible and its potential to evaporate almost an equal amount of liquid in each of the subsequent effects cannot be utilized. So, a part of the external steam is subjected to a sort of wasteful use. On the other hand, if backward feed is adopted, i.e. the cold feed is introduced to the last effect, the feed absorbs heat from steam which does not have the potential of reuse. Although the advantage is partially offset by the fact that the liquid requires to be heated as it passes from one effect to the preceding effect, the net result has been found to be an increase in the capacity. So, backward feed is advantageous if the feed solution is cold.

(iii) If the feed is hot, the forward feed arrangement is likely to offer a better steam economy. It does not absorb any sensible heat when it is introduced to the evaporator (because it is already hot). It rather produces some 'flash steam' when it flows to the effect operating at a lower pressure. In fact, by *vapour flashing* some evaporation in a forward-feed multiple-effect unit occurs in each effect starting from the second. In the backward feed mode, on the contrary, some temperature rise of the liquor occurs in each effect.

The advantages and limitations of different feed modes are given in Table 9.3.

Table 9.3 Advantages and limitations of different modes of feed supply to multiple-effect evaporators

Mode of feed supply	Advantages	Limitations
Forward feed	Simple to operate; less expensive; the liquor flows from one effect to the next driven by the pressure differential between successive effects, and hence no pump is required for transferring the liquor; less chance of deterioration of heat sensitive materials because the more concentrated liquor is vaporized at a lower temperature.	Reduced rate of heat transfer in the second and higher effects; feed should not be below the boiling point because this reduces steam economy by consuming external steam to supply sensible heat.
Backward feed	The most concentrated liquor is in contact with the highest temperature steam and thus lower viscosity and higher heat transfer rate in the first effect as a result.	Inter-effect pumps are necessary; higher risk of damage of the viscous product subjected to a higher temperature; risk of fouling.
Mixed feed	Combines the simplicity of forward feed and economy of backward feed; useful for concentration of a highly viscous feed.	More complex piping and instrumentation which make the arrangement more expensive.
Parallel feed	More suitable for use with crystallizers; allows better control.	More complex arrangement; pumps generally required for each effect.

9.5.3 Effect of Boiling Point Elevation in a Multiple-effect Evaporator

Before going into the method of heat transfer calculations for a multiple-effect evaporator, let us have a look into the phenomena that occur as a result of the elevation of boiling point. We refer to Fig. 9.25. Saturated steam [flow rate = W_s, temperature = T_s, pressure = P_s] enters the steam chest of the first effect. The liquid in effect I boils at a temperature T_{b1}. This is also the

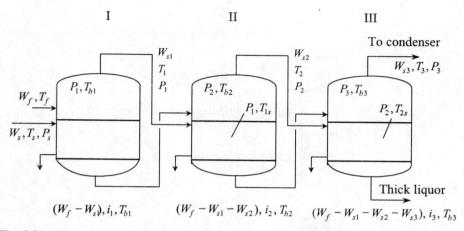

Fig. 9.25 Flow rates, temperatures and pressures in a triple-effect forward-feed evaporator.

temperature of the vapour (or steam) generated in this effect. The pressure in the vapour space is maintained at P_1; the boiling point of the pure solvent (or water) at this pressure is T_{w1}. So, the boiling point elevation in effect I is $T_{b1} - T_{w1}$, and the vapour generated is a superheated vapour. (The superheat is also $T_{b1} - T_{w1}$.) The saturation temperature of this vapour is T_{w1} which is the boiling point of water at the pressure P_1 prevailing in the vapour space.

Solution from effect I flows to effect II where the liquid boils at a temperature T_{b2} and the pressure in the vapour space is maintained at P_2. The boiling point of water at a pressure of P_2 is T_{w2}, and the boiling point elevation is $T_{b2} - T_{w2}$. The vapour from effect I enters the shell or the steam chest of effect II with a degree of superheat, but looses its superheat quickly by transfer of the sensible heat and attains its saturation temperature of T_{w1}. The amount of sensible heat transferred is negligible for all practical purposes. So the temperature driving force for heat transfer to the boiling liquid in effect II is $T_{w1} - T_{b2}$.

Vapour from effect II enters the shell of effect III at a temperature T_{b2}. This vapour has a superheat of $T_{b2} - T_{w2}$, and loses the superheat quickly and attains the saturation temperature of T_{w2}. The temperature driving force in this effect is considered to be $T_{w2} - T_{b3}$, where T_{b3} is the boiling point of the liquid in effect III operating at a pressure of P_3. The boiling temperature of water at a pressure P_3 is T_{w3}, and the boiling point elevation in effect III is $T_{b3} - T_{w3}$. Since we are considering a triple-effect evaporator, effect III is the last effect and the vapour generated in it is discarded in the condenser. Because $T_{b1} > T_{b2} > T_{b3}$, as soon as the liquid from effect I enters the effect II, it flashes, gives off some steam and attains the boiling temperature therein (i.e. T_{b2}). Similar phenomenon occurs when the liquid from effect II enters into effect III.

The boiling point elevations in different effects and the 'effective' temperature driving forces are shown in Fig. 9.26. The 'effective' driving force is the difference between the saturation temperature of steam inside the steam chest or the shell and the boiling temperature of the liquid. Figure 9.26 clearly shows how the effective temperature driving force in each effect is reduced because of the boiling point elevation (BPE). Influence of BPE is the strongest in effect III because the liquid concentration, and, therefore, the BPE, are at the maximum in this effect. It is also easily understandable from Fig. 9.26 that if a large number of effects are used, the driving forces may be too small to make the operation of the unit viable. Further, it will be impossible

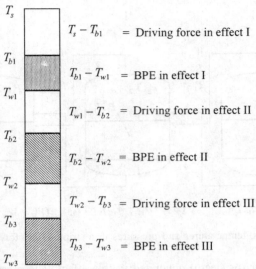

Fig. 9.26   Effective temperature driving force □, and boiling point elevations ▨, ▨, ▧ in a triple-effect forward-feed evaporator.

to run the evaporator if the sum of the BPEs in all the effects is equal to or more than the total available temperature drop across the unit. On the other hand, if we use a single-effect evaporator instead of a triple-effect unit in order to concentrate the feed to the same final concentration, maintaining a pressure P_3, the available temperature driving force will be $T_s - T_{b3}$. As a result, a smaller heat transfer area will be required. However, the steam economy will be much less (at best unity).

9.5.4 Multiple-effect Evaporator Calculations

We can write down the following energy–balance equations for the three effects (neglecting heat losses from the evaporator) on the basis of the flow rates and other data for the various streams (see Fig. 9.25).

Effect I [compare with Eq. (9.6)]

$$W_f i_f + W_s i_s = (W_f - W_{s1})i_1 + W_{s1}i_{s1} + W_s i_{l1} \tag{9.13}$$

where i_{s1} is the enthalpy of the vapour generated; all other terms have significances as stated in Section 9.4.

We can simplify the above energy balance and also similar equations for the latter effects by considering only the latent heats of condensation of the steam in the shells or the steam chests, and *neglecting* the sensible heat of the steam. This will not involve any appreciable error in view of the fact that the amount of sensible heat is very small compared to the latent heat of condensation of steam. With this simplification, we can rewrite Eq. (9.13) as given below.

$$W_f i_f + \underline{W_s \lambda_s} = (W_f - W_{s1})i_1 + W_{s1}i_{s1} \tag{9.14}$$

Similarly, for *Effect II*,

$$(W_f - W_{s1})i_1 + \underline{W_{s1}\lambda_{s1}} = (W_f - W_{s1} - W_{s2})i_2 + W_{s2}i_{s2} \tag{9.15}$$

Effect III

$$(W_f - W_{s1} - W_{s2})i_1 + \underline{W_{s2}\lambda_{s2}} = (W_f - W_{s1} - W_{s2} - W_{s3})i_3 + W_{s3}i_{s3} \qquad (9.16)$$

where

i_k is the enthalpy of the solution leaving the kth effect

W_{sk} is the rate of vapour generation in the kth effect

i_{sk} is the enthalpy of the vapour generated in the kth effect

λ_{s1}, λ_{s2} are the latent heats of steam condensation of steam at pressures P_1 and P_2, respectively.

The underlined quantities in Eqs. (9.14)–(9.16) represent the heat input to the respective effects by steam condensation.

If U_{D1}, U_{D2} and U_{D3} are the corresponding overall heat transfer coefficients and A_1, A_2 and A_3 are the areas of heat transfer required, then we may write

$$\text{Effect 1: } Q_1 = W_s\lambda_s = U_{D1}A_1(T_s - T_{b1}) = U_{D1}A_1\Delta T_1 \qquad (9.17)$$

$$\text{Effect 2: } Q_2 = W_{s1}\lambda_{s1} = U_{D2}A_2(T_{w1} - T_{b2}) = U_{D2}A_2\Delta T_2 \qquad (9.18)$$

$$\text{Effect 3: } Q_3 = W_{s2}\lambda_{s2} = U_{D3}A_3(T_{w2} - T_{b3}) = U_{D3}A_3\Delta T_3 \qquad (9.19)$$

The meanings of the notations are given in Section 9.5.3. Equations (9.14)–(9.19) have to be solved together in order to calculate the heat transfer areas. It needs quite an amount of trial-and-error calculations. As a matter of practice, the heat transfer areas in all the effects are taken to be equal. This keeps the sizes of the effects same and significantly reduces the equipment cost (it is understandable that three effects of the same size will cost substantially less than the three effects of different sizes). The following steps are involved in the calculations.

Step I: In the first step, an initial estimate of the temperature in each effect is made. In making this estimate, it is assumed that the heat transfer rates in the three effects are roughly equal. That is,

$$U_{D1}A_1\Delta T_1 = U_{D2}A_2\Delta T_2 = U_{D3}A_3\Delta T_3 \qquad (9.20)$$

The total available temperature drop is

$$\Sigma\Delta T = \Delta T_1 + \Delta T_2 + \Delta T_3 = (T_s - T_{w3}) - \Sigma\text{BPE}$$

where $T_s - T_{w3}$ is the apparent total temperature drop (see Fig. 9.26), and

$$\Sigma\,\text{BPE} = (T_{b1} - T_{w1}) + (T_{b2} - T_{w2}) + (T_{b3} - T_{w3}) \qquad (9.21)$$

The total boiling point elevation is yet to be determined.

The assumption of equal rate of heat transfer in each of the effects means that the rates of vaporization in the effects are also roughly equal. The total rate of evaporation of water is calculated from the feed and product concentrations and the feed rate. The approximate vaporization rate in each effect is one-third of this (for a triple-effect unit). From this, we calculate the concentration in each effect and estimate the BPE in each effect. Then, we estimate the total available temperature drop, $\Sigma\Delta T$.

The total available temperature drop $\Sigma\Delta T$ is *distributed* among the three effects. As stated before, it is customary to take $A_1 = A_2 = A_3$. Then from Eq. (9.20), we have

$$U_{D1}\Delta T_1 = U_{D2}\Delta T_2 = U_{D3}\Delta T_3 \qquad (9.22)$$

Because ΔTs are inversely proportional to U_Ds, we obtain

$$\frac{\Delta T_1}{\Sigma \, \Delta T} = \frac{1/U_{D1}}{(1/U_{D1}) + (1/U_{D2}) + (1/U_{D3})} \tag{9.23}$$

Thus ΔT_1, ΔT_2, and ΔT_3 can be estimated.

Step II: Once ΔT_1, ΔT_2 and ΔT_3 are estimated, Eqs. (9.14)–(9.19) may be solved to calculate W_s, W_{s1} and W_{s2} and the corresponding rates of heat transfer. We then estimate the areas of the three effects from

$$A_1 = \frac{W_s \lambda_s}{U_{D1} \Delta T_1}; \quad A_2 = \frac{W_{s1} \lambda_{s1}}{U_{D2} \Delta T_2}; \quad A_3 = \frac{W_{s2} \lambda_{s2}}{U_{D3} \Delta T_3} \tag{9.24}$$

These calculated areas, in all probability, will not be equal.

Step III: The average value of these calculated areas is taken as the area of each effect. The ΔT values are *redistributed* in the three effects and the calculations are repeated until the areas turn out to be nearly equal.

The procedure laid down above is applicable to forward-feed evaporators, and will be illustrated by an example. For backward feed (or any other mode of feeding), the energy balance equations can be written accordingly. However, generally the areas of all the effects are kept equal in all cases.

Example 9.4 An aqueous solution of a high molecular weight solute is concentrated from 2% to 35% by weight at the rate of 6000 kg/h in a forward-feed, double-effect evaporator. The dilute feed at 50°C enters the first effect. Saturated steam at an absolute pressure of 2.0 bar is supplied to the first effect, while an absolute pressure of 0.0139 bar is maintained in the second effect. Saturated condensate is removed from each effect. The overall heat transfer coefficients in the first and the second effect are 2000 W/m² K and 1500 W/m² K, respectively. Calculate the evaporator areas and the steam economy. The specific heat of the liquid can be taken as 4.1 kJ/kg K.

SOLUTION Feed rate, W_f = 6000 kg/h, 2% solid

Solid in = (6000)(0.02) = 120 kg/h;   water in = 5880 kg/h
Product out (35% solid), W_p = 120/(0.35) = 343 kg/h
Water out with the product = (343)(1 − 0.35) = 223 kg/h
Total evaporation rate, $W_{s1} + W_{s2}$ = 5880 − 223 = 5657 kg/h   (i)

Boiling temperature in the first effect

Pressure = 2 bar (abs);   temperature, T_s = 120°C; λ_s = 2200 kJ/kg
Absolute pressure in the second (last) effect = 0.0139 bar
Boiling point of the liquid in the second effect, T_2 = 285 K (12°C); λ_s = 2470 kJ/kg
Since the solute has a high molecular weight, the boiling point elevation of the solution is neglected.
Total available temperature drop,

$$\Delta T = \Delta T_1 + \Delta T_2 = T_s - T_2 = 120 - 12 = 108°C \tag{ii}$$

If we allow equal areas to the two effects [Eq. (9.20)], we have

$$U_1 \Delta T_1 = U_2 \Delta T_2$$

or

$$2000 \Delta T_1 = 1500 \Delta T_2 \tag{iii}$$

From Eqs. (ii) and (iii), $\Delta T_1 = 46°C$ and $\Delta T_2 = 62°C$
Temperatures of the saturated vapour (steam) leaving the first and the second effects are:

$$T_1 = 120 - 46 = 74°C; \qquad T_2 = 74 - 62 = 12°C$$

Energy balance over the first effect [see Eq. (9.14)]

Taking T_1 as the reference temperature, $i_f = c_p(T_f - T_1)$; $i_1 = 0$; $i_{s1} = \lambda_{s1}$
At the temperature of 74°C, latent heat of steam, $\lambda_{s1} = 2330$ kJ/kg
Feed temperature, $T_f = 50°C$
Substituting the above values in the energy balance equation,

$$W_f i_f + W_s \lambda_s = (W_f - W_{s1})i_1 + W_{s1} i_{s1}$$

we obtain

$$(6000)(4.1)(50 - 74) + W_s(2200) = (6000 - W_{s1})(0) + W_{s1}(2330) \qquad \text{(iv)}$$

or

$$W_{s1} = 0.9442 W_s - 253.4 \qquad \text{(v)}$$

Energy balance over the second effect [see Eq. (9.15)]

$$(W_f - W_{s1})i_1 + W_{s1}\lambda_{s1} = (W_f - W_{s1} - W_{s2})i_2 + W_{s2} i_{s2}$$

Take T_2 (the boiling point of the liquid in the second effect) as the reference temperature. Thus,

$$i_1 = c_p(T_1 - T_2); \qquad i_2 = 0; \qquad i_{s2} = \lambda_{s2}$$

At the temperature $T_2 = 12°C$, $\lambda_{s2} = 2470$ kJ/kg.
Substituting the values, the energy balance over the second effect is

$$(6000 - W_{s1})(4.1)(74 - 12) + W_{s1}(2330) = W_{s2}(2470)$$

or

$$W_{s2} = 0.8404 W_{s1} + 617.5 \qquad \text{(vi)}$$

Solving Eqs. (i), (v), and (vi), we obtain

$$W_{s1} = 2739 \text{ kg/h}; \quad W_{s2} = 2918 \text{ kg/h}; \quad W_s = 3169 \text{ kg/h}$$

Evaporator area

Area of the first effect,

$$A_1 = \frac{Q_1}{U_1 \Delta T_1} = \frac{W_s \lambda_s}{U_1 \Delta T_1} = \frac{(3169)(2200)}{(2000)(46)} = 75.8 \text{ m}^2$$

Area of the second effect,

$$A_2 = \frac{Q_2}{U_2 \Delta T_2} = \frac{W_{s1} \lambda_{s1}}{U_2 \Delta T_2} = \frac{(2739)(2330)}{(1500)(62)} = 68.6 \text{ m}^2$$

The areas are very close. However, the results may be improved by using better approximations for ΔT_1 and ΔT_2.

Revised calculations

Taking $\Delta T_1 = 48°C$, and $\Delta T_2 = 60°C$, we have

$$T_1 = 120 - 48 = 72°C; \quad T_2 = 72 - 60 = 12°C; \quad \lambda_{s1} = 2335 \text{ kJ/kg}; \quad \lambda_{s2} = 2470 \text{ kJ/kg}$$

Energy balance over the first effect

$$(6000)(4.1)(50 - 72) + W_s(2200) = W_{s1}(2335)$$

or

$$W_{s1} = 0.9422W_s - 231.8 \qquad \text{(vii)}$$

Energy balance over the second effect

$$(6000 - W_{s1})(4.1)(72 - 12) + W_{s1}(2335) = W_{s2}(2470)$$

or

$$W_{s2} = 0.8457W_{s1} + 579.5 \qquad \text{(viii)}$$

Solving Eqs. (i), (vii), and (viii), we obtain

$$W_{s1} = 2751 \text{ kg/h}; \quad W_{s2} = 2906 \text{ kg/h}; \quad W_s = 3166 \text{ kg/h}$$

Evaporator area

Area of the first effect,

$$A_1 = \frac{Q_1}{U_1 \Delta T_1} = \frac{W_s \lambda_s}{U_1 \Delta T_1} = \frac{(3166)(2200)}{(2000)(48)} = 72.5 \text{ m}^2$$

Area of the second effect,

$$A_2 = \frac{Q_2}{U_2 \Delta T_2} = \frac{W_{s1} \lambda_{s1}}{U_2 \Delta T_2} = \frac{(2751)(2335)}{(1500)(60)} = 71.4 \text{ m}^2$$

The recalculated areas are nearly equal. Take $\boxed{72 \text{ m}^2}$ as the area of each effect.

$$\text{Steam economy} = \frac{\text{Water vaporized}}{\text{Steam supplied}} = \frac{W_{s1} + W_{s2}}{W_s} = \frac{5657}{3166} = \boxed{1.79}$$

[In design practice, an approximate excess area (15–25%) may be provided in some cases.]

Note: Because the boiling point elevation is negligible, the vapour generated in an effect is saturated. In such a case, only the latent heat of vaporization of water is needed to be considered as shown in the heat balance Eqs. (9.13)–(9.19). However, if the elevation of the boiling point is substantial, the vapour generated remains superheated, and therefore its enthalpy needs to be considered during heat balance. This is shown in Example 9.6.

Example 9.5 Saturated steam at a pressure of 3.32 bar absolute is fed to the steam chest of the first effect of a multi-effect evaporator. The residual pressure in the condenser attached to the last effect is 0.195 bar. If the sum of the temperature losses because of BPE is 41 K, determine the maximum number of effects to be used. The available temperature driving force in an effect must be at least 8 K.

SOLUTION At a pressure of 3.32 bar, the temperature of saturated steam is 410 K.

Pressure in the last effect = 0.195 bar; corresponding saturation temperature = 333 K

$$\text{Total temperature difference} = 410 - 333 = 77 \text{ K}$$

$$\text{Temperature loss because of BPE} = 41 \text{ K}$$

Available temperature drop across the unit = 77 − 41 = 36 K

Maximum number of effects = 36/8 = | 4 effects |

Example 9.6 A 9.5% solution of caustic soda at 40°C is to be concentrated to 50% in a triple-effect evaporator at the rate of 2.0 tons NaOH per hour. Low pressure steam at 3.3 bar absolute is available. A vacuum of 714 mm Hg is maintained in the last effect of the forward-feed evaporator. Overall heat transfer coefficients of 6000, 3500, and 2500 W/m² °C, corrected for boiling point elevations, may be used for the first, second, and third effects, respectively. Calculate the heat transfer area required (assuming equal area for the three effects), rate of steam consumption, and steam economy.

SOLUTION *Material balance*

Capacity = 2.0 tons NaOH/h; Feed concentration = 9.5%

Feed rate = 2000/0.095 = 21,053 kg/h (= W_f)

Product concentration = 50%; Product rate = 2000/0.5 = 4000 kg/h

Total evaporation rate = 21,053 − 4000 = 17,053 kg/h

Steam

Pressure = 3.3 bar, assumed saturated.

From the steam table, temperature, T_s = 137°C; latent heat of steam, λ_s = 2153 kJ/kg

Pressure in the last effect = 760 − 714 = 46 mm Hg

Boiling point of water at this pressure, T_{w3} = 37°C

Refer to Fig. 9.24.

Step I: The apparent total temperature drop, $T_s - T_{b3}$ = 137 − 37 = 100°C. The *actual* temperature drop will be much less because of substantial boiling point elevation in each effect. In order to calculate the BPEs, it is necessary to know the concentrations of the boiling solutions in each effect. To have initial estimates of concentrations, it is customary to assume *nearly equal evaporation* rates in each effect. In fact, if the feed is cold, the evaporation rate in the first effect becomes appreciably lower than those in other effects. Also, the evaporation rate in the second effect may be somewhat less than that in the third effect, because in the forward-feed system, the temperature of the feed in the third effect remains above the boiling temperature in that effect.

Let us assume the following evaporation rates.

Effect I: 5600 kg/h; Effect II: 5680 kg/h; Effect III: 5773 kg/h

Concentrations in the three effects are

$$\text{Effect I} \quad : C_1 = \frac{2000}{21{,}053 - 5600} = 0.13 \text{ (weight fraction)}$$

$$\text{Effect II} \quad : C_2 = \frac{2000}{21{,}053 - 5600 - 5680} = 0.205$$

Effect III : C_3 = 0.50 (given)

Elevation of the boiling points at these concentrations in the three effects are (using Fig. 9.21 together with the steam table):

$$\text{Effect I} \quad : (BPE)_I = 3.5°C$$
$$\text{Effect II} \quad : (BPE)_{II} = 8°C$$
$$\text{Effect III} : (BPE)_{III} = 39°C$$

Actual total temperature drop available $= \Sigma \Delta T = 100 - (3.5 + 8 + 39) = 49.5°C$

This total available temperature drop is now distributed in the three effects following Eq. (9.23).

$$\frac{\Delta T_1}{\Sigma \Delta T} = \frac{1/6000}{(1/6000) + (1/3500) + (1/2500)} = 0.195$$

Therefore,

$$\Delta T_1 = (49.5)(0.195) = 10°C. \text{ Similarly, } \Delta T_2 = 16°C \text{ and } \Delta T_3 = 23.5°C$$

Step II: Enthalpy balance equations for the three effects are written and solved to calculate the evaporation rates W_{s1}, W_{s2}, and W_{s3}. The relevant enthalpy values and other quantities are given in Table 9.4.

Table 9.4 Enthalpies and other parameters

Effect	I	II	III
Inlet liquor temp.	$T_f = 40°C$	$T_{b1} = 137 - 10 = 127°C$	$T_{b2} = 127 - 3.5 - 16^* = 107.5°C$
Liquor conc. (wt. fraction)	0.13	0.205	0.5
Evaporation rate, kg/h	W_{s1}	W_{s2}	W_{s3}
BPE, (°C; estimated)	3.5	8	39
ΔT, (°C; actual)	10	16	23.5
Vap. temperature, °C (= temp. of the boiling liquid)	127	$127 - 3.5 - 16 = 107.5°C$	$107.5 - 8 - 23.5 = 76°C$
Enthalpy of soln., kJ/kg	486 (= i_1)	385 (= i_2)	460 (= i_3)
Saturated vap. temp.	$127 - 3.5 = 123.5°C$	$107.5 - 8 = 99.5°C$	$76 - 39 = 37°C$
Enthalpy of vap. generated (superheated steam), kJ/kg	$i_{s1} = 2729$	$i_{s2} = 2691$	$i_{s3} = 2646$
Enthalpy of condensate, kJ/kg		$i_{l2} = 519$	$i_{l3} = 418$

Enthalpy of water or steam is taken with reference to water at 0°C and that of the caustic solution is taken with reference to an infinitely dilute solution at 0°C. Enthalpies of solutions are read from Fig. 9.27. Enthalpies of saturated water or steam can be obtained from the steam table. However, these values can be calculated with reasonable accuracy taking the specific heat of water as 4.2 kJ/kg °C, that of steam as 1.97 kJ/kg °C, and taking the heat of vaporization at appropriate temperatures.

Enthalpy balance over Effect I (no subcooling of the condensate)

$$W_f i_f + W_s \lambda_s = W_{s1} i_{s1} + (W_f - W_{s1}) i_1$$

*The steam entering Effect II is at 127°C but has a superheat of 3.5°C [because $(BPE)_1 = 3.5°C$]. It is saturated at $127 - 3.5 = 123.5°C$ (= T_{1s}). The boiling point of the liquid in Effect II = $T_{1s} - \Delta T_2 = 123.5 - 16 = 107.5°C$. This is the temperature of the liquor entering Effect III.

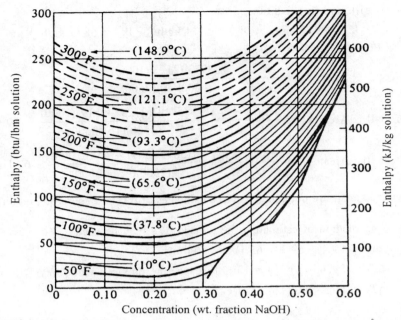

Fig. 9.27 Enthalpy-concentration diagram for the NaOH-water system (Reference state is the liquid water at 0°C)(Example 9.6).

Enthalpy of the feed at 40°C, i_f = 145 kJ/kg

Because the saturated steam fed to the steam chest of Effect I releases only the latent heat, $W_s \lambda_s$ gives the rate of heat supply by steam to this effect.

Putting the values of various quantities, we get

$$(21,053)(145) + W_s(2153) = W_{s1}(2729) + (21,053 - W_{s1})(486)$$

or

$$W_{s1} = 0.96 W_s - 3200 \tag{i}$$

Enthalpy balance over Effect II

$$(W_f - W_{s1})i_1 + W_{s1}i_{s1} = (W_f - W_{s1} - W_{s2})i_2 + W_{s2}i_{s2} + W_{s1}i_{l2}$$

or

$$(21,053 - W_{s1})(486) + W_{s1}(2729) = (21,053 - W_{s1} - W_{s2})(385) + W_{s2}(2691) + W_{s1}(519)$$

or

$$W_{s2} = 0.9146 W_{s1} + 922 \tag{ii}$$

Enthalpy balance over Effect III

$$(W_f - W_{s1} - W_{s2})i_2 + W_{s2}i_{s2} = (W_f - W_{s1} - W_{s2} - W_{s3})i_3 + W_{s3}i_{s3} + W_{s2}i_{l3}$$

or

$$(21,053 - W_{s1} - W_{s2})(385) + W_{s2}(2691) = (21,053 - W_{s1} - W_{s2} - W_{s3})(460) + W_{s3}(2646) + W_{s2}(418)$$

or

$$W_{s3} = 1.073 W_{s2} + 0.0343 W_{s1} - 722 \tag{iii}$$

Total evaporation, $W_{s1} + W_{s2} + W_{s3} = 17,053$ \hfill (iv)

Solving Eqs. (i), (ii), and (iii), we obtain

$$W_s = 8973 \text{ kg/h}; \quad W_{s1} = 5414 \text{ kg/h}; \quad W_{s2} = 5873 \text{ kg/h}$$

Calculation of the heat transfer areas

$$\text{Effect I} \quad : \text{Area, } A_1 = \frac{W_s \lambda_s}{U_{D1} \Delta T_1} = \frac{(8973)(2153)(1000)}{(6000)(10)(3600)} = 89.4 \text{ m}^2$$

$$\text{Effect II} \quad : \text{Area, } A_2 = \frac{W_{s1}(i_{s1} - i_{l2})}{U_{D2} \Delta T_2} = \frac{(5414)(2729 - 519)(1000)}{(3500)(16)(3600)} = 59.3 \text{ m}^2$$

$$\text{Effect III} : \text{Area, } A_3 = \frac{W_{s2}(i_{s2} - i_{l3})}{U_{D3} \Delta T_3} = \frac{(5873)(2691 - 418)(1000)}{(2500)(23.5)(3600)} = 63.1 \text{ m}^2$$

The calculated areas are not equal. Revised estimates of ΔTs are, therefore, required to repeat the calculations.

The following method of estimating the revised ΔTs may be used.

Average area = $(89.4 + 59.3 + 63.1)/3 = 70.6 \text{ m}^2$

$$\Delta T_1 = (89.4/70.6)(10) = 13°C \text{ [because the previous value of } \Delta T_1 \text{ is } 10°C]$$

$$\Delta T_2 = (59.3/70.6)(16) = 14°C$$

$$\Delta T_3 = 49.5 - 13 - 14 = 22.5°C$$

Table 9.5 Enthalpies and other parameters (revised)

Effect	I	II	III
Inlet liquor temp.	$T_f = 40°C$	$T_{b1} = 137 - 13 = 124°C$	$T_{b2} = 124 - 3.5 - 14$ $= 106.5°C$
BPE, (°C, estimated)	3.5	8	39
ΔT, (°C, actual)	13	14	22.5
Vap. temperature, °C (= temp. of the boiling liquid)	124	$124 - 3.5 - 14 = 106.5°C$	$106.5 - 8 - 22.5 = 76°C$
Enthalpy of soln., kJ/kg	470 (= i_1)	380 (= i_2)	460 (= i_3)
Saturated vap. temp.	$124 - 3.5 = 120.5°C$	$106.5 - 8 = 98.5°C$	$76 - 39 = 37°C$
Enthalpy of vap. generated (superheated steam), kJ/kg	$i_{s1} = 2720$	$i_{s2} = 2685$	$i_{s3} = 2646$
Enthalpy of condensate, kJ/kg		$i_{l2} = 513$	$i_{l3} = 412$

There will be consequent changes in the boiling temperatures of the liquids in the three effects. However, because these will not be much, the BPEs in the three effects will remain nearly the same as before. So, the previous values are used.

The revised values of enthalpies and other parameters are shown in Table 9.5. Enthalpy balance equations can now be written for the Effects I, II, and III as before. The solutions of these equations yield 'better' values of W_s, W_{s1}, W_{s2}, and W_{s3}. The calculations are left as an exercise. The revised values are:

$$W_s = 8854 \text{ kg/h}; \quad W_{s1} = 5432 \text{ kg/h}; \quad W_{s2} = 5812 \text{ kg/h}; \quad W_{s3} = 5809 \text{ kg/h}$$

Revised heat transfer areas

$$\text{Effect I} : A_1 = \frac{W_s \lambda_s}{U_{D1} \Delta T_1} = \frac{(8854)(2153)(1000)}{(6000)(13)(3600)} = 68 \text{ m}^2$$

$$\text{Effect II} \; : A_2 = \frac{W_{s1}(i_{s1} - i_{l2})}{U_{D2}\Delta T_2} = \frac{(5432)(2720 - 513)(1000)}{(3500)(14)(3600)} = 68 \text{ m}^2$$

$$\text{Effect III} \; : A_3 = \frac{W_{s2}(i_{s2} - i_{l3})}{U_{D3}\Delta T_3} = \frac{(5812)(2685 - 412)(1000)}{(2500)(22.5)(3600)} = 71 \text{ m}^2$$

The areas are now reasonably close. An average area of 69 m² for each effect is recommended (however, a reasonable excess area may be provided). Thus,

$$\text{Steam rate, } W_s = \boxed{8854 \text{ kg/h}}$$

$$\text{Steam economy} = \frac{\text{Total evaporation}}{\text{Steam fed}} = \frac{17,053}{8854} = \boxed{1.92}$$

Example 9.7   A single-effect evaporator is used to concentrate 3000 kg/h of a dilute feed from 8% to 40% solute content. Besides labour charges, all other operating costs amount to Rs 120,000 per year. Low pressure steam at 4.5 bar absolute is available at a cost of Rs 700 per ton.

It is now planned to increase the capacity of the evaporator by maintaining a vacuum in the evaporator, so as to get the benefit of a larger driving force. Given below are the parameter values and operating conditions for both atmospheric and vacuum operation.

Table 9.6   Parameter values (Example 9.7)

Parameter	Existing	New
Evaporator pressure	1 atm	75 mm Hg
Steam chest pressure	4.5 bar abs.	4.5 bar abs.
Feed concentration	8%	8%
Product concentration	40%	40%
Feed temperature	60°C	60°C
Labour cost	Rs 100/h	Rs 200/h
Other operating costs	Rs 400,000 per year	Rs 600,000 per year
Working days	300 days per year	300 days per year

Assuming that the overall heat transfer coefficient remains constant, calculate the increase in evaporation capacity attainable, and also the percentage change in the cost of concentrating a ton of feed. The specific heat of the dilute liquor is essentially the same as that of water, and the specific heat of the concentrated liquor is 0.764 kcal/kg °C. Because the solute has a high molecular weight, the elevation of boiling point may be neglected. Heat of solution may also be neglected.

SOLUTION   Feed = 3000 kg/h, Feed concentration = 8% solids, Feed temperature = 60°C
  Solids in the feed = (3000)(0.08) = 240 kg
  Product rate = 240/0.4 = 600 kg/h
  Evaporation rate = 3000 − 600 = 2400 kg/h

Existing operating condition

Refer to Eq. (9.6). Taking 0°C as the reference temperature, we have

$$i_f = (1)(60 - 0) = 60 \text{ kcal/kg}$$

Boiling temperature of the solution (BPE = 0) = 100°C = product temperature

$$i_1 = (0.764)(100 - 0) = 76.4 \text{ kcal/kg}$$

i_{s1} = enthalpy of vapour generated at 1 atm pressure = 639 kcal/kg (from the steam table). Latent heat of steam at 4.5 bar, λ_s = 496 kcal/kg, temperature = 425 K

Heat balance,

$$(3000)(60) + W_s(496) = (2400)(639) + (600)(76.4)$$

or

$$W_s = 2821 \text{ kg/h} = 2.821 \text{ ton/h}$$

Heat supplied, $Q = W_s\lambda_s = (2821)(496)$ kJ/h = 1.4 × 10^6 kcal/h

$$U_D A = \frac{Q}{\Delta T} = \frac{(2821)(496)}{(425 - 373)} = 26,910 \text{ kcal/°C}$$

Hourly cost,

$$\text{Steam cost} = (2.821 \text{ ton/h})(\text{Rs } 700/\text{ton}) = \text{Rs } 1975/\text{h}$$

$$\text{Labour cost} = \text{Rs } 100/\text{h}$$

$$\text{Other costs} = \frac{\text{Rs } 400,000/\text{year}}{(300)(24)\text{h}/\text{year}} = \text{Rs } 55.5/\text{h}$$

$$\text{Total cost} = 1975 + 100 + 55.50 = \text{Rs } 2130.50/\text{h}$$

$$\text{Cost per ton of feed} = \frac{\text{Rs } 2130.50/\text{h}}{3.0 \text{ ton}/\text{h}} = \text{Rs } 710 \text{ per ton}$$

Proposed operating condition (under vacuum)

Boiling point of liquor at 75 mm Hg (BPE = 0) = 320 K; ΔT = 425 - 320 = 105 K
Rate of heat supply = $U_D A \Delta T$ (= $W_{s1}\lambda_s$) = (26,910)(425 - 320) = 2.825 × 10^6 kcal/h
Steam rate = Q/λ_{s1} = (2.825 × 10^6)/(496) = 5697 kg/h = 5.697 ton/h
Enthalpy of product (product temp. = 320 K = 47°C) = (0.764)(47 - 0) = 36 kcal/kg
Enthalpy of vapour generated (saturated at 75 mm Hg) = 618 kcal/kg

Heat balance,

$$W_f(60) + 2.825 \times 10^6 = (W_f - W_{s1})(36) + W_{s1}(618)$$

Material balance,

$$W_f(0.08) = (W_f - W_{s1})(0.4)$$

Solving the above two equations, we get

$$W_f = 6373 \text{kg/h} = \text{new capacity}$$

So, the evaporator will now treat 6373 kg feed per hour. This means

$$\frac{6373 - 3000}{3000} \quad \text{or} \quad \boxed{112\%}$$

increase in the capacity.

$$\text{Steam rate, } W_{s1} = 5098 \text{ kg/h} = 5.098 \text{ ton/h}$$

Hourly cost,

$$\text{Steam cost} = (5.098 \text{ ton})(\text{Rs } 700/\text{ton}) = \text{Rs } 3568.6/\text{h}$$

$$\text{Labour cost} = \text{Rs } 200/\text{h}$$

$$\text{Other costs} = \frac{\text{Rs } 600,000/\text{year}}{(300)(24)\text{h}/\text{year}} = \text{Rs } 83.30/\text{h}$$

$$\text{Total cost} = 3568.6 + 200 + 83.30 = \text{Rs } 3851.90/\text{h}$$

$$\text{Cost per ton of feed} = (\text{Rs } 3851.90/\text{h})/(6.373 \text{ ton/h}) = \text{Rs } 604.40/\text{ton}$$

Percentage saving in cost for the proposed method

$$= \frac{\text{Rs } 710 - \text{Rs } 604.4}{\text{Rs } 710} \times 100 = \boxed{15\%}$$

9.6 EVAPORATOR SELECTION

There are a number of factors that govern the selection of an evaporator for a particular service. The more important factors are: (i) solution viscosity, (ii) heat sensitivity of the solute, (iii) scaling or fouling characteristics of the solution, (iv) corrosiveness, (v) presence of suspended solids, and (vi) foaming characteristics. Besides concentration of various solutions and liquors in the chemical process industries, evaporators find wide applications in food processing industries that range from concentration of milk to fruit juices and different types of extracts. The choice of evaporation temperature, vacuum and residence time or contact time is of great importance as these parameters influence the product quality and economy. Loss of aroma compounds and degradation of vitamins during concentration of fruit juices and extracts, degradation of vitamins during milk evaporation to make condensed milk, etc. are a few critical problems which need to be overcome. Any limitations imposed by the auxiliaries and utilities (for example, steam may not be available above a particular pressure, or the vacuum producing device may provide vacuum only up to a given level) also influence the selection. The general guidelines given in Fig. 9.28 may be used for the selection of an evaporator.

9.7 VAPOUR RECOMPRESSION

The basic objective of the multiple-effect evaporator configuration is to achieve a high steam economy by multiple reuse of the heat supplied to the first effect in the form of steam. Similar results can be obtained in a single-effect evaporator if the steam generated by evaporation of the solution is compressed to a higher pressure (simultaneously raising its temperature) using a suitable device, so that the compressed steam can be recirculated to the steam chest. This is the idea behind vapour recompression that essentially pumps heat from a lower level to a higher level. In fact, there are certain applications where the evaporation of a solution has to be done only at a moderate temperature (for example, orange juice is evaporated at less than 55°C), and a multiple-effect equipment cannot be used because of the limitation on the temperature driving force. The vapour recompression system is a practical alternative for achieving good steam economy in such a case.

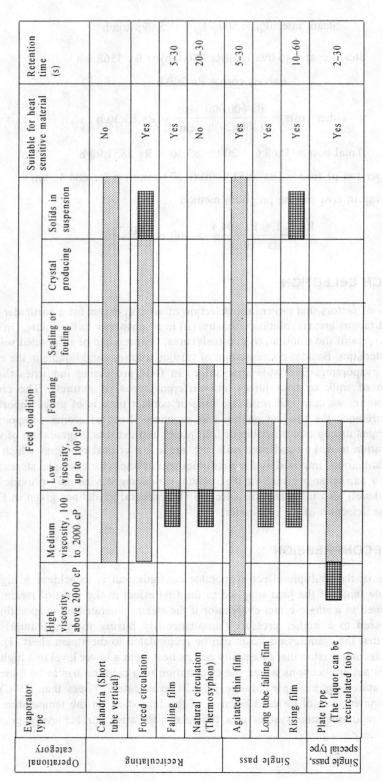

Fig. 9.28 A guide to selection of evaporators (after Parker, 1963).

Recompression of steam may be done in two ways—*thermocompression* and *mechanical recompression*. These processes are schematically shown in Fig. 9.29. Thermocompression of low-pressure steam may be done by using a steam-jet ejector to raise its temperature and pressure. High-pressure steam (the motive fluid) is used in the ejector to compress or 'boost' the temperature of the low-pressure steam from the evaporator or any other equipment. An ejector, as described before in Section 9.2.1, is essentially a fluid-pumping device. The design and construction of a thermocompressor are similar to those of a steam-jet ejector used for producing vacuum. The sectional drawing of a single-nozzle thermocompressor ejector is shown in Fig. 9.29(b).

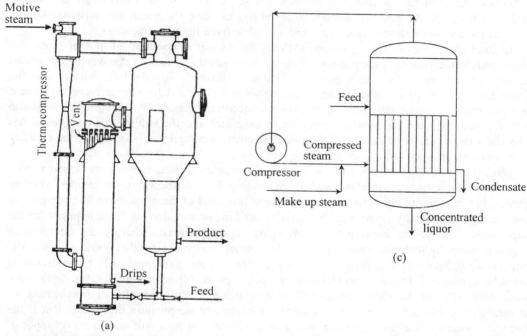

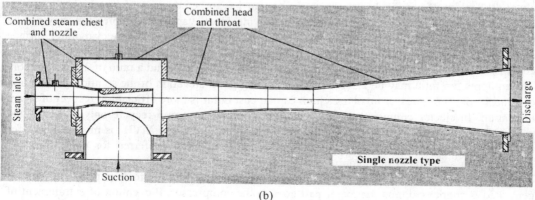

Fig. 9.29 Vapour recompression in evaporators: (a) a single-stage evaporator with thermocompressor, (b) single-nozzle steam-jet ejector for thermocompression (Courtesy: Croll-Reynolds, Westfield, NJ.), and (c) mechanical vapour recompression.

The motive fluid (steam) passes through a nozzle and emerges as a low-pressure, high-velocity jet that moves along the axis of the venturi-shaped diffuser. The low-pressure steam or vapour (called the 'load') enters through the suction and gets mixed with the motive fluid. The mixture flows through the diffuser when the velocity head is converted to the pressure head. The fluid mixture emerges at an intermediate pressure between the motive fluid and the load.

The expansion of the motive fluid as it emerges from the nozzle is almost isentropic. The pressure energy of steam is converted to kinetic energy at about 100% efficiency. The steam velocity may be very high, typically about 1000 m/s or more. However, because of impaction and turbulence in the mixing of the fluids, the efficiency of the diffuser ranges between 65 to 95%. A thermocompressor has its optimal efficiency at a single set of suction, discharge, and motive fluid conditions. At the point of optimal efficiency it has also the maximum capacity. Design charts of thermocompression calculations are available from the manufacturers.

In mechanical vapour recompression (MVR), the vapour is compressed to a pressure that corresponds to the saturation temperature of the steam as per the design. In this way the generated vapour is reused, and only a small quantity of make-up steam is supplied from outside after the system has been stabilized. A centrifugal compressor is well-suited for vapour recompression if the vapour rate is sufficiently high. A positive displacement or axial flow compressor may also be used depending upon the capacity. Since steam has a high specific volume, the compressor has to handle a large volume of vapour. A positive displacement compressor is suitable if the capacity of the evaporator is small.

Much of the energy supplied to the compressor appears as heat in the compressed vapour whose temperature rises considerably and it remains superheated. The vapour is desuperheated by spraying hot condensate into it before it enters the steam chest of the evaporator. The compressor power input is essentially determined by the ultimate temperature driving force required for the evaporation. The power input per unit mass of vapour increases sharply as the required temperature driving force increases. For example, consider an evaporator that operates with a 5°C temperature difference at atmospheric pressure. The vapour generated will be saturated at atmospheric pressure if there is no elevation of boiling point. It will require an ideal work input of about 8 kcal/kg of this vapour for compression to such a pressure that after desuperheating we get saturated steam at 5°C, thereby providing the necessary temperature difference. But if the evaporator works at the same pressure with a 20°C driving force, it will need a power input of about 40 kcal/kg of vapour, because now the vapour has to be compressed to a higher pressure. If the evaporator works at a large temperature driving force, vapour recompression may not be economical. However, if the evaporator works at a small driving force, a large evaporator surface area will be necessary. A trade-off between these opposing factors is attempted in practice. The cost of power for compression depends upon the compression ratio, and a ratio of 2:1 is common. The compression ratio may be 2.5:1, if there is boiling point elevation of the solution. Installation of a mechanical compressor involves a high capital cost but it is more efficient in respect of steam utilization. In ejector compression, it is possible to recompress and recycle only a part of the produced vapour, and the rest is discarded or put to other uses. The MVR is more suitable for use with falling-film and forced-circulation evaporators and less preferred for rising-film and calandria type equipment that require a larger temperature driving force.

The MVR system needs a smaller condenser because only a small part of the vapour generated is condensed, and the major part goes to the compressor. Prevention of entrainment of liquid in the vapour must be ensured so that no impingement of droplets occurs in the compressor. It is desirable that the mist-free vapour is given a little superheat, so that no condensation occurs in the suction pipeline to the compressor.

Example 9.8 Consider Example 9.1. Assume that saturated steam is available at 2.03 kg/cm² absolute. Make a mechanical vapour recompression calculation for the problem, if the compression ratio is 2:1.

SOLUTION Heat of condensation of steam at 2.03 kg/cm² absolute = 2200 kJ/kg
Rate of heat supply necessary (see Example 9.1) = 2.337×10^8 kJ/day
If heating is done exclusively by 'external' steam supply, we have

$$\text{Steam supply rate} = \frac{2.337 \times 10^8 \text{ kJ / day}}{2200 \text{ kJ / kg}} = 1.062 \times 10^5 \text{ kg/day}$$

Now we consider mechanical recompression of the vapour generated in the evaporator operating at 100°C and 1 atm pressure (note that the vapour generated is saturated, because there is no elevation of boiling point). Assuming *isentropic* compression of this vapour to 2 atm (2.03 kg/cm²) pressure and using Molier's chart, we obtain

Enthalpy of compressed vapour = 2800 kJ/kg, temperature = 175.5°C

The saturation temperature of steam at this pressure (2 atm absolute) is 121°C, and its enthalpy at this temperature is 2700 kJ/kg. So, the superheat of the vapour after recompression is 175.5 − 121 = 54.5°C.

Let us assume that this steam is desuperheated by spraying hot water at 100°C (enthalpy = 419 kJ/kg). (This is a standard practice for desuperheating steam.)
If *x* kg of water is supplied per kg of superheated steam, we have

$$x(419) + 2800 = (x + 1)(2700)$$

or

$$x = 0.044 \text{ kg/kg superheated steam}$$

Steam obtained after desuperheating = 1.044 kg/kg superheated steam

Rate of vapour generation in the evaporator = 8.925×10^4 kg/day

Rate of recompressed saturated steam generated = $(8.925 \times 10^4)(1.044)$

$$= 9.318 \times 10^4 \text{ kg/day}$$

This is a little less than the external steam requirement. Therefore,

Make-up steam required = $1.062 \times 10^5 - 9.318 \times 10^4 = 1.302 \times 10^4$ kg/day

SHORT QUESTIONS

1. What do you mean by evaporation? How does evaporation differ from distillation and drying?

2. Make a list of ten examples of industrial evaporation operation.

3. Explain the terms 'forced circulation' and 'natural circulation' in the context of evaporators.

4. Classify different types of evaporators. What are calandria and propeller calandria?

5. What are the advantages and limitations of different types of evaporators?

6. What type of evaporator is suitable for concentrating a highly viscous solution?

7. What are the different types of vacuum equipment used with an evaporator?

8. What is the working principle of a barometric condenser? What is a barometric leg?

9. Name the different types of steam traps. State the working principle of each type.

10. What are the factors that govern the selection of a steam trap?

11. What are the factors that govern the sizing of a steam trap for a particular application?

12. What are the advantages and limitations of different types of steam traps?

13. Describe (i) a direct contact condenser, and (ii) a surface condenser. When would you prefer a surface condenser to a direct contact condenser?

14. Name the different types of entrainment separators or mist eliminators. Discuss their working principles. What is a knockout drum?

15. Define steam economy of an evaporator. Give typical values of steam economy of (i) a single-effect evaporator and (ii) a tripple-effect evaporator.

16. Which evaporator—'natural circulation' or 'forced circulation'—needs more frequent tube cleaning when used for processing the same solution? For which type is the 'downtime' more? ('Downtime' is the loss of production time because of maintenance or repair of an equipment.)

17. What kind of operational problem do you anticipate if the evaporator for concentrating the solution of a solid is highly overdesigned?

18. A new calandria evaporator is to be installed in an existing plant. But the place where it is required to be installed has a low headroom. As a result, the evaporator has been designed with a less than adequate vapour space above the tubes. What are the likely problems associated with an underdesigned vapourhead? What are the remedies?

19. What are the common operational problems of an evaporator?

20. An evaporator of a given capacity has 200 tubes and is expected to have a 'turndown ratio' of 2:1. If, for the same capacity, 50 more tubes are provided, what would be the effect on the turndown ratio?

21. What is meant by 'tube plugging' in an evaporator?

22. What are the consequences of entrainment in a single- and a multiple-effect evaporator?

23. A long tube vertical evaporator having fixed tube sheets and an expansion bellow on the shell had the problem of frequent tube plugging. The operator of the unit developed an ingenious technique of tube cleaning. Instead of using a brush, he used to shoot live steam in the tubes, one-by-one. Though the tubes were fixed to the tube sheets by expansion joints (which is quite dependable), several joints started leaking after a few cleaning cycles. Can you guess why it happened?

24. When do you use a multistage ejector system for creating vacuum?

25. Which components contribute to the cost of a multiple-effect evaporator compared to that of a single-effect one?

26. What type of evaporator do you recommend for the following services?

(a) Concentration of 'black liquor' in the Kraft process of making pulp.

(b) Evaporation of caustic soda from 5% to 45% concentration.

(c) Evaporation of clarified cane-sugar juice.

(d) Evaporation of a solution of sodium dichromate.

(e) Concentration of glycerine solution (purified spent soap lye) in a soap factory.

(f) Evaporation of brine for the manufacture of table salt.

(g) Evaporation of ammonium nitrate from 50% water content to 1.5% water content. (This material is sometimes used as an explosive.)

(h) Evaporation of tomato juice to 30% concentration.

(i) Evaporation of apple juice (maximum allowable product temperature is 60°C).

(j) Concentration of 'spin bath liquid' in a viscose rayon plant.

(k) Evaporation of corn syrup.

(l) Evaporation of milk to produce condensed milk.

27. What type of evaporator is most widely used with a crystallizer?

28. What type of evaporator is suitable for a viscous, heat-sensitive, and highly fouling heat solution?

29. What are the practical operational problems of a falling-film evaporator?

30. What is the approximate heat duty of an evaporator per pound of water evaporated?

31. How do you know or diagnose that the tubes of an evaporator are excessively fouled and need cleaning?

32. What is the typical range of liquid velocity in a forced-circulation evaporator?

33. Single-effect evaporators have a low heat economy. Notwithstanding this, name the types of services for which they are selected.

34. What are the primary instruments necessary for controlling the operation of an evaporator?

35. What are the methods of cleaning an evaporator?

36. Can you use a recirculating-type evaporator for food processing applications? Give reasons.

37. Which one of the film evaporators (falling-film or rising-film) gives a higher heat transfer coefficient?

38. What are the important liquid properties for specifying an evaporator?

39. A lower operating pressure in an evaporator reduces the boiling point, increases the temperature driving force and hence reduces the heat transfer area. Is there any factor that indirectly tends to increase the requirement of heat transfer area as the operating pressure decreases?

PROBLEMS

9.1 (a) Calculate the area of a single-effect evaporator for concentrating 30,000 kg/h of an 8% sucrose solution to 50% using low pressure steam available at 1.8 atm absolute. The feed is at 25°C, and a vacuum of 640 mm Hg is maintained in the evaporator. The corrected heat transfer coefficient is 2000 W/m² °C. The following data are available: specific heat (kcal/kg °C) of the feed is the same as that of weight fraction of water in it; specific heat of concentrated solution = 3.14 kJ/kg°C; boiling point elevation = 5°C.

    (b) At what rate can a 10% solution of a high molecular weight solute be concentrated to 40% in the same evaporator if the other conditions and parameters remain unchanged?

9.2. Consider part (b) of Problem 9.1. At what pressure should saturated steam be fed to the steam chest, if a solution of 5% concentration is required to be concentrated to 50% concentration in the same evaporator at the same rate? Other conditions and parameters remain unchanged.

9.3. A dilute solution is concentrated from 10% solid to 50% solid at the rate of 500 kg solid per hour. The feed is available at 70°C and steam is supplied at 2 atm gauge pressure. The evaporator operates at an absolute pressure of 100 mm Hg.

    (i) Calculate the heat load of the evaporator.

    (ii) Determine the steam consumption rate and the surface area required.

    (iii) If the evaporator pressure is raised to 300 mm Hg, by what percentages would the capacity of the evaporator and the steam rate change?

Data: specific heat of dilute solution = 0.883 kcal/kg °C; specific heat of the product = 0.75 kcal/kg °C; overall heat transfer coefficient = 500 kcal/h m² °C; boiling point elevation may be neglected.

9.4. A 5% aqueous solution of a high molecular weight solute has to be concentrated to 40% in a forward-feed double-effect evaporator at the rate of 8000 kg/h. The feed temperature is 40°C. Saturated steam at 3.5 kg/cm² is available. A vacuum of 600 mm Hg is maintained in the second effect. Calculate the area requirements, if calandrias of equal areas are used. The overall heat transfer coefficients are 550 kcal/h m² °C and 370 kcal/h m² °C in the first and the second effect, respectively. The specific heat of the concentrated liquor is 0.87 kcal/kg °C.

9.5. It is required to concentrate a dilute solution of NaOH from 8% to 45% concentration using a forward-feed triple-effect evaporator. The total evaporation rate is 8000 kg water per hour. The feed enters at 60°C and a vacuum of 0.85 bar is maintained in the last effect. Steam is available at 5 bar absolute pressure. The overall heat transfer coefficients corrected for boiling point elevation are estimated to be 5000, 3400, and 2400 W/m² °C for the first, second, and third effect, respectively. Calculate the evaporation areas, steam rate, and steam economy.

9.6. A solution boils at 106°C at atmospheric pressure in the first effect of a double-effect forward-feed evaporator. The rate of vapour generation is 2000 kg/h, and the liquor with 15% solute enters the second effect. The product leaves the second effect at a temperature of 88°C. A vacuum of 500 mm Hg is maintained in the second effect. The preheated feed,

15,000 kg/h, enters the first effect at a temperature of 70°C after heat exchange with the hot product. Calculate: (i) the temperature of the fresh feed, if the product is cooled to 40°C in the heat exchanger and (ii) the rate and concentration of the feed.

9.7. A solution having 10% dissolved solids is to be concentrated to 30% solids by weight in a triple-effect evaporator at the rate of 10 tons per hour. The feed is at 100°C. Low pressure steam is available at 1.5 kg/cm^2 gauge. A forward-feed configuration is used with a vacuum of 660 mm Hg maintained in the final effect using a single-stage steam-jet ejector. The BPE in the final effect is 10°C and, in the absence of more data , the BPE can be approximately taken to be proportional to the solid concentration in wt.%. If the overall heat transfer coefficients in the effects are 2500, 2000, and 1000 kcal/h m^2 °C, respectively, calculate the evaporator area in each effect.

9.8. A solution having negligible boiling point elevation is concentrated from 10% to 40% solids in a triple-effect evaporator. A vacuum of 650 mm Hg is maintained in the last effect. The feed rate is 800 kg per hour, available at 30°C. The specific heat of the solution is given by $c_p = 0.4 + 0.6w$ kcal/kg °C, where w is the weight fraction of water in solution. The overall heat transfer coefficients in the three effects are estimated to be 2700, 1700, and 975 kcal/h m^2 °C in the first, second, and the third effect, respectively, when operated in the forward feed mode. The corresponding coefficients for the backward feed operation are 2200, 1700, and 1350 kcal/m^2 h °C, respectively. Calculate the evaporation area, steam requirement, and steam economy for both modes of feed supply.

REFERENCES AND FURTHER READING

Bennett, R.C., "Compression evaporation", *Chem. Eng. Progr.*, July 1978, 67– 69.

Capps, R.W., "Properly specify wire-mesh mist eliminators", *Chem. Eng. Progr.*, December 1994, 49–55.

Casten, J.W., "Mechanical compression evaporators", *Chem. Eng. Progr.*, July 1978, 61–66.

Coates J. and B.S. Pressburg, "How heat hransfer occurs in evaporators", *Chem. Eng.*, February 22, 1960, 139–144.

Coates, J. and B.S. Pressburg, "Multiple-effect evaporators", *Chem. Eng.*, March 21, 1960, 157–160.

Croll, S.W., "Keeping steam ejectors on-line", *Chem. Eng.*, April 1998, 108–112.

Fabian, P., R. Cusack, P. Hennessey, and M. Neuman, "Demistifying the selection of mist eliminators", *Chem. Eng.*, November 1993, 148–156.

Ganic, G., On the heat transfer and fluid flow in falling-film shell-and-tube evaporators, in S. Kakac et al. (Eds.), *Heat Exchangers*, Hemisphere Publ. Co., New York, 1981.

Haas, J.H., "Steam traps—key to process heating", *Chem. Eng.*, January 1990, 151–156.

Karuana, G. "A review of evaporators and their applications", *Brit. Chem. Engr.*, July 1965, 466–471.

Lavis, G., "Evaporators—how to make the right choice", *Chem. Eng.*, April 1994, 92–106.

Ludwig, E., *Applied Process Design for Chemical and Petrochemical Plants*, Vol. 1, 2nd ed., Gulf Publ. Co., Houston, 1977.

Mackay, B., "Avoid steam trap problems", *Chem. Eng. Progr.*, January 1992, 41–47.

Mangnall, K., *Vacuum/Pressure Producing Machineries and Associated Equipment*, Hick Hargreaves & Co., Lancashire, 1989.

O'Dell, L.R., "Beware of group trapping", *Chem. Eng. Progr.*, January 1992, 37–40.

Parker, N.H., "How to specify evaporators", *Chem. Eng.*, May 27, 1963, 135–140.

Power, R.B., *Steam Jet Ejectors for the Process Industries*, McGraw-Hill, New York, 1994.

Power, R.B., "Pump-up your energy savings", *Chem. Eng.*, February 1994, 120–126.

Radle, J., "Select the right steam trap", *Chem. Eng. Progr.*, January 1992, 30–36.

Schurter, R.V., "Evaporation—think thin film", *Chem. Eng.*, April 1994, 105–106.

Standiford, F.C., "Evaporation is a unit operation", *Chem. Eng.*, December 9, 1963, 158–176.

Wetherhorn, D., "Guide to trouble-free evaporators", *Chem. Eng.*, June 1, 1970, 187–192.

Yundt, G., "Troubleshooting VC evaporators", *Chem. Eng.*, December 1984, 46–55.

Ziebold, S., "Demistifying mist eliminator selection", *Chem. Eng.*, May 2000, 94–102.

10

Unsteady State and Multidimensional Heat Conduction

In Chapter 2, we discussed *steady state* heat conduction in solid bodies of one-dimensional geometry. However, *unsteady state* heat conduction phenomenon is also pretty common to many industrial processes, and so is heat conduction in bodies of two- or three-dimensional geometry. A few examples may be cited in this connection. In heat treatment of metal parts, the temperature of the body is raised to a desired level followed by cooling at a predetermined rate. The cooling process occurs at *unsteady state*. All kinds of equipment pass through an unsteady state period following the *start-up* until they eventually attain the steady state operating conditions. A furnace may take several days to reach the steady state condition following slow heating after start-up. In the hot rolling process, a hot block of metal or an alloy is forced through the rolls to make a sheet. The desired rate of cooling of the block is a very important factor in fixing the rolling speed. The entire cooling process occurs at unsteady state. The recovery of energy from hot gases is sometimes done by using 'thermal regenerators' (Schmidt and Willmont, 1981). The process has a heating half-cycle during which the hot gases flow through a brick checkerwork that absorbs a large portion of the thermal energy of the gas. This is followed by a cooling half-cycle when a cold gas (for example, combustion air) flows through the hot checkerwork that releases the heat stored in it in the previous half-cycle. This is an example of unsteady state heat transfer (involving both conduction and convection) in a cyclic or periodic mode. A much more sophisticated example of coupled conduction and convection at unsteady state is the heat transfer phenomenon that occurs while a rocket leaves or re-enters the atmosphere. A large amount of heat is generated by viscous dissipation on the outer surface of the rocket raising it to a very high temperature. In an interesting recent article, McGee et al. (1999) have discussed modelling of unsteady state heat conduction during cooking of meat and the strategy of maintaining the food quality. These diverse examples bear ample testimony to the occurrence, importance, and applications of the unsteady state heat transfer processes.

In the analysis of unsteady state heat conduction in a body or system, two types of 'models' are in use—the *lumped parameter model* and the *distributed parameter model*. In the former model, the temperature in a body or system is assumed to depend only on time and not on position. Temperature at any moment is considered to have a value 'lumped' over the entire body or system. For example, if a body has a very high thermal conductivity (or a small volume), the variation in temperature from one point to another in the body is small, i.e. the temperature can reasonably be assumed to depend only on time and not on position in the body. Similar is the picture when a stirred liquid in a tank is heated by passing a hot fluid through a jacket or a heating coil. Because the liquid is kept stirred, its temperature may be taken to be uniform at all points in the tank and dependent only on time. These are situations where the lumped parameter approach is appropriate. We solved a few examples in Chapter 3 using the lumped parameter model (Examples 3.1 and 3.2). In the distributed parameter modelling approach, on the other hand, the temperature (or, for that matter, any other dependent variable, for example, concentration) in a system is assumed to depend upon both position and time. The term 'distributed parameter' implies dependence of a 'parameter' on the position or space coordinates.

Examples of conduction heat transfer in multidimensional geometry are numerous in the engineering practice. Transport of heat within a cylindrical catalyst pallet on which an exothermic reaction occurs, is a practical example. In this chapter, we will discuss the basic mathematical techniques of analytical solution of unsteady state and steady state heat conduction problems. Solutions of relatively simple problems in two- or three-dimensional systems will be taken up as illustrative examples.

10.1 MATHEMATICAL FORMULATION AND INITIAL AND BOUNDARY CONDITIONS

Before taking up the details of analytical solution of unsteady state and multidimensional heat conduction (heat conduction is also called *heat diffusion*) problems, it will be useful to discuss the basic procedure of formulation of such problems. The term 'formulation of a problem' means the process of translating the problem into a mathematical equation, usually a differential equation. Reasonable simplifying assumptions are frequently made in order that the method of solution of the resulting equations does not become too difficult.

Formulation of a few simple problems of heat conduction in one dimension has been discussed in Chapters 2 and 3. The procedure is essentially similar for a higher dimensional problem and is based on differential heat balance over a suitably chosen 'small' volume element in the system. One of the three common coordinate systems—the rectangular Cartesian, the cylindrical polar, and the spherical polar—may be selected depending upon the geometrical shape of the body in which conduction or diffusion of heat is taking place. (The details of these coordinate systems and their transformations are readily available. See, for example, Wylie, 1975.) The rates of heat input, output, generation, and accumulation are written and balanced by the equation of conservation of energy.

$$\text{Rate of accumulation of heat} = (\text{Rate of input}) - (\text{Rate of output}) + (\text{Rate of generation}) \qquad (10.1)$$

Applications of the above principle are illustrated below by a few examples.

Example 10.1 (*Unsteady state heat conduction in a rectangular solid*) Consider a rectangular block of solid having sides a, b, and c. The initial temperature of the body has been T_i throughout. At time $t = 0$, all the six surfaces of the body are raised to a temperature T_s and are maintained at that temperature subsequently. Generation of heat occurs in the body at a volumetric rate (which, in general, may be a function of position and even of time) given by $\psi_v(x, y, z, t)$. It is required to do the mathematical formulation of this unsteady state three-dimensional heat conduction problem, and to specify the initial and boundary conditions.

SOLUTION Let us choose the coordinate axes along the three edges of the rectangular body with the origin placed at a corner of it as shown in Fig. 10.1. An elementary volume of solid of size Δx by Δy by Δz is chosen at a point (x, y, z) for the purpose of heat balance. There will be six heat input and output terms for the six surfaces of the volume element. There may be a generation term and an accumulation term, too. The small volume element is named $ABCDEFGH$. The coordinates of the point A are (x, y, z).

Let us assume that heat input occurs to the element through the surfaces $ABGF$, $ADEF$, and

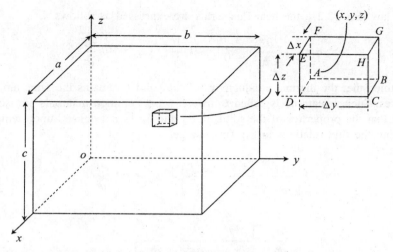

Fig. 10.1 A rectangular block of solid and an elementary volume element (control volume) (Example 10.1).

ABCD (which are normal to the *x*-, *y*- and *z*-axes respectively). The output of heat from the element occurs through the three opposite surfaces. The following are the rates of heat input:

$$\text{Through the surface } ABGF = \Delta y\,\Delta z\,q_x\big|_x \tag{i}$$

$$\text{Through the surface } ADEF = \Delta x\,\Delta z\,q_y\big|_y \tag{ii}$$

$$\text{Through the surface } ABCD = \Delta x\,\Delta y\,q_z\big|_z \tag{iii}$$

Similarly,

$$\text{The rate of heat output through } CDEH = \Delta y\,\Delta z\,q_x\big|_{x+\Delta x} \tag{iv}$$

$$\text{The rate of heat output through } BCHG = \Delta z\,\Delta x\,q_y\big|_{y+\Delta y} \tag{v}$$

$$\text{The rate of heat output through } EFGH = \Delta x\,\Delta y\,q_z\big|_{z+\Delta z} \tag{vi}$$

$$\text{The rate of heat generation in the element} = \Delta x\,\Delta y\,\Delta z\,\psi_v \tag{vii}$$

$$\text{The rate of heat accumulation} = \frac{\partial}{\partial t}\big(\Delta x \Delta y \Delta z\,\rho c_p T\big) \tag{viii}$$

where ψ_v is the volumetric rate of heat generation, ρ is the density of the solid, c_p is its specific heat, and t is the time. Because there is more than one independent variable (namely, x, y, z, and t), the time derivative in Eq. (viii) is taken as the partial derivative.

Following Eq. (10.1), the heat balance equation is written as

$$\Delta y\,\Delta z\,q_x\big|_x + \Delta z\,\Delta x\,q_y\big|_y + \Delta x\,\Delta y\,q_z\big|_z - \Delta y\,\Delta z\,q_x\big|_{x+\Delta x} - \Delta x\,\Delta z\,q_y\big|_{y+\Delta y}$$

$$- \Delta x\,\Delta y\,q_z\big|_{z+\Delta z} + \Delta x\,\Delta y\,\Delta z\,\psi_v = \frac{\partial}{\partial t}\big(\Delta x \Delta y \Delta z\,\rho c_p T\big) = \big(\Delta x \Delta y \Delta z\,\rho c_p\big)\frac{\partial T}{\partial t}$$

Dividing by $\Delta x \Delta y \Delta z$ throughout and taking limits $\Delta x \to 0$, $\Delta y \to 0$, and $\Delta z \to 0$, we have

$$-\frac{\partial q_x}{\partial x} - \frac{\partial q_y}{\partial y} - \frac{\partial q_z}{\partial z} + \psi_v = \rho c_p \frac{\partial T}{\partial t} \tag{ix}$$

By Fourier's law [Eq. (2.2)], the heat flux terms are expressed as follows:

$$q_x = -k\frac{\partial T}{\partial x}; \qquad q_y = -k\frac{\partial T}{\partial y}; \qquad q_z = -k\frac{\partial T}{\partial z}$$

It is assumed that the thermal conductivity of the solid k remains the same along the three coordinate axes. Such a material is called *isotropic*. Most common materials are isotropic. It is also assumed that the properties of the solid, ρ, c_p, and k, do not depend upon temperature.

Substituting the flux relations in Eq. (ix), we get

$$k\left[\frac{\partial^2 T}{\partial x^2} + \frac{\partial^2 T}{\partial y^2} + \frac{\partial^2 T}{\partial z^2}\right] + \psi_v = \rho c_p \frac{\partial T}{\partial t}$$

or

$$\frac{1}{\alpha}\frac{\partial T}{\partial t} = \left[\frac{\partial^2 T}{\partial x^2} + \frac{\partial^2 T}{\partial y^2} + \frac{\partial^2 T}{\partial z^2}\right] + \frac{\psi_v}{k} \tag{x}$$

or

$$\frac{1}{\alpha}\frac{\partial T}{\partial t} = \nabla^2 T + \frac{\psi_v}{k} \tag{xi}$$

where $\alpha = k/\rho c_p$ = thermal diffusivity of the material. Equation (x) or (xi) is the governing partial differential equation for unsteady state three-dimensional heat conduction in the general form. Similar equations for one- or two-dimensional heat conduction can be easily obtained from this general equation. The term $\nabla^2 T$ in Eq. (xi) is called the *Laplacian of temperature*; ∇^2 is the *Laplacian operator* given as

$$\nabla^2 = \left[\frac{\partial^2}{\partial x^2} + \frac{\partial^2}{\partial y^2} + \frac{\partial^2}{\partial z^2}\right]$$

If we put $\partial T/\partial t = 0$ (steady state), and $\psi_v = 0$ (no heat generation), we get the Laplacian equation for temperature

$$\frac{\partial^2 T}{\partial x^2} + \frac{\partial^2 T}{\partial y^2} + \frac{\partial^2 T}{\partial z^2} = 0 \tag{xii}$$

Now, we proceed to write down the *initial* and *boundary conditions*. An initial condition dictates the temperature (or its distribution) within a body at *zero time*. Boundary conditions specify the temperatures at the boundaries enclosing the body. It may be noted that the partial differential equation for temperature in the present case, Eq. (x), has the following highest order partial derivatives: first order with respect to time, and the second order with respect to each of the space variables x, y, and z. We, therefore, need *one initial condition*, and *two boundary conditions* with respect to each of the space variables, i.e. a set of seven conditions.

The initial and boundary conditions are given, explicitly or implicitly, in a problem statement. In the present problem, these are explicitly stated. Given below are the conditions.

Initial condition (I.C.):

$$t = 0, \qquad 0 < x < a, \qquad 0 < y < b, \qquad 0 < z < c; \qquad T = T_i \tag{xiii}$$

Boundary conditions (B.Cs.):

B.C. 1: $x = 0, 0 \leq y \leq b, 0 \leq z \leq c; t \geq 0;$ $T = T_s$ (xiv)

B.C. 2: $x = a, 0 \leq y \leq b, 0 \leq z \leq c; t \geq 0;$ $T = T_s$ (xv)

B.C. 3: $y = 0, 0 \leq x \leq a, 0 \leq z \leq c; t \geq 0;$ $T = T_s$ (xvi)

B.C. 4: $y = b, 0 \leq x \leq a, 0 \leq z \leq c; t \geq 0;$ $T = T_s$ (xvii)

B.C. 5: $z = 0, 0 \leq x \leq a, 0 \leq y \leq b; t \geq 0;$ $T = T_s$ (xviii)

B.C. 6: $z = c, 0 \leq x \leq a, 0 \leq y \leq b; t \geq 0;$ $T = T_s$ (xix)

Here is the explanation of the boundary conditions. Let us consider B.C. 1 as an example. This boundary condition means that the surface $x = 0$ of the rectangular body has a temperature T_s at all time as stated in the problem.

It will be useful to note an important point here. It is not at all necessary to be concerned with the boundary conditions while formulating a problem. We are free to assume heat input at some of the surfaces, and heat output at the rest. However, the assumptions must be *consistent*. Whatever approach is adopted, the process will end up with the same partial differential equation.

Example 10.2 (*Unsteady state heat conduction in a cylinder*) A cylindrical solid body of radius a and height H has an initial temperature distribution given by $T_i = T_i (r, z)$, where r and z represent the radial and axial positions of a point in the cylinder. At time $t = 0$, the surface temperature of the cylinder is brought to T_s and is maintained at this temperature subsequently. Heat generation occurs at a volumetric rate of ψ_v. Develop the partial differential equation governing heat diffusion in the cylinder. Also, specify the initial and boundary conditions.

SOLUTION We will use the cylindrical polar coordinate system for the mathematical formulation of this problem. The choice of the coordinate axes and the cylindrical body is shown in Fig. 10.2. Also shown in the figure is a small annular cylindrical element of inner radius r, thickness Δr and height (or length) Δz. We will make a heat balance over this small element. Because the initial temperature distribution is independent of the polar angle θ and the cylindrical surface temperature is uniform at all time, θ will not appear as a variable in the problem.

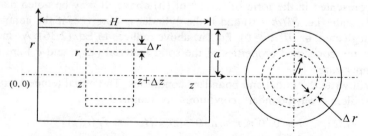

Fig. 10.2 A finite cylinder and an annular cylindrical volume element (Example 10.2).

Diffusion of heat occurs in the cylinder in both radial and axial directions. So there are two terms for heat input to the annular cylindrical element, and two corresponding heat output terms. Besides, we will include heat generation and accumulation terms as usual.

Rate of radial heat input through the curved surface at $r = (2\pi r\Delta z)q_r\big|_r$

Rate of axial heat input through the annular circular surface at $z = (2\pi r\Delta r)q_z\big|_z$

Rate of radial heat output at $r + \Delta r = (2\pi r\Delta z)q_r\big|_{r+\Delta r}$

Rate of axial heat output at $z + \Delta z = (2\pi r\Delta r)q_z\big|_{z+\Delta z}$

Rate of heat generation in the element $= (2\pi r\Delta r\Delta z)\psi_v$

Rate of heat accumulation in the element $= \dfrac{\partial}{\partial t}\Big[2\pi r\Delta r\Delta z\rho c_p T\Big]$

By an unsteady state heat balance, we have

$$(2\pi r\Delta z)q_r\big|_r + (2\pi r\Delta r)q_z\big|_z - (2\pi r\Delta z)q_r\big|_{r+\Delta r} - (2\pi r\Delta r)q_z\big|_{z+\Delta z} + (2\pi r\Delta r\Delta z)\psi_v = \dfrac{\partial}{\partial t}\Big[2\pi r\Delta r\Delta z\rho c_p T\Big]$$

Dividing by $2\pi r\Delta r\Delta z$ throughout and taking limits $\Delta r \to 0$, $\Delta z \to 0$, we get

$$-\dfrac{\partial}{\partial r}(rq_r) - \dfrac{\partial}{\partial z}(rq_z) + r\psi_v = r\rho c_p\dfrac{\partial T}{\partial t}$$

Putting $q_r = -k\dfrac{\partial T}{\partial r}$ and $q_z = -k\dfrac{\partial T}{\partial z}$, we obtain

$$k\dfrac{\partial}{\partial r}\left(r\dfrac{\partial T}{\partial r}\right) + kr\dfrac{\partial^2 T}{\partial z^2} + r\psi_v = r\rho c_p\dfrac{\partial T}{\partial t}$$

or

$$\dfrac{1}{r}\dfrac{\partial}{\partial r}\left(r\dfrac{\partial T}{\partial r}\right) + \dfrac{\partial^2 T}{\partial z^2} + \dfrac{\psi_v}{k} = \dfrac{1}{\alpha}\dfrac{\partial T}{\partial t} \tag{i}$$

or

$$\dfrac{1}{\alpha}\dfrac{\partial T}{\partial t} = \dfrac{\partial^2 T}{\partial r^2} + \dfrac{1}{r}\dfrac{\partial T}{\partial r} + \dfrac{\partial^2 T}{\partial z^2} + \dfrac{\psi_v}{k} \tag{ii}$$

The partial differential equation governing unsteady state heat diffusion in the cylindrical body can be represented in the form of Eq. (i) or (ii) above. It may be noted that if conduction occurs at steady state (i.e. $\partial T/\partial t = 0$) and if the cylinder is *long* (i.e. if the contribution of axial conduction is neglected, or $\partial T/\partial z = 0$), Eq. (ii) above reduces to Eq. (2.36). As in Example 10.1, we assume that the concerned properties of the solid material (k, ρ, and c_p) are independent of position or temperature.

Now we will write the initial and boundary conditions. The initial temperature distribution is given in the problem. Four boundary conditions are required.

I.C.: $t = 0,\ 0 \leq r < a,\ 0 < z < H;\ T = T_i(r, z)$ $\tag{iii}$

B.C. 1: $r = 0,\ 0 \leq z \leq H;\ t \geq 0;$ $T =$ finite or $\partial T/\partial r = 0$ $\tag{iv}$

B.C. 2: $r = a,\ 0 \leq z \leq H;\ t \geq 0;$ $T = T_s$ $\tag{v}$

B.C. 3: $z = 0,\ 0 \leq r \leq a;\ t \geq 0;$ $T = T_s$ $\tag{vi}$

B.C. 4: $z = H,\ 0 \leq r \leq a;\ t \geq 0;$ $T = T_s$ $\tag{vii}$

Boundary condition 1, Eq. (iv), shows that the axis-line temperature may be taken as finite (this is obvious; in fact, the temperature at any position in the solid remains finite), or the temperature gradient at this axis may be taken as zero. This latter condition arises out of *cylindrical symmetry*.

The above initial and boundary conditions can be expressed in a different form (which looks more compact). Let $T = T(r, z, t)$ be the temperature in the solid at any position and time. Then Eqs. (iii)–(vii) can be written in the following forms:

$$\text{I.C.:} \qquad T(r, z, 0) = T_i(r, z) \qquad\qquad\qquad\qquad \text{(viii)}$$

$$\text{B.C. 1:} \qquad T(0, z, t) = \text{finite} \qquad \text{or} \qquad \frac{\partial T(0, z, t)}{\partial r} = 0 \qquad \text{(ix)}$$

$$\text{B.C. 2:} \qquad T(a, z, t) = T_s \qquad\qquad\qquad\qquad\qquad \text{(x)}$$

$$\text{B.C. 3:} \qquad T(r, 0, t) = T_s \qquad\qquad\qquad\qquad\qquad \text{(xi)}$$

$$\text{B.C. 4:} \qquad T(r, H, t) = T_s \qquad\qquad\qquad\qquad\qquad \text{(xii)}$$

Example 10.3 (*Unsteady state heat conduction in a sphere*) Consider a sphere of radius a that has an initial temperature distribution $T = T_i(r)$, where r denotes the radial position in the sphere. At time $t = 0$, the surface temperature of the sphere is brought to T_s and maintained at that value subsequently. It is required to mathematically formulate the unsteady state heat conduction problem and to specify the initial and boundary conditions. Generation of heat occurs in the sphere at a volumetric rate of ψ_v.

SOLUTION Following the approach adopted in Section 2.5.3, we consider a thin spherical shell of inner radius r and thickness Δr as shown in Fig. 10.3 . Because of the symmetric nature of the solid (*spherical symmetry*), the temperature in the sphere will depend upon the radial position r,

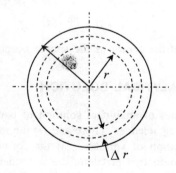

Fig. 10.3 A thin spherical shell (Example 10.3).

but not upon the angles θ and φ of the spherical coordinate system. The rates of heat input, output, generation, and accumulation for the spherical shell are given below.

$$\text{Rate of radial heat input into the shell at } r = 4\pi r^2 q_r \big|_r$$

$$\text{Rate of radial heat output from the shell at } r + \Delta r = 4\pi r^2 q_r \big|_{r + \Delta r}$$

$$\text{Rate of heat generation in the shell} = 4\pi r^2 \Delta r \psi_v$$

$$\text{Rate of heat accumulation in the shell} = \frac{\partial}{\partial t}\Big[4\pi r^2 \Delta r \rho c_p T\Big]$$

The following is the heat balance equation for the spherical shell.

$$4\pi r^2 q_r\big|_r - 4\pi r^2 q_r\big|_{r+\Delta r} + 4\pi r^2 \Delta r \psi_v = \frac{\partial}{\partial t}\Big[4\pi r^2 \Delta r \rho c_p T\Big]$$

Dividing throughout by $4\pi \Delta r$ and taking the limit $\Delta r \rightarrow 0$, we have

$$-\frac{\partial}{\partial r}\left(r^2 q_r\right) + r^2 \psi_v = \rho c_p r^2 \frac{\partial T}{\partial t} \tag{i}$$

Putting $q_r = -k\partial T/\partial r$ and rearranging, we get

$$\frac{1}{\alpha}\frac{\partial T}{\partial t} = \frac{1}{r^2}\frac{\partial}{\partial r}\left(r^2 \frac{\partial T}{\partial r}\right) + \frac{\psi_v}{k} \tag{ii}$$

or

$$\frac{1}{\alpha}\frac{\partial T}{\partial t} = \frac{\partial^2 T}{\partial r^2} + \frac{2}{r}\frac{\partial T}{\partial r} + \frac{\psi_v}{k} \tag{iii}$$

Equation (ii) [or Eq. (iii)] above represents the partial differential equation that governs the temperature distribution in the sphere at any time. If heat conduction occurs at steady state [i.e. if $\partial T/\partial t = 0$], Eq. (ii) reduces to Eq. (2.50) developed in Chapter 2.

We now proceed to specify the initial condition and the two boundary conditions.

I.C.:	$t = 0,\ 0 \le r < a;$	$T = T_i(r)$	(iv)
B.C. 1:	$r = 0,\ t \ge 0;$	$T = \text{finite} \quad \text{or} \quad \partial T/\partial r = 0$	(v)
B.C. 2:	$r = a,\ t \ge 0;$	$T = T_s$	(vi)

In B.C. 1, $\partial T/\partial r$ vanishes at the centre of the sphere because of spherical symmetry.

10.2 TECHNIQUES OF ANALYTICAL SOLUTION

There are two common techniques of analytical solution of partial differential equations which appear in physical and engineering sciences. These are: (i) the method of separation of variables, and (ii) the method of combination of variables. In the former method, the solution for the dependent variable (in a heat conduction problem, the dependent variable is the temperature T) is assumed to be the product of distinct functions of the individual independent variables. When this assumed form of solution is substituted in the governing partial differential equation (PDE), a number of ordinary differential equations (ODE) (same as the number of independent variables) are obtained. In the method of combination of variables, on the other hand, the independent variables are suitably combined to define a new independent variable. On substitution in the governing PDE, an ordinary equation is obtained, the solution of which is the solution of the PDE. This technique is applicable when the physical system has an *internal similitude* (this will be discussed in greater detail in the next section). Application of both the methods will be illustrated with examples. However, it should be remembered that none of these methods is a general one.

Whether one of these techniques is applicable, depends upon the form of the equation and the boundary conditions.

Example 10.4 (*The plane wall transient in one dimension—constant wall temperatures*) Consider unsteady state heat conduction in a plane wall of *large* length and breadth but of finite thickness l. The initial temperature has been T_i throughout the wall. At time $t = 0$, both the surfaces of the wall are brought to a temperature T_s and maintained at this value for all subsequent time. The relevant physical properties of the material of the wall, namely density, specific heat, and thermal conductivity can be assumed to remain independent of temperature. It is required to determine the unsteady state temperature distribution in the wall.

SOLUTION A section of the wall is shown in Fig. 10.4. The x-axis is taken normal to the wall with the origin on a surface of it.

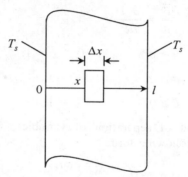

Fig. 10.4 A section of an infinite plane wall (Example 10.4).

In order to formulate the problem mathematically, we consider a small rectangular volume element having one rectangular face of unit area normal to the x-axis and of thickness Δx as shown in the figure. Because the length and breadth of the wall are *large* and the surface temperatures are uniform, heat conduction in the wall will be one-dimensional. In other words, the local temperature will depend upon x but not upon y or z. Following the procedure given in Example 10.1, we make a differential heat balance over the small volume element and arrive at the following second order partial differential equation for the one-dimensional unsteady state temperature distribution in the wall.

$$\frac{\partial T}{\partial t} = \alpha \frac{\partial^2 T}{\partial x^2} \tag{i}$$

where α is the thermal diffusivity of the solid. [Note that Eq. (i) above can be obtained directly from Eq. (x) of Example 10.1, by putting $\partial^2 T/\partial y^2 = 0$, $\partial^2 T/\partial z^2 = 0$ (i.e. no heat conduction in the y- and z-directions), and $\psi_v = 0$ (i.e. no heat generation).

The initial and boundary conditions are:

$$\text{I.C.:} \qquad t = 0, \qquad 0 < x < l; \qquad T = T_i \tag{ii}$$

$$\text{B.C. 1:} \qquad t \geq 0, \qquad x = 0; \qquad T = T_s \tag{iii}$$

$$\text{B.C. 2:} \qquad t \geq 0, \qquad x = l; \qquad T = T_s \tag{iv}$$

Solution of the PDE, Eq.(i), subject to the initial condition (ii) and the boundary conditions (iii) and (iv), gives the unsteady state temperature distribution in the wall. We will use the technique of *separation of variables* as shown below in order to solve the PDE. It is, however, useful to make the equation, the I.C., and the B.Cs. *dimensionless*, before we proceed to solve the equation. Suitable *dimensionless variables* need to be defined for this purpose. Solution of an equation in terms of dimensionless variables is obtained in a form which is readily useful even when the system's parameters and conditions change.

Let us define the following dimensionless variables:

$$\overline{T} = \frac{T - T_s}{T_i - T_s}, \quad \overline{x} = \frac{x}{l}, \quad \theta = \frac{\alpha t}{l^2} \tag{v}$$

Substitution in Eqs. (i)–(iv) yields

$$\text{PDE:} \qquad \frac{\partial \overline{T}}{\partial \theta} = \frac{\partial^2 \overline{T}}{\partial \overline{x}^2} \tag{vi}$$

$$\text{I.C.:} \qquad \theta = 0, \qquad\qquad 0 < \overline{x} < 1; \qquad \overline{T} = 1 \tag{vii}$$

$$\text{B.C. 1:} \qquad \theta \geq 0, \qquad\qquad \overline{x} = 0; \qquad \overline{T} = 0 \tag{viii}$$

$$\text{B.C. 2:} \qquad \theta \geq 0, \qquad\qquad \overline{x} = 1; \qquad \overline{T} = 0 \tag{ix}$$

According to the method of separation of variables, we assume a solution of the dimensionless Eq. (vi) in the following form:

$$\overline{T} = X(\overline{x})\, \Theta(\theta) \tag{x}$$

where X is a function of $\overline{x}$ only, and Θ is a function of θ only. Substituting (x) in (vi), we get

$$\frac{\partial}{\partial \theta}(X\Theta) = \frac{\partial^2}{\partial \overline{x}^2}(X\Theta)$$

or

$$X\frac{d\Theta}{d\theta} = \Theta\frac{d^2 X}{d\overline{x}^2} \tag{xi}$$

Ordinary derivatives, rather than partial derivatives, are written in Eq. (xi), because each of X and Θ is a function of a single variable only. Dividing both sides of Eq. (xi) by $X\Theta$, we get

$$\frac{1}{\Theta}\frac{d\Theta}{d\theta} = \frac{1}{X}\frac{d^2 X}{d\overline{x}^2} \tag{xii}$$

In the above equation, the LHS is a function of θ only, whereas the RHS is a function of $\overline{x}$ only, but they are equal. This is possible only if each side is equal to the same constant. Let us, therefore, write,

$$\frac{1}{\Theta}\frac{d\Theta}{d\theta} = \frac{1}{X}\frac{d^2 X}{d\overline{x}^2} = -\lambda^2 \tag{xiii}$$

The constant, $-\lambda^2$, is deliberately chosen as a negative quantity. This is necessary in order to have *non-trivial eigenfunctions* for X in latter steps. (See the note at the end of this problem.)

From Eq. (xiii), we have

$$\frac{1}{\Theta}\frac{d\Theta}{d\theta} = -\lambda^2$$

or

$$\Theta = Ae^{-\lambda^2\theta} \qquad (A = \text{integration constant}) \qquad \text{(xiv)}$$

Also from Eq. (xii), we get

$$\frac{1}{X}\frac{d^2X}{d\overline{x}^2} = -\lambda^2$$

or

$$\frac{d^2X}{d\overline{x}^2} + \lambda^2 X = 0 \qquad \text{(xv)}$$

Boundary conditions on the function $X(\overline{x})$ can be derived by substituting $\overline{T} = X(\overline{x})\Theta(\theta)$ in Eqs. (viii) and (ix). Substitution in Eq. (viii), for example, gives

$$\theta \geq 0, \quad \overline{x} = 0; \qquad \overline{T} = X(0)\Theta(\theta) = 0$$

i.e.

$$\text{at } \overline{x} = 0, \qquad X = 0 \qquad \text{(xvi)}$$

Similarly,

$$\text{at } \overline{x} = 1, \qquad X = 0 \qquad \text{(xvii)}$$

It is to be noted that Eq. (xv) and B.Cs. (xvi) and (xvii) constitute an eigenvalue problem (or a Sturm-Liouville problem, as discussed in the note below). It is understandable that by selecting $-\lambda^2$ as the constant in Eq. (xiii), we have ensured non-trivial solution of Eq. (xv). The eigenvalues and eigenfunctions (which are infinite in number) of Eq. (xv) are:

$$\lambda_n = n\pi, \qquad\qquad n = 1, 2, 3, \ldots \qquad \text{(xviii)}$$

$$X_n = \sin(n\pi\overline{x}), \qquad\qquad n = 1, 2, 3, \ldots \qquad \text{(xix)}$$

Corresponding to each eigenvalue, we have a solution for Θ given by Eq. (xiv), and a solution for X (eigenfunction) given by Eq. (xix). So, corresponding to each eigenvalue there is a solution for $\overline{T}$ given by Eq. (x). The general solution for $\overline{T}$ will be a *linear combination* (i.e. each solution is multiplied by an arbitrary constant and then added up) of all such solutions. Thus,

$$\overline{T} = \sum_{n=1}^{\infty} B_n \sin(n\pi\overline{x}) \exp(-n^2\pi^2\theta) \qquad \text{(xx)}$$

where B_n ($n = 1, 2, 3, \ldots$) are constants yet to be determined. We will now put the I.C., Eq. (vii), in the above solution to obtain

$$1 = \sum_{n=1}^{\infty} B_n \sin(n\pi\overline{x}) \qquad \text{(xxi)}$$

The constants B_n can be determined by using the *orthogonality property* of the sine functions.

Multiplying both sides of Eq. (xxi) by $\sin(m\pi\bar{x})d\bar{x}$ and integrating from $\bar{x} = 0$ to 1, we get

$$\int_0^1 \sin(m\pi\bar{x})\,d\bar{x} = \int_0^1 \left[\sum_{n=1}^{\infty} B_n (\sin(n\pi\bar{x}))\right] \sin(m\pi\bar{x})\,d\bar{x}$$

$$= \sum_{n=1}^{\infty} B_n \int_0^1 \sin(n\pi\bar{x}) \sin(m\pi\bar{x})\,d\bar{x}$$

All the terms on the RHS of the above equation vanish except the one for which $n = m$ (this happens because of the orthogonality property of the eigenfunctions). Therefore,

$$\int_0^1 \sin(m\pi\bar{x})\,d\bar{x} = B_m \int_0^1 \sin^2(m\pi\bar{x})\,d\bar{x}$$

or

$$\left[-\frac{\cos(m\pi\bar{x})}{m\pi}\right]_{\bar{x}=0}^{1} = B_m \int_0^1 \frac{1}{2}\left[1 - \cos(2m\pi\bar{x})\right]d\bar{x}$$

or

$$\frac{1}{m\pi}\left[1 - (-1)^m\right] = \frac{B_m}{2}\left[\bar{x} - \frac{\sin(2m\pi\bar{x})}{2m\pi}\right]_{\bar{x}=0}^{1} = \frac{B_m}{2}$$

If m is an even number, the LHS = 0, and thus $B_m = 0$. But when m is an odd number, the LHS = $2/m\pi$. Therefore,

$$\frac{2}{m\pi} = \frac{B_m}{2}$$

or

$$B_m = \frac{4}{m\pi} \qquad \text{when } m \text{ is odd.}$$

Let us put $m = 2p + 1$; $p = 0, 1, 2, 3, \ldots$. When

$$B_{2p+1} = \frac{4}{\pi(2p+1)}; \qquad p = 0, 1, 2, 3, \ldots \qquad \text{(xxii)}$$

Substituting for B_n or B_{2p+1} in Eq. (xx), the complete solution for the dimensionless temperature is given by

$$\bar{T} = \frac{4}{\pi} \sum_{p=0}^{\infty} \frac{\sin(2p+1)\pi\bar{x}}{(2p+1)} \exp\left[-(2p+1)^2 \pi^2\theta\right] \qquad \text{(xxiii)}$$

or in terms of the original variables, we have

$$\bar{T} = \frac{T - T_s}{T_i - T_s} = \frac{4}{\pi} \sum_{p=0}^{\infty} \frac{\sin(2p+1)\dfrac{\pi x}{l}}{(2p+1)} \exp\left[-(2p+1)^2 \frac{\pi^2\alpha t}{l^2}\right] \qquad \text{(xxiv)}$$

Equation (xxiv) gives the solution for the dimensionless (or dimensional) temperature distribution in the wall at unsteady state. The temperature in the wall at any position x and time t can be calculated using the above equation. The ranges of variation of the dimensionless variables are: $\overline{T}$ and $\overline{x}$ lie between 0 and 1 $(0 \le \overline{T} \le 1,\ 0 \le \overline{x} \le 1)$, and θ varies from 0 to ∞. The dimensionless time θ is also called the *Fourier number* (Fo).

Note: The basic principles that underlie the methodology of solution of the above problem in terms of a set of sine functions are described here.

Orthogonal functions. To begin with, we define *orthogonal functions*. A sequence of functions $\{\phi_i(x)\}$, $i = 1, 2, 3, \ldots$ defined in an interval (a, b) is said to form an *orthogonal* set if it has the following properties:

$$\int_a^b \phi_m(x)\phi_n(x)\,dx = 0, \qquad m \ne n$$

$$\ne 0, \qquad m = n \tag{10.2}$$

Further, a sequence of functions $\{\psi_i(x)\}$, $i = 1, 2, 3, \ldots$ defined in an interval (a, b) is said to form an orthogonal set with respect to the weight function $w(x)$ if it has the following property:

$$\int_a^b \psi_m(x)\psi_n(x)w(x)\,dx = 0, \qquad m \ne n$$

$$\ne 0, \qquad m = n \tag{10.3}$$

Example If we define $\{\phi_k(x)\} = \sin\dfrac{k\pi x}{l}$, $0 \le x \le l$; $k = 1, 2, 3, \ldots$ it can be easily verified by direct integration that

$$\int_0^l \sin\frac{m\pi x}{l}\sin\frac{n\pi x}{l}\,dx = 0, \qquad m \ne n$$

$$\ne 0, \qquad m = n$$

Therefore, this set of functions forms an orthogonal set in the given interval.

A simple eigenvalue problem. Let us consider the following homogeneous second order ordinary differential equation and the associated homogeneous boundary conditions. [If an equation (or a boundary condition) is satisfied by $y = \phi(x)$ and if it is satisfied also by $y = c\phi(x)$, where c is an arbitrary constant, then the equation (or the boundary condition) is called a homogeneous equation (or boundary condition)]

$$\frac{d^2 y}{dx^2} + \lambda y = 0, \qquad 0 \le x \le l \tag{10.4}$$

$$\text{B.C.: } y(0) = y(l) = 0 \tag{10.5}$$

The solution of the above equation depends upon the nature of the quantity λ, that is, whether λ is negative, zero, or positive. These cases are considered below.

Case 1: λ *is negative* (or $-\lambda$ *is positive*)

The general solution of Eq. (10.4) is

$$y = A \exp\left(\sqrt{-\lambda}\,x\right) + B \exp\left(-\sqrt{-\lambda}\,x\right) \tag{10.6}$$

where $\sqrt{-\lambda}$ is real, and A and B are two constants.

Substitution of the boundary conditions (10.5) in the above general solution (10.6) yields $A = 0$ and $B = 0$. Putting these values of A and B in Eq. (10.6), the solution for y is $y = 0$. Such a solution does not serve any practical purpose and is called a *trivial solution*.

Case 2: λ is zero

In this case, the solution of Eq. (10.4) is $y = Ax + B$. If we use the boundary conditions (10.5), we get $A = 0$, $B = 0$. Then the solution is $y = 0$, which is again *trivial*.

Case 3: λ is positive (or $-\lambda$ is negative)

The solution of Eq.(10.4) may be written in the form:

$$y = A \cos \left(\sqrt{\lambda}\,x\right) + B \sin \left(\sqrt{\lambda}\,x\right)$$

Using the condition $y(0) = 0$, we get

$$0 = A \cos (0) + B \sin (0)$$

or

$$A = 0$$

Using the other boundary condition, $y(l) = 0$, we get

$$0 = B \sin \left(\sqrt{\lambda}\,l\right)$$

If we want a *non-trivial solution* of the problem, we have to put $B \neq 0$, when

$$\sin \left(\sqrt{\lambda}\,l\right) = 0$$

or

$$\sqrt{\lambda}\,l = n\pi$$

or

$$\lambda = \frac{n^2\pi^2}{l^2}, \qquad n = 1, 2, 3, \ldots \tag{10.7}$$

Therefore, the non-trivial solution of Eq. (10.4) subject to the boundary conditions (10.5) is given by

$$y = B \sin \frac{n\pi x}{l}, \qquad n = 1, 2, 3, \ldots \tag{10.8}$$

It is to be noted that the non-trivial solution of the problem is obtained only for a set of discrete values of λ given by Eq. (10.7). These values are called *characteristic values* or *eigenvalues*. The corresponding solutions of the equation as shown in Eq. (10.8) are called the *eigenfunctions*.

If the boundary conditions of the equation are different from those given by Eq. (10.5), the eigenvalues and the eigenfunctions will be expectedly different. For example, if we prescribe the boundary conditions as $y(0) = y'(0)$, it is easy to show that the eigenvalues and the eigenfunctions become

eigenvalues: $\quad \lambda = \dfrac{(2n + 1)^2\pi^2}{4l^2}, \qquad\qquad n = 1, 2, 3, \ldots \tag{10.9}$

eigenfunctions: $\quad y = B \sin\dfrac{(2n + 1)\pi x}{2l}, \qquad n = 0, 1, 2, 3, \ldots \tag{10.10}$

It is also easy to verify that the eigenfunctions given by Eq. (10.8) form an orthogonal set in the interval $[0, l]$. It may be noted that Eqs. (xv)–(xvii) of Example 10.4 are almost identical to Eqs. (10.4) and (10.5) of this section, and this is why the solution to the former could be written in the form of Eq. (xix) of the problem. This clarifies the mathematical basis of the eigenfunctions of Example 10.4 and the technique of determining the constants B_n by using the orthogonality property of the eigenfunctions.

The Sturm-Liouiville problem. Let us now consider the following homogeneous second order equation

$$\frac{d}{dx}[p(x)y'] + [q(x) + \lambda w(x)]y = 0 \tag{10.11}$$

where $y' = dy/dx$; $p(x)$, $q(x)$, and $w(x)$ are continuous functions of x over the interval $[a, b]$, and λ is a parameter. The following homogeneous boundary conditions are prescribed.

$$\alpha_1 y(a) + \beta_1 y'(a) = 0 \tag{10.12}$$

$$\alpha_2 y(b) + \beta_2 y'(b) = 0 \tag{10.13}$$

where $\alpha_1^2 + \beta_1^2 \neq 0$ and $\alpha_2^2 + \beta_2^2 \neq 0$.

If $y = y_i(x)$, $i = 1, 2, 3, \ldots$ are the non-trivial solutions or eigenfunctions of the above equation corresponding to the eigenvalues $\lambda = \lambda_i$, $i = 1, 2, 3, \ldots$ then it can be proved that these eigenfunctions form an orthogonal set with respect to the weight function $w(x)$ in the interval $[a, b]$.

$$\int_a^b y_m(x)y_n(x)w(x)\,dx = 0, \qquad m \neq n$$

$$= 0, \qquad m = n \tag{10.14}$$

Equation (10.11) subject to the boundary conditions (10.12) and (10.13) constitutes what is called the Sturm-Liouville problem. This type of equation frequently appears in the course of solution of second order partial differential equations for diffusional transport. The above mathematical results are of great importance in the solution of such diffusion equations. It is easy to check that Eqs. (10.4) and (10.5) represent a Sturm-Liouville problem. The orthogonality property of the eigenfunctions of a Sturm-Liouville problem will be used in the solution of the transient heat conduction problem for a cylinder.

10.3 TYPES OF BOUNDARY CONDITIONS

In all the previous examples, we assumed a constant surface temperature of each of the bodies considered in the problems. However, other types of conditions may also prevail at the surface (or the boundary) of the solid in which a diffusional transport process occurs.

The various common types of boundary conditions that may arise in diffusional problems may be classified into three kinds. If the temperature at the surface (or the boundary) is given, the resulting boundary condition is called the *Dirichlet condition* or the *boundary condition of first kind*. It can be expressed mathematically as

$$T(l, t) = T_s \tag{10.15}$$

where $T(x, t)$ is the temperature in the body as a function of the position x and the time t. Equation (10.15) means that the temperature of the body at its boundary at $x = l$ is T_s for all time.

In many physical situations, it is possible that the *heat flux* rather than the temperature is specified at the boundary. The resulting boundary condition is mathematically expressed as

$$q_x = -k\frac{\partial T}{\partial x} = -\beta' \qquad \text{at } x = l \tag{10.16}$$

or

$$\frac{\partial T(l, t)}{\partial x} = -\frac{\beta'}{k} = \beta \tag{10.17}$$

where β' is the prescribed heat flux. The boundary condition of the above type, given by Eq. (10.16) or (10.17), is called the *Neuman condition* or a *boundary condition of the second kind*.

Still another type of situation may arise when the body loses (or gains) heat by convection while it is in contact with a fluid at a given temperature T_o. The convective heat transfer coefficient h is also given. The mathematical form of this type of boundary condition is given by

$$x = l, \qquad -k\frac{\partial T}{\partial x} = h(T - T_o) \tag{10.18}$$

The form of the boundary condition given by Eq. (10.18) is called the *Robin condition* or a *boundary condition of the third kind* or a *convection boundary condition*.

The following examples illustrate the techniques of analytical solution of a few unsteady state heat diffusion problems with different types of boundary conditions.

Example 10.5 (*The plane wall transient—finite heat transfer coefficient at the walls*) A 'large' plane wall of thickness l has a uniform initial temperature T_i throughout. At time $t = 0$, both the surfaces are brought in contact with a fluid of bulk temperature T_o. The heat transfer coefficient at both the surfaces is h. It is required to determine the unsteady state temperature distribution in the wall.

SOLUTION A section of the wall is shown in Fig. 10.5. The governing PDE is the same as Eq. (i) of Example 10.4. Here the boundary conditions are of Robin type (in Example 10.4 the boundary conditions were of Dirichlet type).

$$\text{Governing PDE: } \frac{\partial T}{\partial t} = \alpha \frac{\partial^2 T}{\partial x^2} \tag{i}$$

$$\text{I.C.: } T(x, 0) = T_i \tag{ii}$$

$$\text{B.C. 1: } -k\frac{\partial T(0, t)}{\partial x} = h(T_o - T) \tag{iii}$$

$$\text{B.C. 2: } -k\frac{\partial T(l, t)}{\partial x} = h(T - T_o) \tag{iv}$$

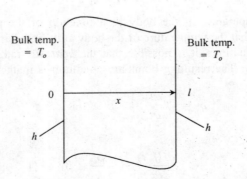

Bulk temp. $= T_o$ Bulk temp. $= T_o$

0   x l

h h

Fig. 10.5 A section of the wall (Example 10.5).

Note the 'signs' used in the equations for the boundary conditions. The heat fluxes in the wall at both the boundaries are based on the assumption that temperature decreases as x increases (it is on this basis that a negative sign is used in the equation for the heat flux). The direct implication of this relation for heat flux along with its negative sign is that heat enters the wall at $x = 0$, and leaves at $x = l$. The temperature driving forces on the RHS of B.C. 1 and B.C. 2 are thus compatible with the assumed direction of the heat flux. It can be verified that the boundary conditions will remain the same even if the heat flow is assumed to occur in the reverse direction.

Introduction of the following dimensionless variables

$$\overline{T} = \frac{T - T_o}{T_i - T_o}; \qquad \overline{x} = \frac{x}{l}; \qquad \theta = \frac{\alpha t}{l^2}; \qquad \text{Bi} = \frac{hl}{k}$$

reduces Eqs. (i)–(iv) to the following forms:

$$\text{PDE:} \qquad \frac{\partial \overline{T}}{\partial \theta} = \frac{\partial^2 \overline{T}}{\partial \overline{x}^2} \tag{v}$$

$$\text{I.C.:} \qquad \overline{T}(\overline{x}, 0) = 1 \tag{vi}$$

$$\text{B.C. 1:} \qquad \frac{\partial \overline{T}(0, \theta)}{\partial \overline{x}} = \text{Bi}\, \overline{T}(0, \theta) \tag{vii}$$

$$\text{B.C. 2:} \qquad \frac{\partial \overline{T}(1, \theta)}{\partial \overline{x}} = -\text{Bi}\, \overline{T}(1, \theta) \tag{viii}$$

where Bi is a dimensionless group called the *Biot number*. We assume a solution of the form (see Example 10.4)

$$\overline{T} = X(\overline{x})\Theta(\theta) \tag{ix}$$

Substituting Eq. (ix) in Eq. (v) and following the procedure of Example 10.4, we get

$$\frac{1}{\Theta}\frac{d\Theta}{d\theta} = -\lambda^2$$

or

$$\Theta = A e^{-\lambda^2 \theta} \qquad \text{where } A = \text{integration constant} \tag{x}$$

And

$$\frac{1}{X}\frac{d^2 X}{d\overline{x}^2} = -\lambda^2$$

or

$$\frac{d^2 X}{d\overline{x}^2} + \lambda^2 X = 0 \tag{xi}$$

Boundary conditions on the function $X(\overline{x})$ are

$$\text{B.C. 1:} \qquad \frac{dX(0)}{d\overline{x}} = \text{Bi}\, X(0) \tag{xii}$$

$$\text{B.C. 2:} \qquad \frac{dX(1)}{d\bar{x}} = \text{Bi}\, X(1) \qquad \qquad \text{(xiii)}$$

Equations (xi)–(xiii) constitute a Sturm-Liouville problem having the following eigenvalues and eigenfunctions (this is left as an exercise).

$$\text{Eigenvalues} = \lambda_n \qquad \text{where } \tan \lambda_n = \frac{2\lambda_n \text{Bi}}{\lambda_n^2 - \text{Bi}^2} \qquad \qquad \text{(xiv)}$$

$$\text{Eigenfunctions,} \ X_n = \cos(\lambda_n \bar{x}) + \frac{\text{Bi}}{\lambda_n} \sin(\lambda_n \bar{x}) \qquad \qquad \text{(xv)}$$

The solution for the PDE can now be written as

$$\bar{T} = \sum_{n=1}^{\infty} A_n \left[\cos(\lambda_n \bar{x}) + \frac{\text{Bi}}{\lambda_n} \sin(\lambda_n \bar{x}) \right] e^{-\lambda_n^2 \theta} \qquad \qquad \text{(xvi)}$$

Now we have to determine the constants A_n by using the I.C., Eq. (vi), and the orthogonality property of the eigenfunctions. The constants are found to be

$$A_n = \frac{\int_0^1 X_n \, d\bar{x}}{\int_0^1 X_n^2 \, d\bar{x}} = \frac{1}{\text{Bi}} \left[\lambda_n \sin \lambda_n + \text{Bi}\, (1 - \cos \lambda_n) \right] \qquad \qquad \text{(xvii)}$$

The complete solution for the dimensionless temperature is given by Eq. (xvi), where the constants A_n are given by Eq. (xvii) and λ_ns are the roots of Eq. (xiv).

Example 10.6 (*Unsteady state heat conduction in a 'long' cylinder with finite wall heat transfer coefficient*) A long cylinder of radius a has a uniform initial temperature T_i throughout. At time $t = 0$, the cylinder is placed in a medium of bulk temperature T_o. If the convection heat transfer coefficient at the cylindrical surface is h, determine the unsteady state temperature distribution in the cylinder.

SOLUTION A *long* cylinder means that its length is much greater than its diameter. Axial conduction effects in such a cylinder are confined to near the ends only. For practical purposes, axial heat conduction can be neglected in regions considerably far from the ends and thus radial conduction is only considered.

The partial differential equation governing the temperature distribution in this case can be obtained from Example 10.2 by putting $\partial^2 T/\partial z^2 = 0$ (no axial conduction), and $\psi_v = 0$ (no heat generation). Thus, we get

$$\frac{1}{\alpha} \frac{\partial T}{\partial t} = \frac{\partial^2 T}{\partial r^2} + \frac{1}{r} \frac{\partial T}{\partial r} \qquad \qquad \text{(i)}$$

The following initial and boundary conditions apply.

$$\text{I.C.:} \qquad t = 0, \qquad 0 \le r < a; \qquad T = T_i \qquad \qquad \text{(ii)}$$

B.C. 1: $t \geq 0$, $r = 0$; $T = \text{finite}$ or $\dfrac{\partial T}{\partial r} = 0$ (iii)

B.C. 2: $t \geq 0$, $r = a$; $-k\dfrac{\partial T}{\partial r} = h(T - T_o)$ (iv)

Introduction of the following dimensionless variables

$$\overline{T} = \frac{T - T_o}{T_i - T_o}; \quad \overline{r} = \frac{r}{a}; \quad \theta = \frac{\alpha t}{a^2}; \quad \text{and} \quad \text{Bi} = \frac{ha}{k}$$

leads to the PDE, I.C. and B.C. in the following forms.

PDE: $\dfrac{\partial \overline{T}}{\partial \theta} = \dfrac{\partial^2 \overline{T}}{\partial \overline{r}^2} + \dfrac{1}{\overline{r}} \dfrac{\partial \overline{T}}{\partial \overline{r}}$ (v)

I.C.: $\overline{T}(\overline{r}, 0) = 1$ (vi)

B.C. 1: $\overline{T}(0, \theta) = \text{finite}$ or $\dfrac{\partial \overline{T}(0, \theta)}{\partial \overline{r}} = 0$ (vii)

B.C. 2: $\dfrac{\partial \overline{T}(1, \theta)}{\partial \overline{r}} = -\text{Bi}\,\overline{T}(1, \theta)$ (viii)

The technique of separation of variables is applicable in this case. Let us assume a solution of the form

$$\overline{T} = R(\overline{r})\Theta(\theta)$$ (ix)

where R and Θ are functions of the respective individual dimensionless variables.

Substituting Eq. (ix) in Eq. (v) and separating the variables, we can write

$$\frac{1}{\Theta}\frac{d\Theta}{d\theta} = \frac{1}{R}\left[\frac{d^2 R}{d\overline{r}^2} + \frac{1}{\overline{r}}\frac{dR}{d\overline{r}}\right] = -\lambda^2$$ (x)

where $-\lambda^2$ is a constant, deliberately taken to be negative in order to ensure non-trivial solution of the resulting Sturm-Liouville problem.

The solution for the function Θ is

$$\Theta = A \exp(-\lambda^2 \theta)$$ (xi)

where A is a constant. The function R is given by the ordinary differential equation

$$\frac{d^2 R}{d\overline{r}^2} + \frac{1}{\overline{r}}\frac{dR}{d\overline{r}} + \lambda^2 R = 0$$ (xii)

This is Bessel equation of order zero having the solution (see Wylie and Barrett, 1982)

$$R = B_1 J_o(\lambda \overline{r}) + B_2 Y_o(\lambda \overline{r})$$ (xiii)

where J_o and Y_o are Bessel functions of first kind and second kind of order zero, respectively. It is known that $Y_o(\lambda \bar{r}) \to -\infty$ as $\bar{r} \to 0$. But the B.C. 1 [Eq. (vii)] demands that $\bar{T}$ must be finite at $\bar{r} = 0$. To make the solution (xiii) compatible with B.C. 1, we put

$$B_2 = 0 \tag{xiv}$$

Therefore,

$$R = B_1 J_o(\lambda \bar{r}) \tag{xv}$$

Substituting $\bar{T} = R\Theta$ in B.C. 2 [Eq. (viii)], we get

$$\frac{dR}{d\bar{r}} = -\mathrm{Bi}\, R \qquad \text{at} \qquad \bar{r} = 1$$

or

$$B_1 \frac{d}{d\bar{r}}[J_o(\lambda \bar{r})] = -B_1 \mathrm{Bi}\, J_o(\lambda \bar{r}) \qquad \text{at} \qquad \bar{r} = 1$$

Applying the differentiation formula of Bessel function (see Wylie and Barrett, 1982) to the above equation, we get

$$\lambda J_1(\lambda) = \mathrm{Bi}\, J_o(\lambda) \tag{xvi}$$

The above equation is the eigencondition and has the roots $\lambda_1,\ \lambda_2,\ \lambda_3,\ \ldots$, which are the eigenvalues; $J_o(\lambda_n \bar{r})$ are the eigenfunctions. The solution for $\bar{T}$ can be written as the linear combination of all possible solutions. Thus, we have

$$\bar{T} = \sum_{n=1}^{\infty} A_n J_o(\lambda_n \bar{r}) \exp\left(-\lambda_n^2 \theta\right) \tag{xvii}$$

where A_n $(n = 1, 2, 3, \ldots)$ are constants to be determined. Using the initial condition, we get

$$1 = \sum_{n=1}^{\infty} A_n J_o(\lambda_n \bar{r})$$

The eigenfunctions $J_o(\lambda_n \bar{r})$ are orthogonal with respect to the weight function $\bar{r}$. Multiplying both sides of Eq. (xvii) by $\bar{r} J_o(\lambda_m \bar{r}) d\bar{r}$ and integrating from 0 to 1, we get

$$\int_0^1 \bar{r} J_o(\lambda_m \bar{r})\, d\bar{r} = \int_0^1 \bar{r} J_o(\lambda_m \bar{r})\, d\bar{r} \left[\sum_{n=1}^{\infty} A_n J_o(\lambda_n \bar{r})\right] = A_m \int_0^1 \bar{r} J_o^2(\lambda_m \bar{r})\, d\bar{r} \tag{xviii}$$

Upon evaluation of the integral on the RHS and using the eigencondition, Eq. (xvi), we get

$$A_m = \frac{2\mathrm{Bi}}{(\lambda_m^2 + \mathrm{Bi}^2) J_o(\lambda_m)} \tag{xix}$$

The complete solution for the temperature distribution is given by

$$\bar{T} = \sum_{m=1}^{\infty} \frac{2\mathrm{Bi}\, J_o(\lambda_m \bar{r})}{(\lambda_m^2 + \mathrm{Bi}^2) J_o(\lambda_m)} \exp(-\lambda_m^2 \theta) \tag{xx}$$

The solution for the 'dimensional' temperature is

$$T = T_o + 2\text{Bi}\,(T_i - T_o) \sum_{m=1}^{\infty} \frac{2\text{Bi}\,J_o(\lambda_m \bar{r})}{(\lambda_m^2 + \text{Bi}^2)\,J_o(\lambda_m)}\,\exp\,(-\lambda_m^2 \theta) \tag{xxi}$$

Example 10.7 (*Unsteady state heat conduction in a sphere*) A sphere of radius a has a uniform initial temperature distribution T_i. At time $t = 0$, the sphere is placed in a fluid of bulk temperature T_o. If the convection heat transfer coefficient at the surface of the sphere is h, determine the unsteady state temperature distribution in the sphere.

SOLUTION The governing partial differential equation in this case may be directly obtained from Eq. (ii) of Example 10.3, if we put $\psi_v = 0$ (no heat generation). The boundary condition at the surface of the sphere is of Robin type.

PDE:
$$\frac{1}{\alpha}\frac{\partial T}{\partial t} = \frac{\partial^2 T}{\partial r^2} + \frac{2}{r}\frac{\partial T}{\partial r} \tag{i}$$

I.C.:
$$T(r,\,0) = T_i \tag{ii}$$

B.C. 1:
$$T(0,\,t) = \text{finite} \quad \text{or} \quad \frac{\partial T(0,\,t)}{\partial r} = 0 \tag{iii}$$

B.C. 2:
$$-k\frac{\partial T(a,\,t)}{\partial r} = h[T(a,\,t) - T_o] \tag{iv}$$

Introduction of the following dimensionless variables

$$\bar{T} = \frac{T - T_o}{T_i - T_o}; \qquad \bar{r} = \frac{r}{a}; \qquad \theta = \frac{\alpha t}{a^2}; \quad \text{and} \quad \text{Bi} = \frac{ha}{k}$$

reduces the above Eqs. (i)–(iv) to the following forms.

PDE:
$$\frac{\partial \bar{T}}{\partial \theta} = \frac{\partial^2 \bar{T}}{\partial \bar{r}^2} + \frac{2}{\bar{r}}\frac{\partial \bar{T}}{\partial \bar{r}} \tag{v}$$

I.C.:
$$\bar{T}(\bar{r},\,0) = 1 \tag{vi}$$

B.C. 1:
$$\bar{T}(0,\,\theta) = \text{finite} \quad \text{or} \quad \frac{\partial \bar{T}(0,\,\theta)}{\partial \bar{r}} = 0 \tag{vii}$$

B.C. 2:
$$\frac{\partial \bar{T}(1,\,\theta)}{\partial \bar{r}} = -\,\text{Bi}\,\bar{T}(1,\,\theta) \tag{viii}$$

The above PDE may be considerably simplified if we use the following transformation of variables.

$$\hat{T}(\bar{r},\,\theta) = \bar{r}\,\bar{T}(\bar{r},\,\theta) \tag{ix}$$

or

$$\frac{\partial \hat{T}}{\partial \bar{r}} = \bar{r}\frac{\partial \bar{T}}{\partial \bar{r}} + \bar{T}$$

or

$$\frac{\partial^2 \hat{T}}{\partial \bar{r}^2} = \bar{r}\frac{\partial^2 \overline{T}}{\partial \bar{r}^2} + \frac{\partial \overline{T}}{\partial \bar{r}} + \frac{\partial \overline{T}}{\partial \bar{r}} = \bar{r}\frac{\partial^2 \overline{T}}{\partial \bar{r}^2} + 2\frac{\partial \overline{T}}{\partial \bar{r}} \tag{x}$$

Also,

$$\frac{\partial \hat{T}}{\partial \theta} = \bar{r}\frac{\partial \overline{T}}{\partial \theta} \tag{xi}$$

Substitution of Eqs. (x) and (xi) in Eq. (v) yields the following equation for $\hat{T}$

$$\text{PDE:} \qquad \frac{\partial \hat{T}}{\partial \theta} = \frac{\partial^2 \hat{T}}{\partial \bar{r}^2} \tag{xii}$$

which is of the form of unsteady state heat conduction equation for a wall [see Example 10.4, Eq. (vi)]. The initial and boundary conditions [Eqs. (vi)–(viii)] reduce to the following form in terms of the new variable $\hat{T}$.

$$\text{I.C.:} \qquad \hat{T}(\bar{r}, 0) = \bar{r}\,\overline{T}(\bar{r}, 0) = \bar{r} \tag{xiii}$$

$$\text{B.C. 1:} \qquad \hat{T}(0, \theta) = \left[\bar{r}\overline{T}(0, \theta)\right]_{\bar{r}=0} = 0 \tag{xiv}$$

$$\text{B.C. 2:} \qquad \frac{\partial \hat{T}(1, \theta)}{\partial \bar{r}} = (1 - \text{Bi})\hat{T}(1, \theta) \tag{xv}$$

Now we apply the technique of separation of variables to Eq. (xii). Let us assume a solution of the form

$$\hat{T} = R(\bar{r})\Theta(\theta) \tag{xvi}$$

where R and Θ are functions of the respective individual variables, $\bar{r}$ and θ. Following the procedure adopted in Example 10.5, we have

$$\Theta = A \exp\left(-\lambda^2\theta\right) \tag{xvii}$$

where A is a constant. The eigenvalues and the eigenfunctions are given by

$$\lambda \cot \lambda = 1 - \text{Bi} \tag{xviii}$$

and

$$R = \sin \lambda_n\bar{r}, \qquad n = 1, 2, 3,\ldots \tag{xix}$$

The solution to the PDE for temperature distribution may be written as

$$\hat{T} = \sum_{n=1}^{\infty} A_n \sin\left(\lambda_n\bar{r}\right) \exp\left(-\lambda_n^2\theta\right) \tag{xx}$$

Using the transformed initial condition (xiii) and also using the orthogonality property of the eigenfunctions, the constants A_n may be evaluated as

$$A_n = \frac{\displaystyle\int_0^1 \bar{r}\sin(\lambda_n\bar{r})\,d\bar{r}}{\displaystyle\int_0^1 \sin^2\left(\lambda_n\bar{r}\right)\,d\bar{r}} = \frac{2\,\text{Bi} \sin \lambda_n}{\lambda_n^2 - (1 - \text{Bi})\sin^2 \lambda_n} \tag{xxi}$$

The complete solution for the dimensionless temperature $\overline{T}$ is given by

$$\overline{T} = \frac{\hat{T}}{\overline{r}} = \frac{2\,\mathrm{Bi}}{\overline{r}} \sum_{n=1}^{\infty} \frac{\sin \lambda_n}{\lambda_n^2 - (1-\mathrm{Bi})\sin^2 \lambda_n} \sin(\lambda_n \overline{r})\exp(-\lambda_n^2 \theta) \qquad \text{(xxii)}$$

10.4 DETERMINATION OF THE AVERAGE TEMPERATURE OF A SOLID

The variation of the local temperature in a body with time or the temperature profile (i.e. the variation of temperature with position) at any instant can be directly calculated from the solution for the dimensionless temperature in the body. The average value of the dimensionless temperature of the body at any time can be determined by using any of the following equations depending upon its geometry.

$$\text{Plane wall:} \qquad \overline{T}_{\mathrm{av}} = \int_0^1 \overline{T}\,d\overline{x} \qquad (10.19)$$

$$\text{Infinite or 'long' cylinder:} \qquad \overline{T}_{\mathrm{av}} = 2\int_0^1 \overline{T}\,\overline{r}\,d\overline{r} \qquad (10.20)$$

$$\text{Sphere:} \qquad \overline{T}_{\mathrm{av}} = 3\int_0^1 \overline{T}\,\overline{r}^2\,d\overline{r} \qquad (10.21)$$

The derivation of these equations is left as an exercise. However, we will work out the equation for the average temperature of a 'long' cylinder undergoing unsteady state heat exchange with an ambient medium with a finite heat transfer coefficient at the surface.

The solution for the temperature distribution is given by Eq. (xx) of Example 10.6. The average temperature is given by

$$\overline{T}_{\mathrm{av}} = 2\int_0^1 \overline{T}\,\overline{r}\,d\overline{r} = 2\int_0^1 \left[\sum_{m=1}^{\infty} \frac{2\,\mathrm{Bi}\,J_o(\lambda_m \overline{r})}{(\lambda_m^2 + \mathrm{Bi}^2)\,J_o(\lambda_m)}\exp(-\lambda_m^2 \theta)\right]\overline{r}\,d\overline{r}$$

Using the known properties of Bessel functions (see Wiley and Barrett, 1982), the above integral can be evaluated to get the following expression for the average temperature of the cylinder.

$$\overline{T}_{\mathrm{av}} = 4\,\mathrm{Bi} \sum_{m=1}^{\infty} \frac{J_1(\lambda_m)}{\lambda_m(\lambda_m^2 + \mathrm{Bi}^2)\,J_o(\lambda_m)}\exp(-\lambda_m^2 \theta) \qquad (10.22)$$

Numerical calculations of temperature, heat flux or amount of heat exchange from a series solution requires the evaluation of the associated eigenvalues and eigenfunctions. In most cases, the eigenvalues are the roots of a transcendental equation and can be evaluated by using a numerical computational method like the Newton-Raphson method. Tabulated roots of a few such transcendental equations are given in Myers (1971), Carslaw and Jaeger (1959).

10.5 NUMERICAL CALCULATION OF UNSTEADY STATE HEAT CONDUCTION

Practical problems on unsteady state heat conduction (or steady state heat conduction in a body of multidimensional geometry) may require calculation of quantities such as the average solid temperature at any instant, central temperature (i.e. temperature at the mid-plane of a wall, at the centre line of a cylinder, or at the centre of a sphere), time required for attaining a prescribed temperature, etc. If a problem admits of an analytical solution, the numerical calculations of the above quantities are not difficult. Analytical solutions frequently occur in the form of infinite series. The accuracy of the calculation increases as more and more number of terms of a series are taken into account. For most practical purposes, truncation of a series solution after a few terms yields sufficiently accurate results. However, there is no general method of ascertaining a priori how many terms of a series should be considered in order to achieve a desired degree of accuracy. A practical approach is to perform the calculations taking n and $n + 1$ terms of the series, to note the difference in results and then to make a judgement on this basis as to how many terms should be taken for the computation. It is rather easy to do this on a computer.

However, the computed results on heat conduction in planar, cylindrical, and spherical geometries are available in the form of charts (Grober et al., 1961; Schneider, 1955) which can be directly used for determining the temperature at any location in a solid or the amount of heat transferred over a period of time. Useful charts for solids of the above geometries are given in Figs. 10.6–10.9. The following notations are used in all the charts:

T_o is the ambient temperature

T_i is the initial temperature of the solid

T_c is the mid-plane, centre-line or centre temperature at any time

T_{av} is the average temperature of the solid at any time

$Q_i = mc_p(T_i - T_o)$ = the initial heat content of the body (taking T_o as the reference temperature)

$Q = mc_p(T_i - T_{av})$ = the heat lost by the body in a given time.

Let us consider the case of a flat plate (i.e. an infinite wall) of thickness $2l$. The value of the dimensionless centre-line temperature, $(T_c - T_o)/(T_i - T_o)$, can be directly obtained from Fig. 10.6(a) for the given values of θ (or the Fourier number, Fo) and the Biot number, Bi. Therefore we can calculate T_c, the mid-plane temperature. The dimensionless temperature at any other location at a distance x (or at a dimensionless distance x/l from the mid-plane), can be read from Fig. 10.6(b). Determination of the fraction of initial heat content of the solid that is lost to the ambient medium is done by using Fig. 10.9. The procedure remains the same whether the solid is heated or cooled.

Examples 10.8 and 10.9 illustrate the use of charts in the case of a slab and of a 'long' cylinder.

Example 10.8 The top surface of a slab of margarine, 50 mm thick, initially at a temperature of 4°C, is kept exposed to an ambient at 25°C. Heat transfer to the slab occurs from the top surface only and the other surfaces may be considered insulated. If the heat transfer coefficient at the open surface is 8 kcal/h m² °C, calculate the bottom surface, mid-plane, and top surface temperatures of the slab after 4 hours. The relevant physical properties of margarine are: ρ = 990 kg/m³, c_p = 0.55 kcal/kg °C, k = 0.143 kcal/h m °C.

SOLUTION The temperature distribution in the slab with an insulated bottom will be the same as that of a slab of double the thickness but having both the faces exposed to the given environment. This is because the mid-plane of this imaginary slab is a plane of symmetry. We shall use Fig. 10.6(a) to calculate the mid-plane temperature of a 0.1 m thick slab first.

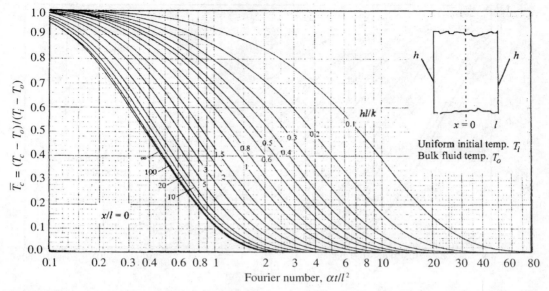

Fig. 10.6(a)   Dimensionless mid-plane temperature of an infinite flat plate with convection boundary condition (for $hl/k = \text{Bi} = \infty$, the surface temperature becomes constant at T_o).

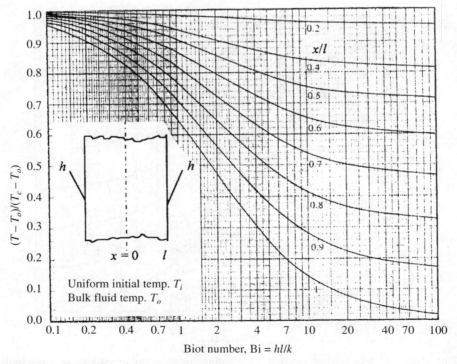

Fig. 10.6(b)   Dimensionless local temperature in an infinite flat plate [to be used with Fig. 10.6 (a)].

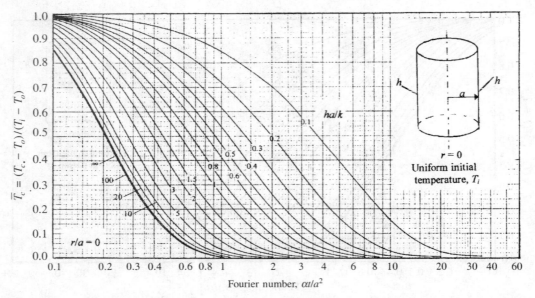

Fig. 10.7(a)   Dimensionless centre-line temperature of an infinite cylinder with convection boundary condition (for $hl/k = \mathrm{Bi} = \infty$, the surface temperature becomes constant at T_o).

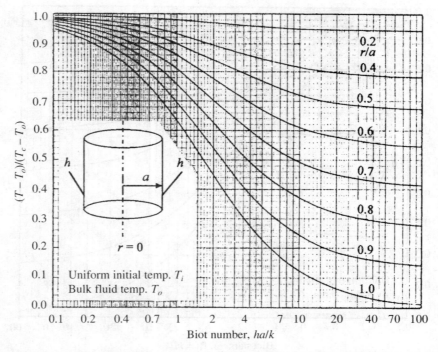

Fig. 10.7(b)   Dimensionless local temperature in an infinite cylinder [to be used along with Fig. 10.7(a)].

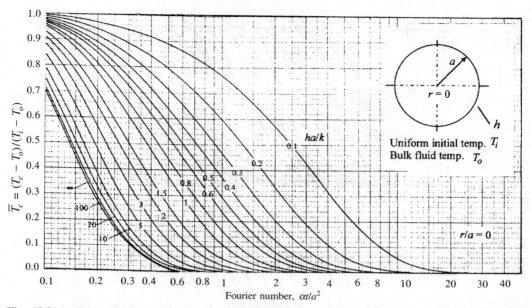

Fig. 10.8(a) Dimensionless centre temperature in a sphere with convection boundary condition (for ha/k = Bi = ∞, the surface temperature becomes constant at T_o).

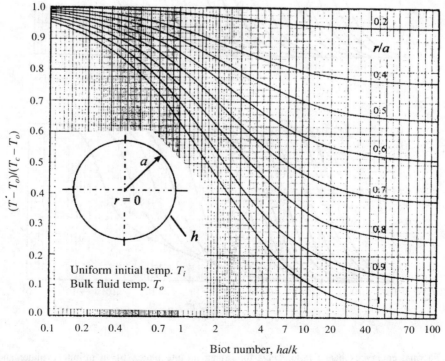

Fig. 10.8(b) Dimensionless local temperature in a sphere [to be used along with Fig. 10.8(a)].

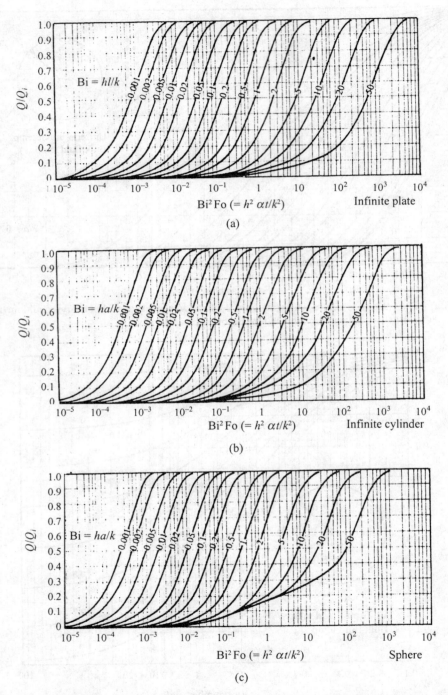

Fig. 10.9 Thermal energy exchange ratio, Q/Q_i, for (a) an infinite plate, (b) an infinite cylinder and (c) a sphere [uniform initial temperature = T_i; convection heat transfer coefficient = h].

Using the notations of Fig. 10.6(a), $l = 0.05$ m, $\rho = 990$ kg/m³, $c_p = 0.55$ kcal/kg °C, and $k = 0.143$ kcal/h m °C, initial temperature, $T_i = 4$°C, and ambient temperature, $T_o = 25$°C. Considering a 0.1 m thick slab, both sides exposed, we have

$$\text{Fourier number, Fo} = \frac{\alpha t}{l^2} = \frac{0.143}{(990)(0.55)} \frac{4.0}{(0.05)^2} = 0.42$$

$$\text{Biot number, Bi} = \frac{hl}{k} = \frac{(8)(0.05)}{0.143} = 2.8$$

For Fo = 0.42 and Bi = 2.8, from Fig. 10.6(a) we get

$$\overline{T_c} = \frac{T_c - T_o}{T_i - T_o} = 0.7$$

Therefore, the centre temperature,

$$T_c = T_o + \overline{T_c}(T_i - T_o) = 25 + (0.7)(4 - 25) = \boxed{10.3\,°C}$$

This is the bottom surface temperature of the given slab, 50 mm thick.

To calculate the temperature of the top surface, we use Fig. 10.6(b).

$$\text{For } x/l = 1.0, \text{ Bi} = 2.8 \text{ and } \overline{T_c} = 0.7, \frac{T - T_o}{T_c - T_o} = 0.382$$

Therefore,

$$T = (0.382)(T_c - T_o) + T_o = (0.382)(10.3 - 25) + 25 = \boxed{19.4°C}$$

To calculate the mid-plane temperature of the given slab, we put

$$x = 0.025, \quad \text{or} \quad x/l = 0.025/0.05 = 0.5, \quad \text{and} \quad \text{Bi} = 2.8$$

Then from Fig. 10.6(b), we get

$$\frac{T - T_o}{T_c - T_o} = 0.842$$

or

$$T = (0.842)(10.3 - 25) + 25 = \boxed{12.6°C}$$

Example 10.9 A 'long' steel shaft of 10 cm diameter is austenised at 870°C and then quenched in a liquid at 30°C. The average surface heat transfer coefficient is 2000 W/m² °C. The relevant thermophysical properties of the shaft material are $k = 20$ W/m °C; $\rho = 7800$ kg/m³; $c_p = 0.46$ kJ/kg °C. Calculate:

(i) the time required for the centre-line temperature to drop down to 200°C (martensite forms at this temperature),

(ii) the temperature at half-radius at that moment, and

(iii) the amount of heat that has been transferred to the liquid by that time per metre length of the shaft.

SOLUTION   We shall use Figs. 10.7–10.9(b) for the calculations.

(i) *Given:* Initial temperature, $T_i = 870°C$, ambient temperature, $T_o = 30°C$, centre-line temperature, $T_c = 200°C$, surface heat transfer coefficient, $h = 2000$ W/m^2 °C, and radius of the cylinder, $a = 0.05$ m.

$$\overline{T}_c = \frac{T_c - T_o}{T_i - T_o} = \frac{200 - 30}{870 - 30} = 0.2$$

$$\text{Biot number, Bi} = \frac{ha}{k} = \frac{(2000)(0.05)}{20} = 5.0$$

$$\text{Thermal diffusivity, } \alpha = \frac{k}{\rho c_p} = \frac{20}{(7800)(460)} = 5.57 \times 10^{-6} \text{ m}^2/\text{s}$$

For $\overline{T}_c = 0.2$ and Bi = 5.0, we get from Fig. 10.7(a),

$$\text{Fourier number, Fo} = \frac{\alpha t}{a^2} = 0.51$$

(i) Hence, the required time,

$$t = \frac{(0.51)(0.05)^2}{5.57 \times 10^{-6}} = \boxed{229 \text{ s}}$$

(ii) At the half-radius, $r/a = 0.5$ and Bi = 5.0, Fig. 10.7(b) reads

$$\frac{T - T_o}{T_c - T_o} = 0.77$$

or

$$T = T_o + 0.77(T_c - T_o) = 30 + 0.77(200 - 30) = \boxed{161°C}$$

(iii) Bi^{2}Fo = $(5.0)^2(0.51) = 12.75$

For Bi^{2}Fo = 12.75 and Bi = 5.0, Fig. 10.9(b) gives

$$Q/Q_i = 0.83$$

Initial amount of heat energy present in 1 m length of the shaft,

$$Q_i = \pi a^2(1)\rho \, c_p(T_i - T_o) = \pi(0.05)^2(7800)(0.46)(870 - 30)$$
$$= 2.367 \times 10^4 \text{ kJ}$$

Amount of heat transferred from the shaft by this time (229 s) is

$$Q = 0.83 Q_i = (0.83)(2.367 \times 10^4 \text{ kJ}) = \boxed{19,646 \text{ J}}$$

10.6 UNSTEADY OR STEADY STATE HEAT CONDUCTION IN A MULTIDIMENSIONAL SOLID

So far we have discussed and analysed heat diffusion problems in systems described by *one space coordinate* only and obtained solutions in simple cases using the technique of separation of variables. The same technique is also applicable to many unsteady state or even steady state problems for systems described by more than one space coordinates. Typical examples of such systems are a 'long' rectangular bar, a rectangular parallelepiped, a finite cylinder, etc. Unsteady state heat conduction in a rectangular bar is analysed in the following problem.

Example 10.10 (*Unsteady state heat conduction in a rectangular bar*) A *long* rectangular bar of sides *a* and *b* has a uniform initial temperature distribution T_i. At time $t = 0$, all the four surfaces of the bar are raised to a temperature T_s and maintained at that value. It is required to determine the unsteady state temperature distribution in the bar.

SOLUTION A cross-section of the bar is shown in Fig. 10.10. It is long in the *z*-direction and has dimensions *a* and *b* along *x*- and *y*-directions, respectively. The temperature in the solid will depend upon the spatial coordinates *x* and *y* and time *t*. That is,

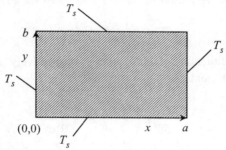

Fig. 10.10 A section of the rectangular bar (Example 10.10).

$$T = T(x, y, t)$$

The partial differential equation governing the temperature distribution in the solid can be obtained from Eq. (xii) of Example 10.1 on putting $\partial^2 T/\partial z^2 = 0$ (no variation of temperature in the *z*-direction) and $\psi_v = 0$ (no heat generation). Thus, we get

$$\text{PDE:} \qquad \frac{1}{\alpha} \frac{\partial T}{\partial t} = \frac{\partial^2 T}{\partial x^2} + \frac{\partial^2 T}{\partial y^2} \qquad \text{(i)}$$

The initial and boundary conditions are:

$$\text{I.C.:} \qquad T(x, y, 0) = T_i \qquad \text{(ii)}$$

$$\text{B.C. 1:} \qquad T(0, y, t) = T_s \qquad \text{(iii)}$$

$$\text{B.C. 2:} \qquad T(a, y, t) = T_s \qquad \text{(iv)}$$

$$\text{B.C. 3:} \qquad T(x, 0, t) = T_s \qquad \text{(v)}$$

$$\text{B.C. 4:} \qquad T(x, b, t) = T_s \qquad \text{(vi)}$$

Let us define the following dimensionless variables:

$$\overline{T} = \frac{T - T_s}{T_i - T_s}; \quad \overline{x} = \frac{x}{a}; \quad \overline{y} = \frac{y}{b}; \quad s = \frac{a}{b}; \quad \theta = \frac{\alpha t}{a^2} \qquad \text{(vii)}$$

The use of these dimensionless variables transforms Eqs. (i)–(vi) to the following forms.

$$\text{PDE:} \qquad \frac{\partial \overline{T}}{\partial \theta} = \frac{\partial^2 \overline{T}}{\partial \overline{x}^2} + s^2 \frac{\partial^2 \overline{T}}{\partial \overline{y}^2} \qquad \text{(viii)}$$

$$\text{I.C.:} \qquad \overline{T}(\overline{x}, \overline{y}, 0) = 1 \qquad \text{(ix)}$$

$$\text{B.C. 1:} \qquad \overline{T}(0, \overline{y}, \theta) = 0 \qquad \text{(x)}$$

$$\text{B.C. 2:} \qquad \overline{T}(1, \overline{y}, \theta) = 0 \qquad \text{(xi)}$$

$$\text{B.C. 3:} \qquad \overline{T}(\overline{x}, 0, \theta) = 0 \qquad \text{(xii)}$$

$$\text{B.C. 4:} \qquad \overline{T}(\overline{x}, 1, \theta) = 0 \qquad \text{(xiii)}$$

We use the technique of separation of variables to solve the PDE and assume a solution of the form

$$\overline{T} = \Theta(\theta) X(\overline{x}) Y(\overline{y}) \qquad \text{(xiv)}$$

where Θ, X, and Y are the functions of θ, $\overline{x}$, and $\overline{y}$, respectively. Substitution of Eq. (xiv) in Eq. (viii) and rearrangement yields

$$\frac{1}{X}\frac{d^2 X}{d\overline{x}^2} + \frac{s^2}{Y}\frac{d^2 Y}{d\overline{y}^2} = \frac{1}{\theta}\frac{d\Theta}{d\theta} \qquad \text{(xv)}$$

or

$$\frac{1}{X}\frac{d^2 X}{d\overline{x}^2} = \frac{1}{\theta}\frac{d\Theta}{d\theta} - \frac{s^2}{Y}\frac{d^2 Y}{d\overline{y}^2} = -\lambda^2 \text{ (say)} \qquad \text{(xvi)}$$

or

$$\frac{d^2 X}{d\overline{x}^2} + \lambda^2 X = 0 \qquad \text{(xvii)}$$

and

$$\frac{1}{\theta}\frac{d\Theta}{d\theta} = \frac{s^2}{Y}\frac{d^2 Y}{d\overline{y}^2} - \lambda^2 = -\mu^2 - \lambda^2 \qquad \text{(xviii)}$$

where

$$\frac{s^2}{Y}\frac{d^2 Y}{d\overline{y}^2} = -\mu^2 \qquad \text{(xix)}$$

Therefore,

$$\frac{1}{\theta}\frac{d\Theta}{d\theta} = -(\mu^2 + \lambda^2) \qquad \text{(xx)}$$

Here λ^2 and μ^2 are constants (preceded by negative sign) introduced on the same argument as put forward in Example 10.4 in order to ensure non-trivial solutions for the function X and Y. Solutions of Eqs. (xvii) and (xix) provide the eigenvalues and eigenfunctions. Boundary conditions on X and Y required for this purpose can be derived from Eqs. (x), (xi), (xii), and (xiii) respectively in the following forms.

$$\text{B.Cs. on } X(\overline{x}): \qquad X(0) = X(1) = 0 \qquad \text{(xxi)}$$

$$\text{B.Cs. on } Y(\overline{y}): \qquad Y(0) = Y(1) = 0 \qquad \text{(xxii)}$$

Solution of Eq. (xvii) subject to the B.Cs. given by Eq. (xxi) is

$$\text{Eigenvalues:} \qquad \lambda = m\pi, \qquad m = 1, 2, 3, \ldots \qquad \text{(xxiii)}$$

$$\text{Eigenfunctions:} \qquad X(\overline{x}) = A \sin (m\pi\overline{x}) \qquad \text{(xxiv)}$$

Solution of Eq. (xviii) subject to the B.Cs. given by Eq. (xxi) is

$$\text{Eigenvalues:} \qquad \mu = sn\pi, \qquad n = 1, 2, 3, \ldots \qquad \text{(xxv)}$$

$$\text{Eigenfunctions:} \qquad Y(\overline{y}) = B \sin (sn\pi\overline{y}) \qquad \text{(xxvi)}$$

Solution to the time-dependent function Θ is obtained from Eq. (xx). Thus,

$$\Theta = C \exp [-(\lambda^2 + \mu^2)\theta] \qquad \text{(xxvii)}$$

The general solution for $\overline{T}$ can be written as a linear combination of individual solutions. Thus,

$$\overline{T} = \sum_{m=1}^{\infty} \sum_{n=1}^{\infty} A_{mn} \sin (m\pi\overline{x}) \sin (n\pi\overline{y}) \exp [-(m^2\pi^2 + s^2n^2\pi^2)\theta]$$

It is now convenient to replace the dimensionless variables by the corresponding dimensional quantities by using Eq. (vii). Thus,

$$(m^2\pi^2 + s^2n^2\pi^2)\theta = \left(m^2\pi^2 + \frac{a^2}{b^2}n^2\pi^2 \right)\frac{\alpha t}{a^2} = \left(\frac{m^2}{a^2} + \frac{n^2}{b^2} \right)\pi^2 \alpha t$$

$$\overline{T} = \sum_{m=1}^{\infty} \sum_{n=1}^{\infty} A_{mn} \sin \left(\frac{m\pi x}{a} \right) \sin \left(\frac{n\pi y}{b} \right) \exp \left[-\left(\frac{m^2}{a^2} + \frac{n^2}{b^2} \right)\pi\alpha^2\,\pi\alpha t \right] \qquad \text{(xxviii)}$$

Putting the I.C., Eq. (ix), in Eq. (xxviii) (at $\theta = 0$ or $t = 0$, $\overline{T} = 1$), we get

$$1 = \sum_{m=1}^{\infty} \sum_{n=1}^{\infty} A_{mn} \sin \left(\frac{m\pi x}{a} \right) \sin \left(\frac{n\pi y}{b} \right) \qquad \text{(xxix)}$$

The functions $\sin \left(\dfrac{m\pi x}{a} \right)$ are orthogonal over $[0, a]$, and $\sin \left(\dfrac{n\pi y}{b} \right)$ are orthogonal over $[0, b]$.

Multiplying both sides of Eq. (xxix) by $\sin \left(\dfrac{p\pi x}{a} \right) \sin \left(\dfrac{q\pi y}{b} \right) dxdy$ and integrating between the respective limits of x and y, we get

$$\int_0^a \int_0^b \sin\left(\frac{p\pi x}{a}\right) \sin\left(\frac{q\pi y}{b}\right) dxdy = A_{pq} \int_0^a \int_0^b \sin^2\left(\frac{p\pi x}{a}\right) \sin^2\left(\frac{q\pi y}{b}\right) dxdy$$

or

$$\int_0^a \sin\left(\frac{p\pi x}{a}\right) dx \int_0^b \sin\left(\frac{q\pi y}{b}\right) dy = A_{pq} \int_0^a \sin^2\left(\frac{p\pi x}{a}\right) dx \int_0^b \sin^2\left(\frac{q\pi y}{b}\right) dy$$

or

$$\left[-\frac{a}{p\pi}\cos\frac{p\pi x}{a}\right]_0^a \left[-\frac{b}{q\pi}\cos\frac{q\pi y}{b}\right]_0^b = A_{pq}\frac{ab}{4}$$

The LHS can be shown to be non-vanishing only for odd values of p and q (see Example 10.4). Therefore,

$$A_{2k+1,\ 2l+1} = \frac{16}{\pi^2(2k+1)(2l+1)}; \quad k = 0, 1, 2, 3, \ldots; \quad l = 1, 2, 3, \ldots \qquad \text{(xxx)}$$

The complete solution for the dimensionless temperature $\overline{T}$ is obtained from Eqs. (xxviii) and (xxx) as

$$\overline{T} = \frac{16}{\pi^2}\sum_{k=0}^{\infty}\sum_{l=0}^{\infty}\frac{\sin\left[(2k+1)\dfrac{\pi x}{a}\right]\sin\left[(2l+1)\dfrac{\pi y}{b}\right]}{(2k+1)(2l+1)}\exp\left[-\left\{\left(\frac{2k+1}{a}\right)^2 + \left(\frac{2l+1}{b}\right)^2\right\}\pi^2\alpha t\right]$$

$$\text{(xxxi)}$$

10.7 THE GRAETZ PROBLEM

The problem of heat transfer to or from a fluid in laminar flow through a circular pipe is well known as the *Graetz problem* after L. Graetz who first solved this problem more than a hundred years ago. If we assume that: (i) the fluid is in fully developed laminar flow (i.e. it has a parabolic velocity distribution), (ii) the fluid properties remain essentially constant, and (iii) heat transfer occurs at steady state, the governing partial differential equation for the local temperature in the fluid may be found to be given by Eq. (10.23) below. This equation may be derived by considering the various heat input and output terms in a small annular cylindrical element of fluid (the depiction of the fluid element and the pipe will look like Fig. 10.2). The input and output terms should take into account the transfer of heat by bulk flow (i.e. owing to the motion of the fluid) in the axial direction and by conduction in the radial direction. The contribution of axial conduction can be neglected in most practical situations without involving appreciable error. Thus, we get

$$2U_o\left[1 - \left(\frac{r}{a}\right)^2\right]\frac{\partial T}{\partial z} = \alpha\left[\frac{\partial^2 T}{\partial r^2} + \frac{1}{r}\frac{\partial T}{\partial r}\right] \qquad (10.23)$$

where z is the axial coordinate and U_o the average fluid velocity. The following boundary conditions apply:

$$z = 0, \qquad 0 \le r \le a; \qquad T = T_o \qquad\qquad (10.24a)$$

$$z > 0, \qquad r = a; \qquad T = T_w \qquad\qquad (10.24b)$$

$$z > 0, \qquad r = 0; \qquad \frac{\partial T}{\partial r} = 0 \qquad\qquad (10.24c)$$

where

T_o is the fluid temperature at the inlet to the heated section of the pipe

T_w is the constant wall temperature

a is the radius of the pipe.

Equation (10.23) subject to the boundary conditions (10.24a, b, c) can be solved by using the method of separation of variables to have the dimensionless temperature distribution in terms of eigenvalues and eigenfunctions in the following form,

$$\frac{T - T_w}{T_o - T_w} = \sum_{n=1}^{\infty} A_n \Theta_n \exp\left(-\lambda_n^2 \bar{z}\right) \qquad\qquad (10.25)$$

where $\bar{z} = 2\alpha z / a^2 U_o$, λ_n are the eigenvalues, and Θ_n are the eigenfunctions given by

$$\frac{d^2\Theta}{d\bar{r}^2} + \frac{1}{\bar{r}} \frac{d\Theta}{d\bar{r}} + \lambda^2 (1 - \bar{r}^2)\Theta = 0 \qquad\qquad (10.26)$$

$$\bar{r} = r/a = 1, \qquad \Theta_n = 0 \qquad\qquad (10.27a)$$

$$\bar{r} = 0, \qquad \frac{d\Theta_n}{d\bar{r}} = 0 \qquad\qquad (10.27b)$$

The rate of heat exchange of the fluid with the wall can be determined by computing the wall temperature gradient and the corresponding wall heat flux. The eigenvalues and eigenfunctions have generally been obtained numerically (for more details, see Longwell, 1966). However, a closed form analytical solution for the temperature distribution can be obtained in terms of confluent hypergeometric functions by adopting the mathematical technique used by Ray et al. (1988) for analysis of gas absorption with a chemical reaction in a laminar falling film. The derivation of Eq. (10.23) and solution of the problem are left as an exercise.

Analyses of heat transfer in laminar flow through conduits of other geometries (for example, triangular, rectangular, etc.) are available in Shah and London (1978).

10.8 SIMILARITY SOLUTION

The partial differential equations arising in some cases of momentum, heat, or mass transfer admit of a special class of solution called the *similarity solution*. Before dealing with the mathematical basis of similarity solution it will be useful to discuss its physical implication. A similarity solution exists for a system if it has an *inherent* or *self-similitude*. To be more down-to-earth, it may be said that the plots of the dependent variables for two particular values of one of the independent variables will look similar except for a certain *scale factor*. The scale factor is a function of the other independent variable. A frequently cited example for illustration of the similarity phenomenon is the flow of an incompressible fluid past a flat plate at zero angle of incidence (also see Chapters 4 and 11).

The two-dimensional velocity profile (which depends upon x and y) at two locations A and B along the plate look similar (see Fig. 11.4) and the profile at B may be obtained by *stretching* that at A. It may be shown that here the local longitudinal velocity in the boundary layer, u, may be written as

$$\frac{u}{V_\infty} = f(\eta), \qquad \eta = \frac{y}{\delta(x)} \tag{10.28}$$

where
  u is the velocity in the x-direction at any point in the boundary layer
  V_∞ is the 'free stream velocity'
  $\delta(x)$ is the boundary layer thickness.

In the definition of η, the transverse coordinate y is scaled by δ which is function of x.

Mathematically, this kind of physical problem (i.e. a problem having a self-similitude) is governed by a partial differential equation that remains *invariant* under a group of transformation called the *similarity transformation*. The use of such a transformation, in effect, leads to a reduction in the number of independent variables. If there are only two independent variables in the problem (for example, boundary layer flow discussed above or unsteady heating of a semi-infinite solid), the partial differential equation reduces to an ordinary differential equation.

Example 10.11 Let us consider unsteady state heating (or cooling) of a semi-infinite solid.* The solid is initially at a uniform temperature T_i, throughout. At time $t = 0$, the flat surface $x = 0$ is raised to a temperature T_s and is maintained at this value at all subsequent time. It is required to determine the unsteady state temperature distribution in the solid.

SOLUTION The sketch of the system and the coordinate system are shown in Fig. 10.11. This is a problem of one-dimensional unsteady state heat diffusion. The governing PDE is

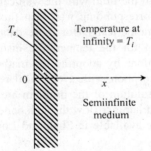

Fig. 10.11 A semi-infinite solid medium (Example 10.11).

PDE: $$\frac{\partial T}{\partial t} = \alpha \frac{\partial^2 T}{\partial x^2} \tag{i}$$

I.C.: $$T(0, x) = T_i \tag{ii}$$

B.C. 1: $$T(t, 0) = T_s \tag{iii}$$

B.C. 2: $$T(t, \infty) = T_i \tag{iv}$$

---

*A solid medium, only one surface of which is at our reach and it, otherwise, extends to infinity is called a 'semi-infinite solid'.

The B.C. 1 implies that, since the thermal conductivity of the solid is finite, the temperature at 'infinite' distance from the surface remains equal to the initial temperature for all finite time. Let us introduce a dimensionless temperature and a modified time variable

$$\overline{T} = \frac{T - T_i}{T_s - T_i} \quad \text{and} \quad \tau = \alpha t$$

The PDE, the I.C., and the B.Cs. get transformed to the following forms.

PDE: $$\frac{\partial \overline{T}}{\partial \tau} = \frac{\partial^2 \overline{T}}{\partial x^2} \tag{v}$$

I.C.: $$\overline{T}(0, x) = 0 \tag{vi}$$

B.C. 1: $$\overline{T}(\tau, 0) = 1 \tag{vii}$$

B.C. 2: $$\overline{T}(\tau, \infty) = 0 \tag{viii}$$

Now, we try the following one-parameter group of transformation,

$$T' = a^{\beta_1} \overline{T}; \qquad x' = a^{\beta_2} x; \qquad \tau' = a^{\beta_3} \tau \tag{ix}$$

where a is the parameter of transformation. Substituting in Eq. (v), we get

$$\frac{a^{-\beta_1}}{a^{-\beta_3}} \frac{\partial T'}{\partial \tau'} = \frac{a^{-\beta_1}}{a^{-2\beta_2}} \frac{\partial^2 T'}{\partial x'^2} \tag{x}$$

Equation (v) is said to be 'invariant' with respect to the transformation (ix) if its form remains the same after the transformation. This happens if

$$\frac{a^{-\beta_1}}{a^{-\beta_3}} = \frac{a^{-\beta_1}}{a^{-2\beta_2}}$$

or

$$a^{(\beta_3 - \beta_1)} = a^{(2\beta_2 - \beta_1)}$$

or

$$\beta_3 - \beta_1 = 2\beta_2 - \beta_1$$

Thus β_1 can be chosen arbitrarily, and $\beta_3 = 2\beta_2$. Let us put

$$\beta_1 = 0 \quad \text{and} \quad \beta_2/\beta_3 = \delta = 1/2$$

So, a *similarity transformation* for this problem exists. The quantities that remain invariant can now be determined. From Eq. (ix), we have

$$a = \left(\frac{x'}{x}\right)^{1/\beta_2} = \left(\frac{\tau'}{\tau}\right)^{1/\beta_3}$$

or

$$\frac{x'}{x} = \left(\frac{\tau'}{\tau}\right)^{\beta_2/\beta_3} = \left(\frac{\tau'}{\tau}\right)^{\delta}$$

Therefore,

$$\frac{x'}{\tau'^{\delta}} = \frac{x}{\tau^{\delta}}$$

Thus

$$\eta' = \frac{x}{\tau^{\delta}} = \frac{x}{\sqrt{\tau}} \qquad \text{(because } \delta = 1/2) \qquad \text{(xi)}$$

is also an invariant of this system. This variable η' is called the *similarity variable*. However, for solution of this problem it is more convenient to use the variable

$$\eta = \frac{\eta'}{2} = \frac{x}{2\sqrt{\tau}} \qquad \text{(xii)}$$

Introduction of the variable, η or η', reduces Eq. (v) to an ordinary differential equation. Also, upon using this transformation, the conditions (vi) and (viii) reduce to the same condition with respect to η. The given problem is now solved by using this technique.

Putting $\eta = x/2\sqrt{\tau}$ and using the chain rule of differentiation, we get

$$\frac{\partial \overline{T}}{\partial \tau} = \frac{\partial \overline{T}}{\partial \eta}\frac{\partial \eta}{\partial \tau} = \frac{\partial \overline{T}}{\partial \eta}\left(-\frac{x}{4\tau\sqrt{\tau}}\right)$$

$$\frac{\partial \overline{T}}{\partial x} = \frac{\partial \overline{T}}{\partial \eta}\frac{\partial \eta}{\partial x} = \frac{\partial \overline{T}}{\partial \eta}\left(\frac{1}{2\sqrt{\tau}}\right)$$

$$\frac{\partial^2 \overline{T}}{\partial x^2} = \frac{\partial}{\partial x}\left(\frac{\partial \overline{T}}{\partial x}\right) = \frac{\partial}{\partial \eta}\left(\frac{\partial \overline{T}}{\partial \eta}\frac{1}{2\sqrt{\tau}}\right)\frac{\partial \eta}{\partial x} = \frac{\partial^2 \overline{T}}{\partial \eta^2}\frac{1}{4\tau}$$

Substituting in Eq. (v), we get

$$\frac{\partial \overline{T}}{\partial \eta}\left(-\frac{x}{4\tau\sqrt{\tau}}\right) = \frac{\partial^2 \overline{T}}{\partial \eta^2}\frac{1}{4\tau}$$

or

$$\frac{\partial^2 \overline{T}}{\partial \eta^2} + 2\eta\frac{\partial \overline{T}}{\partial \eta} = 0 \qquad \text{(xiii)}$$

The above equation contains only one independent variable, η. So the partial derivatives may be replaced by ordinary derivatives. Using the similarity transformation (xii), Eq. (v) has been transformed to an ordinary differential equation

$$\frac{d^2 \overline{T}}{d\eta^2} + 2\eta\frac{d\overline{T}}{d\eta} = 0 \qquad \text{(xiv)}$$

The initial condition (vi) reduces to the following form for the variable $\overline{T} = \overline{T}(\eta)$.

$$\eta = \infty, \qquad \overline{T} = 0 \qquad \text{(xv)}$$

The boundary condition (vii) reduces to

$$\eta = 0, \qquad \overline{T} = 1 \qquad \text{(xvi)}$$

Note that the other boundary condition (viii) reduces to the same form as (xv). Thus, the original set of three conditions now provide two boundary conditions with respect to η, which are adequate for solving the ordinary differential Eq. (xiv).

Putting $d\overline{T}/d\eta = \zeta$ in Eq. (xiv) and integrating, we get

$$\frac{d\zeta}{d\eta} + 2\eta\zeta = 0$$

or

$$\zeta = Ae^{-\eta^2}$$

or

$$\frac{d\overline{T}}{d\eta} = Ae^{-\eta^2} \qquad \text{(xvii)}$$

Integrating again, we obtain

$$\overline{T} = A \int_0^{\eta} e^{-\eta^2} \, d\eta + B \qquad \text{(xviii)}$$

where A and B are two integration constants.

Using the boundary condition (xvi), we get

$$1 = A \int_0^0 e^{-\eta^2} \, d\eta + B$$

or

$$B = 1 \qquad \text{(xix)}$$

Using the condition (xv), we get

$$0 = A \int_0^{\infty} e^{-\eta^2} \, d\eta + 1$$

Because $\int_0^{\infty} e^{-\eta^2} \, d\eta = \dfrac{\sqrt{\pi}}{2}$, $A = -\dfrac{2}{\sqrt{\pi}}$ the final solution is obtained from Eq. (xviii).

$$\overline{T} = \frac{T - T_i}{T_s - T_i} = 1 - \frac{2}{\sqrt{\pi}} \int_0^{\eta} e^{-\eta^2} \, d\eta \qquad \text{(xx)}$$

The integral in the above equation is called the *error function*. Thus the solution of Eq. (xiv) may also be written as

$$\overline{T} = 1 - \text{erf} \, (\eta) = \text{erfc} \, (\eta) \qquad \text{(xxi)}$$

It will be interesting to see how the heat flux at the boundary depends upon time. Heat flux q_x at the boundary is given by

$$[q_x]_{x=0} = -k \left[\frac{dT}{dx} \right]_{x=0} = h(T_s - T_i)$$

where h is the corresponding heat transfer coefficient at the free surface of the semi-infinite solid.

Performing differentiation of both sides of Eq. (xx) and simplifying, we get

$$[q_x]_{x=0} = (T_s - T_i)\sqrt{\frac{k\rho c_p}{\pi t}} = h(T_s - T_i) \tag{xxii}$$

where

$$h = \sqrt{\frac{k\rho c_p}{\pi t}}$$

The above equation indicates that the heat flux as well as the heat transfer coefficient is infinitely 'large' initially, but becomes very small at 'large' times.

A brief list of values of the error function is given in Table 10.1. More elaborate tables are available (Carslaw and Jaeger, 1959; Myers, 1971).

Table 10.1   Values of the error function

η	erf (η)	η	erf (η)	η	erf (η)
0.0	0.0000	0.45	0.47548	0.90	0.79791
0.05	0.05637	0.50	0.52050	0.95	0.82089
0.1	0.11246	0.55	0.56332	1.00	0.84270
0.15	0.16800	0.60	0.60386	1.10	0.88020
0.20	0.22270	0.65	0.64203	1.30	0.93401
0.25	0.27633	0.70	0.67780	1.50	0.96610
0.30	0.32863	0.75	0.71116	1.80	0.98909
0.35	0.37938	0.80	0.74210	2.00	0.99532
0.40	0.42839	0.85	0.77067	3.00	0.99998

10.9   NUMERICAL SOLUTION OF HEAT CONDUCTION PROBLEMS

Analytical solutions to heat conduction problems involving more than one independent variable (time and space coordinates) can be obtained for solids of simple geometries and simple boundary conditions. Besides the techniques used in the previous examples, integral transforms (e.g. Laplace transform, Fourier transform, etc.) constitute powerful mathematical tools for analytical solutions of heat diffusion and other diffusional problems. Analytical solutions to many conduction or diffusion problems are available in the literature (see, e.g. Carslaw and Jaeger, 1959; Crank, 1974; Mikhailov and Ozisik, 1984).

Numerical methods, based on the finite difference or finite element techniques are frequently used to solve heat conduction and diffusional problems when analytical methods fail. A large number of books are available which deal with numerical methods and application to a wide variety of problems including conduction heat transfer (for example, Myers, 1971; Adams and Rogers, 1973; Carnahan et al. 1969).

SHORT QUESTIONS

1. Qualitatively state the condition under which a cylindrical shell can be considered similar to an infinite plane wall for the purpose of analysis of heat conduction.

2. What are the different types of boundary conditions encountered in conduction heat transfer?

3. Consider unsteady state heat conduction in a semi-infinite solid medium. Discuss how the relative importance of heat transfer resistance, owing to convection and that owing to conduction in the solid, changes with time.

4. What is the physical meaning of a semi-infinite solid? Under what condition can a relatively small block of solid be approximated to a semi-infinite solid?
   [*Hint:* If the time of contact of the solid with the medium (hot or cold) is small.]

5. What do you mean by a 'long' cylinder? Discuss the limitations of using the solution for the transient temperature in a 'long' cylinder to analyse heat conduction in a finite rod.

6. Why is heat conduction in fins considered one-dimensional in most cases? Explain qualitatively:

PROBLEMS

10.1 A large plane wall of thickness l has a uniform initial temperature distribution T_i throughout. At time $t = 0$, one of the surfaces of the wall is brought to a temperature T_1 and the other surface to a temperature T_2, and both surfaces are maintained at these temperatures thereafter. It is required to determine the unsteady state temperature distribution in the wall.

Hints: Because two surfaces are maintained at different temperatures, the boundary conditions are *not* homogeneous and the technique of separation of variables will not be directly applicable. However, the problem may be made homogeneous by expressing the non-dimensional temperature as the sum of a steady state part and an unsteady state part. That is,

$$\overline{T} = \frac{T - T_1}{T_2 - T_1} = \hat{T}_s(\overline{x}) + \hat{T}_u(\theta, \overline{x})$$

where $\hat{T}_s(\overline{x})$ is the steady state part and $\hat{T}_u(\theta, \overline{x})$ the unsteady state part. On substitution in the governing PDE, an ODE for $\hat{T}_s(\overline{x})$ [with boundary conditions $\hat{T}_s(0) = 0$, and $\hat{T}_s(1) = 1$] and a PDE for $\hat{T}_u(\theta, \overline{x})$ [I.C.: $\hat{T}_u(0, \overline{x}) = (T_i - T_1)/(T_2 - T_1) - \overline{x}$; B.C. 1: $\hat{T}_u(\theta, 0) = 0$; B.C. 2: $\hat{T}_u(\theta, 1) = 0$] are obtained. The final solution is

$$\overline{T} = \frac{T - T_1}{T_2 - T_1} = \overline{x} + \frac{2}{\pi(T_2 - T)} \sum_{n=1}^{\infty} \left[\frac{(T_i - T_1)}{n} + \frac{(-1)^n (T_2 - T_i)}{n} \right] \sin(n\pi\overline{x}) e^{-n^2\pi^2\theta}$$

10.2 The temperatures at the ends $x = 0$ and $x = l$ of a rod having perfectly insulated curved surface are held at temperatures T_1 and T_2, respectively, until the steady state condition is attained. Then the temperatures at the two ends are interchanged. If this interchange occurs at time $t = 0$, determine the resultant unsteady state temperature distribution in the rod.

Hint: Determine the initial steady state temperature of the rod which is $T = T_1 + (T_2 - T_1)x/l$. As there is no heat loss from the curved surface of the rod, the equation of one-dimensional heat conduction in an infinite wall applies. One boundary condition is non-homogeneous which may be homogenized by assuming the solution to consist of steady state and unsteady state parts (see Problem 10.1). The solution is

$$\overline{T} = \frac{T - T_1}{T_2 - T_1} = (1 - \overline{x}) - \frac{2}{\pi}\sum_{n=1}^{\infty}\frac{1}{n}\sin(2n\pi\overline{x})e^{-4n^2\pi^2\theta}$$

10.3 Rework Example 10.5 assuming that one surface of the wall ($x = 0$) is in contact with a fluid of bulk temperature T_o, the surface heat transfer coefficient being h, whereas the other surface ($x = l$) is held at temperature T_o.

10.4 A thin circular disk of radius a and width w has a uniform initial temperature T_i. At time $t = 0$, the cylindrical edge of the disk is brought to a temperature T_o, and maintained at that temperature thereafter. Heat exchange starts simultaneously between the two circular faces of the disk and the ambient medium which is also at a temperature T_o. The surface heat transfer coefficient is h. Determine the unsteady state temperature distribution in the disk. What is the temperature of the disk after a large time?

Hints: Neglect temperature variation along the thickness of the disk because it is very thin. Obtain the following PDE by making a heat balance over a thin annular cylindrical element of thickness Δr and height w at a distance r from the centre line of the disk.

$$\frac{\partial \overline{T}}{\partial \theta} = \frac{\partial^2 \overline{T}}{\partial \overline{r}^2} + \frac{1}{\overline{r}}\frac{\partial \overline{T}}{\partial \overline{r}} - \beta\overline{T}; \qquad \overline{T} = \frac{T - T_o}{T_i - T_o}; \qquad \overline{r} = \frac{r}{a}; \qquad \beta = \frac{2ha^2}{wk}$$

Use the transformation $\hat{T} = \overline{T}e^{-\beta\theta}$, and obtain the PDE, the I.C. and the B.Cs. for $\hat{T}$.

10.5 A wooden board, 2.5 cm thick, is dried at 120°C and then cooled in a flowing stream of air. If the air temperature is 25°C and the surface heat transfer coefficient is 70 W/m² °C, calculate

(i) the maximum temperature in the board after 10 minutes, and
(ii) the amount of heat lost by 1 m² of the board during this time.

Given: $\rho = 700$ kg/m³, $c_p = 1300$ J/kg °C, $k = 0.15$ W/m K.

10.6 A large steel plate, 5 cm thick, is hot rolled at 650°C and then placed on an insulating brick floor. Heat loss occurs from the exposed top surface to the flowing air at 25°C. The heat transfer coefficient owing to combined forced and natural convection is 100 W/m² °C. Heat loss from the bottom surface may be neglected. Calculate the time required for the bottom surface to reach 100°C. What is the mid-plane temperature of the plate at that moment? *Given:* $\rho = 7800$ kg/m³, $c_p = 460$ J/kg °C, $k = 30$ W/m °C.

[*Hints:* Consider a plate of double the given thickness , i.e. 10 cm (or $l = 5$ cm) for the purpose of using Fig. 10.6. Because $\partial T/\partial x = 0$ at the bottom surface (where $x = 0$), the mid-plane temperature of a 10 cm thick plate will be the same as the bottom surface temperature of the given plate.]

Note that the variation in the plate temperature with position is small, because the air-film resistance mainly governs the rate of cooling. With a lumped parameter analysis of the problem, it may be seen that it takes 1 h 3 min for the plate to attain a temperature of 100°C.

10.7 A manganese steel ball, 4 cm in diameter, is heated to 500°C in a furnace and then quenched in an oil at 80°C for 'surface hardening'. The ball is removed from the bath after the temperature at 3 mm below the surface has dropped to 250°C. If the surface heat transfer coefficient is 200 W/m² °C, calculate the quench time required. What is the maximum temperature in the ball at that instant? *Given:* $\rho = 7800$ kg/m³, $c_p = 450$ J/kg °C, $k = 45$ W/m °C.

10.8 A long steel rod, 10 cm in diameter, initially at 40°C, is drawn through a continuous annealing furnace in which the temperature is maintained at 800°C. The centre-line temperature of the rod must not be below 650°C when the rod leaves the furnace. If the heat transfer coefficient because of combined convection and radiation is 200 W/m² °C and the rod moves at 0.5 m/min, calculate the length of the furnace using Fig. 10.7. Repeat the calculation using the exact solution determined in Example 10.6 truncated after the second term of the series. *Given:* $\rho = 7830$ kg/m³, $c_p = 440$ J/kg °C, $k = 42$ W/m °C.

10.9 A thick concrete slab at 130°C is suddenly exposed to a cold environment so as to reduce its surface temperature to 15°C. Calculate the time required for the temperature 5 m below the surface to drop down to 50°C. What is the rate of heat loss from the surface at that moment? (Use an error function table for the calculation.)

10.10 A plastic sheet, 25 cm thick, moulded at a temperature of 160°C is cooled by a high velocity air stream at 25°C. If the surface heat transfer coefficient is 200 W/m² °C, calculate the time required for the average temperature of the sheet to drop down to 40°C. What is the maximum temperature of the sheet at that time. Use the analytical solution truncated after the first two terms and compare the results with those obtained by using the charts. *Given:* $\rho = 1050$ kg/m³; $c_p = 2.1$ kJ/kg °C; $k = 0.15$ W/m °C.

10.11 A cylindrical furnace of 3 m internal diameter has a 25 cm thick wall of thermal conductivity 2.5 kcal/h m °C. The furnace is lit at zero time and the flue gas temperature inside the furnace becomes 1400°C immediately. The wall receives heat from the hot flue gas and the heat transfer coefficient (convection and radiation combined) is 300 kcal/h m² °C. Because the outer wall heat transfer coefficient is very small during the start-up, the outer surface may be considered to be perfectly insulated. If the initial temperature of the wall is 30°C, calculate the time required for the inner surface temperature to reach 1200°C. *Given:* $\rho = 3000$ kg/m³ and $c_p = 0.35$ kcal/kg °C.

[*Hint:* The wall can be approximated by an infinite flat plate.]

10.12 Pipelines for municipal water supply in cold countries are laid deep in the ground so that the water does not freeze in the winter. At a certain place, the average temperature during the winter is estimated to be 3°C and the minimum temperature goes down to −15°C and continues for about a month. Calculate the depth at which the pipe should be laid in order to prevent freezing of water. For soil, $\rho = 2000$ kg/m³, $c_p = 0.5$ kcal/kg °C, and $k = 0.5$ kcal/h m °C.

10.13 A long cylindrical rod is to be annealed by putting it in a furnace. The initial temperature of the rod is 30°C and its average temperature increases to 700°C in half an hour. If the surface heat transfer coefficient is 90 kcal/h m² °C, calculate the diameter of the rod and the temperature of the furnace. *Given:* $\rho = 7900$ kg/m³, $c_p = 0.048$ kcal/kg °C, and $k = 0.17$ kcal/h m °C.

10.14 A *long* wooden bar of rectangular cross-section (100 mm by 25 mm) at a uniform initial temperature of 35°C suddenly comes in contact with a flame at 700°C. If the ignition temperature of the wood is 500°C, calculate the time after which the block starts burning.

  Given: surface heat transfer coefficient, $h = 15$ kcal/h m² °C, $\rho = 1090$ kg/m³, $c_p = 0.58$ kcal/kg °C, and $k = 0.31$ kcal/h m °C. Neglect any chemical decomposition process that may occur in the wood prior to ignition.

  [*Hint:* Rework Example 10.10 with convective boundary condition.]

REFERENCES AND FURTHER READING

Adams, J.A. and D.F. Rogers, *Computer-aided Heat Transfer Analysis*, McGraw-Hill, New York, 1973.

Carnahan, B., H.A. Luther, and J.O. Wilkes, *Applied Numerical Analysis*, John Wiley, New York, 1969.

Carslaw, H.S. and J.C. Jaeger, *Conduction of Heat in Solids*, Oxford University Press, London, 1959.

Crank, J., *The Mathematics of Diffusion*, Claredon Press, Oxford, 1974.

Grober, H., S. Erk, and U. Grigull, *Fundamentals of Heat Transfer*, McGraw-Hill, New York, 1961.

Longwell, P.A., *Mechanics of Fluid Flow*, McGraw-Hill, New York, 1966.

McGee, H., J. McInerney, and A. Harrus, "The virtual cook: modeling heat transfer in the kitchen", *Physics Today*, November 1999, 30–36.

Mikhailov, M.D. and M.N. Ozisik, *Unified Analysis and Solutions of Heat and Mass Diffusion*, John Wiley, New York, 1984.

Myers, G.E., *Analytical Methods in Conduction Heat Transfer*, McGraw-Hill, New York, 1971.

Ray, P., S.K. Bayen, B.K. Dutta, and A.S. Gupta, *Warme-und Stoffubertragung*, **22**, (1988), 195–199.

Schmidt, F.W. and A.J. Willmont, *Thermal Energy Storage and Regeneration*, Hemisphere Publ. Co., New York, 1981.

Schneider, P.J., *Conduction Heat Transfer*, Addison-Wesley, Reading, MA, 1955.

Shah, R.K. and A.L. London, *Laminar Flow Forced Convection in Ducts*, Academic Press, New York, 1978.

Wylie, C.R. and L.C. Barrett, *Advanced Engineering Mathematics*, 5th ed., McGraw-Hill, New York, 1982.

11

Boundary Layer Heat Transfer

Formation of hydrodynamic and thermal boundary layers in flow past an immersed surface has been introduced in Chapter 4. The rate of convective heat transfer and, therefore, the convective heat transfer coefficient are strongly governed by the velocity field near the immersed surface. In simple situations and geometries, convective heat transfer through a boundary layer is amenable to theoretical analysis and the rate of heat transfer can be calculated. The theoretical principles underlying boundary layer heat transfer and a few simple applications will be discussed in this chapter. As the theoretical calculations of the rate of heat transfer involve the solution of both hydrodynamic and thermal boundary layer equations, the relevant transport equations will be developed first. The hydrodynamic boundary layer equations originate from the equation of continuity and the equation of motion of a fluid.

11.1 THE EQUATION OF CONTINUITY

One of the fundamental assumptions underlying the mechanics of solids or fluids is that the material is a *continuum*, that is, there is no discontinuity or vacant space in the material or the medium. This assumption is not realistic if we consider a medium at molecular level, because it is known that every medium has intermolecular or interatomic vacant spaces which are the discontinuities in the medium. But, in a macroscopic sense, a common solid or fluid medium can be considered to be a continuum.

The equation of continuity has its basis in the concept of continuum and in the law of conservation of mass. It can be derived by writing a mass balance over a differential volume in a fluid.

Rate of input of mass – Rate of output of mass = Rate of accumulation of mass   (11.1)

Figure 11.1 shows a differential volume or *control volume* in a continuum. The control volume has the shape of a tiny rectangular parallelepiped having sides Δx, Δy, and Δz, respectively, in the rectangular Cartesian coordinate system. One corner of the control volume has the coordinates (x, y, z) as shown in the figure. The components of the fluid velocity at any location are denoted by u, v, and w along the x-, y-, and z-directions, respectively. Accordingly, we may visualize that the fluid enters the control volume through each of the three faces normal to the axes of coordinates and leaves through the respective opposite faces. Thus if we consider the two surfaces of the small volume element normal to the x-axis, we may write

Rate of input of mass at x through the area $\Delta y \Delta z = (u\rho \Delta y \Delta z)|_x$

and

Rate of output of mass at $x + \Delta x$ through the area $\Delta y \Delta z = (u\rho \Delta y \Delta z)|_{x+\Delta x}$

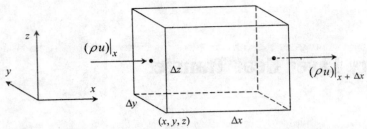

Fig. 11.1 A volume element of sides Δx, Δy and Δz selected for making the mass balance.

Similarly, considering the velocity component in the y-direction, the rate of input at $y = (v\rho\Delta x\Delta z)|_y$, and the rate of output of mass through the opposite surface $= (v\rho\Delta x\Delta z)|_{y + \Delta y}$.

Also in the z-direction, the rate of input at z through the area $\Delta x\Delta y = (w\rho\Delta x\Delta y)|_z$, and the rate of output of mass through the opposite surface $= (w\rho\Delta x\Delta y)|_{z + \Delta z}$.

The mass of the fluid contained in the element at any time $= \Delta x\Delta y\Delta z\rho$, and

$$\text{the rate of accumulation of mass in it} = \frac{\partial}{\partial t}(\Delta x\Delta y\Delta z\rho) = \Delta x\Delta y\Delta z\frac{\partial\rho}{\partial t}$$

Substituting the various quantities in the general mass balance Eq. (11.1), we have

$$(u\rho\Delta y\Delta z)|_x - (u\rho\Delta y\Delta z)|_{x + \Delta x} + (v\rho\Delta x\Delta z)|_y - (v\rho\Delta x\Delta z)|_{y + \Delta y}$$

$$+ (w\rho\Delta x\Delta y)|_z - (w\rho\Delta x\Delta y)|_{z + \Delta z} = \Delta x\Delta y\Delta z\frac{\partial\rho}{\partial t}$$

Dividing both sides by $\Delta x\Delta y\Delta z$, and taking limits $\Delta x \to 0$, $\Delta y \to 0$, and $\Delta z \to 0$, we get

$$\lim_{\Delta x \to 0} \frac{(u\rho)|_x - (u\rho)|_{x + \Delta x}}{\Delta x} + \lim_{\Delta y \to 0} \frac{(v\rho)|_y - (v\rho)|_{y + \Delta y}}{\Delta y} + \lim_{\Delta z \to 0} \frac{(w\rho)|_z - (w\rho)|_{z + \Delta z}}{\Delta z} = \frac{\partial\rho}{\partial t}$$

or

$$\frac{\partial\rho}{\partial t} = -\left[\frac{\partial}{\partial x}(u\rho) + \frac{\partial}{\partial y}(v\rho) + \frac{\partial}{\partial z}(w\rho)\right] \tag{11.2}$$

The above equation is called the *equation of continuity*. It relates the time rate of change in the fluid density with the spatial rates of changes in the mass velocities. If we use the vector notation, Eq. (11.2) reduces to the form

$$\frac{\partial\rho}{\partial t} = -\nabla \cdot \rho\mathbf{V} \tag{11.3}$$

where $\mathbf{V}$ is the velocity vector ($\mathbf{V} = iu + jv + kw$; i, j and k are unit vectors), and

$$\nabla = i\frac{\partial}{\partial x} + j\frac{\partial}{\partial y} + k\frac{\partial}{\partial z}$$

In Eq. (11.3), $\mathbf{V} \cdot \rho\mathbf{V}$ is called the *divergence* of the vector $\rho\mathbf{V}$. It is a vectorial operation. Besides, it has a physical significance. Divergence of the mass velocity vector means the *net rate* at which the fluid mass flows out (that is, 'diverges' out) of the volume element.

On differentiation of the RHS, Eq. (11.2) can be written in the following form:

$$\frac{\partial\rho}{\partial t} + u\frac{\partial\rho}{\partial x} + v\frac{\partial\rho}{\partial y} + w\frac{\partial\rho}{\partial z} = -\rho\left[\frac{\partial u}{\partial x} + \frac{\partial v}{\partial y} + \frac{\partial w}{\partial z}\right] \qquad (11.4)$$

The LHS of the above equation is called the *substantial derivative* of density or the rate of change of density along the path or flow line of a fluid element. From Eq. (11.4), we have

$$\frac{D\rho}{Dt} = -\rho\mathbf{V}\cdot\mathbf{V}, \ (\mathbf{V} = iu + jv + kw) \qquad (11.5)$$

where $D\rho/Dt$ is the substantial derivative and $\mathbf{V}\cdot\mathbf{V}$ is the divergence of the velocity vector $\mathbf{V}$. If the fluid is *incompressible*, the LHS of Eq. (11.4) [or Eq. (11.5)] is zero. Therefore,

$$\frac{\partial u}{\partial x} + \frac{\partial v}{\partial y} + \frac{\partial w}{\partial z} = \mathbf{V}\cdot\mathbf{V} = 0 \qquad (11.6)$$

11.2 THE EQUATION OF MOTION

Materials (including fluids) are deformed when acted upon by a force. The extent of deformation deperds upon the force applied per unit area, that is, upon the *stress*. The stresses that may act on a fluid may be *normal stress*, which is another name for static pressure, and shear stresses that act tangentially to the surface.

The forces that may act on a fluid are broadly of two types—*body forces*, which are proportional to the volume of fluid considered, and *surface forces* which depend upon the area. Gravitational and magnetic (in the case of an electrically conducting fluid) forces are examples of body forces. The static pressure applied to a fluid or tangential forces (viscous stresses) are surface forces. The equation of motion of a fluid is obtained by balancing the rate of change of momentum of a fluid element with the surface and the body forces acting on it. Before proceeding to develop these equations, it will be useful to discuss the stress components that act on a fluid element in motion.

The stresses on a small volume element (or control volume) which act along the *x*-direction are shown in Fig. 11.2. The stress acting on the surface normal to the *x*-direction (*y-z* plane) is

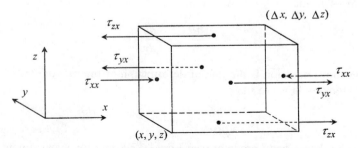

Fig. 11.2 Stress components acting in the *x*-direction.

denoted by τ_{xx}. It is the normal stress. The stresses acting on the z-x and x-y planes (parallel to the x-axis) are called shear stresses denoted by τ_{zx} and τ_{yx}, respectively. It may be noted that there are two subscripts on τ. The second subscript denotes the direction of action of the stress.

A fluid at rest can withstand the normal stress. But common fluids fail to withstand shear stress. They deform under shear stress, and the rate of deformation is proportional to the component of shear stress concerned. Figure 11.3 shows the deformation of a fluid element in one-dimensional flow. The element, having a rectangular cross-section $ABCD$ (Δx by Δy) at time

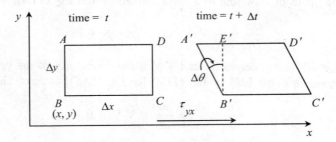

Fig. 11.3 Deformation in flow between two parallel plates.

t, undergoes an angular deformation $\Delta\theta$ after a time Δt. Over the time Δt, the point A moves to A' and the point B to B'; the surface $ABCD$ deforms to $A'B'C'D'$. This is because, in presence of the existing velocity gradient, the point A slows down, thus, deforming the cross-section. It is assumed that flow occurs in the x-direction only. The fluid velocity at $B = u|_y$, and that at $A = u|_{y+\Delta y}$. Therefore, we have

$$BB' = u|_y\,\Delta t; \qquad AA' = u|_{y+\Delta y}\,\Delta t; \qquad A'E' = u|_y\,\Delta t - u|_{y+\Delta y}\,\Delta t$$

and

$$\tan(\Delta\theta) = \frac{A'E'}{B'E'} = \frac{(u|_y - u|_{y+\Delta y})\Delta t}{\Delta y}$$

or

$$\Delta\theta = \tan^{-1}\frac{(u|_y - u|_{y+\Delta y})\Delta t}{\Delta y} \approx \frac{(u|_y - u|_{y+\Delta y})\Delta t}{\Delta y}$$

Taking limits $\Delta t \to 0$ and $\Delta y \to 0$, we have

$$\frac{d\theta}{dt} = -\frac{du}{dy} \tag{11.7}$$

The LHS of the above equation is the rate of deformation or the *rate of strain*. If τ_{yx} is the shear stress, then by Hooke's law, we have

$$\tau_{yx} \propto \frac{d\theta}{dt} \qquad \text{or} \qquad \tau_{yx} = G\frac{d\theta}{dt} \tag{11.8}$$

Replacing $d\theta/dt$ by the velocity gradient [see Eq. (11.7)], and the shear modulus G by μ in the case of a fluid, we have

$$\tau_{yx} = -\mu\frac{du}{dy} \tag{11.9}$$

The above equation is the well-known *Newton's law of viscosity*. A fluid which obeys this law is a *Newtonian fluid*.

When there is a velocity gradient in a fluid, the faster-moving layer pulls the adjacent slower-moving layer. It is in this way that momentum is transported from the former layer to the latter. The shear stress (which is equal to the force per unit area) acting on the surface causes this transport of momentum. The stress tends to increase the momentum of the layer and is regarded as the *momentum flux by diffusion*.*

We will now make a momentum balance over the control volume shown in Fig. 11.2 in order to develop the equation of motion. Momentum balance is based on the law of conservation of momentum which can be expressed in the following form [compare with Eq. (11.1) for mass balance over the control volume].

Rate of momentum input − Rate of momentum output + Sum of the forces
acting on the system = Rate of momentum accumulation (11.10)

The above general form can be used to balance each of the components of momentum. Two types of 'input' and 'output' terms should be considered—one accounting for *molecular transport* (i.e. momentum transport owing to the velocity gradient) or *diffusive transport* and the other accounting for *convective transport* or *bulk flow*. We now proceed to balance the x-component of momentum over the control volume in Fig. 11.2 (in which the rectangular Cartesian coordinate system is used).

Convective momentum input and output terms

The rate of input of fluid mass through the surface of area $\Delta y\Delta z$ at the position x on the x-axis is $(\rho u\Delta y\Delta z)|_x$ and the corresponding rate of input of x-momentum is $(\rho u\Delta y\Delta z \cdot u)|_x = (\rho u^2\Delta y\Delta z)|_x$. The rate of output of x-momentum by bulk flow through the opposite surface at $x + \Delta x$ (which is also of area $\Delta y\Delta z$) is $(\rho u^2\Delta y\Delta z)|_{x + \Delta x}$.

Now if we consider the surface of area $\Delta x\Delta z$ at the position y on the y-axis, the rate of input of fluid mass $= (\rho v\Delta x\Delta z)|_y$, and the corresponding rate of input of x-momentum is $(\rho v\Delta x\Delta z \cdot u)|_y$. The rate of output of x-momentum through the opposite surface (also of area $\Delta x\Delta z$) $= (\rho v\Delta x\Delta z \cdot u)|_{y + \Delta y}$. Similarly, considering the surface of area $\Delta x\Delta y$, the rate of input of x-momentum is $(\rho w\Delta x\Delta y \cdot u)|_z$, and the rate of output through the opposite surface is $(\rho w\Delta x\Delta y \cdot u)|_{z + \Delta z}$.

Diffusive momentum input and output terms

There are three stress components acting on the three surfaces of the control volume, namely, τ_{xx}, τ_{yx} and τ_{zx}. The first one (τ_{xx}) is the normal stress, and the latter two are the shear stresses. These

---

*Equation (11.9) can be written as $\tau_{yx} = -\dfrac{\mu}{\rho}\dfrac{d(u\rho)}{dy} = -v\dfrac{d(u\rho)}{dy}$, where $u\rho$ is the volumetric concentration of momentum. So, this equation of transport of momentum follows the law of diffusional transport.

stresses cause diffusional momentum transport as mentioned before. The rates of input of the x-component of momentum are:

$$\text{Through the surface } \Delta y \Delta z: (\tau_{xx} \cdot \Delta y \Delta z)|_x$$

$$\text{Through the surface } \Delta x \Delta z: (\tau_{yx} \cdot \Delta x \Delta z)|_y$$

$$\text{Through the surface } \Delta x \Delta y: (\tau_{zx} \cdot \Delta x \Delta y)|_z$$

The corresponding rates of output of the x-component of momentum through the opposite surfaces are

$$(\tau_{xx} \cdot \Delta y \Delta z)|_{x+\Delta x}, \ (\tau_{yx} \cdot \Delta x \Delta z)|_{y+\Delta y} \text{ and } (\tau_{zx} \cdot \Delta x \Delta y)|_{z+\Delta z}, \text{ respectively.}$$

Forces acting on the system

The forces acting on a system may be of various types. Here we shall consider only the pressure force and the gravitational force. If p is the local pressure, the net force on the element acting in the x-direction because of the fluid pressure is $(\Delta y \Delta z)p|_x - (\Delta y \Delta z)p|_{x+\Delta x}$. The gravitational force is given by $\Delta x \Delta y \Delta z \rho g_x$, where g_x is the component of g (acceleration due to gravity) acting in the x-direction.

Rate of momentum accumulation and momentum balance

The mass of the volume element or control volume being $\Delta x \Delta y \Delta z \rho$, its x-momentum is $(\Delta x \Delta y \Delta z \rho)u$. Hence, the rate of accumulation of x-momentum is the time rate of change in this quantity. All the input, output, applied force, and accumulation terms are now put in Eq. (11.10). Thus,

$$(\Delta x \Delta y \Delta z)\frac{\partial}{\partial t}(\rho u) = \Delta y \Delta z \left(\rho u^2|_x - \rho u^2|_{x+\Delta x}\right) + \Delta x \Delta z \left(\rho v u|_y - \rho v u|_{y+\Delta y}\right)$$

$$+ \Delta x \Delta y \left(\rho w u|_z - \rho w u|_{z+\Delta z}\right) + \Delta y \Delta z \left(\tau_{xx}|_x - \tau_{xx}|_{x+\Delta x}\right)$$

$$+ \Delta x \Delta z \left(\tau_{yx}|_y - \tau_{yx}|_{y+\Delta y}\right) + \Delta x \Delta y \left(\tau_{zx}|_z - \tau_{zx}|_{z+\Delta z}\right)$$

$$+ \Delta x \Delta y (p|_x - p|_{x+\Delta x}) + \Delta x \Delta y \Delta z \rho g_x$$

Dividing throughout by $\Delta x \Delta y \Delta z$ and taking limits $\Delta x \to 0$, $\Delta y \to 0$, and $\Delta z \to 0$, we get

$$\frac{\partial}{\partial t}(\rho u) = -\left[\frac{\partial}{\partial x}(\rho u^2) + \frac{\partial}{\partial y}(\rho u v) + \frac{\partial}{\partial z}(\rho u w)\right] - \left(\frac{\partial \tau_{xx}}{\partial x} + \frac{\partial \tau_{yx}}{\partial y} + \frac{\partial \tau_{zx}}{\partial z}\right) - \frac{\partial p}{\partial x} + \rho g_x$$

$$(11.11)$$

Equation (11.11) is the x-component of the equation of motion of a fluid. The y- and z-components of the equation can be easily obtained by balancing the respective momentum components according to the above procedure. These equations are

$$\frac{\partial}{\partial t}(\rho v) = -\left[\frac{\partial}{\partial x}(\rho u v) + \frac{\partial}{\partial y}(\rho v^2) + \frac{\partial}{\partial z}(\rho v w)\right] - \left(\frac{\partial \tau_{xy}}{\partial x} + \frac{\partial \tau_{yy}}{\partial y} + \frac{\partial \tau_{zy}}{\partial z}\right) - \frac{\partial p}{\partial y} + \rho g_y$$

$$(11.12)$$

$$\frac{\partial}{\partial t}(\rho w) = -\left[\frac{\partial}{\partial x}(\rho u w) + \frac{\partial}{\partial y}(\rho v w) + \frac{\partial}{\partial z}(\rho w^2)\right] - \left(\frac{\partial \tau_{xz}}{\partial x} + \frac{\partial \tau_{yz}}{\partial y} + \frac{\partial \tau_{zz}}{\partial z}\right) - \frac{\partial p}{\partial z} + \rho g_z$$

(11.13)

To complete the development of the equation of motion it is necessary to express the stresses (τ_{ij}; $i, j = x, y, z$) in terms of the velocity gradients. Such expressions can be derived by relating the deformation of a fluid element with the stresses to which it is subjected. These derivations are available in standard texts on fluid mechanics (see, for example, Longwell, 1966; Schlichting, 1968; Churchill, 1988.). Given below are the relations among the stress components and the velocity gradients.

Shear stress

$$\tau_{xy} = -\mu\left(\frac{\partial u}{\partial y} + \frac{\partial v}{\partial x}\right); \qquad \tau_{yz} = -\mu\left(\frac{\partial v}{\partial z} + \frac{\partial w}{\partial y}\right); \qquad \tau_{zx} = -\mu\left(\frac{\partial w}{\partial x} + \frac{\partial u}{\partial z}\right) \quad (11.14)$$

and

$$\tau_{xy} = \tau_{yx}; \qquad \tau_{yz} = \tau_{zy}; \qquad \tau_{zx} = \tau_{xz}$$

Normal stress

$$\tau_{xx} = p - \mu\left(2\frac{\partial u}{\partial x} - \frac{2}{3}\nabla \cdot \mathbf{V}\right) - \eta\nabla \cdot \mathbf{V} \tag{11.15}$$

where p is the static pressure, $\nabla \cdot \mathbf{V}$ is the divergence of the velocity vector (see Section 11.1), and η is called the *bulk viscosity*. Similarly, we can write the other two normal stress components (τ_{yy} and τ_{zz}).

If we consider the motion of an incompressible fluid, then $\nabla \cdot \mathbf{V} = 0$ [see Eq. (11.6)], and ρ = constant. Substituting Eqs. (11.14) and (11.15) in Eq. (11.11), we obtain the following form of the x-component of the equation of motion.

$$\frac{\partial u}{\partial t} + u\frac{\partial u}{\partial x} + v\frac{\partial u}{\partial y} + w\frac{\partial u}{\partial z} = -\frac{1}{\rho}\frac{\partial p}{\partial x} + \frac{\mu}{\rho}\left[\frac{\partial^2 u}{\partial x^2} + \frac{\partial^2 u}{\partial y^2} + \frac{\partial^2 u}{\partial z^2}\right] + g_x \tag{11.16}$$

The y- and z-components of the equation are

$$\frac{\partial v}{\partial t} + u\frac{\partial v}{\partial x} + v\frac{\partial v}{\partial y} + w\frac{\partial v}{\partial z} = -\frac{1}{\rho}\frac{\partial p}{\partial y} + \frac{\mu}{\rho}\left[\frac{\partial^2 v}{\partial x^2} + \frac{\partial^2 v}{\partial y^2} + \frac{\partial^2 v}{\partial z^2}\right] + g_y \tag{11.17}$$

$$\frac{\partial w}{\partial t} + u\frac{\partial w}{\partial x} + v\frac{\partial w}{\partial y} + w\frac{\partial w}{\partial z} = -\frac{1}{\rho}\frac{\partial p}{\partial z} + \frac{\mu}{\rho}\left[\frac{\partial^2 w}{\partial x^2} + \frac{\partial^2 w}{\partial y^2} + \frac{\partial^2 w}{\partial z^2}\right] + g_z \tag{11.18}$$

The equation of motion in any other coordinate system can be obtained from the above equations by using the corresponding rules of coordinate transformation. The equations of motions in the cylindrical and the spherical polar coordinate system can be found in Bird et al. (1960).

11.3 BOUNDARY LAYER FLOW OVER A FLAT PLATE

The simplest case of boundary layer flow is the flow of a fluid past a flat plate (Fig. 11.4). Despite its simplicity, this flow situation appears in a number of engineering applications. Furthermore, the flow over a slightly curved surface can be approximated by that over a flat plate with reasonable accuracy. The physical description of the boundary layer flow over a flat plate has been given in Chapter 4. Here we shall write down the simplified equation of motion for boundary layer flow of a *Newtonian fluid* over a flat plate. This will be followed by solution of the equation in order to obtain the velocity distribution (or velocity fields) within the boundary layer and also the boundary layer thickness.

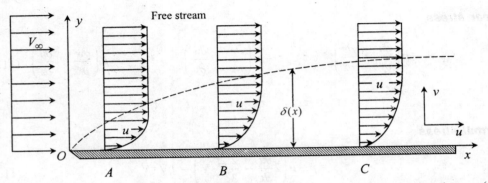

Fig. 11.4   Laminar boundary on a flat plate (the region between the plate and the dotted curve is the boundary layer).

11.3.1 Differential Equations for Laminar Boundary Layer Flow

The concept of boundary layer and the related theory were originally introduced by L. Prandtl in 1904. Prandtl suggested that the flow of a fluid, whose viscosity is not high, past an immersed surface consists of two zones—a thin region near the boundary where the effects of viscosity are prominent, and an outer region away from the surface where the influence of viscosity is very small and therefore negligible. Most of the fluid flows through this outer zone and is called the *bulk fluid*. The bulk flow is virtually inviscid (or non-viscous). It may be noted that the above kind of boundary layer flow occurs at a pretty high Reynolds number.

Laminar boundary layer flow over a flat plate is 'two-dimensional', that is, there is a velocity component u along the plate (i.e. the x-direction), and another velocity component v normal to the plate (along the y-axis). In order to develop simplified equations of boundary layer flow, the following assumptions are made:

(i) The fluid approaches the plate at *zero angle of incidence*, that is , the flow direction of the bulk fluid is along the plate.

(ii) The plate is wide. So the velocity component in the z-direction (across the plate), w, is zero.

(iii) Steady state flow occurs (i.e. $\partial u/\partial t = \partial v/\partial t = 0$)

(iv) There is no body force.

Only the x- and the y-components of the equation of motion need to be considered. From Eqs. (11.16) and (11.17), we may write

$$u\frac{\partial u}{\partial x} + v\frac{\partial u}{\partial y} = -\frac{1}{\rho}\frac{\partial p}{\partial x} + \frac{\mu}{\rho}\left[\frac{\partial^2 u}{\partial x^2} + \frac{\partial^2 u}{\partial y^2}\right] \tag{11.19}$$

$$u\frac{\partial v}{\partial x} + v\frac{\partial v}{\partial y} = -\frac{1}{\rho}\frac{\partial p}{\partial y} + \frac{\mu}{\rho}\left[\frac{\partial^2 v}{\partial x^2} + \frac{\partial^2 v}{\partial y^2}\right] \tag{11.20}$$

The equations are still not simple. In order to make further simplifications, Prandtl adopted an *order-of-magnitude analysis* of the terms of the above equations. The basic idea is that the terms which are much smaller than others in each of the equations can be neglected. The details of the order-of-magnitude analysis (Longwell, 1966, Schlichting, 1968, Bejan, 1984) will not be given here; only the salient points are mentioned.

  (i) The velocity component u is much larger than the component v inside the boundary layer (i.e. $u \gg v$).

 (ii) The transverse gradient of u is much larger than the other velocity gradients. That is,

$$\frac{\partial u}{\partial y} \gg \frac{\partial u}{\partial x}, \frac{\partial v}{\partial x}, \frac{\partial v}{\partial y} \qquad \text{and} \qquad \frac{\partial^2 u}{\partial y^2} \gg \frac{\partial^2 u}{\partial x^2}$$

(iii) All the terms in Eq. (11.20) are negligibly small, and this equation need not be considered in the solution of the boundary layer flow problem.

The simplified form of Eq. (11.19) obtained on the above basis is given below.

$$u\frac{\partial u}{\partial x} + v\frac{\partial u}{\partial y} = -\frac{1}{\rho}\frac{\partial p}{\partial x} + \frac{\mu}{\rho}\frac{\partial^2 u}{\partial y^2} \tag{11.21}$$

Further, if there is no externally imposed pressure gradient (i.e. if $\partial p/\partial x = 0$), we have

$$u\frac{\partial u}{\partial x} + v\frac{\partial u}{\partial y} = v\frac{\partial^2 u}{\partial y^2} \tag{11.22}$$

and

$$\frac{\partial u}{\partial x} + \frac{\partial v}{\partial y} = 0 \tag{11.23}$$

where $v = \mu/\rho$ is the kinematic viscosity or momentum diffusivity; Eq. (11.23) is the equation of continuity for two-dimensional steady state flow of an incompressible fluid.

The following boundary conditions are applicable:

$$u = v = 0 \qquad \text{at } y = 0 \tag{11.24}$$

$$u = V_\infty \qquad \text{at } y \to \infty \tag{11.25}$$

The first condition indicates that the fluid velocities are zero at the plate surface (*no-slip condition*); the second one means that the fluid velocity is the same as the *free-stream velocity* V_∞, at a point 'far away' from the surface. The free stream velocity is, in fact, the bulk fluid velocity. The flow outside the boundary layer is also called *potential flow*.

11.3.2 Solution of the Boundary Layer Equations

The Prandtl boundary layer equation (Eq. 11.22) together with the equation of continuity (Eq. 11.23) was first solved by Blasius in 1908 in his doctoral dissertation. Blasius used the principle called the *similarity principle* (see Section 10.8) to solve these equations. The solution is called the *similarity solution*. Details of the principle of similarity are available in the literature (Schlichting, 1968; Bejan, 1984). A similarity solution exists for a system if it has an internal or self-similitude. It means that the plots of the dependent variable at two locations in the system will look similar except for a certain scale factor which is a function of another independent variable. In the case of laminar boundary layer flow, the profiles of longitudinal velocity u (which depend on the independent variables x and y) at the locations A, B, and C (Fig. 11.4) look similar in the sense that the profile at B (or at C) may be obtained by *stretching* the profile at A. The stretching is done by using a scale factor. Here the transverse coordinate y is *scaled* by $\delta(x)$, which is the local boundary layer thickness and is a function of x. To this effect, Blasius defined a dimensionless variable η such that the velocity profile is uniquely given as

$$\frac{u}{V_\infty} = \phi(\eta), \qquad \eta = \frac{y}{\delta(x)} \qquad (11.26)$$

To start the procedure of solution, we first define a stream function $\Psi(x, y)$ which satisfies the equation of continuity, Eq. (11.23) [see Schlichting (1968) for further details]. The partial derivatives of the stream function give the respective velocity components.

$$u = \frac{\partial \Psi}{\partial y}, \qquad v = -\frac{\partial \Psi}{\partial x} \qquad (11.27)$$

where

$$\Psi = (vxV_\infty)^{1/2} f(\eta) \qquad (11.28)$$

with

$$\eta = \frac{y}{(xv/V_\infty)^{1/2}} \qquad (11.29)$$

The above definition of η follows from Eq. (11.26) and also from the fact that by an order-of-magnitude analysis the boundary layer thickness can be found to be $\delta(x) \sim (xv/V_\infty)^{1/2}$ [the sign '$\sim$' means "is of the order of"]. The dimensionless variable η is called the *similarity variable*. From Eqs. (11.27) through (11.29), we get

$$u = \frac{\partial \Psi}{\partial y} = \frac{\partial \Psi}{\partial \eta}\frac{\partial \eta}{\partial y} = (vxV_\infty)^{1/2}\frac{df}{d\eta}\left(\frac{V_\infty}{vx}\right)^{1/2} = V_\infty \frac{df}{d\eta} \qquad (11.30)$$

$$v = -\frac{\partial \Psi}{\partial x} = -\frac{\partial \Psi}{\partial \eta}\frac{\partial \eta}{\partial x} = -\left[(vxV_\infty)^{1/2}\frac{df}{d\eta}\frac{d\eta}{dx} + f\frac{d}{dx}(vxV_\infty)^{1/2}\right]$$

or

$$v = \frac{1}{2}\frac{yV_\infty}{x}\frac{df}{d\eta} - \frac{1}{2}\left(\frac{V_\infty v}{x}\right)^{1/2} f = \frac{1}{2}\left(\frac{V_\infty v}{x}\right)^{1/2}\left[\eta\frac{df}{d\eta} - f\right] \tag{11.31}$$

On further differentiation and simplification, we obtain

$$\frac{\partial u}{\partial x} = -\frac{V_\infty}{2x}\eta\frac{d^2 f}{d\eta^2} \tag{11.32}$$

$$\frac{\partial u}{\partial y} = V_\infty\left(\frac{V_\infty}{vx}\right)^{1/2}\frac{d^2 f}{d\eta^2} \tag{11.33}$$

$$\frac{\partial^2 u}{\partial y^2} = \frac{V_\infty^2}{vx}\frac{d^3 f}{d\eta^3} \tag{11.34}$$

Substituting the expressions for u, v, $\partial u/\partial x$, $\partial u/\partial y$, and $\partial^2 u/\partial y^2$ in Eq. (11.22), we obtain a third order nonlinear ordinary differential equation for f. It is to be noted that if the similarity variable exists, the partial differential equation representing the system can be reduced to an ordinary differential equation

$$\frac{1}{2}f\frac{d^2 f}{d\eta^2} + \frac{d^3 f}{d\eta^3} = 0 \tag{11.35}$$

The boundary conditions on the function $f(\eta)$ can be obtained from Eqs. (11.24) and (11.25) together with Eqs. (11.30) and (11.31). Thus from Eqs. (11.24) and (11.30), we have

$$\text{at } y = 0 \quad \text{or} \quad \eta = 0, \quad u = 0, \quad \text{i.e.} \quad df/d\eta = 0 \tag{11.36}$$

From Eqs. (11.24), (11.31), and (11.36), we have

$$\text{at } y = 0 \quad \text{or} \quad \eta = 0, \quad v = 0, \quad \text{i.e.} \quad \eta df/d\eta - f = 0 \quad \text{or} \quad f = 0 \tag{11.37}$$

From Eqs. (11.25) and (11.30), we have

$$\text{at } y \to \infty \quad \text{or} \quad \eta \to \infty, \quad u = V_\infty, \quad \text{i.e.} \quad df/d\eta = 1 \tag{11.38}$$

Equation (11.35) subject to the boundary conditions (11.36), (11.37), and (11.38) does not admit of an analytical solution. Numerical solutions for f, f' ($= df/d\eta$), and f'' ($= d^2f/d\eta^2$) have been obtained by a number of researchers. Representative results are given in Table 11.1, and a plot of the dimensionless longitudinal velocity distribution (u/V_∞ vs. η) is shown in Fig. 11.5. Table 11.1 can be used to determine the velocities u and v at any location in the boundary layer.

Longitudinal velocity in the laminar boundary layer over a flat plate has been measured experimentally by many researchers (using instruments like a hot-wire anemometer). A set of experimental data is shown in Fig. 11.5, and the Blasius solution is also plotted for comparison. The very good match between the theoretical results and the experimental data validates the assumptions and approach of Blasius.

Table 11.1 The Blasius function and its derivatives for boundary layer flow over a flat plate

$\eta = \dfrac{y}{(xv/V_\infty)^{1/2}}$	f	$f' = u/V_\infty$	f''
0.0	0.0	0.0	0.33206
0.4	0.02656	0.06641	0.33147
0.8	0.10611	0.26471	0.32730
1.2	0.23795	0.38378	0.31659
1.6	0.42032	0.51676	0.29667
2.0	0.65003	0.62977	0.26675
2.4	0.92230	0.72899	0.22809
2.8	1.23099	0.81152	0.18401
3.2	1.56911	0.87609	0.13913
3.6	1.92954	0.92333	0.09809
4.0	2.30576	0.95552	0.06424
4.4	2.69238	0.97587	0.03897
4.8	3.08534	0.98779	0.02187
5.0	3.28329	0.99155	0.01591
5.4	3.68094	0.99616	0.00793
6.0	4.27964	0.99898	0.00240

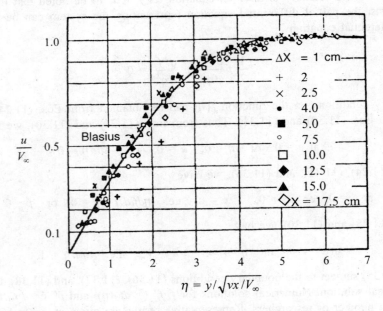

Fig. 11.5 Dimensionless longitudinal velocity distribution in the laminar boundary layer over a flat plate—Blasius solution and the experimental data ($V_\infty = 8$ m/s).

11.3.3 Boundary Layer Thickness and Drag Coefficient

Qualitatively, the boundary layer thickness is the distance from the plate within which the longitudinal velocity u changes from zero at the plate surface to the free stream velocity V_∞ at the edge of the boundary layer. However, the condition given by Eq. (11.25) shows that this distance is infinite in the theoretical sense. To make it practically meaningful, the boundary layer thickness is defined as the distance from the plate where 99% of the free stream velocity is attained*. Referring to Fig. 11.5 and Table 11.1, it is seen that $f'(\eta)$ or u/V_∞ attains a value of at 0.99 $\eta \approx 5.0$. So, we take the boundary layer thickness $y = \delta$ when $\eta = 5.0$. Thus,

$$\eta = y(V_\infty/vx)^{1/2} = \delta(V_\infty/vx)^{1/2} = 5.0$$

or

$$\delta = \delta(x) = 5.0\,(vx/V_\infty)^{1/2} = 5.0\,x(\mathrm{Re}_x)^{-1/2} \tag{11.39}$$

It appears from the above equation that the thickness of the boundary at any point on the plate is proportional to the square root of the distance of the point from the leading edge of the plate.

The frictional drag on the plate is determined from the shear stress at the plate surface (i.e. at $y = 0$ or $\eta = 0$). Thus,

$$\left[\tau_{yx}\right]_{y=0} = \tau_0 = \mu\left[\frac{\partial u}{\partial y}\right]_{y=0} \tag{11.40}**$$

From Eq. (11.33), we get

$$\left[\frac{\partial u}{\partial y}\right]_{\eta=0} = V_\infty\left(\frac{V_\infty}{vx}\right)^{1/2} f''(0) \tag{11.41}$$

Putting the above result in Eq. (11.40) and using the value of $f''(0)$ from Table 11.1, we obtain

$$\tau_0 = 0.332\mu V_\infty\left(\frac{V_\infty}{vx}\right)^{1/2} = 0.332\rho V_\infty^2(\mathrm{Re}_x)^{-1/2} \tag{11.42}$$

where Re_x is the local Reynolds number defined as

$$\mathrm{Re}_x = \frac{xV_\infty}{v} \tag{11.43}$$

If the length of the plate (oriented in the direction of flow) is l, the total drag force F_D per unit breadth of the plate (considering flow on one side of the plate) is obtained from Eq. (11.42) as

$$F_D = \int_0^l \tau_0\,dx = 0.332\,\rho V_\infty^2\left(\frac{V_\infty}{v}\right)^{1/2}\int_0^l x^{-1/2}\,dx = 0.664\mu V_\infty(\mathrm{Re}_l)^{1/2} \tag{11.44}$$

where Re_l = plate Reynolds number = $\dfrac{lV_\infty}{v}$ $\tag{11.45}$

The drag coefficient C_D for skin friction is defined by the equation

$$F_D = C_D A\left[\frac{1}{2}\rho V_\infty^2\right]$$

---

* However, this definition, though generally accepted, is arbitrary. One may very well take the boundary layer thickness as the distance from the plate where 99.9% of the free stream velocity is attained.

** No negative sign in included because u increases with y.

or

$$C_D = \frac{\text{Drag force}}{(\text{Area})(\text{Kinetic energy per unit volume})}$$

$$= \frac{F_D}{(l)(\rho V_\infty^2/2)} = \frac{0.664\, \mu V_\infty (\text{Re}_l)^{1/2}}{l(\rho V_\infty^2/2)} = 1.328\,(\text{Re}_l)^{-1/2} \qquad (11.46)$$

The boundary layer thickness $\delta(x)$ defined here is more specifically called the *momentum thickness*. Two other types of boundary layer thicknesses have also been defined. These are called the *displacement thickness* and the *kinetic energy thickness* (see Schlichting, 1968).

Example 11.1 Water at 25°C flows at a velocity of 1 m/s on both sides of a flat plate, 1 m long 0.5 m wide, at zero angle of incidence. Calculate:

 (i) the boundary layer thickness midway of the plate, at $x = 0.5$ m,

 (ii) the local drag coefficient at $x = 0.5$ m,

 (iii) the force required to hold the plate in position, and

 (iv) the shear stress at a plane, distant $\delta/2$ from the surface at $x = 0.5$ m.

SOLUTION For water at 25°C, the momentum diffusivity, $\nu = 8.63 \times 10^{-7}$ m²/s
Given: $V_\infty = 1$ m/s. The local Reynolds number at $x = 0.5$ m is

$$\text{Re}_x = \frac{xV_\infty}{\nu} = \frac{(0.5)(1.0)}{8.63 \times 10^{-7}} = 5.8 \times 10^5$$

 (i) From Eq. (11.39), the boundary layer thickness is

$$\delta = 5.0 \frac{x}{(\text{Re}_x)^{1/2}} = \frac{(5)(0.5)}{(5.8 \times 10^5)^{1/2}} = 0.0033 \text{ m} = \boxed{3.3 \text{ mm}}$$

 (ii) The local drag coefficient can be expressed as

$$C_D = \frac{\text{Local drag force per unit area}}{\text{Kinetic energy per unit volume}} = \frac{[\tau_0]_x}{(1/2)\rho V_\infty^2}$$

Using Eq. (11.42) for the wall shear stress, we get

$$C_D = \frac{0.332\rho V_\infty^2 (\text{Re}_x)^{-1/2}}{(1/2)\rho V_\infty^2} = 0.664(5.8 \times 10^5)^{-1/2} = \boxed{8.72 \times 10^{-4}}$$

(iii) From Eq. (11.44), the drag force acting on one side of the plate is

$$F_D = 0.664\mu V_\infty (\mathrm{Re}_l)^{1/2}.w$$
$$= (0.664)(8.6 \times 10^{-4})(1)(1.16 \times 10^6)^{1/2}(0.5)$$
$$= 0.307 \text{ N}$$

where

w = plate breadth = 0.5 m
$\mu = 8.6 \times 10^{-4}$ kg/m s
$\mathrm{Re}_l = lV_\infty/v = 1.16 \times 10^6$

The total drag force acting on both sides of the plate

$$= (2)(0.307) = \boxed{0.614 \text{ N}} = \text{force required to hold the plate in position.}$$

(iv) Shear stress at any point in the boundary layer

$$\tau_{yx} = \mu \frac{\partial u}{\partial y} = \mu \frac{\partial u}{\partial \eta} \frac{\partial \eta}{\partial y}$$

$$= \mu V_\infty (V_\infty/vx)^{1/2} f''(\eta) \qquad \text{[see Eq. (11.33)}$$

At a point halfway in the boundary layer, i.e. at $x = 0.5$ m,

$$y = \delta/2 = 0.0033/2 = 0.00165 \text{ m}$$

$$\eta = \frac{y}{(vx/V_\infty)^{1/2}} = \frac{(0.00165)(1)^{1/2}}{[(8.63 \times 10^{-7})(0.5)]^{1/2}} = 2.5$$

From Table 11.1, at $\eta = 2.5$, $f''(\eta) = 0.218$. Therefore,

$$\tau_{xy} = (8.6 \times 10^{-4})(1) \left[\frac{1}{(8.63 \times 10^{-7})(0.5)} \right]^{1/2} (0.218) = \boxed{0.285 \text{ N/m}^2}$$

11.3.4 The Momentum Integral Equation

von Karman (1921) and Pohlhausen (1921) developed an approximate, nonetheless much simpler, method of solution of the boundary layer equations. This method, called the *integral method*, gives a solution that satisfies the differential equations (11.22) and (11.23), *on the average*, over the boundary layer thickness. The first step is to integrate Eq. (11.22) over the boundary layer thickness to obtain the *momentum integral equation*. It is again assumed that there is no externally imposed pressure gradient (i.e. $\partial p/\partial x = 0$).

$$\int_0^\delta u \frac{\partial u}{\partial x} dy + \int_0^\delta v \frac{\partial u}{\partial y} dy = v \int_0^\delta \frac{\partial^2 u}{\partial y^2} dy \tag{11.47}$$

The continuity equation, (11.23), is also integrated from $y = 0$ to $y = \delta$.

$$\int_0^\delta \frac{\partial u}{\partial x} dy + \int_0^\delta \frac{\partial v}{\partial y} dy = 0 \tag{11.48}$$

In the following mathematical treatment of the problem, the boundary condition (11.24) will be used as such. Instead of Eq. (11.25), the following conditions are used at the edge of the boundary layer.

$$y = \delta: \qquad u = V_\infty, \quad v = v_\delta, \quad \partial u/\partial y = 0 \qquad\qquad (11.49)$$

The condition $\partial u/\partial y = 0$ at $y = \delta$ is necessary for the continuity of shear stress at the edge of the boundary layer.

From Eq. (11.48), we have

$$\int_0^\delta \frac{\partial u}{\partial x} \, dy + [v]_{y=0}^{y=\delta} = 0$$

or

$$v_\delta = -\int_0^\delta \frac{\partial u}{\partial x} dy \qquad\qquad (11.50)$$

Because,

$$(uv) = \int \frac{\partial}{\partial y}(uv) \, dy = \int u \frac{\partial v}{\partial y} \, dy + \int v \frac{\partial u}{\partial y} \, dy$$

using the equation of continuity, we have

$$\int_0^\delta v \frac{\partial u}{\partial y} \, dy = [uv]_{y=0}^\delta - \int_0^\delta u \frac{\partial v}{\partial y} \, dy = V_\infty v_\delta - \int_0^\delta u \left(-\frac{\partial u}{\partial x}\right) dy$$

$$= -V_\infty \int_0^\delta \frac{\partial u}{\partial x} \, dy + \int_0^\delta u \frac{\partial u}{\partial x} \, dy$$

Substituting the above result in Eq. (11.47), we obtain

$$\int_0^\delta u \frac{\partial u}{\partial x} dy - V_\infty \int_0^\delta \frac{\partial u}{\partial x} dy + \int_0^\delta u \frac{\partial u}{\partial x} dy = v \left[\frac{\partial u}{\partial y}\right]_{y=0}^\delta$$

Using the condition (11.49) and Eq. (11.40) on the RHS of the above equation, we get

$$\int_0^\delta 2u \frac{\partial u}{\partial x} dy - V_\infty \int_0^\delta \frac{\partial u}{\partial x} dy = -v \left[\frac{\partial u}{\partial y}\right]_{y=0} = -\frac{\tau_0}{\rho}$$

or

$$\int_0^\delta \frac{\partial}{\partial x}(V_\infty u) \, dy - \int_0^\delta \frac{\partial u^2}{\partial x} dy = \frac{\tau_0}{\rho}$$

Changing the order of differentiation and integration, we get

$$\frac{d}{dx} \int_0^\delta (V_\infty u) \, dy - \frac{d}{dx} \int_0^\delta u^2 dy = \frac{\tau_0}{\rho}$$

or

$$\frac{d}{dx} \int_0^\delta (V_\infty - u) u \, dy = \frac{\tau_0}{\rho} \tag{11.51}$$

or

$$\frac{d}{dx} \int_0^\delta (V_\infty - u) \rho u \, dy = \tau_0 \tag{11.52}$$

Equation (11.51) [or (11.52)] is called the *von Karman momentum integral equation*. The equation has a physical importance; it indicates that the relative reduction of the total rate of momentum input at any section of the boundary layer is balanced by the shear stress at the plate surface. This equation can be used to obtain the velocity distribution, boundary layer thickness, and drag coefficient by starting with an assumed functional form of the dimensionless velocity, u/V_∞, in terms of the dimensionless variable, y/δ. As an illustration, let us choose a third degree polynomial function for the velocity profile.

$$\frac{u}{V_\infty} = a_0 + a_1 \left(\frac{y}{\delta}\right) + a_2 \left(\frac{y}{\delta}\right)^2 + a_3 \left(\frac{y}{\delta}\right)^3 \tag{11.53}$$

The following boundary conditions are satisfied by u.

$$y = 0: \quad u = 0, \quad \frac{\partial^2 u}{\partial y^2} = 0^* \tag{11.54}$$

$$y = \delta: \quad u = V_\infty, \quad \frac{\partial u}{\partial y} = 0 \quad \frac{\partial^2 u}{\partial y^2} = 0, \text{ etc.} \tag{11.55}$$

From Eqs. (11.53) and (11.54),

$$a_0 = 0$$

From Eqs. (11.53) and (11.55), we get

$$1 = a_1 + a_2 + a_3 \tag{11.56}$$
$$0 = a_1 + 2a_2 + 3a_3 \tag{11.57}$$
$$0 = 2a_2$$

Solving the above equations, we get

$$a_1 = 3/2; \quad a_2 = 0; \quad a_3 = -1/2 \tag{11.58}$$

Substituting the values of these coefficients in Eq. (11.53), the velocity distribution is given by

$$\frac{u}{V_\infty} = \frac{3}{2}\left(\frac{y}{\delta}\right) - \frac{1}{2}\left(\frac{y}{\delta}\right)^3 \tag{11.59}$$

The following expression for the wall shear stress can be obtained from the above equation.

$$\tau_0 = \mu \left[\frac{\partial u}{\partial y}\right]_{y=0} = \frac{3\mu V_\infty}{2\delta} \tag{11.60}$$

---

*This condition follows from Eq. (11.22), because $u = v = 0$ at $y = 0$.

From Eq. (11.51), we obtain

$$V_\infty^2 \frac{d}{dx} \int_0^\delta \left(1 - \frac{u}{V_\infty}\right) \frac{u}{V_\infty} dy = \frac{3\mu V_\infty}{2\rho\delta}$$

Using Eqs. (11.59) and (11.60), and substituting $\zeta = y/\delta$, we have

$$V_\infty^2 \frac{d}{dx} \int_{\zeta=0}^1 \delta \left(1 - \frac{3}{2}\zeta + \frac{1}{2}\zeta^3\right)\left(\frac{3}{2}\zeta - \frac{1}{2}\zeta^3\right) d\zeta = \frac{3\mu V_\infty}{2\rho\delta}$$

Integrating with respect to ζ, we get

$$\frac{39}{280} V_\infty^2 \frac{d\delta}{dx} = \frac{3\mu V_\infty}{2\rho\delta} \tag{11.61}$$

The above equation can be integrated using the condition: $x = 0$, $\delta = 0$.

$$\int_{\delta=0}^\delta \delta \, d\delta = \frac{140}{13} \frac{\mu}{\rho V_\infty} \int_{x=0}^x dx$$

or

$$\frac{1}{2}\delta^2 = \frac{140}{13} \frac{\mu x}{\rho V_\infty}$$

or

$$\delta = \left(\frac{280}{13} \frac{\mu x}{\rho V_\infty}\right)^{1/2} = 4.64 \frac{x}{\sqrt{x V_\infty / v}} = 4.64 \frac{x}{(\mathrm{Re}_x)^{1/2}} \tag{11.62}$$

This relation for the boundary layer thickness δ compares very well with that obtained from the exact solution as given by Eq. (11.39). The local wall shear stress and the average drag coefficient can be determined following the procedure given before.

$$\tau_0 = \mu \left[\frac{\partial u}{\partial y}\right]_{y=0} = \frac{3\mu V_\infty}{2\delta} = 0.323 \, \rho V_\infty^2 (\mathrm{Re}_x)^{-1/2} \tag{11.63}$$

$$C_D = \frac{\int_0^l \tau_0 \, dx}{(l)(\rho V_\infty^2 / 2)} = 1.292 \, (\mathrm{Re}_l)^{-1/2} \tag{11.64}$$

The relations for τ_0 and C_D match closely with Eqs. (11.42) and (11.46) obtained from the exact solution. Thus, it is seen that the momentum integral method is not only simple but also quite accurate for solution of the laminar boundary layer equations. As mentioned in Chapter 4, the boundary layer over a flat plate remains laminar up to a Reynolds number of about 10^5 to 5×10^5 (depending upon the roughness of the surface) and beyond which the flow gradually turns turbulent.

11.4 FORCED CONVECTION HEAT TRANSFER IN LAMINAR BOUNDARY LAYER FLOW OVER A FLAT PLATE

If a cold fluid flows past a hot flat plate at a reasonably high velocity, a thermal boundary layer is formed on the plate, in addition to the momentum boundary layer. The phenomenon has been qualitatively explained in Chapter 4. The rate of heat flow is greatly influenced by the flow field in the boundary layer. The temperature of the fluid in contact with the plate remains the same as that of the plate. The fluid temperature drops down to the bulk temperature or the 'free stream temperature' at the edge of the boundary layer. In general, the thickness of the thermal boundary layer will be different from that of the velocity boundary layer; the ratio of the thicknesses of the two boundary layers depends upon the Prandtl number.

11.4.1 Boundary Layer Temperature Equation

The differential equation that governs the temperature distribution in the two-dimensional boundary layer can be derived by making a heat balance over a small control volume within the boundary layer. Heat input to and output from the control volume occurs by both convection and conduction (or 'diffusion') along the x- and the y-directions. Let the control volume have a size Δx by Δy by 1, and let q_x and q_y denote the conduction heat fluxes in the respective directions. The various heat input and output terms are written below assuming steady state conditions (see Fig. 11.6). Heat generation by viscous dissipation of momentum is neglected.

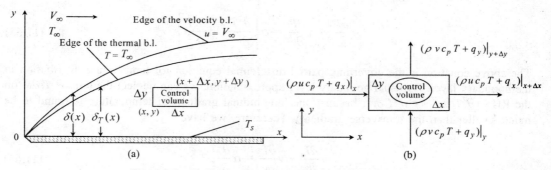

Fig. 11.6 Control volume in a thermal boundary layer.

	Convection terms	*Conduction terms*
Rate of heat input at x	$= (\Delta y \cdot \rho u c_p T)\big|_x$	$+ (\Delta y \cdot q_x)\big|_x$
Rate of heat output at $x + \Delta x$	$= (\Delta y \cdot \rho u c_p T)\big|_{x+\Delta x}$	$+ (\Delta y \cdot q_x)\big|_{x+\Delta x}$
Rate of heat input at y	$= (\Delta x \cdot \rho v c_p T)\big|_y$	$+ (\Delta x \cdot q_y)\big|_y$
Rate of heat output at $y + \Delta y$	$= (\Delta x \cdot \rho v c_p T)\big|_{y+\Delta y}$	$+ (\Delta x \cdot q_y)\big|_{y+\Delta y}$
Rate of heat accumulation	$= 0$ (at steady state)	

By a steady state heat balance, we get

$$(\Delta y \cdot \rho u c_p T)\big|_x + (\Delta y \cdot q_x)\big|_x + (\Delta x \cdot \rho v c_p T)\big|_y + (\Delta x \cdot q_y)\big|_y$$
$$= (\Delta y \cdot \rho u c_p T)\big|_{x+\Delta x} + (\Delta y \cdot q_x)\big|_{x+\Delta x} + (\Delta x \cdot \rho v c_p T)\big|_{y+\Delta y} + (\Delta x \cdot q_y)\big|_{y+\Delta y}$$

Dividing both sides by $\Delta x \Delta y$ and rearranging, we obtain

$$\frac{(uT)|_{x+\Delta x} - (uT)|_x}{\Delta x} + \frac{(vT)|_{y+\Delta y} - (vT)|_y}{\Delta y} = \frac{1}{\rho c_p}\left[\frac{q_x|_x - q_x|_{x+\Delta x}}{\Delta x} + \frac{q_y|_y - q_y|_{y+\Delta y}}{\Delta y}\right]$$

Taking limits $\Delta x \to 0$, $\Delta y \to 0$, and putting $q_x = -k(\partial T/\partial x)$ and $q_y = -k(\partial T/\partial y)$, we get

$$\frac{\partial}{\partial x}(uT) + \frac{\partial}{\partial y}(vT) = -\frac{1}{\rho c_p}\left[\frac{\partial q_x}{\partial x} + \frac{\partial q_y}{\partial y}\right] = \alpha\left[\frac{\partial^2 T}{\partial x^2} + \frac{\partial^2 T}{\partial y^2}\right]$$

or

$$T\frac{\partial u}{\partial x} + u\frac{\partial T}{\partial x} + T\frac{\partial v}{\partial y} + v\frac{\partial T}{\partial y} = \alpha\left[\frac{\partial^2 T}{\partial x^2} + \frac{\partial^2 T}{\partial y^2}\right]$$

or

$$T\left[\frac{\partial u}{\partial x} + \frac{\partial v}{\partial y}\right] + u\frac{\partial T}{\partial x} + v\frac{\partial T}{\partial y} = \alpha\left[\frac{\partial^2 T}{\partial x^2} + \frac{\partial^2 T}{\partial y^2}\right]$$

For an incompressible fluid, the term in the bracket on the LHS is zero (by using the equation of continuity). Therefore, we have

$$u\frac{\partial T}{\partial x} + v\frac{\partial T}{\partial y} = \alpha\left[\frac{\partial^2 T}{\partial x^2} + \frac{\partial^2 T}{\partial y^2}\right] \tag{11.65}$$

The above equation is the governing partial differential equation for temperature distribution in the boundary layer. By usual boundary layer approximation, we can neglect the term $\partial^2 T/\partial x^2$ on the RHS ($\partial^2 T/\partial x^2 \ll \partial^2 T/\partial y^2$, because the longitudinal gradient of temperature is found to be much smaller than the transverse gradient). Therefore, we have

$$u\frac{\partial T}{\partial x} + v\frac{\partial T}{\partial y} = \alpha\frac{\partial^2 T}{\partial y^2} \tag{11.66}$$

Equation (11.66) is the governing equation for boundary layer temperature. Let T_s be the plate temperature and T_∞ the temperature of the free stream. The boundary conditions on temperature can now be written as

$$y = 0, \qquad T = T_s \tag{11.67}$$

$$y \to \infty, \qquad T = T_\infty \tag{11.68}$$

$$x = 0, \qquad T = T_\infty \tag{11.69}$$

The following dimensionless variables are introduced.

$$\eta = \frac{y}{(xv/V_\infty)^{1/2}} \quad \text{and} \quad \overline{T} = \frac{T - T_s}{T_\infty - T_s} \tag{11.70}$$

The gradients of temperature may be transformed as follows.

$$\frac{\partial T}{\partial x} = \frac{\partial T}{\partial \eta} \frac{\partial \eta}{\partial x} = \frac{\partial T}{\partial \eta} \cdot y \left(\frac{V_\infty}{\nu}\right)^{1/2} \left[-\frac{1}{2} x^{-3/2}\right] \tag{11.71}$$

$$\frac{\partial T}{\partial y} = \frac{\partial T}{\partial \eta} \frac{\partial \eta}{\partial y} = \frac{\partial T}{\partial \eta} \left(\frac{V_\infty}{\nu x}\right)^{1/2} \tag{11.72}$$

$$\frac{\partial^2 T}{\partial y^2} = \frac{\partial^2 T}{\partial \eta^2} \left(\frac{V_\infty}{\nu x}\right) \tag{11.73}$$

Substituting for u and v from Eqs. (11.30) and (11.31), and for $\partial T/\partial x$, $\partial T/\partial y$, and $\partial^2 T/\partial y^2$ from Eqs. (11.71)–(11.73) in Eq. (11.66), we get

$$V_\infty \frac{df}{d\eta} \frac{dT}{d\eta} y \left(\frac{V_\infty}{\nu}\right)^{1/2} \left[-\frac{1}{2} x^{-3/2}\right] + \frac{1}{2} \left(\frac{V_\infty \nu}{x}\right)^{1/2} \left[\eta \frac{df}{d\eta} - f\right] \frac{\partial T}{\partial \eta} \left(\frac{V_\infty}{\nu x}\right)^{1/2} = \alpha \frac{\partial^2 T}{\partial \eta^2} \left[\frac{V_\infty}{\nu x}\right] \tag{11.74}$$

On simplification, we obtain

$$\frac{d^2 T}{d\eta^2} + \frac{1}{2} \mathrm{Pr}\, f \frac{dT}{d\eta} = 0 \tag{11.75}$$

Putting the dimensionless temperature defined by Eq. (11.70), we get

$$\frac{d^2 \overline{T}}{d\eta^2} + \frac{1}{2} \mathrm{Pr}\, f \frac{d\overline{T}}{d\eta} = 0 \tag{11.76}$$

where $\mathrm{Pr} = \nu/\alpha$ is the Prandtl number. The partial derivatives in Eq. (11.74) can be replaced by the ordinary derivatives because the equation ultimately involves η as the sole independent variable. The function $f(\eta)$ is the solution of Eq. (11.35).

The partial differential equation (11.66) can be reduced to an ordinary differential equation (11.76) because *a similarity solution for the boundary layer temperature exists* (as in the case of boundary layer velocity field). The boundary conditions on the dimensionless temperature $\overline{T}$ can be obtained from Eqs. (11.67)–(11.69).

$$\eta = 0, \qquad \overline{T} = 0 \tag{11.77}$$
$$\eta = 1, \qquad \overline{T} = 1 \tag{11.78}$$

Let us now attempt to solve Eq. (11.76) subject to the above boundary conditions. Replacing f by $2f'''/f''$ [see Eq. (11.35)] and substituting $d\overline{T}/d\eta = \xi$, Eq. (11.76) is reduced to

$$\frac{d\xi}{d\eta} = \mathrm{Pr} \frac{f'''}{f''} \xi \qquad \left[\text{where } f' = \frac{df}{d\eta}\right] \tag{11.79}$$

or

$$\int \frac{d\xi}{\xi} = \mathrm{Pr} \int \frac{f'' d\eta}{f'''} + \ln C_1$$

or

$$\xi = C_1 (f'')^{\mathrm{Pr}} \qquad [C_1 \text{ is an integration constant.}]$$

or

$$\frac{d\overline{T}}{d\eta} = C_1 (f'')^{\mathrm{Pr}} \tag{11.80}$$

Integrating again, we get

$$\overline{T} = C_1 \int_0^\eta (f'')^{\mathrm{Pr}} d\eta + C_2 \qquad [C_2 \text{ is another integration constant.}] \tag{11.81}$$

Using the boundary condition (11.77), we get $C_2 = 0$. Therefore,

$$\overline{T} = C_1 \int_0^\eta (f'')^{\mathrm{Pr}} d\eta \tag{11.82}$$

Now, we use the boundary condition (11.78) to determine the constant C_1.

$$1 = \overline{T} = C_1 \int_0^\infty (f'')^{\mathrm{Pr}} d\eta$$

or

$$C_1 = \frac{1}{\displaystyle\int_0^\infty (f'')^{\mathrm{Pr}} d\eta} \tag{11.83}$$

Substituting for C_1 in Eq. (11.82), we get the solution for the non-dimensional boundary layer temperature distribution.

$$\frac{T - T_s}{T_\infty - T_s} = \overline{T} = \frac{\displaystyle\int_0^\eta (f'')^{\mathrm{Pr}} d\eta}{\displaystyle\int_0^\infty (f'')^{\mathrm{Pr}} d\eta} \tag{11.84}$$

Making use of the numerical solution of the momentum boundary layer equation (see Section 11.3.2) we can compute the non-dimensional boundary layer temperature $\overline{T}$ as a function of η for different values of the Prandtl number. A plot of such computed results is shown in Fig. 11.7.

An interesting result can be obtained from Eq. (11.84) if the Prandtl number of the fluid is unity. Putting Pr = 1 in Eq. (11.84) and using the condition given by (11.36) and (11.38) [i.e. $f' = u/V_\infty = 0$ at $\eta = 0$; $f' = 1$ at $\eta = \infty$], we have

$$\overline{T} = \frac{\displaystyle\int_0^\eta f'' d\eta}{\displaystyle\int_0^\infty f'' d\eta} = \frac{[f']_0^\eta}{[f']_0^\infty} = f'(\eta) = \frac{u}{V_\infty} \tag{11.85}$$

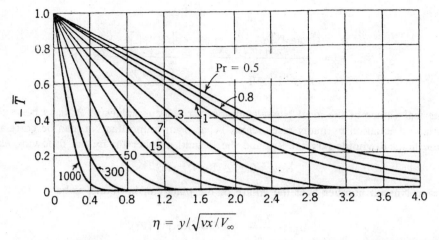

Fig. 11.7 Dimensionless temperature distribution in the thermal boundary layer over a heated flat plate.

Equation (11.85) shows that if Pr = 1, the distribution of the dimensionless velocity and of the temperature in the boundary layer become identical.

The boundary layer heat transfer coefficient or the *wall heat transfer coefficient* is practically more important than the temperature distribution. The local heat transfer coefficient is defined as

$$-k\left[\frac{\partial T}{\partial y}\right]_{y=0} = \left[q_0(x)\right]_{y=0} = h_x(T_s - T_\infty) \qquad (11.86)$$

or

$$h_x = -\frac{k}{(T_s - T_\infty)}\left[\frac{\partial T}{\partial y}\right]_{y=0} \qquad (11.87)$$

The *local* Nusselt number is defined as

$$\mathrm{Nu}_x = \frac{h_x x}{k} = -\frac{x}{(T_s - T_\infty)}\left[\frac{\partial T}{\partial y}\right]_{y=0} \qquad (11.88)$$

From Eq. (11.84), we have

$$\frac{x}{(T_\infty - T_s)}\left[\frac{\partial T}{\partial y}\right]_{y=0} = x\left[\frac{\partial \overline{T}}{\partial y}\right]_{y=0}$$

$$= x\left[\frac{d\overline{T}}{d\eta}\right]_{\eta=0}\left[\frac{\partial \eta}{\partial y}\right]_{y=0} = x\frac{\left[(f'')^{\mathrm{Pr}}\right]_{\eta=0}}{\displaystyle\int_0^\infty (f'')^{\mathrm{Pr}}\, d\eta}\left(\frac{V_\infty}{vx}\right)^{1/2}$$

or

$$\text{Nu}_x = = \frac{\left[(f'')^{\text{Pr}}\right]_{\eta=0}}{\int\limits_0^{\infty} (f'')^{\text{Pr}} \, d\eta} \left(\frac{V_{\infty} x}{\nu}\right)^{1/2} = \frac{\left[(f'')^{\text{Pr}}\right]_{\eta=0}}{\int\limits_0^{\infty} (f'')^{\text{Pr}} \, d\eta} (\text{Re}_x)^{1/2} \qquad (11.89)$$

The numerical values of the local Nusselt number for various values of Pr have been computed from the above equation by many researchers. It has been found that Nu_x can be correlated very well to the local Reynolds number Re_x and the Prandtl number Pr, by the following equation.

$$\text{Nu}_x = 0.332 \, (\text{Re}_x)^{1/2} \, (\text{Pr})^{1/3} \qquad (11.90a)$$

The average Nusselt number over the length of the plate can be found to be

$$\text{Nu}_{\text{av}} = \frac{l \, h_{\text{av}}}{k} = 0.664 \, (\text{Re}_l)^{1/2} \, (\text{Pr})^{1/3} \qquad (11.90b)$$

Equations (11.90) are exact at Pr = 1. The error involved is less than two per cent for $0.5 \leq \text{Pr} \leq 50$. Beyond this range of the Prandtl number, the error is higher.

Equation (11.90a) has been cited already in Chapter 4 [Eq. (4.14)], where the method of calculation of the average Nusselt number over the length of the plate has also been shown. While this equation is applicable to heat transfer in a laminar boundary layer, heat transfer calculations for a turbulent boundary layer flow over a flat plate may be done using Eq. (4.15).

The thermal boundary layer thickness δ_T is typically defined as the distance from the plate where 99% of the dimensionless free stream temperature is attained. It can be calculated from Eq. (11.84) by evaluating the value of η at which $\overline{T} = 0.99$ for a given value of Pr. The ratio of the momentum and thermal boundary layer thicknesses are well correlated by

$$\frac{\delta}{\delta_T} = \text{Pr}^{1/3} \qquad (11.91)$$

The above correlation also shows that the momentum and thermal boundary layer thicknesses are equal for Pr = 1. If Pr > 1 (i.e. $\nu > \alpha$), the thermal boundary layer lies within the momentum boundary layer. The reverse is true if Pr < 1 (i.e. if $\nu < \alpha$).

Example 11.2 Air at 30°C flows over a flat surface at a velocity of 8 m/s. The temperature of the surface is 200°C. Calculate the thermal boundary layer thickness and the local heat transfer coefficient 0.75 m from the leading edge.

SOLUTION The relevant thermophysical properties of air should preferably be taken at the mean temperature of the boundary layer, i.e. at (200 + 30)/2 = 115°C. The following values are used: $\nu = 2.5 \times 10^{-5}$ m²/s; Pr = 0.69; k = 0.036 W/m °C.

From Eqs. (11.91) and (11.39), the thermal boundary layer thickness,

$$\delta_T = \delta(\text{Pr})^{-1/3} = 5.0 \frac{x}{(\text{Re}_x)^{1/2}} (\text{Pr})^{-1/3}$$

Given: x = 0.75 m, V_{∞} = 8 m/s

$$\mathrm{Re}_x = \frac{xV_\infty}{\nu} = \frac{(0.75)(8)}{2.5 \times 10^{-5}} = 2.4 \times 10^5$$

Therefore,

$$\delta_T = \frac{(5.0)(0.75)}{(2.4 \times 10^5)^{1/2}(0.69)^{1/3}} = 0.0086 \text{ m} = \boxed{8.6 \text{ mm}}$$

The local heat transfer coefficient can be calculated from Eq. (11.90a).

$$\mathrm{Nu}_x = 0.332 \, (\mathrm{Re}_x)^{1/2} \, (\mathrm{Pr})^{1/3} = (0.332)(2.4 \times 10^5)^{1/2} \, (0.69)^{1/3} = 144$$

$$h_x = \frac{k}{x}\mathrm{Nu}_x = \frac{0.036}{0.75}144 = \boxed{6.9 \text{ W/m}^2 \text{ °C}}$$

Example 11.3   A thin metal plate separates two air streams flowing on either side of it at zero angle of incidence. Free stream velocity is 6 m/s and the temperature of the air on one side of the plate is 150°C, and the corresponding values on the other side are 3 m/s and 50°C. Calculate the local rate of heat exchange between the two air streams and also the plate temperature at a distance 0.7 m from the leading edge. Assume the boundary layers to be laminar. Axial conduction in the plate may be neglected.

SOLUTION   The thermophysical properties of the air on either side of the plate are needed at the respective mean air temperatures, which are not known. As an approximation, the plate temperature is assumed to be equal to the mean of the bulk air temperatures on the two sides of the plate. So, $T_s = (150 + 50)/2 = 100$°C.

Let the bulk temperatures of the two air streams be denoted by $T_{1\infty}$ (= 150°C) and $T_{2\infty}$ (= 50°C), respectively. The distance of the point from the leading edge, $x = 0.7$ m.

Side 1 *of the plate*

Mean air temperature = $(T_s + T_{1\infty})/2 = (100 + 150)/2 = 125$°C; $V_{1\infty} = 6$ m/s
Properties: $\nu_1 = 2.6 \times 10^{-5}$ m²/s; $\mathrm{Pr}_1 = 0.69$; $k_1 = 0.0336$ W/m °C

$$\mathrm{Re}_{x1} = \frac{xV_{1\infty}}{\nu_1} = \frac{(0.7)(6)}{2.6 \times 10^{-5}} = 1.615 \times 10^5$$

$$\mathrm{Nu}_{x1} = 0.332 \, (\mathrm{Re}_{x1})^{1/2} \, (\mathrm{Pr}_1)^{1/3}$$

$$= (0.332)(1.615 \times 10^5)^{1/2} \, (0.69)^{1/3} = 118$$

$$h_{x1} = \frac{k_1}{x} \, \mathrm{Nu}_{x1} = \frac{0.0336}{0.7}(118) = 5.66 \text{ W/m}^2 \text{ °C}$$

Side 2 *of the plate*

Mean air temperature = $(T_s + T_{2\infty})/2 = (100 + 50)/2 = 75$°C; $V_{2\infty} = 3$ m/s
Properties: $\nu_2 = 2.076 \times 10^{-5}$ m²/s; $\mathrm{Pr}_2 = 0.70$; $k_2 = 0.03$ W/m °C

$$\mathrm{Re}_{x2} = \frac{xV_{2\infty}}{\nu_2} = \frac{(0.7)(3)}{2.076 \times 10^{-5}} = 1.012 \times 10^5$$

$$Nu_{x2} = 0.332 \ (Re_{x2})^{1/2} \ (Pr_2)^{1/3}$$

$$= (0.332)(1.012 \times 10^5)^{1/2} (0.70)^{1/3} = 93.8$$

$$h_{x2} = \frac{k_2}{x} Nu_{x2} = \frac{0.03}{0.7} 93.8 = 4.02 \ \text{W/m}^2 \ ^\circ\text{C}$$

The overall heat transfer coefficient U is given by

$$\frac{1}{U} = \frac{1}{h_{x1}} + \frac{1}{h_{x2}} = \frac{1}{5.66} + \frac{1}{4.02} = 0.425$$

or

$$U = 2.353 \ \text{W/m}^2 \ ^\circ\text{C}$$

The local rate of heat exchange $= U(T_{1\infty} - T_{2\infty}) = (2.353)(150 - 50) = \boxed{235 \ \text{W/m}^2}$

The plate temperature is given by

$$T_s = T_{2\infty} + (T_{1\infty} - T_{2\infty}) \left[\frac{1/h_{x2}}{1/U} \right] = 50 + (150 - 50) \left[\frac{1/4.02}{0.425} \right] = \boxed{108 \ ^\circ\text{C}}$$

(The calculated plate temperature is close enough to the value assumed for the evaluation of the mean temperatures at which the properties of air were used. No further calculation is necessary.)

Example 11.4 Air at 25°C flows over both sides of a 0.4 × 0.3 m copper plate, 25.4 mm thick (density = 8880 kg/m^3; specific heat = 0.385 kJ/kg K). If the air velocity is 1 m/s and the plate temperature is assumed to remain uniform (in view of the high thermal conductivity of copper), calculate the temperature of the plate after 1 hour if its initial temperature is 120°C.

SOLUTION The mean temperature of the boundary layer is $T_f = (120 + 25)/2 = 72.5°C$ at the start. This value will decrease with time. But the final value is not yet known. So the properties of the air at 72.5°C are used.

$$\nu = 2.07 \times 10^{-5} \ \text{m}^2/\text{s}; \qquad Pr = 0.70; \qquad k = 0.03 \ \text{W/m} \ ^\circ\text{C}$$

$$\text{Length of the plate, } l = 0.4 \ \text{m}; \quad V_\infty = 1 \ \text{m/s}$$

Therefore,

$$Re_l = \frac{lV_\infty}{\nu} = \frac{(0.4)(1)}{2.07 \times 10^{-5}} = 1.932 \times 10^4$$

The average heat transfer coefficient is calculated from Eq. (11.90b).

$$\frac{lh_{av}}{k} = (0.664)(1.932 \times 10^4)^{1/2} (0.70)^{1/3} = 82$$

or

$$h_{av} = (82)(0.03)/(0.4) = 6.15 \ \text{W/m}^2 \ ^\circ\text{C}$$

As the thermal conductivity of copper is high, we do a 'lumped parameter' approximation for the plate temperature. The rate of change of this temperature is given by the equation

$$-\frac{d}{dt}(mc_p T) = Ah_{av}(T - T_\infty); \qquad (A = \text{plate area})$$

Applying the boundary condition

$$\text{At } t = 0, \qquad T = T_i$$

and integrating, we get

$$\frac{T - T_\infty}{T_i - T_\infty} = \exp\left[-\frac{Ah_{\text{av}}}{m\,c_p}\,t\right]$$

Given: $m = (0.4)(0.3)(0.0254)(8880) = 27.07$ kg; $A = (0.4)(0.3)(2) = 0.24$ m² (considering both sides of the plate); $c_p = 0.385$ kJ/kg K; $h_{\text{av}} = 6.15$ W/m² °C; $T_i = 120$°C; $T_\infty = 25$°C; time, $t = 1\ h = 3600$ s

Substituting the above values in the integrated equation, we get

$$\frac{T - 25}{120 - 25} = \exp\left[-\frac{(0.24)(6.15)}{(27.07)(385)}(3600)\right] = 0.60$$

or

$$T = \boxed{82°C}$$

Note: Although the plate temperature is changing with time, we have used the equations for steady state heat transfer through the boundary layer. This is a pseudo-steady state approximation.

Once the final temperature is calculated, the properties of air may be evaluated at the revised mean temperature, $[(120 + 25)/2 + (81.6 + 25)/2]/2$. However, the revised calculations will not change the results appreciably in the present case.

In reality, the plate will cool much faster because of natural convection effects which have not been considered in the above analysis.

11.4.2 The Energy Integral Equation

The momentum integral method of analysis of boundary layer flow (see Section 11.3.4) can be extended to analyze the phenomenon of convective heat transfer in a laminar boundary layer. Let us integrate both sides of Eq. (11.66) from the surface of the plate (i.e. $y = 0$) to the edge of the thermal boundary layer (i.e. $y = \delta_T$). Thus, we get

$$\int_0^{\delta_T} u\frac{\partial T}{\partial x}\,dy + \int_0^{\delta_T} v\frac{\partial T}{\partial y}\,dy = \alpha\int_0^{\delta_T} \frac{\partial^2 T}{\partial y^2}\,dy \qquad (11.92)$$

Because

$$\frac{d}{dx}\int_0^{\delta_T} uT\,dy = \int_0^{\delta_T} \frac{\partial}{\partial x}(uT)\,dy = \int_0^{\delta_T} T\frac{\partial u}{\partial x}\,dy + \int_0^{\delta_T} u\frac{\partial T}{\partial x}\,dy$$

the first term on the LHS of Eq. (11.92) becomes

$$\int_0^{\delta_T} u\frac{\partial T}{\partial x}\,dy = \frac{d}{dx}\int_0^{\delta_T} uT\,dy - \int_0^{\delta_T} T\frac{\partial u}{\partial x}\,dy \qquad (11.93)$$

Similarly, the second term on the LHS of Eq. (11.92) yields

$$\int_0^{\delta_T} v \frac{\partial T}{\partial y} dy = [vT]_{y=0}^{\delta_T} - \int_0^{\delta_T} T \frac{\partial v}{\partial y} dy = [vT]_{\delta_T} + \int_0^{\delta_T} \frac{\partial u}{\partial x} T dy$$

$$= -T_\infty \int_0^{\delta_T} \frac{\partial u}{\partial x} dy + \int_0^{\delta_T} \frac{\partial u}{\partial x} T dy$$

$$= -\frac{d}{dx} \int_0^{\delta_T} u T_\infty dy + \int_0^{\delta_T} \frac{\partial u}{\partial x} T dy \qquad (11.94)*$$

Using Eqs. (11.93) and (11.94) in Eq. (11.92), we get

$$\frac{d}{dx} \int_0^{\delta_T} u T dy - \frac{d}{dx} \int_0^{\delta_T} u T_\infty dy = \alpha \left[\frac{\partial T}{\partial y}\right]_{y=0}^{\delta_T} = -\alpha \left[\frac{\partial T}{\partial y}\right]_{y=0}$$

or

$$\frac{d}{dx} \int_0^{\delta_T} u(T - T_\infty) dy = -\alpha \left[\frac{\partial T}{\partial y}\right]_{y=0} \qquad (11.95)**$$

Equation (11.95) is the *energy integral equation* for heat transfer in laminar boundary layer flow.

In order to solve Eq. (11.95) for the boundary layer temperature distribution, we assume [following the procedure for solution of the momentum integral equation (11.51)] that the dimensionless temperature profile can be expressed as a polynomial in y/δ_T. Let us take this as a third degree polynomial.

$$\frac{T - T_s}{T_\infty - T_s} = b_0 + b_1\left(\frac{y}{\delta_T}\right) + b_2\left(\frac{y}{\delta_T}\right)^2 + b_3\left(\frac{y}{\delta_T}\right)^3 \qquad (11.96)$$

The following boundary conditions apply.

$$y = 0; \qquad T = T_s, \qquad \partial^2 T/\partial y^2 = 0 \qquad (11.97)$$

$$y = \delta_T; \qquad T = T_\infty, \qquad \partial T/\partial y = 0 \qquad (11.98)$$

The boundary condition (11.97) follows from Eq. (11.66) because at $y = 0$ (i.e. at the surface of

---

* Because at $y = 0$, $v = 0$; and from the equation of continuity (11.23), $\dfrac{\partial v}{\partial y} = -\dfrac{\partial u}{\partial x}$.

At $y = \delta_T$, $T = T_\infty$ and $[vT]_{\delta_T} = T_\infty[v]_{\delta_T} = T_\infty \int_0^{\delta_T} \frac{\partial v}{\partial y} dy = -T_\infty \int_0^{\delta_T} \frac{\partial u}{\partial x} dy$

** By physical reasoning we may put $\partial T/\partial y = 0$ at $y = \delta_T$, i.e. at the edge of the boundary layer. This is because the rate of change in temperature with y becomes extremely small there. Another argument may be put forward to justify this boundary condition. Beyond the boundary layer (i.e. for $y > \delta_T$) the temperature is uniform, or $\partial T/\partial y = 0$. Because there must not be a discontinuity in the temperature gradient in the boundary layer, $\partial T/\partial y$ should be zero at the edge of the boundary layer.

the plate), the velocity components are zero ($u = 0$, $v = 0$). Putting the conditions (11.97) and (11.98) in Eq. (11.96), we obtain

$$0 = b_0 \tag{11.99}$$

$$0 = 2b_2/\delta_T^2 \tag{11.100}$$

$$1 = b_0 + b_1 + b_2 + b_3 \tag{11.101}$$

$$0 = b_1/\delta_T + 2b_2/\delta_T + 3b_3/\delta_T \tag{11.102}$$

Solving the above equations, we get

$$b_0 = b_2 = 0, \qquad b_1 = 3/2, \qquad b_3 = -1/2 \tag{11.103}$$

The corresponding expression for the temperature profile is given by

$$\frac{T - T_s}{T_\infty - T_s} = \frac{3}{2}\left(\frac{y}{\delta_T}\right) - \frac{1}{2}\left(\frac{y}{\delta_T}\right)^3 \tag{11.104}$$

in which the thermal boundary layer thickness δ_T is yet to be determined.

Substituting for u [from Eq. (11.59)] and for T [from Eq. (11.104)] in Eq. (11.95), we get

$$\frac{d}{dx}\int_0^{\delta_T} V_\infty \left[\frac{3}{2}\left(\frac{y}{\delta}\right) - \frac{1}{2}\left(\frac{y}{\delta}\right)^3\right](T_\infty - T_s)\left[-1 + \frac{3}{2}\left(\frac{y}{\delta_T}\right) - \frac{1}{2}\left(\frac{y}{\delta_T}\right)^3\right]dy$$

$$= -\alpha(T_\infty - T_s)\frac{\partial}{\partial y}\left[\frac{3}{2}\left(\frac{y}{\delta_T}\right) - \frac{1}{2}\left(\frac{y}{\delta_T}\right)^3\right]_{y=0} = -\alpha(T_\infty - T_s)\frac{3}{2\delta_T}$$

If we substitute $\chi = \delta_T/\delta$ and $\xi = y/\delta_T$, the above equation reduces to

$$V_\infty \frac{d}{dx}\int_{\xi=0}^1 \delta_T\left(\frac{3}{2}\chi\xi - \frac{1}{2}\chi^3\xi^3\right)\left(1 - \frac{3}{2}\xi + \frac{1}{2}\xi^3\right)d\xi = \frac{3\alpha}{2\chi\delta}$$

Integrating, we obtain

$$V_\infty \frac{d}{dx}\left[\delta\left(\frac{3}{20}\chi^2 - \frac{3}{280}\chi^4\right)\right] = \frac{3\alpha}{2\chi\delta} \tag{11.105}$$

An analytical solution of the above equation can be obtained if $\chi = \delta_T/\delta < 1$ (i.e. the thermal boundary layer remains within the velocity boundary layer). This happens when Pr < 1. Under this condition the term $(3/280)\chi^4$ in Eq. (11.105) can be neglected when compared with $(3/20)\chi^2$. With this simplification, the equation reduces to

$$\frac{d}{dx}(\delta\chi^2) = \frac{10\alpha}{V_\infty\chi\delta} \tag{11.106}$$

If we substitute $\delta = (280\mu x/13\rho V_\infty)^{1/2}$ from Eq. (11.62) in the above equation, we get

$$\frac{d\chi^3}{dx} + \frac{4}{3x}\chi^3 = \frac{39}{56}\frac{\alpha}{\nu}\frac{1}{x}$$

or

$$\frac{d\zeta}{dx} + \frac{4}{3x}\zeta = \frac{39}{56x}\frac{1}{\text{Pr}}, \quad \text{where } \chi^3 = \zeta$$

Integrating, we get

$$\zeta = \frac{13}{14}\frac{1}{\text{Pr}} + Cx^{-3/4}$$

where C is an integration constant. From physical reasoning, ζ is finite at $x = 0$. This is possible only if we put $C = 0$. Therefore,

$$\zeta = \chi^3 = \frac{13}{14}\frac{1}{\text{Pr}}$$

i.e.

$$\chi = \frac{\delta_T}{\delta} = \left[\frac{13}{14}\frac{1}{\text{Pr}}\right]^{1/3} \approx (\text{Pr})^{-1/3} \tag{11.107}$$

or

$$\delta_T = \delta\,\text{Pr}^{-1/3} = 4.64\,\frac{x}{(\text{Re}_x)^{1/2}(\text{Pr})^{1/3}} \tag{11.108}$$

Equation (11.108) gives the relation for the thermal boundary layer thickness at any distance from the leading edge of the plate. The relative thickness of the thermal boundary layer compared to the velocity boundary layer can be determined from the value of the Prandtl number using Eq. (11.107) which is identical to Eq. (11.91) derived before.

The temperature distribution in the boundary layer can be obtained from Eq. (11.104) after substituting δ_T given by Eq. (11.108). The local temperature gradient at the plate surface can be obtained by differentiating Eq. (11.104). Thus,

$$\left[\frac{\partial T}{\partial y}\right]_{y=0} = (T_\infty - T_s) \cdot \frac{3}{2\delta_T} \tag{11.109}$$

The local heat transfer coefficient [as expressed by Eq. (11.87)] can now be determined as

$$h_x = -\frac{k}{(T_s - T_\infty)}\left[\frac{\partial T}{\partial y}\right]_{y=0} = -\frac{k}{(T_s - T_\infty)} \cdot (T_\infty - T_s)\frac{3}{2\delta_T}$$

Substituting δ_T from Eq. (11.108), we get

$$h_x = \frac{3k}{9.28x}(\text{Re}_x)^{1/2}(\text{Pr})^{1/3} = 0.323\,\frac{k}{x}(\text{Re}_x)^{1/2}(\text{Pr})^{1/3} \tag{11.110}$$

The local Nusselt number is obtained as

$$\text{Nu}_x = \frac{xh_x}{k} = 0.323\,(\text{Re}_x)^{1/2}(\text{Pr})^{1/3} \tag{11.111}$$

The above relation for Nusselt number is nearly identical to that given by Eq. (11.90a). This shows that the *integral method* is a very good technique for the analysis of boundary layer transport. Figure 11.8 shows a comparison of experimental heat transfer data with that predicted by Eq. (11.111) or (11.90b).

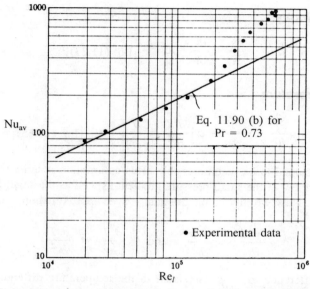

Fig. 11.8   Experimental results on Nusselt number for laminar boundary layer heat transfer for flow of air past a flat plate.

Although the method of analysis given in Section 11.4.1 is *exact*, it is useful only in simple flow situations. On the other hand, the integral method, though approximate, can be used in a greater number of cases. In many situations of engineering importance, however, none of these methods can be used, and empirical correlations for the Nusselt number as a function of Reynolds number, Prandtl number and other parameters, if any, remain to be the answer to the problems of heat transfer calculation. As discussed in Chapter 4, a pretty large number of such correlations have been reported in the literature and may be selected for specific uses.

11.5   APPLICATION TO FREE CONVECTION HEAT TRANSFER

The phenomenon of free convection heat transfer has been discussed in Chapter 5 and a number of correlations for the Nusselt number in different situations have been cited. Fluid motion in free convection in an otherwise stagnant fluid remains confined to a rather narrow zone near the surface. This suggests that the boundary layer approximation is applicable. Let us consider steady state natural convection heat transfer from a vertical flat plate. If the plate is sufficiently wide, the motion near the plate will be *two-dimensional*. Using the coordinates as shown in Fig. 5.2(a), the equation of motion can be derived from Eq. (11.16). It should, however, be noted that the body force in the equation of motion for the x-component of the velocity arises here, because of the density difference, $\rho_\infty - \rho$, between the cold bulk fluid and the warm fluid near the surface. Also the component of g in the vertical or the x-direction is $g_x = g$. Therefore, at steady state,

Equation of motion

$$u\frac{\partial u}{\partial x} + v\frac{\partial u}{\partial y} = -\frac{1}{\rho}\frac{\partial p}{\partial x} + v\left(\frac{\partial^2 u}{\partial x^2} + \frac{\partial^2 u}{\partial y^2}\right) + \left(\frac{\rho_\infty - \rho}{\rho}\right)g \qquad (11.112)$$

Equation of continuity

$$\frac{\partial u}{\partial x} + \frac{\partial v}{\partial v} = 0 \qquad [\text{Eq. (11.23)}]$$

Equation of temperature

$$u\frac{\partial T}{\partial x} + v\frac{\partial T}{\partial y} = \alpha\left(\frac{\partial^2 T}{\partial x^2} + \frac{\partial^2 T}{\partial y^2}\right) \qquad [\text{Eq. (11.65)}]$$

Here the fluid is assumed to be incompressible. This assumption is quite reasonable even for a gaseous medium in view of the small pressure gradient as well as the small length of the surface under consideration. Now we invoke the boundary layer approximation, namely

$$\frac{\partial^2 u}{\partial x^2} \ll \frac{\partial^2 u}{\partial y^2} \quad \text{and} \quad \frac{\partial^2 T}{\partial x^2} \ll \frac{\partial^2 T}{\partial y^2}.$$

Also the density difference, $\rho_\infty - \rho$, is related to the temperature difference, $T_\infty - T$, and the coefficient of volumetric expansion β as follows.

$$\rho_\infty - \rho = \beta\rho(T_\infty - T) \qquad (11.113)$$

With the above boundary layer approximations, if we use Eq. (11.113) and neglect the pressure gradient (i.e. putting $\partial p/\partial x = 0$), Eqs. (11.112) and (11.65) can be rewritten as

$$u\frac{\partial u}{\partial x} + v\frac{\partial u}{\partial y} = v\frac{\partial^2 u}{\partial y^2} + g\beta(T - T_\infty) \qquad (11.114)$$

$$u\frac{\partial T}{\partial x} + v\frac{\partial T}{\partial y} = \alpha\frac{\partial^2 T}{\partial y^2} \qquad [\text{Eq. 11.66}]$$

The following boundary conditions apply.

$$y = 0; \qquad u = 0, \qquad T = T_s \qquad (11.115)$$

$$y = \infty; \qquad u = 0, \qquad T = T_\infty \qquad (11.116)$$

A *similarity solution* to the above equation was suggested by Pohlhausen in 1921, who defined the following similarity variable ξ and the stream function Ψ.

Similarity variable: $\xi = C\dfrac{y}{x^{1/4}}$ (11.117)

Stream function: $\Psi = 4vCx^{3/4}F(\xi)$ (11.118)

where

$$C = \left[\frac{g\beta(T_s - T_\infty)}{4\nu^2} \right]^{1/4}$$

The above choice of stream function satisfies the equation of continuity, Eq. (11.23). Differentiating Ψ, we get

$$u = \frac{\partial \Psi}{\partial x} = 4\nu C^2 x^{1/2} \frac{dF}{d\xi} \tag{11.119}$$

$$v = -\frac{\partial \Psi}{\partial y} = \nu C x^{-1/4} \left[\xi \frac{dF}{d\xi} - 3F \right] \tag{11.120}$$

Let us define the following dimensionless temperature.

$$\overline{T} = \frac{T - T_\infty}{T_s - T_\infty} \tag{11.121}$$

If we put Eqs. (11.119) and (11.120) in Eqs. (11.114) and (11.66), we get

$$\frac{d^3F}{d\xi^3} + 3F\frac{d^2F}{d\xi^2} - 2\left(\frac{dF}{d\xi}\right)^2 + \overline{T} = 0 \tag{11.122}$$

$$\frac{d^2\overline{T}}{d\xi^2} + 3\mathrm{Pr}F\frac{d\overline{T}}{d\xi} = 0 \tag{11.123}$$

The following are the boundary conditions on F and $\overline{T}$ which are derived from Eqs. (11.115), (11.116), and (11.119)–(11.121).

$$\xi = 0; \qquad F = 0, \quad \frac{dF}{d\xi} = 0, \quad \overline{T} = 1 \tag{11.124}$$

$$\xi = \infty, \qquad \frac{dF}{d\xi} = 0, \quad \overline{T} = 0 \tag{11.125}$$

Equations (11.122) and (11.123) constitute two simultaneous non-linear ordinary differential equations in F and $\overline{T}$. These do not admit of analytical solutions, though numerical solutions have been obtained by many researchers.

The local heat flux at the surface of the plate can be written as

$$q = -k\left[\frac{\partial T}{\partial y}\right]_{y=0} = -k(T_s - T_\infty)\left[\frac{d\overline{T}}{d\xi}\right]_{\xi=0} \frac{d\xi}{dy} \tag{11.126}$$

or

$$h(T_s - T_\infty) = -\frac{Ck}{x^{1/4}}(T_s - T_\infty)\left[\frac{d\overline{T}}{d\xi}\right]_{\xi=0}$$

The local Nusselt number at x is given by

$$\text{Nu}_x = \frac{hx}{k} = -Cx^{3/4}\left[\frac{d\overline{T}}{d\xi}\right]_{\xi=0} = -\left[\frac{\text{Gr}_x}{4}\right]^{1/4}\left[\frac{d\overline{T}}{d\xi}\right]_{\xi=0} \tag{11.127}$$

where Gr_x is the *local* Grashof number (see Chapter 5) defined as

$$\text{Gr}_x = \frac{g\beta(T_s - T_\infty)x^3}{\nu^2} \tag{11.128}$$

The average Nusselt number, hL/k (where L is the height of the plate), can be obtained from Eq. (11.127). Thus,

$$\text{Nu}_L = \frac{1}{L}\int_0^L \text{Nu}_x\, dx = -\frac{4}{3}\left[\frac{\text{Gr}_L}{4}\right]^{1/4}\left[\frac{d\overline{T}}{d\xi}\right]_{\xi=0} \tag{11.129}$$

Equations (11.127) and (11.129) can be used to calculate the local and the average heat transfer coefficients. But the gradient of dimensionless temperature at the plate surface, that is, $[d\overline{T}/d\xi]_{\xi=0}$ should be known. The values of this quantity, which is a function of Prandtl number [as evident from Eq. (11.123)], can be obtained from the numerical solution to the equations. Table 11.2 gives a short list of the values of the surface temperature gradient.

Table 11.2 Numerical values of the surface temperature gradient

Pr	0.01	0.733	1	2	10	100	1000
$-\left[\dfrac{d\overline{T}}{d\xi}\right]_{\xi=0}$	0.0812	0.05080	0.5671	0.7165	1.1694	2.191	3.968

If the fluid is air (Pr = 0.7), Eq. (11.129) reduces to

$$\text{Nu}_L = 0.47(\text{Gr}_L)^{1/4} \tag{11.130}$$

Let us compare the prediction of the above equation with that by Eq. (5.4) cited before. Taking $\text{Gr}_L = 10^8$ (so that the free convection boundary layer remains laminar), we have

from Eq. (11.130), $\text{Nu}_L = 47$
from Eq. (5.4), taking Pr = 0.7, $\text{Nu}_L = 52$

This again shows that in such cases, experimental heat transfer data are in good agreement with the predictions of the boundary layer theory.

11.6   HEAT TRANSFER IN TURBULENT BOUNDARY LAYER FLOW

In 1883 Osborne Reynolds established, by his 'dye experiment' on the flow of a liquid through a circular tube, that the flow remains *streamline* or *laminar* if the Reynolds number is less than 2100. The flow becomes *turbulent* (i.e. the flow is characterized by turbulent fluctuations in velocity and pressure) when the Reynolds number goes above 4000. In the intermediate range of Reynolds number, the flow is called *transitional*. Analysis of turbulent flow is extremely important because in a large variety of practical situations, the flow of fluids remains turbulent. In the case of laminar flow, shear stresses and velocity gradients are related by simple laws and these relations lead to the equations of motion (Section 11.2). For turbulent flow, there are no such fundamental laws and, therefore, the exact quantitative analysis of the turbulent flow phenomenon is not possible.

Reynolds' dye experiment, and more sophisticated photographic studies of turbulent flow carried out later after sprinkling fine powders on water flowing through a channel indicated that in turbulent flow secondary motions of the fluid are superimposed on the main flow. Turbulent flow is characterized by irregular fluctuations in local velocity and by mixing through eddies. The superimposed secondary motions have tremendous influence on the drag force and also on convective heat and mass transfer rates.

There are two important approaches towards explaining the phenomenon of turbulent motion. The first approach, which is also the older one, is an empirical approach. This is based upon the temporal mean (or time-mean) of the flow quantities like velocity components and pressure. The empirical approach or theories were developed during the period 1900 to 1915, and Ludwig Prandtl was the foremost contributor. The second approach is based upon the 'statistical theory of turbulence' proposed by G.I. Taylor In 1921, Taylor introduced the concept that the local velocity of a turbulent fluid is a random continuous function of time and position.

11.6.1 Mean and Fluctuating Quantities

According to the empirical theories of turbulence, each flow quantity, such as velocity components and pressure, is assumed to consist of a mean and a fluctuating component. The instantaneous velocity at any point fluctuates with time, and a flow measuring device like a pitot tube measures only the time-average of the local velocity (Fig. 11.9). But an instrument like a hot-wire anemometer can be used to measure the instantaneous velocity. The instantaneous x-component of velocity u_i can be written as (the subscript i means instantaneous).

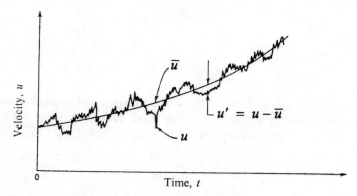

Fig. 11.9 Fluctuation of a velocity component about the mean value.

$$u_i = u + u' \tag{11.131}$$

where u_i is the instantaneous x-component of velocity, u is the mean and u' is the fluctuating part. The other two velocity components and the local pressure can correspondingly be written as

$$v_i = v + v'; \qquad w_i = w + w'; \qquad p_i = p + p' \qquad (11.132)$$

If $\hat{t}$ is a time larger than the time scale of fluctuation of velocity, then the mean velocity, say u, is related to the instantaneous velocity as

$$u = \frac{1}{\hat{t}} \int_0^{\hat{t}} u_i \, dt = \bar{u}_i \qquad (11.133)$$

Taking the time-mean of both sides of Eq. (11.131), we get

$$\frac{1}{\hat{t}} \int_0^{\hat{t}} u_i \, dt = \frac{1}{\hat{t}} \int_0^{\hat{t}} u \, dt + \frac{1}{\hat{t}} \int_0^{\hat{t}} u' \, dt = u + \frac{1}{\hat{t}} \int_0^{\hat{t}} u' \, dt \qquad (11.134)$$

Therefore, from Eqs. (11.133) and (11.134), we have

$$\frac{1}{\hat{t}} \int_0^{\hat{t}} u' \, dt = \bar{u}' = 0 \qquad (11.135)$$

Thus the time-mean (or the temporal mean) of the fluctuating velocity component u' is zero. Similarly,

$$\int_0^{\hat{t}} v' \, dt = \int_0^{\hat{t}} w' \, dt = \int_0^{\hat{t}} p' \, dt = 0 \qquad (11.136)$$

However, the mean square (or root mean square) of the fluctuating velocity (or pressure) is not zero. These are defined as

$$\frac{1}{\hat{t}} \int_0^{\hat{t}} u'^2 \, dt = \bar{u}'^2$$

$$\frac{1}{\hat{t}} \int_0^{\hat{t}} v'^2 \, dt = \bar{v}'^2 \qquad (11.137)$$

$$\frac{1}{\hat{t}} \int_0^{\hat{t}} w'^2 \, dt = \bar{w}'^2$$

The concept of time-averaging can be used to write the equation of continuity and the equations of motion for turbulent flow of an incompressible fluid. If we use a bar (–) to mark a time-averaged quantity, the equation of continuity follows from Eq. (11.6). Thus,

$$\frac{\partial \bar{u}}{\partial x} + \frac{\partial \bar{v}}{\partial y} + \frac{\partial \bar{w}}{\partial z} = 0$$

But

$$\bar{u} = \frac{1}{\hat{t}} \int_0^{\hat{t}} u \, dt = \frac{1}{\hat{t}} \int_0^{\hat{t}} (u + u') \, dt = \frac{u}{\hat{t}} \int_0^{\hat{t}} dt + \frac{1}{\hat{t}} \int_0^{\hat{t}} u' \, dt$$

or

$$\bar{u} = u$$

Similarly,

$$\bar{v} = v, \qquad \bar{w} = w$$

The equation of continuity then becomes

$$\frac{\partial u}{\partial x} + \frac{\partial v}{\partial y} + \frac{\partial w}{\partial z} = 0$$

which is same as Eq. (11.6).

The time-averaged x-component of equation of motion can be written from Eq. (11.11). Assuming the density ρ to be constant, and neglecting the body force term, we get

$$\rho \frac{\partial \bar{u}_i}{\partial t} = -\rho \left[\frac{\partial \overline{u_i^2}}{\partial x} + \frac{\partial \overline{u_i v_i}}{\partial y} + \frac{\partial \overline{u_i w_i}}{\partial z} \right] + \mu \left[\frac{\partial^2 \bar{u}_i}{\partial x^2} + \frac{\partial^2 \bar{u}_i}{\partial y^2} + \frac{\partial^2 \bar{u}_i}{\partial z^2} \right] - \frac{\partial \bar{p}_i}{\partial x} \qquad (11.138)$$

But

$$\bar{u}_i = u \quad \text{and} \quad \bar{p}_i = p$$

Therefore,

$$\overline{u_i^2} = \frac{1}{\hat{t}} \int_0^{\hat{t}} (u + u')^2 \, dt = \frac{1}{\hat{t}} \int_0^{\hat{t}} (u^2 + 2uu' + u'^2) dt$$

$$= \frac{u^2}{\hat{t}} \int_0^{\hat{t}} dt + \frac{2u}{\hat{t}} \int_0^{\hat{t}} u' dt + \frac{1}{\hat{t}} \int_0^{\hat{t}} u'^2 dt = u^2 + \overline{u'^2}$$

and

$$\overline{u_i v_i} = \frac{1}{\hat{t}} \int_0^{\hat{t}} (u + u')(v + v') \, dt = \frac{uv}{\hat{t}} \int_0^{\hat{t}} dt + \frac{v}{\hat{t}} \int_0^{\hat{t}} u' dt + \frac{u}{\hat{t}} \int_0^{\hat{t}} v' dt + \frac{1}{\hat{t}} \int_0^{\hat{t}} u'v' dt$$

$$= uv + \overline{u'v'}$$

Similarly,

$$\overline{u_i w_i} = uv + \overline{u'w'}$$

Putting the above results in Eq. (11.138) and using Eq. (11.6), we get

$$\rho \frac{\partial u}{\partial t} + u \frac{\partial u}{\partial x} + v \frac{\partial u}{\partial y} + w \frac{\partial u}{\partial z} = -\frac{\partial p}{\partial x} + \mu \left[\frac{\partial^2 u}{\partial x^2} + \frac{\partial^2 u}{\partial y^2} + \frac{\partial^2 u}{\partial z^2} \right]$$

$$- \frac{\partial}{\partial x} \left[(\rho \overline{u'^2}) + \frac{\partial}{\partial y} (\rho \overline{v'u'}) + \frac{\partial}{\partial z} (\rho \overline{u'w'}) \right] \qquad (11.139)$$

A comparison of Eq. (11.139) with Eq. (11.16) reveals that in the former equation there are three additional stress terms, involving $\rho \overline{u'^2}$, $\rho \overline{u'v'}$, and $\rho \overline{u'w'}$ arising out of the contributions of the fluctuating velocity components. These stresses are called *Reynolds stresses*.

11.6.2 The Concept of Eddy Viscosity

Boussinesq, in 1877, proposed that the Reynolds stresses can be related to the gradients of the mean velocity components in turbulent flow in a way similar to Eq. (11.14). But the viscosity of the fluid has to be replaced by a *turbulent* or *eddy viscosity*, μ_M. For example,

$$\rho\overline{u'v'} = -\mu_M\left(\frac{\partial u}{\partial y} + \frac{\partial v}{\partial x}\right) = \tau_{xy}^{(t)} \tag{11.140}$$

or

$$\overline{u'v'} = -\varepsilon_M\left(\frac{\partial u}{\partial y} + \frac{\partial v}{\partial x}\right) \tag{11.141}$$

Here $\varepsilon_M = \mu_M/\rho$ is the turbulent or eddy diffusivity of momentum. Originally Boussinesq assumed ε_M to be a scalar quantity (i.e. independent of direction). But this assumption is not consistent with the fundamental laws of fluid motion (see Hinze, 1975). Thus, ε_M varies with direction as well as position. For one-dimensional flow through a parallel plate channel (or a circular tube), only the eddy diffusion of momentum in the transverse (or radial) direction is important. In such a situation the turbulent shear stress, $\tau_{xy}^{(t)}$ may be expressed as

$$\tau_{yx}^{(t)} = -\mu_M\frac{\partial u}{\partial y} \tag{11.142}$$

It may be noted that the eddy viscosity μ_M is much larger (by even 1000 times) than viscosity. The 'total' shear stress (taking into account the contributions of both molecular, or laminar, and eddy transport of momentum) is expressed as [see Eq. (4.50)]

$$\tau_{yx} = \tau_{yx}^{(l)} + \tau_{yx}^{(t)} = -\rho(\nu + \varepsilon_M)\frac{\partial u}{\partial y} \tag{11.143}$$

Here $\tau_{xy}^{(l)}$ accounts for molecular transport of momentum. A similar equation for the total rate of heat transfer from the surface to the turbulent bulk fluid has been given before [see Eq. (4.49)].

$$q = -(\alpha + \varepsilon_H)\rho c_p\frac{\partial T}{\partial y} \tag{11.144}$$

where ε_H is the *eddy diffusivity of heat*.

11.6.3 The Prandtl Mixing-Length Theory

Prandtl proposed his mixing-length theory in 1925 in order to describe turbulent momentum transport. This is probably the most important of the empirical theories of turbulence. Prandtl visualized that in a fluid in turbulent motion, the fluid 'particles' (or clusters of molecules) or 'elements' or 'eddies' undergo random motion. It is through the motion of these particles or eddies that turbulent transport of momentum from one fluid layer to an adjacent layer occurs. If a slow particle enters into a faster moving layer, it is dragged forward, but a faster particle entering into a slower layer tends to speed it up. In an attempt to quantify this phenomenon, Prandtl introduced the concept of *mixing-length*. The Prandtl mixing-length is physically described as the average distance travelled by a fluid particle in a direction transverse to the mean flow before it gets mixed with other particles and loses its identity. It is akin to the mean free path in the kinetic theory of gases.

Turbulent velocity profile near a wall is shown in Fig. 11.10. The mean flow occurs in the x-direction only, and in this sense the flow is one-dimensional. In the y- and z-directions, there are fluctuating components only and the mean values of these components, v and w, are obviously zero. Let us assume that at section (1) in Fig. 11.10, the mean velocity in the x-direction is u, and that at section (2), at a distance l, is $u + l(\partial u/\partial y)$. Prandtl postulated that the local fluctuating velocity in the x-direction is the same as the difference between the mean velocities of the two layers separated by a distance l. Thus,

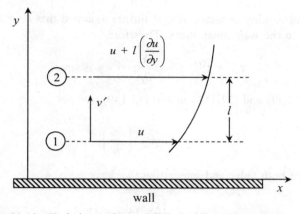

Fig. 11.10 Turbulent velocity profile and Prandtl mixing-length.

$$u' = u + \left(l\frac{\partial u}{\partial y}\right) - u = l\left(\frac{\partial u}{\partial y}\right) \tag{11.145}$$

The distance l is called the *Prandtl mixing-length*. It is the distance (on an average) over which a fluid element (or particle or eddy) retains its identity. Prandtl further assumed that the turbulent fluctuation in velocity in the y-direction, i.e. v', is of the same order as u'. This means

$$v' \sim l\left(\frac{\partial u}{\partial y}\right) \tag{11.146}$$

The stress component for turbulent fluctuations (or *Reynolds stress*) can be written as

$$\tau_{yx}^{(t)} = \rho\overline{u'v'} \sim \rho l^2\left(\frac{\partial u}{\partial y}\right)^2 \tag{11.147}$$

If we define l, such that

$$\tau_{yx}^{(t)} = \rho l^2\left(\frac{\partial u}{\partial y}\right)^2 \tag{11.148}$$

we may correspondingly define the eddy viscosity or eddy momentum diffusivity ε_M as follows. (This term has been introduced in Section 4.8 as well.)

$$\tau_{yx}^{(t)} = \rho \varepsilon_M \left(\frac{\partial u}{\partial y} \right), \qquad \varepsilon_M = l^2 \left(\frac{\partial u}{\partial y} \right) \qquad (11.149)$$

While Boussinesq introduced the concept of eddy diffusivity, a quantitative expression for the same is obtained as Eq. (11.149) following the Prandtl mixing-length theory. Prandtl also developed relations for the velocity distribution in turbulent fluid by using this theory. He postulated that the mixing length is proportional to the distance from the wall, y. Thus,

$$l = my \qquad (11.150)$$

where m is the proportionality constant. It was further assumed that *near the wall* the turbulent shear stress is equal to the wall shear stress. Therefore,

$$\tau_w = \tau_{yx}^{(t)} = -\mu \frac{\partial u}{\partial y}, \qquad y \text{ is small.} \qquad (11.151)$$

Substituting Eqs. (11.150) and (11.151) in Eq. (11.149), we get

$$\tau_w = \rho m^2 y^2 \left(\frac{\partial u}{\partial y} \right)^2$$

Taking square root of both sides and integrating, we have

$$u = \frac{1}{m} \sqrt{\frac{\tau_w}{\rho}} \ln y + C' \qquad (11.152)$$

where C' is the integration constant. Let us now take $u^* = \sqrt{\tau_w/\rho}$ as the reference velocity, called the *friction velocity* or *shear velocity*, $u^+ = u/u^*$ as the dimensionless velocity, and $y^+ = (y/v)\sqrt{\tau_w/\rho}$ as the dimensionless distance from the plate. In terms of these dimensionless variables, Eq. (11.152) becomes

$$u^+ = \frac{1}{m} \ln y^+ + \frac{1}{m} \ln \left[v \sqrt{\frac{\rho}{\tau_w}} \right] + C' \sqrt{\frac{\rho}{\tau_w}}$$

or

$$u^+ = \frac{1}{m} \ln y^+ + C'' \qquad (11.153)$$

Equation (11.153), derived on the basis of the Prandtl mixing-length theory, gives the velocity profile near the wall for turbulent flow of a fluid. Very close to the wall, however, there exists a *laminar sub-layer* [see Fig. 4.1(a)] in which the flow is of purely viscous nature and, therefore, no effect of turbulence is experienced in this layer. Because this layer is thin and the flow is laminar, the velocity profile in it is linear in distance from the wall. Following Newton's law of viscosity, we can write the following equation for this layer.

$$\tau = \tau_w = \mu \frac{\partial u}{\partial y}$$

or

$$u = \frac{\tau_w}{\rho} y \tag{11.154}*$$

The above velocity distribution is linear and satisfies the boundary condition at the wall ($u = 0$ at $y = 0$). In terms of the dimensionless variables, the above equation reduces to the following form

$$u^+ = y^+ \tag{11.155}$$

It is not possible to theoretically predict the range of y^+ over which Eqs. (11.153) and (11.155) are applicable. Also, it is not possible to theoretically calculate the values of the coefficients m and C'' in Eq. (11.153). However, careful experimental measurements of local velocities at different distances from the wall, and also the wall shear stresses were carried out by many researchers (for example, Nikuradse, 1932). It was found that the experimental data could be fitted remarkably well by using Eq. (11.153) for the *buffer zone* and the *turbulent zone*, and Eq. (11.155) for the laminar sub-layer. Estimating the coefficients from the experimental data, von Karman (1939) proposed the following equations for the different zones.

Laminar sub-layer : $\quad 0 < y^+ < 5, \quad\quad u^+ = y^+ \tag{11.156a}$

Buffer zone : $\quad\quad 5 < y^+ < 30, \quad\quad u^+ = 5.0 \ln y^+ + 3.05 \tag{11.156b}$

Fully turbulent zone : $\quad 30 < y^+ < 400, \quad\quad u^+ = 2.5 \ln y^+ + 5.5 \tag{11.156c}$

The above velocity distribution is well-known as universal velocity distribution. A comparison with experimental data is shown in Fig. 11.11. The excellent match testifies that the

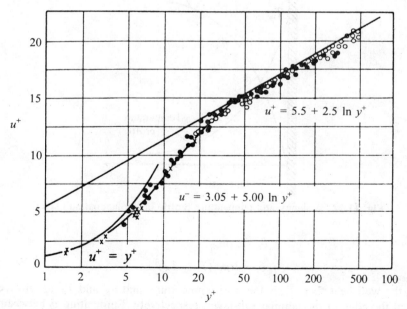

Fig. 11.11   Universal velocity distribution—experimental results.

---

*In this equation we have not used a negative sign to express the shear stress, since the velocity increases with y, the distance from the wall.

Prandtl mixing-length theory provides a satisfactory empirical picture of turbulent flow. Other equations for velocity distribution in turbulent flow are available in Knudsen and Katz (1958), Longwell (1966), and Bejan (1984).

The universal velocity profile Eqs. (11.156a, b, c) also describe the velocity distribution for turbulent flow of a fluid through a smooth tube (slightly different values of the coefficients have been suggested by other workers). The equation may be used to obtain a relation for the *Fanning friction factor f* in terms of the wall shear stress, that is,

$$\tau_w = \frac{1}{2} f \rho u_m^2 \tag{11.157}$$

where u_m is the average velocity.

11.6.4 The Prandtl Analogy

We are now in a position to develop the Prandtl analogy given before in Eq. (4.56). It is assumed that like velocity profile, the temperature profile is also linear in the laminar sub-layer in the close vicinity of the tube wall (Fig. 11.12). As stated before, the contribution of eddy transport is negligible in this layer [$\varepsilon_M = 0$, $\varepsilon_H = 0$]. Integrating Eqs. (4.48) and (4.49) over the laminar sub-layer of thickness δ_l (with slight changes in the notations, putting $\tau = \tau_w$, $q = q_w$; see Fig. 11.12), we get

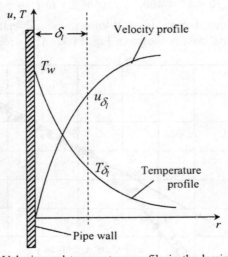

Fig. 11.12 Velocity and temperature profile in the laminar sub-layer.

$$\tau_w \delta_l = \nu \rho u_{\delta_l} \tag{11.158}$$

$$q_w \delta_l = \alpha \rho c_p (T_w - T_{\delta_l}) \tag{11.159}$$

where q_w is the wall heat flux, T_w is the wall temperature, and u_{δ_l} and T_{δ_l} are the velocity and temperature at the edge of the laminar sub-layer, respectively. Eliminating δ_l between the above equations, we obtain

$$\frac{\nu u_{\delta_l}}{\tau_w} = \frac{\alpha c_p}{q_w} (T_w - T_{\delta_l}) \tag{11.160}$$

Now we shall consider momentum and heat transfer in the turbulent core beyond $r = \delta_l$, where r is the radial position in the tube. The mean velocity of the fluid relative to that at $r = \delta_l$ is $(u_m - u_{\delta_l})$ and the temperature driving force in the turbulent core is $(T_{\delta_l} - T_b)$, where T_b is the bulk fluid temperature and u_m is the mean velocity. Now we go back to Eq. (4.54). We replace **V** by $(u_m - u_{\delta_l})$, $h^{\#}$ by $q_w/(T_{\delta_l} - T_b)$ and put $\nu + \varepsilon_M = \alpha + \varepsilon_H$. Also from Eq. (4.51) we can write $f = \dfrac{2\tau_w}{\rho(u_m - u_{\delta_l})^2}$. With these changes, Eq. (4.54) reduces to

$$\frac{(u_m - u_{\delta_l})}{\tau_w} = \frac{c_p(T_{\delta_l} - T_b)}{q_w} \tag{11.161}$$

Eliminating T_{δ_l} from Eqs. (11.160) and (11.161), we get

$$\frac{q_w}{c_p \tau_w}\left[\frac{\nu u_{\delta_l}}{\alpha} + (u_m - u_{\delta_l})\right] = T_w - T_b \tag{11.162}$$

The velocity u_{δ_l} can be expressed in terms of τ_w and u_m by using the universal velocity distribution. It is known from Eq. (11.156a), that the laminar sub-layer extends up to $y^+ = 5$, and the velocity at the edge of it is given by

$$u_{\delta_l}^+ = y^+ = 5$$

That is,

$$u_{\delta_l} = 5u^* = 5\sqrt{\frac{\tau_w}{\rho}} = 5u_m\sqrt{\frac{f}{2}} \tag{11.163}$$

Substituting for u_{δ_l} from Eq. (11.163) in Eq. (11.162) and putting $q_w = h(T_w - T_b)$, $\tau_w = \dfrac{1}{2}f\,\rho u_m^2$ and $\dfrac{\nu}{\alpha} = \text{Pr}$, we get

$$\frac{h}{c_p\,(f/2)\,\rho u_m^2}\left[(\text{Pr})(5u_m)\sqrt{\frac{f}{2}} + u_m\left(1 - 5\sqrt{\frac{f}{2}}\right)\right] = 1$$

or

$$\frac{h}{\rho u_m c_p} = \text{St} = \frac{f/2}{1 + 5\sqrt{\dfrac{f}{2}}(\text{Pr} - 1)} \tag{11.164}$$

Equation (11.164) is the *Prandtl analogy* which has been given before in Eq. (4.56). The analogy can be used to estimate the Stanton number, and hence the heat transfer coefficient, from a knowledge of the Fanning friction factor.

[#]It is to be noted that $q_w/(T_{\delta_l} - T_b)$ is equal to the heat transfer coefficient for the turbulent core only. This coefficient is different from that for heat transfer to or from the tube.

The essence of Prandtl analogy lies in the assumption that purely viscous flow occurs in the laminar sub-layer within which the eddy transport parameters (ε_M and ε_H) are zero. In the turbulent core, on the other hand, molecular transport parameters (ν and α) have negligible contribution compared to their eddy counterparts (i.e. ε_M and ε_H). Also, the eddy diffusivities are equal in the turbulent core ($\varepsilon_M = \varepsilon_H$). Experimental data reported by different researchers, however, show that ε_M and ε_H are not quite equal. However, their ratio becomes nearly unity at very high Reynolds number.

11.6.5 The von Karman Analogy

While Prandtl developed the analogy, Eq. (11.164), by considering the variations in velocity and temperature of the fluid over the laminar sub-layer only, von Karman (1939) improved upon the Prandtl analogy by considering the variations of these quantities in the buffer zone as well and making use of the universal velocity profile, Eq. (11.156). The von Karman analogy, in terms of the friction factor, is given below.

$$\mathrm{Nu} = \frac{(f/2)(\mathrm{Re})(\mathrm{Pr})}{1 + 5\sqrt{f/2}\,\{(\mathrm{Pr} - 1) + \ln[1 + (5/6)(\mathrm{Pr} - 1)]\}} \qquad (11.165)$$

Several other analyses of turbulent heat transfer carried out by Martinelli, Lyon, Deissler, etc. have led to a number of other forms of 'analogies'. These have been discussed in some detail by Knudsen and Katz (1958) and Bejan (1984).

The heat transfer analogies prove to be very useful for estimation of the heat transfer coefficient if a suitable correlation to fit the particular geometry or flow configuration is not available. However, a correlation for the friction factor or the drag coefficient may be available. So the heat transfer coefficient can be calculated from a knowledge of the friction factor by using a heat transfer analogy.

Example 11.5   Calculate the water-side Nusselt number in Example 4.12 using the von Karman analogy.

SOLUTION   Here, Reynolds number, $\mathrm{Re} = 7.44 \times 10^4$; friction factor, $f = 0.00485$; Prandtl number, $\mathrm{Pr} = 5.12$ (see Example 4.12).

Putting these values in Eq. (11.165), we get

$$\mathrm{Nu} = \frac{(0.00485/2)(7.44 \times 10^4)(5.12)}{1 + 5\sqrt{0.00485/2}\,\{(5.12 - 1) + \ln[1 + (5/6)(5.12 - 1)]\}}$$

$$= \boxed{388}$$

The Prandtl analogy predicts $\mathrm{Nu} = 458.7$.

SHORT QUESTIONS

1. What do you mean by boundary layer? How is the boundary layer thickness defined? At which point does the boundary layer turn turbulent?

2. How is the local Nusselt number related to the surface temperature gradient?

3. How can the momentum (or velocity) boundary layer thickness be related to that of the thermal boundary layer?

4. Write down three relations that may be used to calculate the heat transfer coefficient for turbulent pipe flow from the value of the friction factor.

5. What is Reynolds stress? How does its value compare with purely viscous stress?

6. What is eddy diffusivity of momentum? How does its typical value compare with the value of momentum diffusivity?

7. Explain Prandtl mixing-length. What are the basic assumptions underlying the definition of this quantity? What is the order-of-magnitude of its value?

8. For two-dimensional steady state flow of an incompressible fluid, the gradient of longitudinal velocity at a point is $\partial u/\partial x = -0.2$ s^{-1}. What is the value of the transverse velocity gradient $\partial v/\partial y$ at the same point?

9. Consider boundary layer flow of a fluid over a flat plate. Where does the velocity gradient $\partial u/\partial y$ has the maximum value? What is its minimum value and where does it occur?

10. Air flows over a flat plate at 100°C. The boundary layer thickness at a certain location on the plate is 7 mm. If the Prandtl number of air is 0.69, calculate the thickness of the thermal boundary layer at the same location.

11. Water flows over a flat plate, 1 m long, at steady state. If the local Reynolds number halfway the plate is 7×10^4, calculate the boundary layer thickness at that point. What is the maximum thickness of the boundary layer for this plate?

12. If the local drag coefficient is given by $C_D = 0.664$ (Re)$^{-1/2}$, calculate the total drag force because of laminar boundary layer flow of water at a velocity of 2 m/s over a flat surface 1.2 m long and 0.7 m wide.

13. Heat transfer from a flat plate at 90°C occurs to a fluid through a laminar boundary layer. The free stream velocity of the fluid is 4 m/s, and its bulk temperature is 25°C. If the longitudinal velocity at a point in the boundary layer is 1.2 m/s, what is the temperature at that point? Prandtl number of the fluid is unity.
    Hint: If Pr = 1, $u/U_\infty = (T - T_s)/(T_\infty - T_s)$.

14. If a fluid has a Prandtl number Pr = 5, how does the velocity boundary layer thickness compare with the thermal boundary layer thickness?

PROBLEMS

11.1 Air at 30°C flows over a flat surface, 1.5 m long and 0.5 m wide, at a velocity of 5 m/s. Calculate: (i) the boundary layer thickness at a point 1 m from the leading edge, (ii) the shear stress at the surface and at a point 5 mm above the surface, 1 m from the leading edge, (iii) the total drag force acting on the surface, and (iv) the percentage increase in the local boundary layer thickness, if the free stream velocity is reduced by 50%.

11.2 Water at 25°C flows over a flat plate at a velocity of 1 m/s. Calculate the rate of flow of water through a section of the boundary layer 0.5 m from the leading edge. Also, calculate the velocity components, u and v, at a point in the boundary layer where $x = 0.3$ m and $y = 1$ mm (for water, at 25°C, $v = 8.63 \times 10^{-7}$ m^2/s).

11.3 Using the momentum integral technique, obtain the relations for the boundary layer thickness and the drag coefficient assuming the following forms of velocity profile:

(i) a linear profile, $u/U_\infty = \eta$, and

(ii) a sine function profile, $u/U_\infty = a \sin (b\eta)$, $\eta = y/\delta$.

Compare these expressions with those obtained from the exact solution.

11.4 The following approximate form of the function $f(\eta)$ was suggested by Churchill (1988).

$$f(\eta) = \eta \left[1 + \left(\frac{6}{\eta} \right)^{7/3} \right]^{-3/7}$$

(i) Show that the function f satisfies the boundary conditions given by Eqs. (11.36), (11.37), and (11.38).

(ii) Determine a relation for the boundary layer thickness using the above function and comment on its accuracy.

11.5 Cold air at 5°C flows at a velocity of 10 m/s over a flat surface of size 1 m × 1 m. If the surface temperature is 35°C, calculate:

(i) the velocity and the thermal boundary layer thicknesses midway of the surface using the exact solution as well as the approximate solution (i.e. the solution obtained by the integral technique),

(ii) the local heat transfer coefficient and heat flux at a point 0.2 m from the leading edge, and

(iii) the average heat transfer coefficient and the rate of heat loss from the surface.

11.6 At what rate must heat be removed from the bottom side of a 0.7 m × 0.7 m flat plate, if hot air at 60°C flows over the top of the plate at 8.5 m/s in order to maintain the plate at 20°C?

11.7 In Example 11.6, calculate the thickness of a stagnant film of air that offers the same resistance to heat transfer, on an average, as the thermal boundary layer.

11.8 Hot air at 90°C flows over one surface of a 1 m × 1 m thin plate at 8 m/s. Cold air at 10°C flows over the other surface at 5 m/s. If the plate material has a high thermal conductivity, calculate the plate temperature and the rate of heat flow across the plate.

11.9 A 0.2 m × 0.2 m metal plate has embedded heating wires so as to generate heat at the rate of 250 W. The plate is placed in a wind tunnel at 10 m/s air velocity. Heat loss occurs from both sides of the plate. At steady state condition the plate temperature is found to be 130°C. If the tunnel air temperature is 30°C, calculate the percentage deviation of the calculated average heat transfer coefficient from the experimental value obtained by using the above plate temperature. What are the probable causes of this deviation?

11.10 The vertical wall of a furnace, 2 m in height, has an outer surface temperature of 60°C. Using the exact solution of the natural convection heat transfer equation, calculate the local heat transfer coefficient midway of the wall. Also, calculate the rate of heat loss to the ambient at 25°C. Does the boundary layer on the wall remain laminar all along the height of the wall?

REFERENCES AND FURTHER READING

Bird, R.B., W.E. Stewart, and E.N. Lightfoot, *Transport Phenomena*, John Wiley, New York, 1960.

Bejan, A., *Convection Heat Transfer,* John Wiley, New York, 1984.

Churchill, S.W., *Viscous Flows — The Practical Use of Theory*, Butterworths, Boston, 1988.

Hinze, J.O., *Turbulence,* 2nd ed., McGraw-Hill, New York, 1975.

Knudsen, H.G. and D.L. Katz, *Fluid Dynamics and Heat Transfer*, McGraw-Hill, New York, 1958.

Longwell, P.A., *Mechanics of Fluid Flow*, McGraw-Hill, New York, 1966.

Nikuradse, J., *Forsch. Ingenieurw*, **356** (1932).

Pohlhausen, K., *Z. Angew Math. Mech.*, **1** (1921), 252.

Schlichting, H., *Boundary Layer Theory*, McGraw-Hill, New York, 1968.

von Karman, Th., *Z. Angew Math. Mech.*, **1** (1921), 223.

von Karman, Th., *Trans. ASME*, **61** (1939), 705.

Answers to Selected Problems

CHAPTER 2

Short Questions

6. W/m $(°C)^3$;   1.106 W/m °C;   0.827 W/m °C
7. 229°C

Problems

2.1 26.1 kW;   239°C
2.2 859 W;   0.0156%;   yes

2.3 $\dfrac{(T - T_i) + (\beta/2)(T^2 - T_i^2)}{(T_o - T_i) + (\beta/2)(T_o^2 - T_i^2)}$ ln $(r_o/r_i) = \ln(r/r_i)$; β in $°C^{-1}$

2.4 (a) 659 W/m²;   (b) 46°C;   (c) 12.6%
2.5 618°C
2.6 7.85 kg/h
2.7 24.8 h
2.8 890°C
2.9 75% thicker;   0.25
2.10 0.0473;   0.0955 Btu/h ft °F
2.11 0.6 W/m °C
2.12 decreases by 15.7%
2.13 267°C
2.14 0.69 m
2.15 (a) $T_1 = 241.2 - (41.205)/(1 - 1.5872x)$; $T_2 = 331.74 - (103)/(1 - 1.5872x)$;
   (b) 33.3 W;   (c) 1655 W/m²;   (d) 181°C,   702 W/m²
2.16 (a) 0.2577 °C/W;   (b) 427 W;   (c) $-10°C$,   13.1 W/m²
2.17 4.2 kW
2.18 $1 \leq (r_o/r_i) \leq 1.02$;   yes, it overestimates
2.19 0.442 W/m °C
2.20 17.2 W
2.21 (i) 300°C,   313.02°C;   (ii) 1 MW/m³;   (iii) 306.5°C
2.22 (a) $T = 250 + 4687.5(0.6x - x^2)$;   (b) $T_{max} = 672°C$ at $x = 0.3$ m;   (c) 531°C

2.23 $T_1 = T_s + \dfrac{\psi_{v2}}{6k_2}\left(r_2^2 - r_1^2\right) + \dfrac{r_1^2}{3k_2}(\psi_{v1} - \psi_{v2})\left(1 - \dfrac{r_1}{r_2}\right) + \dfrac{\psi_{v1}}{6k_1}\left(r_1^2 - r^2\right)$;

$T_2 = T_s + \dfrac{\psi_{v2}}{6k_2}\left(r_2^2 - r^2\right) + \dfrac{r_1^3}{3k_2}(\psi_{v1} - \psi_{v2})\left(\dfrac{1}{r} - \dfrac{1}{r_2}\right)$

CHAPTER 3

Short Questions

20. 21 W/m^2 °C
21. 80°C; 11.3 W/m^2 °C

Problems

3.1 40 W/m^2 °C
3.2 2.3 h
3.3 (a) 338 K; (b) 12.6 W/m^2 °C; (c) 2.16 mm
3.4 43.1°C; 118.6 s
3.6 1.26 h; 0.62 mm; 1040 W/m^2 °C
3.7 (a) 101.39°C; (b) yes, 100 kW/m^3; (c) 24.5 W/m^2 °C
3.10 50.4 °C/W
3.12 61°C
3.14 381°C
3.15 313.6 A; 60.02°C; 50.8°C
3.17 8.2 W
3.18 (a) yes, 0.000653°C/W; (b) 25°C; (c) 11.4%
3.20 56.6 W
3.24 $r_{o,\,\text{crit}} = 2k/h_o$

CHAPTER 4

Problems

4.2 55.7°C; 30.4 m
4.3 32.6°C
4.7 (a) 30.5°C; (b) 2.8 turns (4.45 m)
4.10 121 W/m^2 K
4.11 h_{av} = 4080 W/m^2 K, Q = 183.6 kW per metre width
4.13 0.281 kg/h
4.14 (a) laminar; (b) being heated; (e) (i) 411 W/m^2; (ii) 6.85 W/m^2 °C; (iii) 2.48
4.18 32 m
4.21 49.4 kW

CHAPTER 5

Problems

5.1 Q = 4371 W (total); 0.838
5.3 67 W; 59%
5.5 6.5 h
5.9 The coil is overdesigned
5.12 36.4 km
5.13 42 W; k_{eff} = 0.06 W/m °C; 31.5 W
5.16 1.085 kW

CHAPTER 6

Problems

6.3 233.7°F
6.5 25 W (taking $C_{sf} = 0.006$; $n = 1$)
6.7 31.45 kW/m²
6.10 (a) 0.161 m, 0.107 mm; (b) 1200 W/m² °C; (c) 1180 W/m² °C
6.11 (a) 0.146 mm; (b) 0.245 m/s; (c) 4610 W/m² °C, 5180 W/m² °C

6.14 $\bar{h} = \left[\dfrac{g \cos \theta \, \rho_l(\rho_l - \rho_v)k_l^3 L_v'}{L \mu_l(T_v - T_w)} \right]^{1/4}$

6.20 (a) $\bar{h}$ = 1642 W/m² °C, 34.4%; (b) 1.63 m

CHAPTER 7

7.1 λ_{max} = 1.16 µm; $E_{gray, \lambda max}$ = 3.131 × 10⁶ W/m² µm
7.3 379.5 W/m²; 189.7 W/m²; 130.8 W/m²
7.4 (a) 2.276 µm; (b) 1.489 × 10⁵ W/m²; (c) 0.5422; (d) 28%; (e) 1.489 × 10⁵ W/m²
7.5 (a) 205 K; (b) (i) 20 µm, (ii) 31.6 µm; (c) 14.14 µm
7.6 Case 1: rate of heat loss = 203 W (radiation), and 227 W (free convection); rate of steam condensation = 0.734 kg/h
Case 2: reduction in heat loss = 26%; radiation heat transfer coefficient, h_r = 3.56 W/m² °C
7.7 (a) 2.523 × 10⁴ W/m²; (b) 0.445; (c) 2.898 µm
7.10 (iia) 0.25; (iib) 0.106
7.12 329 W
7.14 813 W; 674 K
7.16 350°C
7.20 539.4°C; 328°C; 81.6%
7.26 726.5 W; 6.2 W/m² °C

CHAPTER 8

Hints to Short Questions

1. Fouling, leakage at the pass partition plate. **2.** A larger clearance between a tube and a baffle hole. **3.** Erosion of tubes that are directly exposed to the entering shell fluid and which may lead to tube failure/leakage; an impingement plate should be installed. **4.** Hydraulic test for any leakage and the capability of withstanding the desired pressure. **7.** Erosion at the junction of the pass partition plate and the tube sheet. **8.** To allow gases to escape during start-up or during operation of a condenser; to empty the shell completely. **9.** Non-condensables will accumulate leading to a drastic fall in the capacity of the condenser. **10.** Provide the shell with a rupture disk. **11.** Corrosive service. **12.** Increase the cooling water velocity; increased turbulence reduces scaling. **13.** Vibration and tube damage. **14.** 25% **15.** By increasing the tube length.

16. Just plug the tube.   **17.** Subcooling of the condensate; a little increase in the reflux ratio and thereby in the purity of the top product.   **18.** It will cause excessive scaling.   **19.** By introducing some non-condensables.   **20.** Vapour loss through the vent, high vent temperature, less subcooling of the condensate than the designed value.   **21.** Axial nozzle entry velocity much higher than the tube velocity; high radial nozzle entry velocity.   **22.** Erosion/corrosion at the tube inlet, tube sheet erosion, erosion damage at the shell inlet, vibration damage, tube-end stress corrosion cracking, gasket leak at the floating head, etc.   **23.** Inside of the tubes: poking by a wire brush; outside of the tubes: by wire brush (floating-head, U-tube bundle exchanger), chemical cleaning (for fixed head).   **24.** Lifting lug.   **25.** Capacity is likely to fall because condensation coefficient on a vertical tube is less than that on a horizontal tube (however, if the coolant-side coefficient virtually controls, there may not be any appreciable effect).   **26.** Since water boiling heat transfer coefficient is very high and tube-side (i.e. gas-side) coefficient is low, tube-wall temperature remains close to the boiling water temperature. If the hot gases flow through the shell, the tube wall temperature will remain nearly the same as in the previous case. But the shell will now be exposed to hot gases entering at a very high temperature which the shell may not stand. So the hot gases are never passed through the shell.   **28.** All the liquids may not drain out. **29.** The tube temperature over this small empty region will be much higher. This will cause thermal stress which may have damaging effect.

CHAPTER 9

Hints to Short Questions

6. Agitated film (for viscosity over 1000 cP), calandria or LTV (below 1000 cP).   **11.** Condensate load, the quantum of non-condensables to be discharged, the allowable maximum quantity of condensate accumulation in the steam chest.   **13.** If the vapour is valuable and recovery is necessary; if water is not in abundant supply.   **15.** 0.8, 2.4.   **16.** Natural circulation. **17.** Less liquid velocity and greater fouling.   **18.** Entrainment.   **19.** Tube fouling, foaming, entrainment.   **20.** Lower turndown ratio.   **21.** Choking of tubes because of solid deposition over a period of time.   **24.** When live steam is forced through a particular tube, it expands in length putting great stress at the tube joint leading to failure.   **25.** Vacuum device, piping and pumps. **26.** (a) LTV, (b) LTV, forced circulation, (c) Calandria, (d) Calandria, forced circulation, (e) Natural circulation (external calandria), (f) Calandria, (g) Rising film, (h) Forced circulation, agitated thin film, (i) Agitated thin film (under vacuum), (j) Natural circulation, (k) Rising film (l) Falling film.   **27.** Forced circulation.   **29.** Proper liquid distribution in the tubes.   **30.** About 1000 Btu.   **32.** 1.5 to 7 m/s.   **33.** Small size units, if the evaporator is not required to operate at an elevated temperature in order to avoid degradation of the product (like fruit juice, pharmaceuticals, etc.), if the construction of the evaporator is very expensive, like a wiped or agitated film unit.   **34.** Flow, pressure, temperature and level recorders and controllers. **35.** Mechanical cleaning of the tubes, boiling water or dilute feed in the tubes to dissolve the scales. **39.** Yes, viscosity of the solution. As the boiling temperature is reduced (by operating at a vacuum), the solution viscosity increases. This reduces the heat transfer coefficient and tends to increase the requirement of heat transfer area.

Problems

Note: In the absence of data the specific heat of a solution has been estimated by c_{pl} = wt. fraction of the solvent × specific heat of the solvent.

9.1 (a) 156 m²; (b) 36,600 kg/h
9.2 Steam pressure = 2.6 bar abs. (129°C)
9.3 (i) 2.152 × 10⁶ kcal/h; (ii) 4226 kg/h, 52 m²; (iii) capacity reduced by 33%; steam rate = 2915 kg/h
9.4 105 m²
9.6 (i) 36°C; 10,640 kg/h

CHAPTER 10

Problems

10.3 $\bar{T} = \dfrac{T - T_o}{T_i - T_o} = 2\displaystyle\sum_{n=1}^{\infty} \dfrac{\lambda_n + \mathrm{Bi}\,\sin\lambda_n}{\lambda_n^2 + \mathrm{Bi}\,\sin^2\lambda_n}\,\sin(\lambda_n\bar{x})e^{-\lambda_n^2\theta}$

10.5 (i) $T_c = 64°C$; $Q = 153.5$ kJ
10.6 75 min; 98.5°C
10.7 87 s; 255°C
10.8 6 m
10.12 1.6 m

CHAPTER 11

Short Questions

8. 0.2 s⁻¹
9. At the plate surface; zero; at the edge of the boundary layer
10. 7.9 mm
11. 9.4 mm; 13.3 mm
12. 1.44 N
13. 70.5°C
14. $\delta/\delta_T = 1.71$

Problems

11.1 (i) 9.2 mm; (ii) 0.0178 N/m²; (iii) 0.0218 N; (iv) 41.4%
11.2 2.157 × 10⁻³ m³/s per metre breadth of the plate; 0.618 m/s; 5 × 10⁻⁴ m/s
11.3 (i) $\delta = 3.46(vx/U_\infty)^{1/2}$; (ii) $\delta = 4.79(vx/U_\infty)^{1/2}$
11.5 (i) 4.5 mm, 5.0 mm; (ii) 13.6 W/m² °C; (iii) 12.16 W/m² °C, 365 W (one side)
11.6 275 W
11.7 2 mm

A Note on Thermophysical Properties

In some of the exercise problems, the required thermophysical properties are not enumerated. These may be obtained from a few common data sources available. Thermophysical properties of air and water are necessary for the solution of some of the problems and these data are given here over a limited temperature range as a ready reference. Given below are a few common references that provide thermophysical properties of a large number of substances. Reference 5 is extremely useful for estimating the properties of pure substances and mixtures.

1. *CRC Handbook of Chemistry and Physics*, CRC Press, Washington, D.C.

2. *International Critical Tables*, McGraw-Hill, New York, 1926–1933.

3. *Lange's Handbook of Chemistry*, McGraw-Hill, New York.

4. *Perry's Chemical Engineers' Handbook*, McGraw-Hill.

5. *The Properties of Gases and Liquids*, R.C. Reid, J.M. Prausnitz, and B.E. Poling, 4th ed., McGraw-Hill, New York, 1987.

References 1, 3 and 4 are revised from time to time. The latest edition available should preferably be consulted.

Thermophysical Properties of Air

Temp.	Density	Specific heat	Viscosity	Momentum diffusivity	Thermal conductivity	P_r
K	kg/m^3	kJ/kg K	10^{-5} N s/m^2	10^{-6} m^2/s	10^{-2} W/m K	
250	1.3947	1.006	1.596	1.144	2.23	0.720
300	1.1614	1.007	1.846	1.589	2.63	0.707
350	0.9950	1.009	2.082	2.092	3.01	0.700
400	0.8711	1.014	2.301	2.641	3.38	0.690
450	0.7740	1.021	2.507	3.240	3.73	0.686
500	0.6964	1.030	2.701	3.879	4.07	0.684
550	0.6329	1.040	2.884	4.557	4.39	0.683
600	0.5804	1.051	3.058	5.269	4.69	0.685
650	0.5356	1.063	3.225	6.021	4.97	0.690
700	0.4975	1.075	3.388	6.810	5.24	0.695

Thermophysical Properties of Saturated Water

T K	P bar	ρ kg/m³	h_v kJ/kg	c_p kJ/kg K	μ 10^{-3} N s/m²	k W/m K	P_r	σ 10^{-2} N/m	β 10^{-6} K
273.16	0.00611	1000	2502	4.217	1.75	0.659	13.0	7.55	68.05
280	0.00990	1000	2497	4.198	1.422	0.574	10.26	7.48	− 32.74
290	0.01917	999	2461	4.184	1.080	0.590	7.56	7.37	174
300	0.03531	997	2438	4.179	0.855	0.613	5.83	7.17	276.1
310	0.06221	993	2414	4.178	0.695	0.628	4.62	7.00	361.9
320	0.1053	989	2390	4.180	0.577	0.640	3.77	6.83	436.7
330	0.1719	984	2366	4.184	0.489	0.650	3.15	6.66	504
340	0.2713	979	2342	4.188	0.420	0.660	2.66	6.49	566
350	0.4163	974	2317	4.195	0.365	0.668	2.29	6.32	624.2
360	0.6209	967	2291	4.203	0.324	0.674	2.02	6.14	698
370	0.9040	961	2265	4.214	0.289	0.679	1.80	5.95	728.7
373.16	1.0133	958	2257	4.217	0.279	0.68	1.76	5.89	750
380	1.2869	953	2239	4.226	0.260	0.683	1.61	5.76	788
390	1.794	945	2212	4.239	0.237	0.686	1.47	5.56	841
400	2.455	937	2183	4.256	0.217	0.688	1.34	6.36	896
420	4.370	919	2123	4.302	0.185	0.688	1.16	4.94	1010
440	7.333	901	2059	4.36	0.162	0.682	1.04	4.51	
460	11.71	879	1989	4.44	0.143	0.673	0.95	4.07	
480	17.90	857	1912	4.53	0.129	0.660	0.89	3.62	
500	26.40	831	1825	4.66	0.118	0.642	0.86	3.16	
520	37.7	804	1730	4.84	0.108	0.621	0.84	2.69	

h_v = heat of vaporization; ρ = surface tension; β = volume expansion coefficient (see Chapter 5) Other terms have usual significances.

Index